This book presents a diverse collection of chapters on basic research at the molecular level using the Lepidoptera as model systems. This volume, however, is more than just a compendium of information about insect systems in general or the Lepidoptera in particular. Each chapter is a self-contained treatment of a broad subject area, providing sufficient background to give readers a sense of the guiding principles and central questions associated with each topic, in addition to major methodologies and findings. Comparisons with other major model systems are emphasized, with special attention given to the fruit fly, *Drosophila melanogaster.* Topics include a historical overview of research using lepidopteran models, silkworm genetics, mobile elements of lepidopteran genomes, lepidopteran phylogeny, experimental embryogenesis and homeotic genes, chorion gene regulation and evolution, regulation of silk protein and homeobox genes in the silk gland, control of transcription by RNA polymerase III, hormonal regulation of gene expression during development, hormone action in the central nervous system, the molecular genetics of moth olfaction, the immune response, and use of engineered baculoviruses for basic biological studies and insect pest control.

MOLECULAR MODEL SYSTEMS IN THE LEPIDOPTERA

MOLECULAR MODEL SYSTEMS IN THE LEPIDOPTERA

Edited by

MARIAN R. GOLDSMITH
University of Rhode Island

ADAM S. WILKINS
Company of Biologists, Ltd., U.K.

CAMBRIDGE
UNIVERSITY PRESS

CAMBRIDGE UNIVERSITY PRESS
Cambridge, New York, Melbourne, Madrid, Cape Town, Singapore, São Paulo

Cambridge University Press
The Edinburgh Building, Cambridge CB2 2RU, UK

Published in the United States of America by Cambridge University Press, New York

www.cambridge.org
Information on this title: www.cambridge.org/9780521402491

First published 1995
This digitally printed first paperback version 2006

A catalogue record for this publication is available from the British Library

Library of Congress Cataloguing in Publication data

Molecular model systems in the Lepidoptera / edited by Marian R.
Goldsmith, Adam S. Wilkins.
p. cm.
Topics originated from 2 workshops held in 1988 and 1991.
Includes bibliographical references (p.) and index.
ISBN 0–521–40249–2 (hc)
1. Lepidoptera – Molecular aspects. 2. Molecular biology –
Methodology. I. Goldsmith, Marian R. II. Wilkins, A. S. (Adam
S.), 1945- .
QL562.2.M64 1994
595.78′0487322–dc20

93–47979
CIP

ISBN-13 978-0-521-40249-1 hardback
ISBN-10 0-521-40249-2 hardback

ISBN-13 978-0-521-02827-1 paperback
ISBN-10 0-521-02827-2 paperback

Dedicated to Carroll M. Williams, whose contributions and inspiration played such a large part in the development of lepidopteran model systems

Contents

Contributors

PETER E. DUNN Department of Entomology, Purdue University, 1158 Entomology Hall, West Lafayette, IN 47907

THOMAS H. EICKBUSH Department of Biology, University of Rochester, Rochester, NY 14627

TIMOTHY FRIEDLANDER Center for Agricultural Biotechnology, University of Maryland Biotechnology Institute, College Park, MD 20742-

MARIAN R. GOLDSMITH Department of Zoology, University of Rhode Island, Kingston, RI 02881

CHI-CHUNG HUI Department of Developmental Biology, National Institute for Basic Biology, Okazaki 444, Japan

KOSTAS IATROU Department of Medical Biochemistry, Faculty of Medicine, University of Calgary, 3330 Hospital Drive N.W., Calgary, Alberta T2N 4N1, Canada

JOHN A. IZZO Intramural Research Program, National Institute of Aging, Gerontology Research Center, Baltimore, MD 21224

FOTIS C. KAFATOS European Molecular Biology Laboratory, Meyerhofstrasse 1, D69117, Heidelberg, Germany

ROBERT F. LECLERC Center for Agricultural Biotechnology, University of Maryland Biotechnology Institute, College Park, MD 20742

CHARLES MITTER Department of Entomology, University of Maryland at College Park, College Park, MD 20742

AMY B. MULNIX Department of Biology, Earlham College, Richmond, IN 47374-4095

TOSHIFUMI NAGATA Department of Developmental Biology, National Institute for Basic Biology, Okazaki 444, Japan

LISA M. NAGY Laboratory of Molecular Biology, University of Wisconsin-Madison, Madison, WI 53706

HANH T. NGUYEN Department of Cardiology, Children's Hospital, Harvard Medical School, Boston, MA 02115

JEROME C. REGIER Center for Agricultural Biotechnology, University of Maryland Biotechnology Institute, College Park, MD 20742

LYNN M. RIDDIFORD Department of Zoology, University of Washington, Seattle, WA 98195

NIKOLAUS A. SPOEREL Department of Biochemistry, University of Connecticut Health Center, Farmington, CT 06032

KAREN U. SPRAGUE Institute of Molecular Biology, University of Oregon, Eugene, OR 97403

YOSHIAKI SUZUKI Department of Developmental Biology, National Institute for Basic Biology, Okazaki 444, Japan

JAMES W. TRUMAN Department of Zoology, University of Washington, Seattle, WA 98195

GEORGE TZERTZINIS The Biological Laboratories, Harvard University, 16 Divinity Avenue, Cambridge, MA 02138

KOHJI UENO Department of Developmental Biology, National Institute for Basic Biology, Okazaki 444, Japan

RICHARD G. VOGT Department of Biological Sciences, University of South Carolina, Columbia, SC 29208

BRIAN M. WIEGMANN Department of Entomology, University of Maryland at College Park, College Park, MD 20742

ADAM S. WILKINS Company of Biologists, Ltd., Austin Building, Room S14, University of Cambridge, New Museum Site, Pembroke Street, Cambridge, CB2 3ED, U.K.

JUDITH H. WILLIS Department of Zoology, University of Georgia, Athens, GA 30602

Preface

The Lepidoptera present something of a puzzle in biological science. Although represented by the only insect to be fully domesticated – the silkworm, *Bombyx mori* – and providing experimental subjects for much ground-breaking work in physiology and genetics, their virtues as model systems for molecular studies are still comparatively little recognized. This seems particularly striking when one notes the attention lavished in recent years on one diminutive distant cousin of the butterflies and moths, namely, the fruit fly, *Drosophila melanogaster*. Yet the Lepidoptera continue to have much to offer, both in their own right as experimental subjects and in a wide range of comparative studies. Our deliberate intention in organizing the present volume was to help focus attention on this diverse, fascinating, and immensely useful group of organisms and, if possible, to increase their relative appreciation.

The germ for the idea of this book grew out of the first workshop on Molecular Genetics and Molecular Biology of the Lepidoptera, held at the Orthodox Academy in Kolymbari, Crete, in September 1988. This meeting – an outgrowth of several earlier silkworm meetings held in France and the United States – was organized by Fotis Kafatos and Marian Goldsmith in response to the needs of a growing community of lepidopteran researchers. For this group, there seemed to be no forum to share data, compare notes, and exchange ideas in the company of people familiar with the advantages and idiosyncrasies of their own experimental organisms. As impressed by the diversity of systems being studied at the molecular level in these insects and by the high quality of the work presented at Kolymbari as were the other participants, we conceived of this book while still buzzing with excitement from the meeting. Rather than publishing a series of potentially short-lived reports on the actual presentations we had just heard, we and the prospective authors agreed that there was the greater need for a collection of broad overviews of selected subject areas, which would convey the nature and excitement of current work on the molecular biology of the Lepidoptera. To give the book a coherent framework, we took evolution, development, and regulation as unifying themes.

The sixteen chapters that follow will, we hope, achieve these aims. Each is intended to present a contemporary picture of its subject, both for workers in these different areas and for other biologists who would like to know more about the use of Lepidoptera as molecular model systems. Accordingly, each chapter has been written as a self-contained treatment of its subject, and we have tried to provide enough background to give readers a sense of the guiding principles and central questions associated with each of the fields represented, in addition to major methodologies and findings. Furthermore, acknowledging the importance of work in *Drosophila melanogaster* in biology today, the authors have also consciously framed the articles presented here in relation to what is known about the fruit fly, which often presents either an informative evolutionary parallel or a contrast to what has been found in lepidopteran systems. Finally, we unashamedly have tried to bring out the reasons for using lepidopteran models, including both their experimental advantages and their historical importance in the development of many lines of research.

Although the gestation period for this volume was lengthy, the delay was, in some respects, useful. While many of the topics covered in this volume originated with the first Lepidoptera workshop, others were added after the second workshop, held in 1991 under the leadership of Karen Sprague and Gerard Chavancy. In our intention to highlight certain well-developed or emerging lines of work, others were unavoidably omitted. We sincerely apologize for our omissions, and hope that, at the least, we have provided enough signposts to enable interested readers to fill in these gaps on their own.

We would like to extend our thanks to our authors, who met our deadlines and delivered their manuscripts to us on schedule, and to our editor, Robin Smith, for his patience and strong support of this project throughout. Whatever omissions there may be in the coverage presented here, we believe that the chapters in this book will provide a useful view of the state and breadth of lepidopteran molecular biology in the mid-1990s and that the reader will come away with not only a better idea of the fascinating biology of the Lepidoptera but a feeling for the directions that research in lepidopteran systems is likely to take in the second half of this decade.

1

A brief history of Lepidoptera as model systems

JUDITH H. WILLIS, ADAM S. WILKINS, and
MARIAN R. GOLDSMITH

Introduction

Throughout the twentieth century, the Lepidoptera, especially moths, have played important roles in fundamental studies in physiology, especially endocrinologic phenomena, and in biochemistry. Increasingly, they are also finding use in studies of biological development. Unfortunately, the recent resurgence of *Drosophila melanogaster* as the insect species par excellence for developmental analyses (see Lawrence, 1992) has seemingly eclipsed both past and current work on Lepidoptera. This book is intended, therefore, in part, to act as a counter, even an antidote, to the *Drosophila*-centered view of insect biology. In particular, this introductory chapter is designed both to remind the reader of the special value of the Lepidoptera as model systems and to help preserve our knowledge of key discoveries in this field, discoveries whose general implications extend far beyond the particular organisms under study. The chapters that follow emphasize the streams of contemporary research that have flowed from these beginnings and, we hope, convey the breadth and intrinsic interest of lepidopteran biology today.

In taking stock of the origins of lepidopteran research, one can discern three distinct kinds of initial interest. The first was inspired directly by the beauty and variety of lepidopteran species, in particular, by the variety of wing color patterns. This variety has been and continues to be important in studies of systematics and evolution and in the elaboration of ideas about pattern formation (reviewed in Nijhout, 1991). The detailed analysis of the pigments that decorate Lepidoptera was also important in stimulating research in another direction, in spawning major areas of bio-organic chemistry (reviewed in Kayser, 1985).

The second aspect of Lepidoptera that has favored their use as biological model systems is the large body size of certain species and their amenability to surgical manipulation. Much of what we know about how hormones regulate the lives of all insects had its origins in ligations, transplantations, and parabioses in diverse Lepidoptera. The large size of the caterpillars and adults of certain species has also been significant in many biochemical studies. For instance, most of the insect hemolymph products were first characterized in species of moths with large larvae and pupae that can be bled easily (reviewed in Kanost, Kawooya, Law, Ryan, Van Heusden, and Ziegler, 1990).

The third major impetus to lepidopteran research was economic, and here one can subdivide the kinds of interest into two sorts. On the one hand, several species have direct positive economic value, through their production of silk. The silkworm, *Bombyx mori,* whose cultivation began more than five thousand years ago in China (Kuhn, 1988), is the principal species in this respect. Early work with *Bombyx,* connected to the obviously desirable goals of increasing the quantity and quality of silk, produced some major findings of fundamental import. These included the first identification of microorganisms as a cause of disease, first a fungus in 1835 (Bassi, 1958) and, later, two bacterial species (Pasteur, 1870). In addition, some of the first carefully studied homeotic transformations were reported in *Bombyx* in the late 1930s and early 1940s (Sakata, 1938; Hashimoto, 1941). In fact, the modern genetic analysis of *Bombyx* goes back to the turn of the century (Toyama, 1912; see also Tazima, 1964) and provided some of the first modern accounts of mutations. With respect to fundamental research today, work on silk moths has several major strands: the molecular biology of silk production, whose analysis includes the special transcriptional and translational mechanisms employed by the silk gland; the study of the organization, evolution, and regulation of the large, complex chorion multigene family; the recently begun investigation of homeotic genes and mutant effects at the molecular level; and the combined molecular and genetic analyses of embryogenesis. The development of the genetics of *Bombyx* continues to play a key part in all of this work, and the different facets of gene regulation in the silk glands and the chorion system are described in several of the chapters in this book.

The other, and contrary, economic importance of Lepidoptera concerns their major roles as pests, when they act as consumers of vegetables, orchards, forests, shade trees, and clothing. Our knowledge of pheromone signaling by virgin females arose from the need to identify some vulnerability in these remarkably successful pests. One such potential weak spot concerns the sex signaling system by which females attract males (reviewed in Tamaki, 1985), sometimes over considerable distances; the first sex pheromone to be identified chemically was bombykol from *B. mori* (Butenandt, Beckman, Stamm, and Hecker, 1959). This research has led into the detailed characterization of the neural response to pheromones, a subject reviewed by Vogt in this volume. In addition to research strategies based on pheromones, much modern research on lepidopteran control strategies involves recombinant DNA techniques, using baculoviruses, natural insect infectious agents, as vectors. More recently, emphasis in baculovirus research has shifted, in part, toward their employment in basic research on lepidopteran biology. Both of these areas are reviewed by Iatrou in this volume.

In this chapter, we look more closely at the origins of some of these lines of research under, admittedly, somewhat arbitrary headings. As we hope will become apparent, the current molecular studies are not only adding much of direct and central interest to our understanding of particular phenomena but are serving to bring together previously disparate areas, in particular those of hormonal physiology and gene action in development.

Endocrinology

Early endocrinologic studies

One of the pioneers of modern endocrinologic research in Lepidoptera, Carroll Williams, once remarked at the start of his graduate course in insect development that some scientists are problem-driven and select the species most suitable for analysis of their specific problems, whereas others become enamored of a particular species and analyze diverse problems as they are revealed in their favorite model system. Williams himself was probably in the second group. His Ph.D. thesis was on *Drosophila* flight, but most of his research was directed toward solving interesting problems that arose as he studied *Hyalophora cecropia,* though later he also worked on *Manduca sexta.*

Williams's prime research organism, *H. cecropia,* in fact, has had strong attractions for both problem- and organism-centered biologists and was important in some of the earliest grafting experiments to investigate developmental and physiological problems. Thus, a pioneer series of grafting protocols was carried out on the large pupae of this species by the Harvard biologist H. E. Crampton (1900) to test the suggestions of his colleague, A. G. Mayer (1896), that hemolymph contributed to wing pigmentation. Crampton found that in most instances of interspecies grafts, wing pigmentation was independent of hemolymph composition. Mayer's work, it should be noted, had involved a histological analysis of wing development in several Lepidoptera, including *H. cecropia,* and was focused on the steps involved in producing a patterned wing from an imaginal disc. (Mayer's paper, incidentally, contains a concluding section entitled "Summary of Conclusions Believed to Be New to Science." Would that such conclusions were a requirement of today's journals!)

Though Crampton's procedure was, in principal, a crucial breakthrough in methodology, there is no evidence indicating that his experiments were known to the next worker to employ them. This was the Polish biologist S. Kopec (1922), whose surgical manipulations were conducted on a far smaller species, *Lymantria dispar,* the gypsy moth. Kopec's papers established the field of insect endocrinology and, indeed, also that of invertebrate endocrinology. Kopec described the motivation for his research as a "wish to investigate the relation which probably exists between the nervous system and the processes of metamorphosis in insects." Kopec took the progeny of a single gypsy moth female, ablated their brains or ganglia, placed ligations, transplanted tissues, and studied the resulting effects on molting. Reading his paper (published in English) in the *Biological Bulletin* is a humbling experience. Berta Scharrer (1987) has summarized the impact of these experiments as "not only the first demonstration of an endocrine activity in any invertebrate, but . . . also the first indication anywhere in the animal kingdom that the nervous tissue is capable of producing hormones."

All students of insect endocrinology appreciate that these discoveries were confirmed and elaborated by V. B. Wigglesworth in the bug *Rhodnius prolixus* (Wigglesworth, 1934, 1940). Nevertheless, lepidopterans continued to play a key role in permitting generalizations of Wigglesworth's schemes and in expanding and clarifying key issues. Wigglesworth's discovery of a diffusible factor (i.e., juvenile hormone) that inhibited metamorphosis (Wigglesworth, 1936) was followed quickly by two sets of experiments in *B. mori.* In the first, the French worker J. J. Bounhiol (1938) confirmed Wigglesworth's hypothesis that the corpora allata (CA) are the source of juvenile hormone. He showed that ablation of the CA from early instar *Bombyx* larvae resulted in premature metamorphosis, an event signaled by the premature spinning of a diminutive cocoon. Inside the cocoon, the tiny pupa metamorphosed into a tiny adult, the first evidence that metamorphosis in holometabolous insects is also influenced by the CA. Ablation, however, provides only half of the proof needed to establish an endocrine source definitively.

The second half was produced in a set of experiments by a Japanese scientist, S. Fukuda (1944), who implanted CA from early instars into a final instar *Bombyx* larva; the host molted to a supernumerary giant larva that subsequently yielded a giant cocoon, giant pupa, and giant moth. (A drawing immortalizing these experiments was subsequently published in *Scientific American* [Williams, 1958], and slides with this figure have since been projected on countless lecture room screens.)

Yet, more than two decades were to pass before the hormonal activity now known as juvenile hormone (JH) was first partially purified, this preparation being described as a "golden oil" (Williams, 1956a). It is a remarkable coincidence that adult males of *H. cecropia,* Williams's favorite experimental insect, have the most abundant source of juvenoids yet discovered. That the first extracts were produced when Williams was visiting Wigglesworth might not appear to be so remarkable, except that Williams was studying longevity in adults at that time, not the control of metamorphosis. His finding set off a fierce competition to purify JH and identify its chemical nature, turning former collaborators into competitors. It was H. Röller, a scientist previously unknown to the American "moth club," who won the race by purifying the active principle from Cecropia and elucidating the structure of what is now designated JH I (Röller, Dahm, Sweely, and Trost, 1967). Röller had been a student of Piepho and had worked on the influence of the CA in adult insects, so he was well qualified to enter the competition. He was also fortunate in his collaboration with the chemist K. H. Dahm, another German émigré. Their publications were followed by the discovery of JH II from *H. cecropia* (Meyer, Schneiderman, Hanzmann, and Ko, 1968); this form of JH was missed by Röller because he had relied on a beetle, *Tenebrio,* for his bioassay (Röller and Bjerke, 1965). *Tenebrio,* as it happens, is not as sensitive to JH II as the lepidopteran (*Galleria*) that Meyer et al. used for their bioassays.

The first suggestion that juvenoids might have a polyisoprenoid structure with an epoxide component was based on theoretical considerations by Bowers, Thompson, and Uebel, (1965). Only a few years later, the hypothesized compound was isolated from a lepidopteran, *M. sexta,* and christened JH III (Judy, Schooley, Dunham, Hall, Bergot, and Siddall, 1973). It is only within the lepidopterans that a single species appears to synthesize more than one form of juvenoid; indeed, five are now recognized in *Manduca* (Schooley, Baker, Tsai, Miller, and Jamieson, 1984). All other orders of insect tested to date use only JH III, with the exception of flies, which make a *bis*-epoxide of JH III (Richard et al., 1989), and hemipterans, which have JH I (Numata et al., 1992).

The elaborate steps that were first needed to purify JH would not have been necessary had Williams done one additional experiment: to pinpoint the location of JH in *H. cecropia* males. The original isolation was from the abdomens of males, and it took 11 years to separate active juvenoids from abdominal lipids. In 1976, two decades after JH had first been isolated from *H. cecropia* abdomens, Röller's group discovered that all of the juvenoids in the abdomen of the *H. cecropia* male are found in the tiny accessory gland, nicely sequestered from the lipid-rich fat body (Shirk, Dahm, and Röller, 1976). Indeed, we now know that adult male CA secrete JH acid, and that the methyl transferase that converts it to JH is in the accessory gland (Weirich and Culver, 1979). The physiological function of this storage, however, remains obscure; males that are allatectomized as pupae mate normally, and their progeny develop unimpaired (Williams, 1959).

In the 1940s, the ideas of Kopec, who was executed by the Nazis in 1941, were continued by others, resulting in the discovery of ecdysone, the "molting hormone." Fukuda (1940) and Williams (1947) independently showed that molting was initiated by the prothoracic gland. Williams, using isolated abdomens of *H. cecropia* as his assay material, showed that the brain, acting as an endocrine organ, stimulated the prothoracic glands to produce molting hormone. Thus, by the 1950s, the fundamentals of insect endocrinology were established – a brain hormone drives prothoracic glands, and their secretion, in turn, causes molting. The CA control the nature of the molt: When juvenoids are abundant, the molt is larval-to-larval; when their titer is low or absent, the molt is larval-to-pupal or pupal-to-adult.

As with JH, Lepidoptera were the first source of the ecdysteroids, produced by the prothoracic gland. Ecdysone was isolated from 500 kg of male *B. mori* pupae, using a fly bioassay discovered by Frankel, perfected by Baker and Plagg, and exploited with brilliant success by Butenandt and Karlson (1954). (The females from the same shipment of *Bombyx* yielded the adults that provided the pheromone for Butenandt; see Horn, 1989.) Concern that ecdysone might be no more than a tanning agent (given that is what the assay measured) was dispelled by Williams, who demonstrated that it causes adult development and molting in abdomens isolated from diapausing *H. cecropia* pupae (see Butenandt and Karlson, 1954). It took a further 1,000 kg of *Bombyx* pupae to yield 250 mg of ecdysone, sufficient to provide analytic and crystallographic evidence that at least some invertebrates use steroids as hormones (Huber and Hoppe, 1965; Karlson, Hoffmeister, Hummel, Hocks, and Spiteller, 1965).

Later endocrinologic studies

Williams (1967) speculated that hormones could be third-generation pesticides (following the first generation – kerosene, arsenates, nicotine, and rotenone – and the second – organics such as DDT). He predicted that insects could not mount resistance to hormonally active materials. Even though this prediction proved naive, industry and governments were not deterred from becoming interested in insect biochemistry. Research teams were established to identify hormone analogues, especially those that might have specificity for particular pests. Plants turned out to be repositories of phytoecdysteroids, juvenoids, and antijuvenoids (Bowers, 1985; Horn, 1989). Not surprisingly, much of this work relied on lepidopteran assay systems. Lepidopterans also provided the material and assay system that led to the amino acid sequence for the prothoracicotropic hormone (PTTH), the brain factor that initiates insect endocrinology and triggers each molt (Kawakami et al., 1990).

One of the most useful lepidopteran model systems for endocrinologic studies has proven to be *M. sexta,* although fruitful results have been primarily in basic research rather than insect control. It was Lawrence Gilbert who first recognized its potential to be the "white rat" of insect physiology (see Gilbert, 1989, for review). He learned that *Manduca,* unlike native American silk moths, can be raised in the laboratory on an artificial diet with ease, speed, and considerable synchrony. It has a facultative, photoperiodically controlled pupal diapause.

An instance of the usefulness of *Manduca* for bioassays involved the fortuitous discovery of a few black larvae in the colony that Lynn Riddiford and Carroll Williams were using at Harvard. A stock of these larvae was established and subsequent analyses revealed that this abnormal pigmentation could be prevented if fourth instar larvae were treated with juvenoids at the time of head capsule slippage. This led to a rapid bioassay for juvenoids (Truman, Riddiford, and Safranek, 1973).

The special feature of *Manduca,* however, which is responsible for its prominence, is that key events in larval development can be timed with great precision. Using classic techniques of ligation, Truman and Riddiford (1974; Truman, Riddiford, and Safranek, 1974; Fain and Riddiford, 1976, and references therein) established the critical periods for release of the major hormones. Their work established that if close attention were paid to rearing conditions, one could predict precisely when PTTH, juvenoids, and ecdysteroids would be released in the final instar.

As techniques became available, hormone titers were measured. Small blips of ecdysteroids found early in the last instar had been seen in radioimmunoassays of many species. It was the synchrony of *Manduca* development, and Riddiford's development and use of protocols to monitor epidermal development with tissue culture and transplantation, that revealed that these blips constituted a "commitment peak." Larval cells become committed to pupal development as judged by their becoming insensitive to juvenoids (Mitsui and Riddiford, 1978; see also Riddiford, this volume).

The programming of metamorphosis has also been studied in *Galleria mellonella,* where it has been shown that allatectomized larvae can bypass the pupal stage and form adult cuticle (Sehnal, 1972). In terms of the now better understood role of ecdysteroids in triggering the transcriptional activation of particular genes (see the section "Molecular studies"), this finding is now less mysterious than it seemed at first. A provocative account of other studies on *Galleria* has been provided by Kumaran (1991).

An important aspect of current endocrinologic studies involves the characterization of particular cellular effects with hormonal changes. One of these phenomena is programmed cell death. Finlayson's (1956) precise account of which muscles break down at adult eclosion in several moths was followed by Lockshin's extensive studies (see Lockshin, 1985, for review) and those of the Truman group (Schwartz and Truman, 1982; see also Truman, this volume). Studies of programmed cell death have culminated in the discovery that baculoviruses produce proteins that curtail apoptosis (a particular scenario of cell death) in infected lepidopteran cells (Clem, Fechheimer, and Miller, 1991).

Diapause and eclosion

Two other hormonally triggered phenomena that have been investigated intensively in Lepidoptera are diapause and eclosion. Lepidoptera, indeed, were the key subjects for studies that defined diapause in insects, elucidated the basis for entry and exit, and characterized its metabolism. Duclaux, in 1869 (cited in Lees, 1955), recognized that diapausing *Bombyx* embryos would not develop at room temperature unless they had been subjected to a period of chilling. Diapause, as a term, was first applied to grasshopper embryonic movements, but by 1904, was restricted to describe arrested development (Lees, 1955).

The phenomenology of diapause has been studied extensively, with Williams (1956b) defining critical periods for chilling and Danilevskii (1965) providing detailed life histories that reveal the genetic plasticity of photoperiodic response. The cause of arrest in the pupae of *H. cecropia* was pinpointed by Williams (1952) as a quiescence of the brain/prothoracic gland axis, while at the same time the basis of diapause in the embryos of *Bombyx,* which involves the brain and subesophageal gland, was being clarified by Fukuda (1951) and Hasegawa (1951). The first indications, however, that the phenomenon involves an endocrinologic signal came from a much earlier study of Umeya (1926, cited in Denlinger, 1985), who showed that hemolymph transfusions alter the incidence of diapause. The *Bombyx* diapause factor, which leads to embryologic arrest, has recently been purified and sequenced (Imai et al., 1991). The active agent in some larval diapauses was identified as JH when Chippendale and Yin (1973) used ligations and hormone injections with the corn borer, *Diatraea grandiosella,* to show that juvenoids are the principal regulatory agents for entry and maintenance of the diapause state.

Eclosion, the emergence of an insect from the cuticle of the previous stage, is also under hormonal control, but in this instance, the control involves a novel peptide hormone; research on eclosion mechanisms is, in contrast to the other studies, fairly recent. Truman first established that eclosion of silk moths occurs at the appropriate day in development and at a certain photophase when an eclosion hormone is released (Truman and Riddiford, 1970; Truman, 1985). His laboratory shifted to *Manduca* and went on to purify eclosion hormone, obtain its sequence (Marti, Takio, Walsh, Terzi, and Truman, 1987), and isolate and characterize its gene (Horodyski, Riddiford, and Truman, 1989). Other researchers have also begun to work on eclosion hormone, simultaneously obtaining its sequence from *Manduca* (Kataoka, Troetschler, Kramer, Cesarin, and Schooley, 1987) and *Bombyx* (Kono, Nagasawa, Isogai, Fugo, and Suzuki, 1987).

One of the targets of eclosion hormone is the insect's central nervous system (CNS); the molecular events that accompany acquisition of developmental competence in the CNS to respond to the hormone have also been described (Morton and Truman, 1986). In silk moths that lack valves in their cocoons, eclosion is facilitated by the secretion of a trypsinlike enzyme termed, simply, "cocoonase" (Kafatos and Williams, 1964), which served as one of the early insect models for studying the regulation of gene activity.

Nonendocrinologic physiology: biochemistry of the hemolymph

Although a major function of the hemolymph is the carrying of endocrinologic agents (in addition, of course, to basic nutrients for cell growth and metabolism), it has also proven to be a rich source of other substances. In particular, the characterization of various protein families of biological interest was often first possible in the Lepidoptera because of their large size. Products first isolated in the Lepidoptera were then subsequently sought and found in other insects, including *Drosophila.* One group of note are the vitellogenins, the yolk proteins. Telfer (1954), studying *H. cecropia,* was the first to recognize female-specific proteins in the hemolymph of insects and subsequently showed that the vitellogenins are taken up by the ovaries (Telfer, 1960) by passing between follicle cells (Telfer, 1961).

In addition to the yolk proteins, there are numerous insect storage and transport proteins of significance that were first isolated and characterized in the Lepidoptera. These include the first insect hemolymph protein of known function, apart from hemoglobin, namely, lipoprotein I (Chino, Murakami, and Harashima, 1969); the first methionine-rich storage proteins (Tojo, Betachaku, Ziccardi, and Wyatt, 1978); the recognition of arylophorins as a class of proteins with high content of aromatic amino acids (Telfer, Keim, and Law, 1983); first recognition of flavoproteins with high histidine, associated copper, and no cystine (Telfer and Massey, 1987); the first insect iron-binding proteins (Huebers et al., 1988); and the first insect serine protease inhibitors (serpins) (Kanost, Prasad, and Wells, 1989). For hemolymph biochemistry, the reader should consult reviews by Kanost et al. (1990) and Telfer and Kunkel (1991). The utility of lepidopterans as models for understanding lipid transport is summarized by Law and Wells (1989). Diverse tissues synthesize hemolymph proteins (Palli and Locke, 1987, 1988; Sass, Kiss, and Locke, 1993, 1994).

A further aspect of the biochemistry of the hemolymph that is of great importance and deserves mention is the system of humoral immunity mediated by special proteins carried in the hemolymph. It had long been known that moth hemolymph carries lysozyme (Chadwick, 1970) and it had been thought that this was the chief agent of protection against infection, given that its levels increase following pathogenic challenge. The story, however, has proven far more interesting.

The pioneering figure in this work was H. Boman, who, stimulated by a conversation with Williams, began working with *H. cecropia* on humoral defenses against bacteria in diapausing pupae. The outcome of this work was the purification and characterization of four classes of peptides that function as defense molecules against pathogens and parasites (Boman and Hultmark, 1987; Boman, Faye, Gudmundsson, Lee, and Lidholm, 1991). Bacterial infection, which is signaled by the peptidoglycans of the bacterial cell surface (Dunn, Dai, Kanost, and Geng, 1985), induces, in addition to lysozyme, cecropins, attacins, and hemolins. The cecropins are small peptides that work via amphipathic alpha-helices, a mode of action confirmed by the finding that peptides built with D amino acids are totally active. Thus the need to postulate any enzymatic or other specific recognition process has been ruled out (Wade et al., 1990). Boman speculated that cecropins might be used by vertebrates; he targeted a relatively sterile region of the intestine as a place where there was no obvious explanation for its paucity of microbes and succeeded in isolating cecropinlike molecules from pigs. An entirely new class of molecules in the mammalian defense repertoire was thus discovered (Agerberth et al., 1991). The attacins, with molecular weights of about 20,000 daltons, have antibacterial activity only against a limited array of bacteria, and appear to act by disrupting the outer bacterial membrane (reviewed in Boman and Hultmark, 1987).

An important addition to the defense system is hemolin. Although recognized as the most abundant of the pathogen-induced proteins in *H. cecropia* (Rasmuson and Boman, 1979), its significance was obscure initially because of its lack of bactericidal activity. However, when the amino acid sequence of hemolin from *H. cecropia* and its *Manduca* homologue were obtained, both were found to share considerable sequence similarity to vertebrate immunoglobulins (Sun, Lindström, Boman, Faye, and Schmidt, 1990; Ladendorff and Kanost, 1991). The evidence of a connection between vertebrate and invertebrate immune systems was buttressed by the recent discovery that genes for cecropin contain response elements and binding factors clearly homologous to the NF–κB system, which coordinates immunologic responses in vertebrates (Sun and Faye, 1992a).

The current status of immunologic studies on the Lepidoptera is described by Mulnix and Dunn in this volume. One of the central questions concerns the specificity of the immune response, and it is becoming clear that Lepidoptera, and by extension, other insects, possess considerably more specificity of immune response than was once thought.

Genes, genetics, and development

Gene action and biochemistry: early studies

In one of the pioneering studies on the nature of gene action, Caspari (1933) investigated the pleiotropic action of a single gene on the pigmentation of larval cuticle and on adult eyes and testes in the moth *Ephestia.* He was the first to perform transplantation experiments designed "to attack the question as to the mode of action of genes via the method elsewhere most customary in the physiological study of development, namely by means of experimental surgery" (translation from Bodenstein, 1971). The clear interpretation of the results obtained was that some diffusible factor could affect pigmentation of implant or host. Caspari's experiments were cited by Beadle and Ephrussi (1936) in their work on eye disc distransplantation in *Drosophila.* Subsequently, Beadle acknowledged in his Nobel lecture (Beadle, 1977) that it was his and Ephrussi's *Drosophila* studies, not the much-cited and honored *Neurospora* work with Tatum, that led to the development of the one-gene–one-enzyme hypothesis. Hence, the one-gene–one-enzyme hypothesis had its origins, at least in part, in work with a small moth.

Early genetic studies

Though little known outside the world of lepidopteran research, the first reported studies of mutants in animal systems, and hence the first genetic analyses of silk moths, go back to the first decade of the twentieth century (Coutagne, 1902; Toyama, 1906, in Tazima, 1964). *Bombyx* thus ranks with the laboratory mouse as one of the first animal species to receive attention from modern geneticists, following the rediscovery of Mendel's findings in 1900. Indeed, the silkworm remains second only to the fruit fly, *Drosophila melanogaster,* as an insect model for genetic studies; the current status of silkworm genetics is reviewed by Goldsmith in this volume. One of the early mutants to be found, Gray egg (*Gr*), which makes an altered chorion, provided the first indications of a possible developmental and comparative genetics of the chorion; such studies are described in this book by Goldsmith and by Regier, Friedlander, Leclerc, Mitter, and Wiegmann. The molecular aspects are briefly described here as well as by Eickbush and Izzo and by Kafatos, Tzertzinis, Spoerel, and Nguyen in this volume.

Another especially important area in genetics research in *Bombyx* has proved to be that of the homeotic mutants of the *E* locus. Several of the first detailed studies were undertaken in the thirties and forties (for early reviews of this work, consult Tazima, 1964). Some of this work, on mutants that cause thoracic-abdominal and abdominal-thoracic changes, thus preceded the ground-breaking experiments by E. B. Lewis on the bithorax complex of *Drosophila* (Lewis, 1963, 1978) by several decades. Early lepidopteran development is described by Nagy in this volume, and molecular analysis of the homeotic genes of *Bombyx* by Ueno, Nagata, and Suzuki.

The silk gland

A principal motivation for developing a genetics of *Bombyx* was its use in sericulture. Japan, with its large silk industry, has been the primary site of this research. Given that the high capacity for silk production by *Bombyx* is a function of the special attributes of the silk gland, which is a modified salivary gland, it is not surprising that at some point biochemists and molecular biologists would begin to focus attention on the silk gland and its workings. The principal pioneer in bringing this research to the gene level was Yoshiaki Suzuki, whose contributions date from the late 1960s (see Suzuki, 1977, for a review of early work on silk gland differentiation). Another major early contributor was J. P. Garel, who showed that the unusual amino acid composition of the chief silk protein, fibroin, was matched by the presence in the silk gland of specialized tRNAs (Garel, Mandel, Chavancy, and Daillie, 1970, 1971).

Indeed, the silk gland is a factory for the production of two classes of protein, fibroin (the main structural protein of silk) and the sericins (which coat fibroin). Not surprisingly, the silk gland is not only highly enriched for the tRNAs required in abundance for the special codon requirements of these proteins but for the mRNAs that encode these proteins as well (Suzuki and Brown, 1972). This mRNA enrichment is achieved primarily by substantially reduced rates of degradation rather than by enhanced rates of transcription or translation (Suzuki and Giza, 1976). A parallel to this aspect of silk gland development was found in the galea, a modified mouthpart that differentiates during adult development to produce cocoonase, the eclosion enzyme mentioned earlier. Again, relatively high stability enables cocoonase message to accumulate to extremely high levels despite average rates of transcription (Kafatos, 1972).

The unusually glycine-rich nature of fibroin claims attention not only for its intrinsic interest but because it also permitted one of the only direct gene isolations ever achieved. The method employed fractionation, based on the GC-content of these genes rather than on standard cloning techniques (Tsujimoto and Suzuki, 1984). This study helped to confirm that the selective activation of the fibroin gene occurs without altering DNA structure by chemical modifications such as methylation.

These findings, and others pertaining to chorion composition and structure (Kafatos et al., 1977; Mazur, Regier, and Kafatos, 1980, 1982), were achieved essentially at the dawn of the present molecular era, dominated by gene cloning and gene analysis studies. It is to this contemporary work that we now briefly turn.

Molecular studies

Endocrinology

The prelude to present molecular approaches to the study of hormone action was an early demonstration that the site of ecdysone action is at the level of the gene. Clever and Karlson (1960) transferred ecdysone isolated from moths to the midge, *Chironomus,* and provided evidence that the initial site of action is the chromosome. Within 20 minutes of receiving an injection of ecdysone isolated from *Bombyx,* puffing (RNA synthesis) ensued at a few precise sites on the salivary gland chromosomes of the midge. This was the first evidence, in any system, that hormones could have genes as their primary target. The subsequent elucidation of steroid hormones as direct effectors of gene expression in vertebrates (reviewed in Rories and Spelsberg, 1991, and Carson-Jurica, Schrader, and

O'Malley, 1990) gives particular historical significance to the early results with lepidopteran ecdysone.

Subsequent studies with *Drosophila* have paved the way for comparable analyses in key lepidopteran systems, as reviewed by Riddiford in this volume, including, not least, the discovery of multiple ecdysone receptor forms. Ultimately, the depth of prior characterization of physiologic responses in the Lepidoptera should permit an even more thorough analysis of the whole endocrinologic and gene regulatory system. The molecular characterization should, in particular, help to elucidate the known cellular changes that are triggered by hormonal change, such as those occurring in the nervous system during metamorphosis, which are described by Truman in this volume.

Still awaiting elucidation is how juvenoids act to prevent ecdysteroids from initiating metamorphosis. The action is clearly not one of simple antagonism, for juvenoids do not interfere with the molting process, but only with its outcome. Analyses of cuticular proteins (see the next section) have permitted structural correlates of metamorphosis to be examined at the molecular level. The "gene set" paradigm for metamorphosis, in which the distinctive features of each stage were postulated to be underwritten by a set of stage-specific genes, has been found to be far too simplistic (Willis, 1986). Major cuticular proteins are present in more than one metamorphic stage, and each protein is clearly encoded by a single gene (Riddiford, 1987).

Genomes, genes, and gene products

The ability to isolate and characterize innumerable DNA sequences also began to provide findings of significance for evolutionary studies. The first emerged from comparative genome analyses, which allowed extensive characterization of mobile elements and their arrangement in the context of overall genome structure. The wild silk moth, *Antheraea pernyi,* was the first insect shown to have the characteristic short repeated sequence interspersion length found in most animals (Efstratiadis, Crain, Britten, Davidson, and Kafatos, 1976), in contrast to the long sequence interspersion arrangement then known for *Drosophila* and the honeybee. The consequence of this early work has been a wealth of studies of interest not only for those working on Lepidoptera but for insect biologists in general and even vertebrate biologists. This work is reviewed by Eickbush in this volume.

The second area has been in comparative studies of various structural gene products, in particular those of the chorion and the cuticle. Work with chorion (eggshell) proteins revealed one of the most extensive multigene families recognized to date. Kafatos and his students began their analyses with species of *Antheraea*. Later they turned to *Bombyx* to reduce the complications that population polymorphism contributed to the heterogeneity of the chorion protein array and to take advantage of chorion mutants in studying cell differentiation and morphogenesis of the eggshell.

The foundation of the evolutionary chorion work has been the intensive analysis of chorion gene regulation in *Bombyx* and the differences and similarities shown in this regulatory system to that of *Drosophila*. Analysis of *Bombyx* chorion protein variants extended the genetic map to the chromosome level (Goldsmith and Basehoar, 1978; Goldsmith and Clermont-Rattner, 1979) and, together with studies of cloned DNA (Eickbush, Jones, and Kafatos, 1981), revealed a giant locus, described by Eickbush and Izzo in this volume. In silk moth species, most chorion genes are arranged in pairs with a short noncoding region between. Each member belongs to a different family, but both genes are expressed at the same time; transcription proceeds in opposite directions. Despite major differences in the *Drosophila* system – where the chorion genes are not arranged in pairs but in two large clusters of independently transcribed genes, clusters that undergo gene amplification during choriogenesis (Digan, Spradling, Waring, and Mahowald, 1979; Levine and Spradling, 1985) – Mitsialis and Kafatos (1985) were able to show, by making transgenic *Drosophila*, that certain *cis*-controlling elements in chorion gene regulation have been conserved over a period of more than 200 million years of divergent evolution between the Diptera and the Lepidoptera. Current studies and understanding of chorion gene regulation are reviewed by Kafatos et al. in this volume.

The chorion evolutionary studies have taken off in two directions. On the one hand, much fruitful comparison at the level of gene structure has revealed the importance of gene conversion hotspots (Eickbush and Burke, 1986) and raised questions about the nature of the forces that shaped the evolution of the chorion gene cluster in lepidopterans, reviewed by Eickbush and Izzo in this volume. On the other, the ways in which selective forces and ecological niches have affected chorion composition, chorion gene regulation, and ultimately, chorion morphology throughout the Lepidoptera is of great importance for understanding the evolution of phenotype and development, a subject discussed by Regier et al. in this volume.

The cuticle provides an intellectual challenge comparable to that of the chorion, for here, too, we need to learn how molecules contribute to the form and function of a structure that is necessary for the success of all insects. Unfortunately, the analysis of cuticular proteins has not yet reached the same level of sophistication as the chorion, despite the greater abundance of starting material. Although the number of proteins may be comparable, and they can be assigned to families, their assembly into cuticle is complicated by the presence of chitin and the diverse forms of sclerotizing agents. Chorion proteins, in contrast, need only assemble with each other, and for a given species, a single cross-linking agent appears to be employed.

Still, much progress has been made. We have learned from studies with *H. cecropia* cuticular proteins (Cox and Willis, 1985; Willis, 1987, 1989), as well as those with locusts (Andersen, 1979), that the physical properties of cuticles are reflected in the electrophoretic properties of cuticular proteins. Originally, mechanical properties of cuticle had been correlated only with chitin and water content and degree of sclerotization; now we recognize that the more acidic proteins are extracted from flexible cuticles and that across orders one can distinguish a cuticular protein sequence as coming either from flexible or rigid cuticles (Willis, 1987, 1989; Lampe and Willis, 1994). Recent work with another lepidopteran, *Calpodes ethlius,* has demonstrated elegantly the sites of synthesis and precise cuticle localization of diverse integumentary and hemocyte proteins (Sass, Kiss, and Locke, 1993, 1994).

In addition to the molecular developmental analyses of structural proteins, a major, though still comparatively recent, growth area in lepidopteran research has concerned early development. As with the molecular endocrinologic research, a major impetus has come from *Drosophila* studies, with the use of *Drosophila* homeotic and segmentation genes to clone their lepidopteran homologues; this work and its relationship to what is known in Lepidoptera are reviewed by Nagy and by Ueno et al. in this volume. Satisfyingly, two of the regions identified by molecular means are found to correspond to previously known – and in the case of the *E* locus, one exceptionally well-characterized – genetic regions. The comparative aspects possible both within the Lepidoptera and between the Lepidoptera and the Diptera promise an enriched understanding of developmental mechanisms not possible from the fruit fly work alone. In particular, the relationship between segmentation and the imposition of segment identity should show some interesting differences between the long germ band form of *Drosophila* and the forms of germ band development seen in the Lepidoptera. The developmental and evolutionary issues associated with these phenomena are discussed by Nagy and by Regier et al. in this volume.

Lastly, but by no means least, is the work on the silk gland, the area of lepidopteran research that was the first to benefit from modern molecular techniques. Recent findings significantly extend the earlier work. In particular, the novel aspects of RNA polymerase III in the transcription of tRNA genes have produced some major surprises, including the involvement of an RNA molecule as part of the catalytic equipment, which is discussed by Sprague in this volume, and the presence and apparent activity of several homeodomain proteins homologous to known *Drosophila* homeodomain proteins potentially implicated in differentiation of the silk gland, described by Hui and Suzuki in this volume.

Insect tissue culture and virology

The development of methodology in two areas has contributed substantially to lepidopteran research in the past and promises to continue to do so in the future, particularly in connection with the use of baculoviruses as cloning vectors. These areas are insect tissue culture and the related study of baculoviruses. Insect tissue culture, indeed, had its origins in lepidopteran research, and this history is intertwined with studies on baculoviruses (see Benz, 1986). Although a disease of silkworms now known to be caused by baculoviruses had been described in the sixteenth century, progress in elucidating the cause of the disease was only made in the twentieth century and was laborious. It was complicated by the presence of visible occlusion bodies that kept investigators from recognizing that the causal agent was viral.

The first to culture insect tissue was the geneticist Richard Goldschmidt. He placed cysts taken from testes of *H. cecropia* into hanging drops of hemolymph and found that spermatogenesis could ensue under these tissue culture conditions (Goldschmidt, 1915). Two decades later, another advance was made by Trager, who multiplied baculoviruses in tissue culture (Trager, 1935), but substantial progress in this area had to wait another two decades, when S. S. Wyatt designed a complex culture medium based on the levels of organic compounds and amino acids found in *Bombyx* hemolymph (Wyatt, 1956). Indeed, it was collaboration in this study that led G. R. Wyatt and Kalf (1957) to discover that trehalose is the principal sugar in insect hemolymph.

Subsequent modifications of Wyatt's medium by Grace (1962) allowed the successful culture of diverse lepidopteran tissues and the establishment of numerous immortal cell lines from insect tissues. Interestingly, a Chinese investigator, Za-Yin Gaw, had earlier developed the first immortal insect cell line from *Bombyx* tissues. Unfortunately, his achievement remained little known despite having been published in English (Gaw, Lin, and Zia, 1959).

Such cell lines made possible the long-term cultivation of baculoviruses. Although the initial interest in baculovirus culture was in their use as agents of insect control, more recent work has emphasized their potential as cloning vectors. The high rates of synthesis of the viral coat protein, polyhedrin, led to the development of genetically engineered viruses. These viruses contain a recombinant gene in which the polyhedrin promoter is fused to coding sequences of products of interest to commerce and basic research. The discovery that lepidopteran cells infected with these viruses carry out core glycosylation made it possible to produce active products, while constructs expressed in bacteria had failed (Smith, Summers, and Fraser, 1983; Maeda, 1989a). The current state of the use of baculoviruses as cloning vehicles is described by Iatrou in this volume.

The improvement in culture media driven by viral research soon attracted those interested in using tissue and cell culture to probe the physiology of insects (Oberlander and Miller, 1987). *Galleria* wing imaginal discs were used in the first demonstration that ecdysteroids could stimulate imaginal disc development in vitro (Oberlander and Fulco, 1967). Wyatt and Wyatt (1971) demonstrated that cultured wing epidermis from diapausing silk moth pupae would respond to ecdysteroids by increasing its rate of protein synthesis. Willis and Hollowell (1976) were successful in obtaining longer term wing cultures to form second pupal or adult cuticle and demonstrated that ecdysteroids and juvenoids could act directly on pupal wing epidermis to regulate metamorphosis in a completely defined medium. Later experiments with larval integument from *H. cecropia* revealed that the epidermis was a site of synthesis of the vast majority of the larval cuticular proteins (Willis, Regier, and Debrunner, 1981), and more recent studies by Sass, Kiss, and Locke (1993) define diverse secretory pathways for epidermal proteins. New approaches to studying insect endocrinology have been made possible with the availability of ecdysteroid-responsive cell lines from imaginal discs, first isolated from Lepidoptera (Lynn, Miller, and Oberlander, 1982). One may reasonably anticipate that with the extended understanding of mechanisms in endocrinologic response and in development, these in vitro culture systems will lend themselves to further intensive analysis of basic cellular phenomena.

Conclusions

This introduction has documented but a few of the diverse systems where research with lepidopterans has provided the foundation for insect endocrinology and biochemistry; the impact of these model systems on developmental studies is only just beginning to be felt. Insights gained from work with Lepidoptera and their molecules have served to advance our understanding of basic biological processes, including those of mammals. The lesson from this should be that an understanding of complex biological phenomena will occur only when diverse systems are studied. It is our hope that, as the usefulness of lepidopteran systems comes to be more widely appreciated, reviewers of grants will be far less likely to challenge proposals from students of Lepidoptera and other insects with the question, "Why not *Drosophila?*"

Acknowledgments

Thanks to John Willis, Lisa Johnson, and Roger Lebrun for their help with the manuscript.

2

Genetics of the silkworm: revisiting an ancient model system

MARIAN R. GOLDSMITH

Introduction

The earliest known silk textile is nearly five thousand years old (Kuhn, 1988). The condition and weave of the fiber indicate that the insect that produced it, *Bombyx mori*, was already domesticated, implying that it had long been subjected to inbreeding and selection. In modern times, *Bombyx* has been used as a model for genetic studies since the birth of genetics as a formal science in the early 1900s. As early as 1905, Toyama, one of the founders of silkworm genetics, was breeding genetic hybrids between Thai and Japanese silkworms for improved vigor and silk production (Yokoyama, 1968). He first reported discovery of a chorion mutation that affects the shape and transparency of the eggshell in 1910 (Tazima, 1964), the same year as the publication of Morgan's famous white-eyed mutant of *Drosophila melanogaster*. Japanese geneticists maintained active research into the fundamental principles of genetics, keeping pace with the field and often leading with early reports of such phenomena as dominance (Toyama, 1906, cited in Tazima, 1964), sex linkage (Tanaka, 1914, cited in Sturtevant, 1915), maternal inheritance (Toyama, 1912), homeotic mutants (Suzuki, 1929, Sasaki, 1930, and Suzuki and Ohta, 1930 – all cited in Tazima, 1964), enzyme polymorphisms (Matsumura, 1934, cited in Kikkawa, 1953), and unstable mutations (Hatamura, 1939). Although studies with many Lepidoptera have made important contributions to genetics, today, with more than two hundred mutations mapped, the silkworm stands as the only member of this taxonomic group whose genetic system is well-enough established to consider adopting it as a molecular genetic model for solving a broad range of fundamental biological problems.

The recent explosive rate of progress in developmental and molecular biology in the fruit fly, *Drosophila melanogaster,* has shown that concentrating efforts on a single member of a taxonomic group for which there is a strong genetic foundation is an effective way to accumulate a large body of information quickly and to develop new experimental approaches that can then be brought to bear successfully on answering questions in less well-characterized and less tractable species. Study of mutants enables identification of critical steps in biological processes, often uncovering features that may be inaccessible or unsuspected when studying the wild type alone. Genetic maps provide a framework for identifying and cataloging genes with related functions, even if their products are unknown, for their direct isolation, and, where transformation is available, for their reinsertion into the germline to investigate their functions in situ (Ashburner, 1992). The availability of genetic maps in a variety of related species makes it possible to study the dynamics of genome structure and organization in an evolutionary context (Heckel, 1993). Moreover, because of its long history of use in physiological, developmental, and biochemical, as well as genetic studies, the silkworm can serve as a reference for lepidopterans as a whole that will greatly strengthen and accelerate research efforts in all of these areas.

A major purpose of this chapter is to call attention to *Bombyx* as a genetic resource. To provide a context, the first part gives an overview of the genetic system of the Lepidoptera, calling attention to unusual or idiosyncratic features of their inheritance mechanisms that set them apart from other major animal model systems and that warrant further investigation in their own right. This section also provides specific details about the silkworm as a representative species where this information is available. The second part describes the current status of traditional silkworm genetics and highlights a number of available mutants that have been studied on the molecular level or are relevant to currently active areas of research. The final section describes new approaches being developed to exploit existing silkworm resources on the molecular level in order to widen the scope of genetic analysis in this organism.

A number of excellent volumes and reviews are available for additional background on silkworm and lepidopteran genetics. Tazima's *Genetics of the Silkworm* (1964) gives a comprehensive overview of the status of silkworm genetics through the early 1960s and is especially valuable for its historical background and detailed descriptions of experimental findings in many well-established genetic systems. Kikkawa's earlier review (1953) emphasizes biochemical genetics and covers additional details to Tazima's volume. Strunnikov's *Control of Silkworm Reproduction, Development and Sex* (1983) gives extensive experimental data on work carried out in the former Soviet Union and not readily available elsewhere. Discussion of many other topics touched on here are found in Robinson's comprehensive *Lepidoptera Genetics* (1971), which also includes a detailed summary of Tazima's volume. Finally, Sheppard's review (1961) covers studies using Lepidoptera as models in population genetics.

Genetic architecture: an overview

Lepidopteran chromosomes are small (1–4 μm long in mitotic spreads), holocentric (see "Chromosome structure"), and relatively numerous, with a modal haploid number (*n*) around 30 (Robinson, 1971). In *B. mori, n* = 28 and the haploid genome size is 530 million base pairs (0.5 pg DNA); (Gage, 1974a; Rasch, 1974); this is approximately three and a half times the size of the fruit fly genome (140 million base pairs; Rasch, Barr, and Rasch, 1971) and one-sixth the size of the human genome (3,000 million base pairs; Stephens, Cavanaugh, Gradie, Mador, and Kidd, 1990). Based on a 2.5-fold variation in relative chromosome length in pachytene (Traut, 1976), individual chromosomes measure 8–20 million pairs. The present genetic linkage map in the silkworm measures 1,000 centiMorgan (cM), with linkage groups carrying five or more markers ranging from 20 to 55 cM (Doira, 1992; Figure 2.1).

Lepidopteran females are heterogametic in sex chromosome constitution; males are homogametic. Based on the convention that designates the female sex chromosome as W and its counterpart as Z (see discussion in Robinson, 1971), *B. mori* females are ZW and males are ZZ; females of other lepidopteran species may be ZW, ZO, ZZW, or ZZWW (the last being a case of presumed fragmentation of the W; Traut and Mosbacher, 1968; Suomalainen, 1969; see also Robinson, 1971). Hasimoto first suggested that the W has a female-determining role in the silkworm, using polyploids (1933; cited in Tazima, 1964). By producing strains carrying extra chromosome fragments or translocations derived from the W and Z chromosomes, Tazima confirmed this model of sex determination in *Bombyx* and showed that the sex determinants are localized at one end

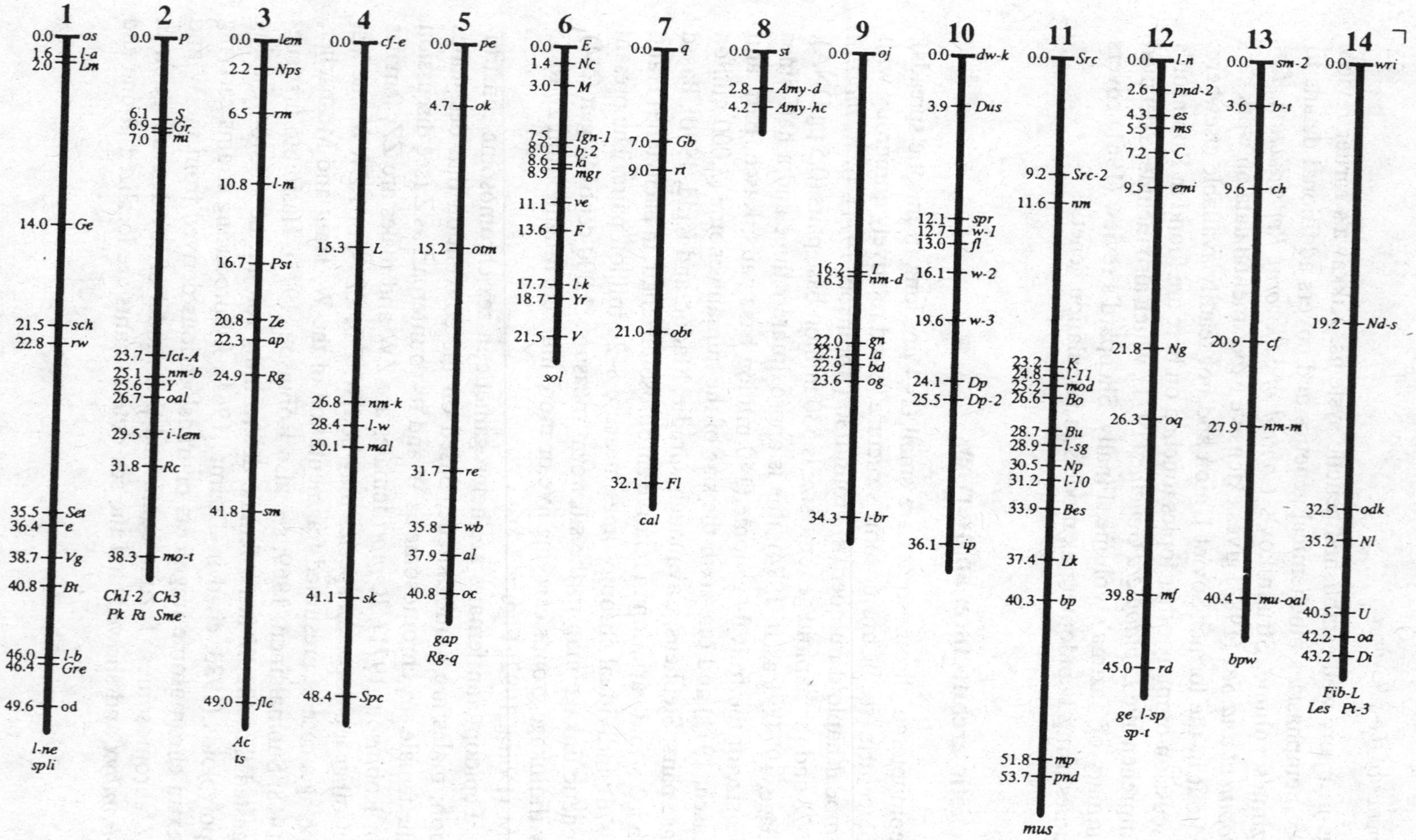
1
0.0 os
1.6 l-a
2.0 Lm
14.0 Ge
21.5 sch
22.8 rw
35.5 Set
36.4 e
38.7 Vg
40.8 Bt
46.0 l-b
46.4 Gre
49.6 od
l-ne
spli
2
0.0 p
6.1 S
6.9 Gr
7.0 mi
23.7 Ict-A
25.1 nm-b
25.6 Y
26.7 oal
29.5 i-lem
31.8 Rc
38.3 mo-t
Ch1·2 Ch3
Pk Rt Sme
3
0.0 lem
2.2 Nps
6.5 rm
10.8 l-m
16.7 Pst
20.8 Ze
22.3 ap
24.9 Rg
41.8 sm
49.0 flc
Ac
ts
4
0.0 cf-e
15.3 L
26.8 nm-k
28.4 l-w
30.1 mal
41.1 sk
48.4 Spc
5
0.0 pe
4.7 ok
15.2 otm
31.7 re
35.8 wb
37.9 al
40.8 oc
gap
Rg-q
6
0.0 E
1.4 Nc
3.0 M
7.5 Ign-1
8.0 b-2
8.6 ka
8.9 mgr
11.1 ve
13.6 F
17.7 l-k
18.7 Yr
21.5 V
sol
7
0.0 q
7.0 Gb
9.0 rt
21.0 obt
32.1 Fl
cal
8
0.0 st
2.8 Amy-d
4.2 Amy-hc
9
0.0 oj
16.2 I
16.3 nm-d
22.0 gn
22.1 la
22.9 bd
23.6 og
34.3 l-br
10
0.0 dw-k
3.9 Dus
12.1 spr
12.7 w-1
13.0 fl
16.1 w-2
19.6 w-3
24.1 Dp
25.5 Dp-2
36.1 ip
11
0.0 Src
9.2 Src-2
11.6 nm
23.2 K
24.8 l-11
25.2 mod
26.6 Bo
28.7 Bu
28.9 l-sg
30.5 Np
31.2 l-10
33.9 Bes
37.4 Lk
40.3 bp
51.8 mp
53.7 pnd
mus
12
0.0 l-n
2.6 pnd-2
4.3 es
5.5 ms
7.2 C
9.5 emi
21.8 Ng
26.3 oq
39.8 mf
45.0 rd
ge l-sp
sp-t
13
0.0 sm-2
3.6 b-t
9.6 ch
20.9 cf
27.9 nm-m
40.4 mu-oal
bpw
14
0.0 wri
19.2 Nd-s
32.5 odk
35.2 Nl
40.5 U
42.2 oa
43.2 Di
Fib-L
Les Pt-3

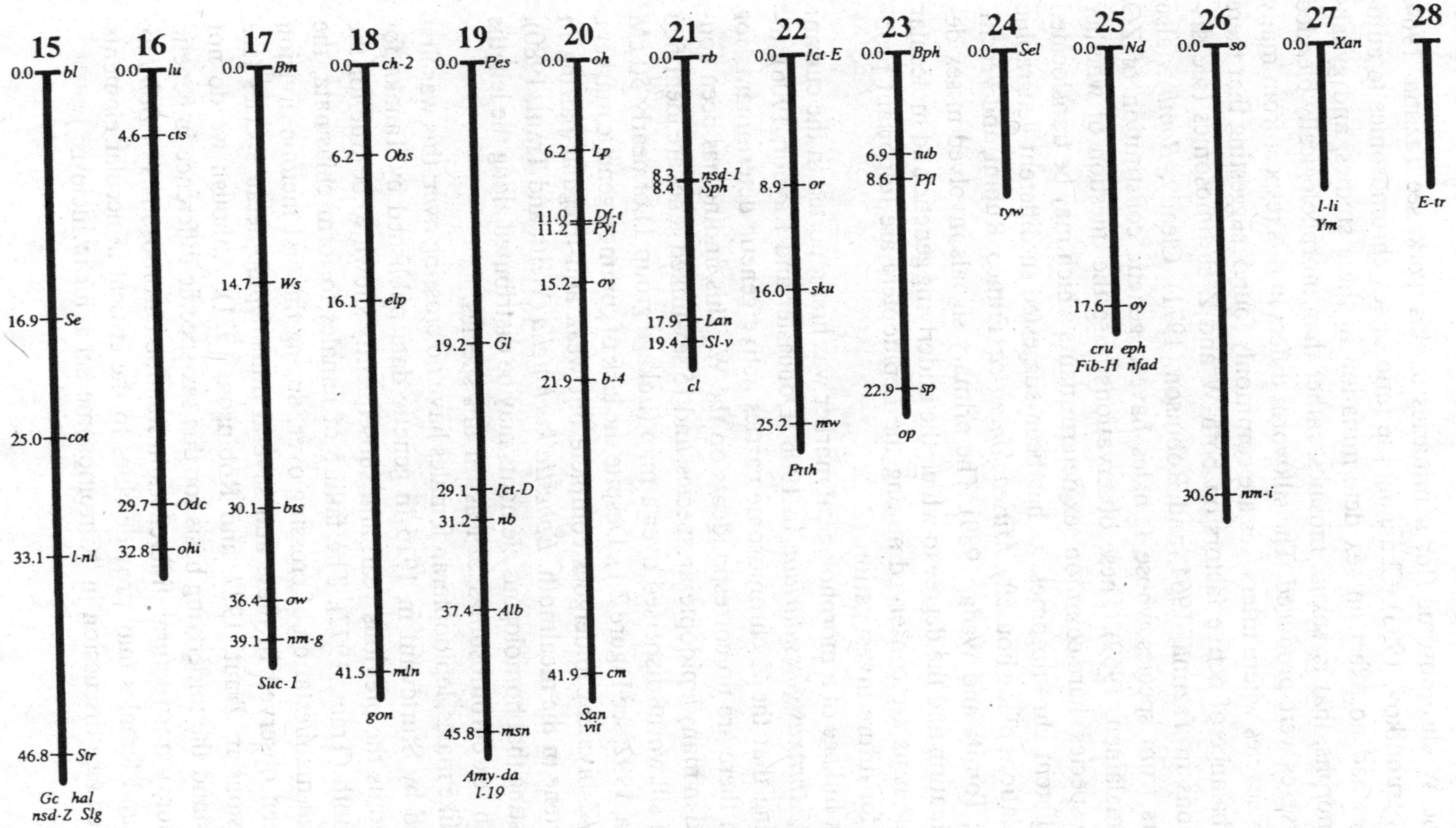

Figure 2.1. Genetic linkage maps in *Bombyx mori*. (Doira, 1992, with permission.)

of the W chromosome (for summaries of this work, see Tazima, 1964, and Strunnikov, 1983). Changing the ratio of sex chromosomes to autosomes had no effect on sex determination in these studies, and gynandromorphs, that is, sexual mosaics, rather than morphologically blended intersexes were produced. The silkworm differs in this respect from many other species where intersexes are commonly found, suggesting that there is a balance of active factors on both W and Z chromosomes (see discussions in Tazima, 1964, and Robinson, 1971). Clearly, *Bombyx* also differs from species whose females have a genetic constitution of ZO (Suomalainen, 1969). These observations raise the question of whether these species share common sex determinants which may be translocated to different chromosomes, as has been suggested in different geographic populations of the housefly, *Musca domestica* (Franco, Rubini, and Vecchi, 1982; Tomita and Wada, 1989). The primary signals involved in sex determination and the degree to which the underlying genetic and molecular mechanisms are conserved among the Lepidoptera are important problems for future investigation.

Sex linkage of a morphological marker was first reported in the currant moth, *Abraxas glossulariata,* in 1906 (Doncaster and Raynor, 1906), indicating that the Z chromosome carries active genetic determinants for which there are no correspondents on the W. This finding has been confirmed in many lepidopteran species, and 15 sex-linked genes are identified in the silkworm dispersed over a map (linkage group 1) of nearly 50 cM (Doira, 1992; see Figure 2.1). Despite the lack of common genetic markers, the WZ bivalent undergoes complete but weak and irregular pairing in pachytene in the mealmoth, *Ephestia kuehniella* (Weith and Traut, 1980), suggesting that homologous elements may be distributed along the lengths of both sex chromosomes, at least in this species.

Unlike males, lepidopteran females have no crossing over; this was first noticed by Sturtevant in 1915 in genetic data published by Tanaka for silk moths the previous year, and subsequently shown to be true for the mealmoth (Traut, 1977). The failure of females to form chiasmata, the physical manifestation of crossing over, during the first meiotic division has been observed in these and several other lepidopteran species (see discussions in Traut, 1977, and Robinson, 1971). Although we do not understand the underlying basis for this sex-specific difference in genetic behavior, ultrastructural differences in meiotic chromosomes of *Bombyx* males and females may provide clues to the critical molecular elements involved (see discussion in "Chromosome behavior in meiosis").

There appears to be no dosage compensation in Lepidoptera; this is most convincingly demonstrated in a study of a sex (Z)-linked gene for 6-phosphogluonate dehydrogenase in two species of *Heliconis* butterflies in which males showed twice the levels of enzyme activity as females (Johnson and Turner, 1979). Unlike dosage-compensated species such as mammals, in which one copy of the homogametic chromosome pair (XX) is inactivated in somatic tissues by heterochromatization (Lyon, 1992), in many lepidopteran species (as well as in other taxa with heterogametic females, like birds; see Cock, 1964), there is ample evidence that it is the W chromosome that becomes inactivated. Thus, for example, in females (but not males) a heterochromatic body may be found in tissues as diverse as Malpighian tubules, silk glands, mandibular glands, and epidermal cells (Traut and Mosbacher, 1968; Suomalainen, 1969; Traut, 1976). The sex chromatin body has been correlated with the W chromosome in endomitotic tissue (Perdrix-Gillot, 1979), in artificially induced polyploids (Ito, 1977), and most tellingly, in mutants carrying translocations between the W and various autosomes in both *Ephestia* (Traut and Scholz, 1978; Schulz and Traut, 1979) and *Bombyx* (Qingzhong, Sugai, and Osiki, personal communication). Sex heterochromatin differs from heterochromatin in polytene chromosomes in that it replicates in proportion to the degree of cellular polyploidy in terminally differentiated tissues, but later than the bulk of the genome (Traut and Scholz, 1978; Perdrix-Gillot, 1979).

That lepidopteran sex heterochromatin is transcriptionally inactive has been shown by its failure to label in the presence of tritiated uridine in silk gland cells of *Ephestia* (Traut and Scholz, 1978). Consistent with this observation is the finding of condensed tangles of undispersed, globular particles on the lateral element of the synaptonemal complex associated with the W chromosome in the WZ bivalent and on the W segment of W-autosome translocations in chromatin spreads of pachytene nuclei examined in the electron microscope in the mealmoth (Weith and Traut, 1980, 1986), the silkworm (Wang and Xu, 1990), and the waxmoth, *Galleria mellonella* (Wang, Marec, and Traut, 1993). Nevertheless, uridine incorporation is associated with the heterochromatin in nurse cells, indicating that at least part of the W chromosome remains active in tissues involved in sex determination (Guelin, cited in Traut and Scholz, 1978). These kinds of observations have led to the suggestion that heterochromatization serves to reduce or eliminate the function of the W chromosome in selected tissues, analogous to the role of chromosome diminution in *Ascaris, Cyclops,* and the Sciaridae (Traut and Scholz, 1978).

Even in the absence of dosage compensation, it would be of interest to find out whether heterochromatization of the W chromosome shares any features with the better known phenomenon of X chromosome inactivation in mammals (Lyon, 1992; Riggs and Pfeifer, 1992). The identification of W-specific sequences would facilitate further investigation of this process at the molecular level.

As in most insects and in many other animals and plants, terminally differentiated tissues in the Lepidoptera undergo DNA endoreduplication without cell division, the levels depending on the tissue (D'Amato, 1989). Numerous tiny chromosomes may condense during this process (Perdrix-Gillot, 1979), but, as in most insect orders except Diptera and Hymenoptera, they are never seen to polytenize (Nagl, 1978). Chromosomes undergo 18 to 20 replication cycles in the middle and posterior sections of the silk gland in *Bombyx* (Gage, 1974; Perdrix-Gillot, 1979), the highest of any known tissue (D'Amato, 1989); replication cycles are highly synchronous, with most variability occurring during the G or resting phase (Perdrix-Gillot, 1979). The synchrony, large size of silk gland nuclei, and abundant tissue make this a promising system for studying the control of endomitosis, especially in light of recent advances in our understanding of the regulation of the cell cycle in eukaryotic systems (Norbury and Nurse, 1992).

A curious feature of lepidopteran genetics is that males produce both eupyrene or true sperm and significant numbers of apyrene or nonnucleated sperm. The former mature during larval development and participate in fertilization, whereas the latter differentiate during the pupal to adult metamorphosis, and though they are physiologically capable of activation and migration to the sperm storage organ in females during copulation, they subsequently degenerate (Silbergleid, Shepherd, and Dickinson, 1984). Studies in the codling moth, *Cydia pomonella,* have shown that formation of apyrene sperm is induced by a factor found in the hemolymph (Jans, Benz, and Friedländer, 1984), and that the abnormal meioses that give rise to them are delayed and have a shortened prophase relative to those of eupyrene sperm (Friedländer and Hauschteck-Jungen, 1986). Control of this intriguing phenomenon bears further investigation on the molecular level, particularly in light of its possible significance for population dynamics and evolution. For an extensive bibliography and discussion of the possible function of apyrene sperm, see Silbergleid, Shepherd, and Dickinson (1984).

Chromosome structure

As indicated earlier, a distinctive property of lepidopteran chromosomes is that, instead of a single, localized centromere and associated microtubule attachment apparatus, they possess multiple or diffuse kinetochores. The evidence for this is largely indirect, including the lack of well-defined primary constrictions in mitotic and meiotic chromosome spreads, complete terminalization of chiasmata at diakinesis in males of some species, and parallel alignment of homologues during their migration to the poles (Murakami and Imai, 1974; Maeki, 1981b). A diffuse kinetochore is also indicated by the presence of supernumerary chromosomes in natural populations (Robinson, 1971; Bigger, 1976) and the stability of chromosome fragments following irradiation (Tazima, 1964; Bauer, 1967; Maeki, 1981b; see discussions in Gupta and Narang, 1981, and Friedländer and Wahrman, 1970). The best evidence for dispersed centromeres in lepidopteran material is provided by published electron micrographs of transverse sections of mitotic chromosomes showing individual microtubules embedded directly along the face of dense chromatin masses in the absence of well-defined kinetochores (Friedländer and Wahrman, 1970; Maeki, 1981b).

The exact nature of the dispersed centromere, the extent of its evolutionary diversification, and whether it changes functionally at various stages of cell division are still open questions (see discussion in Rieder, 1982). Localized chromosome constrictions resembling unitary centromeres have been demonstrated in some species at the level of light microscopy (Bigger, 1975, 1976). It is not clear whether these are true monokinetic chromosomes, or appear so because of the techniques used for spreading (Fontana, 1976; Gupta and Narang, 1981). Differences in chromosome orientation during meiosis I and II suggest that the structures responsible for centromere function may change, with, perhaps, partial transfer to the telomeres (Maeki, 1981b); this coincides with the observation of centromerelike structures in staged and sectioned meiotic chromosomes of *Bombyx* examined by transmission electron microscopy (see Rasmussen and Holm, 1982, for additional references and discussion). With the isolation of molecular probes for highly conserved centromeric DNA sequences (Rattner, 1991) and characterization of kinetochore proteins from other eukaryotes (Brinkley, Valdivia, Tousson, and Balczon, 1989; Rattner, 1991), a search for the structural homologues of these components has begun in the silkworm (Maekawa and Tsuchida, personal

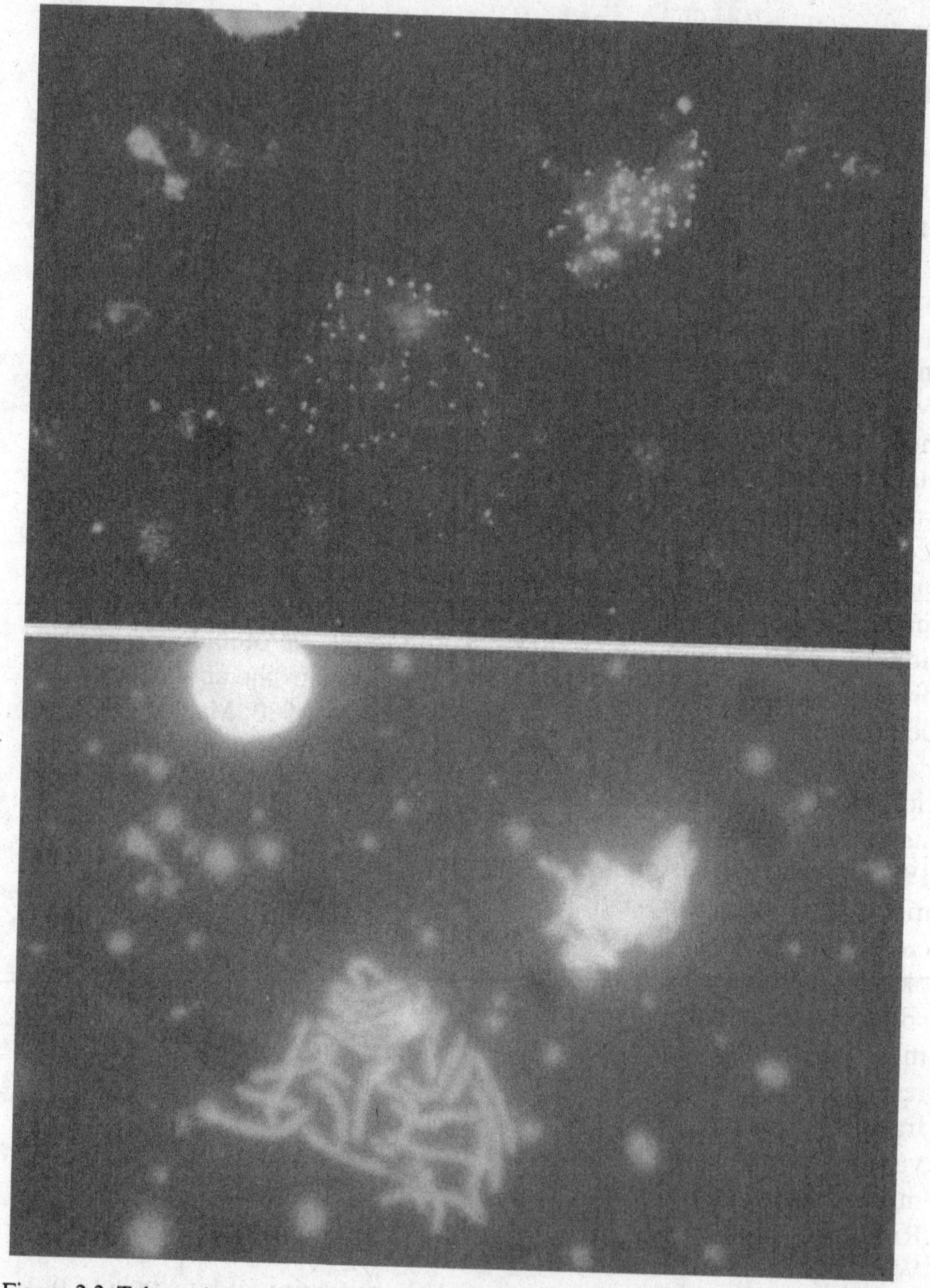

Figure 2.2 Telomeric sequences in *Bombyx mori* prometaphase chromosomes revealed by fluorescence in situ hybridization using $(TTAGG)_{25}$ labeled with fluorescein isothiocyanate (FITC) as probe. (*Top*) FITC fluorescence; (*bottom*) counterstaining with propidium iodide. (Photograph courtesy of Kozo Tsuchida, Haruhiko Fujiwara, and Hideaki Maekawa, National Institute of Health and Tokyo University; see also Okazaki et al., 1993.)

communication). This work should lay a foundation to clarify the unique features of the diffuse centromere and provide probes for future work in other lepidopteran species.

Apart from the centromere, lepidopteran chromosomes display features similar to those of other eukaryotes. Meiotic chromosomes from the silkworm and other moth species form synaptonemal complexes (von Wettstein, Rasmussen, and Holm, 1984; see "Chromosome behavior in meiosis") and, upon spreading, display characteristic loop domains attached to an electron dense scaffolding (Rattner, Goldsmith, and Hamkalo, 1980, 1981; Weith and Traut, 1980, 1986). Telomeric DNA was recently isolated from the silkworm by homology to a widely conserved 6 base pair telomeric repeat unit (TTNGGG) and mapped to chromosome ends by preferential digestion with Bal 31 nuclease digestion and in situ hybridization (Okazaki, Tsuchida, Maekawa, Ishikawa, and Fujiwara, 1993; see Figure 2.2). The telomeric sequence in silkworms differs from that of vertebrates and many invertebrates in being a pentanucleotide (TTAGG). Based on the intensity of hybridization in genomic Southern blots using a pentameric versus a hexameric probe, a similar, short repeat unit appears to be present in wild silk moths and in several other orders of insects, although not in *Drosophila* (Okazaki et al., 1993), whose telomeric region contains a complex array of longer and more diverse sequence elements which are also found in heterochromatin (Young, Pession, Traverse, French, and Pardue, 1983; Valgeirsdottir, Traverse, and Pardue, 1990).

Chromosome behavior in meiosis

The absence of crossing over in lepidopteran females is an intriguing phenomenon that raises the question of what elements of the crossover mechanism differ from those in males. A brief description of the chromosomal events in meiosis may help focus attention on the relevant stages in the two sexes for future studies on the molecular level.

Using serial three-dimensional reconstructions of oocytes and spermatocytes, Rasmussen's and Holm's pioneering and elegant ultrastructural studies of meiosis in the silkworm (summarized in Rasmussen and Holm, 1982) showed that both sexes form normal synaptonemal complexes, the specialized structures elaborated during alignment and pairing of homologous chromosomes in prophase of the first meiotic division, and that they are nearly indistinguishable, except for a small difference in spacing of the lateral components (100–120 nm for males versus 70–80 nm for females). Synaptonemal complexes are also present in females of *Ephestia* (Weith and Traut, 1980) and other moths (see von Wettstein et al., 1984, for details and discussion). This contrasts dramatically with the situation in the fruit fly, where the lack of crossing over in males, the heterogametic sex, can be readily understood by the absence of a synaptonemal complex (Rasmussen, 1973).

Chromosome behavior in silkworms is essentially equivalent between the sexes from initial pairing at zygotene and formation of the synaptonemal complex through the bouquet stage, when the bivalents, which become attached randomly to the nuclear envelope by their telomeres, draw together to a small region with chromosome bodies looped out. During this time numerous chromosome entanglements or interlockings are resolved, presumably by chromatid breakage and reunion, which is postulated to be under the control of topoisomerase II (von Wettstein et al., 1984). Although fewer interlockings are found in oocytes, clearly both sexes are capable of undergoing these kinds of precise chromosome repair processes (Rasmussen and Holm, 1982; von Wettstein et al., 1984). Following pachytene, chromosome structure diverges significantly between the sexes, with the disappearance of the synaptonemal complex in males and its ultimate conversion into elimination chromatin in females. These events are thought to be a consequence rather than a cause of differential crossing over in order to maintain the pairing of chromosomes to ensure their orderly movement to the poles at anaphase I (Rasmussen, 1977; von Wettstein et al., 1984; Carpenter, 1987).

The single important structural feature that distinguishes male and female meiotic figures at the stage when crossing over is expected to occur is the recombination nodule (Rasmussen and Holm, 1982). First observed by Carpenter (1975) in serial sections of *Drosophila* ovaries, these small, nearly spherical electron dense bodies are adjacent to the central elements of the synaptonemal complex and thought to mediate reciprocal exchanges (Carpenter, 1987). Considerable evidence from several organisms indicates that "late" recombination nodules that persist to pachytene play a direct role in crossing over (Carpenter, 1987), including the fact that they are abnormal in structure or missing from certain meiotic mutants in fruit flies (Carpenter, 1979). The number of recombination nodules and their spatial distribution along the chromosomes are consistent with the expected number of crossovers in silkworm males, but these structures are entirely missing from silkworm females (Rasmussen and Holm, 1982). Thus, recombination nodules are the major cytological element that reflects the differential ability of males and females to undergo crossing over.

Silkworm strains with heritable variation in recombination frequencies for a given chromosome are known (Turner, 1979; Ebinuma and Yoshitake, 1981); however, none thus far shows abnormal segregation patterns implicating them in fundamental processes of meiosis. This suggests that using a genetic approach to uncover the distinctive, sex-specific elements underlying differential genetic recombination in the Lepidoptera may have to await progress in more experimentally tractable organisms. Although relatively little is known about the biochemistry of meiosis (see Risley, 1986, and Stern and Hotta, 1987), a few distinctive features of chromatin structure and a small number of recombination-specific proteins have been found in meiotic tissue of a lily (and, to a lesser extent, in mice), some of which are absent or reduced in an ameiotic lily cultivar (Hotta, Bennett, Toledo, and Stern, 1979; Hotta and Stern, 1981). These observations might provide good starting points for a reexamination of meiosis in silkworm males and females.

Cytogenetics

Small chromosome size, dispersed centromeres, and lack of other distinctive cytological features have limited the usefulness of cytogenetics in the Lepidoptera. Even in the silkworm, one of the most intensely studied species, few chromosomes have been reported to show consistent morphological features that can be used for routine identification (Traut, 1976; Kawazoé, 1987). Although it is relatively easy to obtain well-separated mitotic chromosomes for light microscopy, with the exception of early embryonic cells (Maeki, 1982; Kawazoé, 1987), most published mitotic figures are highly compacted and uniformly stained, providing few identifying landmarks. Meiotic chromosomes show significant improvement because of their extended morphology and complex chromomere patterns (Traut, 1976), and under favorable cytological conditions a heterogametic pair of chromosomes may be distinguished in lepidopteran females by different sizes and/or staining patterns (*B. mori:* Kawazoé, 1987; Kawamura and Niino, 1991; *B. mandarina:* Kawazoé, 1987; *E. kuehniella,* Traut and Rathjens, 1973; Schulz and Traut, 1979; *Graphium sarpedon:* Maeki, 1981a, 1982; and other species: Traut and Mosbacher, 1968; Suomalainen, 1969). These features have been enhanced using variations of standard cytological procedures, such as treatment with colchicine and hypotonic medium (Fontana, 1976; Goodpasture, 1976), G banding (Bigger, 1975), and C banding (Maeki, 1981a), as well as spreading of silver stained synoptonemal complexes for examination by electron microscopy (Weith and Traut, 1986; Bhagirath, Kundu, and Ibotombi, 1988; Ibotombi, Bhagirath, and Kundu, 1988). As yet, there have been no reports of systematic testing of fluorescent dyes used for mammalian karyotyping; perhaps this should also be attempted.

To what extent the failure to obtain clear banding patterns is a result of intrinsic chromosome structure is unclear. Other organisms having small chromosomes with diffuse kinetochores such as hemipterans, nematodes such as *Ascaris lumbricoides* and *Caenorhabditis elegans,* and the wood rush, *Luzula* (Godward, 1985), show similarly undistinguished karyotypes. Dispersed centromeres may affect the distribution of sequences that have been implicated in G, Q, and R banding in mammalian chromosomes, such as constitutive heterochromatin (Godward, 1985), AT- and GC-rich regions (Comings, 1978), and repeated sequences like the Alu and L1 families (Korenberg and Rykowski, 1988). Along these lines, it is of interest to note that BMC1, a sequence family related to L1 (see Eickbush, this volume), was recently shown by in situ hybridization to be distributed fairly uniformly in silkworm chromosomes (Tsuchida and Maekawa, personal communication). Future work on the nature and arrangement of sequences potentially able differentially to bind stain or to influence local chromosome domain folding (Comings, 1978) may help explain the lack of morphological features for differentiating lepidopteran karyotypes and provide additional molecular approaches for specific chromosome identification (see "Molecular approaches to cytogenetics").

Although no attempt has been made to cover the genome systematically in *Bombyx* with chromosome aberrations as has been done with *D. melanogaster* (Lindsley et al., 1972), many stocks containing translocations and supernumerary chromosome fragments have been used to advantage for cytogenetic analysis, even with conventional staining techniques. Visualization of sex-limited translocations was first reported in *Ephestia* by the presence of unpaired chromosome loops in aberrant bivalents (Schulz and Traut, 1979); the ZW chromosome pair has been identified similarly in silkworm strains carrying large genetically marked autosomal fragments translocated to the W chromosome (Wang and Xu, 1990; Kawamura and Niino, 1991). Autosomal translocations have been detected cytologically in *Bombyx* (Murakami and Imai, 1974); and even with poorly differentiated banding or chromomere patterns, this kind of material has been used to correlate aberrations with known linkage groups involving autosomes and sex chromosomes. For example, Banno and Doira recently detected a Robertsonian fusion in an aberrant stock showing novel linkage relationships by the presence of an abnormally elongated chromosome paired with two smaller ones whose banding patterns matched the longer chromosome's "arms" (Banno and Doira, personal communication). It has also been possible to trace some of the karyotypic changes between the domesticated silkworm and the Japanese population of its nearest wild ancestor, *B. mandarina*, which has a haploid chromosome number of 27 (Kawazoé, 1987). A particularly elegant series of studies in which Kawamura identified an egg size–determining gene (*Esd*) on the W chromosome of the silkworm by a combination of cytogenetic and genetic approaches is described in a later section ("Oogenesis – egg morphology and chorion mutants").

Despite this success, routine use of lepidopteran chromosomes for conventional cytogenetic analysis is difficult at best, and the situation calls for a radical change in technique. This may be forthcoming with the development of new molecular approaches in the silkworm (described at the end of this chapter).

Traditional genetic resources

Genetics of the silkworm has evolved along two lines. On the one hand are stocks carrying distinctive morphological, developmental, or behavioral mutations and on the other are large collections of inbred geographic and improved races, called *zairaishu* and *kairyoshu* in Japanese, used for practical breeding. One or more alleles are still available for most mapped loci (see Figure 2.1) as well-defined genetic stocks, with the largest collection at the Institute of Genetic Resources at Kyushu University in Fukuoka, Japan (Doira, 1992). In addition, many of the geographic races (which may be found in most countries that have been involved historically with sericulture in Asia and Europe) have been characterized in terms of quantitative or complex traits useful for practical breeding, as well as for some known mutations and biochemical markers. Both kinds of stock collections remain as largely untapped resources for modern genetic and molecular studies.

Most markers assigned to the genetic linkage maps are spontaneous, involving visible characters found during mass rearing for silk production, annual maintenance of genetic strains, and practical breeding. Mutations have also been picked up by mutagenesis, primarily using X-ray and gamma irradiation (see Tazima, 1964, for a description of early work in radiation genetics), but also by chemical induction (Datta, Sengupta, and Das, 1978; Kawaguchi, Doira, Banno, and Fujii, 1985), by screening the *zairaishu* for characters of importance for sericulture, and by crossbreeding with *B. mandarina,* with which the silkworm is able to produce partially fertile hybrids. As with other organisms, irradiation is effective for obtaining chromosome aberrations, and many autosomal and sex-limited translocations as well as supernumerary chromosome fragments carrying visible markers have been used to advantage in breeding and experimental studies (see Tazima, 1964, and Strunnikov, 1983, for details). Despite considerable success of mutant hunts using traditional genetic approaches (see Tazima, 1964), mutagenesis does not seem to have been used extensively for this purpose, perhaps because of the expense and labor involved in rearing on the scale needed to obtain practical yields of narrowly defined mutations.

Polygenic or quantitative traits of economic importance are well segregated in the geographic races. Standard measures of genetic variance such as specific or general combining ability (Kanaratanakul et al., 1987a; Datta and Pershad, 1988; Jolly et al., 1989) show high values in interstrain crosses, indicating that there is substantial divergence in genes controlling characters associated with silk production (Kanaratanakul et al., 1987b; Udupa and Gowda, 1988), growth rate, survival, fecundity, fertility, and disease resistance (Chinnaswamy and Devaiah, 1987; Kremky and Michalska, 1988; Tayade, 1989). Many of these characters are of interest not only for sericulture but also for understanding fundamental biological processes in lepidopteran development and physiology and for devising means of controlling lepidopteran pests. Work has begun on construction of molecular linkage maps in the silkworm, in part aimed at being able to carry out genetic studies with these kinds of traits (see "A molecular linkage map").

Silkworm mutants

Silkworm mutations affect a broad range of developmental, physiological, biochemical, and behavioral characters. These include stage-specific lethals, abnormalities in larval body shape, segment identity, organ and tissue formation, pigmentation patterns and biosynthetic pathways, variant isozyme and hemolymph proteins, life history traits such as stage duration, number of molts, and entry into diapause, resistance to fungal and viral disease and airborne pollutants, altered food discrimination, and aberrant cocoon spinning. In the following sections I have selected a few classes of silkworm mutants to describe in some detail, based partly on the major subjects treated in this volume and partly on the availability of especially large classes of mutations that promise to uncover new areas of investigation by taking advantage of recent rapid advances in technology for isolating genes and studying their structure and function on the molecular level. These are presented in a loose developmental sequence according to the metamorphic stage of the most prominent phenotypic effect, although it will be evident that this may not correspond to the critical period of gene action.

This report focuses primarily on findings in silkworm genetics by Japanese investigators. Given the diverse nature of the subjects touched upon here and that much of the extensive literature is written in Japanese or difficult to obtain outside Japan, I have not attempted a comprehensive review. It must be noted that information on silkworm genetics is published in many other countries, including China, India, Korea, Thailand, and the former Soviet Union, but much of this work is not included here, largely because of its relative inaccessibility. For a comprehensive overview of the early decades of silkworm genetics, Tazima's *Genetics of the Silkworm* (1964) remains the most informative reference available in English; brief descriptions of mutations and pictures of mutants held in Japanese stock centers may be found in Tazima's volume and in publications by Chikushi (1972) and Doira (1978, 1983, 1992).

Oogenesis – egg morphology and chorion mutants

A small number of maternally acting mutations affect egg size and shape and chorion morphology. Of the former, few have been studied biochemically; those listed here appear to be involved with general processes of egg formation and have no specific effects on early embryogenesis (Table

Table 2.1. *Mutations affecting egg formation*

Character	Gene[a]	Locus	Phenotype	Process, product or mechanism affected	References[b]
Egg size	*sm*, small egg	3–41.8	oocytes degenerate; female sterile	abnormal follicle cell differentiation; reduced yolk uptake, RNA synthesis, transport, accumulation, protein synthesis	Kawaguchi and Doira, 1973; Kawaguchi and Fujii, 1983, 1984
	sm-2, small egg 2	13–0.0	small; female sterile; autonomous	follicle formation; reduced yolk uptake; egg-specific protein present	Kawaguchi et al., 1988a, 1988b, 1990
	vit, scanty of vitellin[M]	20–?	small, white egg; female sterile	extraovarian yolk protein synthesis reduced; egg-specific protein normal	
	emi, miniature egg	12–9.5	small, partially fertile, viable		
	Ge, Giant egg	1–14.0	40% size increase; yolk composition normal; autonomous	possible translocation of *Esd* locus from W chromosome	Kawaguchi et al., 1987, 1991; Kawamura, 1990
	Esd, Egg size–determining gene	W–?	40% size increase in polyploids; autonomous		Kawamura and Nakada, 1981; Kawamura, 1988; Sahara, Kawamura, and Iizuka, 1990

Egg shape	*rd*, clumpy	12–45.0	irregular, variable		
	l-sp, lethal spindle	12–?	ellipsoidal; poor hatchability	double fertilization mosaics; polyploidy common	
	sp-t, spindle egg of Tamazawa[Lt]	12–?	long, narrow		
	elp, ellipsoid egg	18–16.1	ellipsoidal		
	sp, spindle-shaped egg	23–22.9	long, narrow eggs; ends attenuated		
Chorion	*Gr*, gray egg[S,X]	2–6.9	elliptical egg; chorion transparent in homozygote, opaque in heterozygote	framework formation abnormal	Weare, Paul, and Goldsmith, unpublished observations
	Gr[col], collapsing egg	2–6.9	thin, collapsing chorion	reduced secretion of early, early-middle chorion proteins; synthesis normal	Nadel, Goldsmith, Goplerud, and Kafatos, 1980
	Gr[16], European 16 gray egg	2–6.9	opaque chorion; homozygote wrinkled	altered posttranslational modification of putative framework proteins	Gautreau et al., 1993
	Gr[B], Bird-eye egg	2 < 6.9	thin, collapsing chorion; transparent center, opaque rim in heterozygote	lacks Hc, many A, B proteins; deletion in *Ch1-2* encompassing A/B.L1–8, Hc.1–15, A/B.R	Nadel, Thireos, and Kafatos, 1980; Durnin-Goodman and Iatrou, 1989
	mgr, mottled gray egg	6–8.9	transparent and opaque areas in chorion		
	Se, White-sided egg	15–16.9	chorion opaque, wrinkled; homozygote sterile		

[a]Mutations spontaneous unless noted. X = X-ray induced; M = MNU induced; Lt = low temperature induced; S = both spontaneous and induced.
[b]characteristics from Doira (1992) unless indicated.

2.1). Silk moth follicle development and choriogenesis are described in this volume in the chapters by Kafatos et al. and by Regier et al.

Two unlinked small egg mutations, *sm* and *sm-2,* give rise to small, infertile eggs, if any, which contain all expected major yolk proteins but in markedly reduced amounts, suggesting that they are defective in yolk uptake. In both mutants the protein composition of larval and pupal hemolymph indicates normal fat body function, but levels of yolk proteins remain high following the stage in pupal development when they should be taken up by the oocyte via the surrounding follicular epithelium (Kawaguchi and Doira, 1973; Kawaguchi and Fujii, 1984; Kawaguchi et al., 1988a, 1990). It is notable that *sm-2* follicles synthesize egg- (female) specific protein (see Riddiford, this volume), but it is found only in trace amounts in the yolk (Kawaguchi et al., 1990).

For both mutants, reciprocal transplants of normal and mutant ovarian imaginal disks into fifth instar larvae develop autonomously and express the donor genotype, consistent with the likelihood that the primary lesions involve the function of cells in the follicle complex (Kawaguchi and Doira, 1973; Kawaguchi et al., 1990). In females carrying the allele sm^n, nurse cells and follicular epithelium appear normal histologically and contain a normal array of bulk RNA species, but labeling with radioactive uridine detectable by autoradiography is reduced throughout pupal development. Further, there is no evidence for movement of labeled material from the nurse cells to the oocytes via intercellular bridges, as would be expected for newly synthesized ribosomes (Kawaguchi and Fujii, 1983). These follicles also label poorly with radioactive amino acids in vitro, showing that their ability to carry out protein synthesis is significantly diminished, although it is not yet clear whether this is a specific or general effect of impaired RNA synthesis (Kawaguchi and Fujii, 1984). In *sm-2,* the size and number of follicular epithelial cells are reduced up to 30 percent, suggesting that this gene may be involved in the initial formation of the follicle complex (Kawaguchi et al., 1988b).

Two additional mutations, scanty of vitellin (*vit*) and miniature egg (*emi*), also produce small eggs (see Table 2.1). Unlike *sm* and *sm-2,* eggs laid by homozygous *vit* females are sterile but contain normal amounts of egg-specific protein, indicating that both this aspect of follicle cell function and protein uptake by the oocyte are normal; however, vitellins are found only in trace amounts (Doira, personal communication). Thus it is possible that this mutation involves the fat body, the major site of synthesis of yolk proteins (see Riddiford, this volume). *emi* eggs are fertile to varying degrees, and larvae may be viable (Doira, personal communication).

As a group, the egg morphology mutations provide a useful model for studying genes affecting formation and function of the follicle complex. Further characterization of the small and miniature egg mutants with the aid of the ultrastructural and biochemical techniques used for examining the mechanism of yolk uptake in *Hyalophora cecropia, Manduca,* mosquitoes, *Drosophila,* and other insects (Raikhel and Dhadialla, 1992) may help answer some of the outstanding questions remaining in this well-studied process.

Giant egg (*Ge*) is a sex-linked mutation in which females produce eggs that are 20 percent longer on the long axis and weigh 44 percent more than wild type, but have normal yolk composition (Kawaguchi, Shito, Fujii, and Doira, 1987). Wild-type tetraploid animals (ZZWW) produce eggs equivalent in size to those laid by *Ge* diploids, and *Ge* tetraploids produce eggs that are even larger (Kawamura and Nakada, 1981). These observations are explained by postulating the presence of a quantitative Egg size–determining gene, *Esd,* on the W chromosome, which is supported by the dose-dependent egg size increases found in ZWW triploid and ZZWW tetraploid females (Kawamura, 1988). Ovary transplantation experiments demonstrate that both *Ge* and *Esd* act autonomously (Kawaguchi, Banno, Koga, Doira, and Fujii, 1991; Sahara, Kawamura, and Iizuka, 1990) and support the idea that egg size and number are limited, perhaps by the female's nutritional status at pupation (Kawaguchi et al., 1987; Sahara et al., 1990). Additional observations of the eggs laid by the females used in these studies give hints of the presence of Z-linked gene(s) that exert a negative effect on egg formation (Kawamura, 1988).

The similar modes of action of *Ge* and *Esd* and the linkage of these genes to the sex chromosomes led to the idea that *Ge* is a translocation of the *Esd* locus (Kawamura, 1988). This hypothesis is strongly supported by cytological examination of the lengths of ZW bivalents in normal and mutant stocks, showing that the Z lateral component derived from the *Ge* strain is significantly longer than the wild type (Kawamura, 1990). In this study, the W chromosome was identified by a large translocation of chromosome 2 that formed a distinctive unpaired loop in late pachytene. *Esd* is the only gene known on the W chromosome. Once positional cloning becomes available in the silkworm (see "A molecular linkage map"), the fortuitous transfer of this gene to the Z chromosome opens up the possibility of being able to design a strategy to identify and isolate a portion of the W chromosome, which plays an active role in sex determination and whose function has been at least partially characterized.

Chorion mutations have proven to be a fertile area for genetic analysis since Toyama's first report (1912) of a mutant with an opaque eggshell (*Gr*). Mutations at several loci affect chorion morphology (see Table 2.1); only the gray egg mutations have been studied in molecular detail and are described here, along with studies of chromosome abnormalities involving the chorion structural genes.

Based on the graded appearance and close linear arrangement of several spontaneous and X-ray–induced *Gr* "alleles," Takasaki recognized in the late 1950s that this chromosome region may be a complex locus (Takasaki, 1962; see also Tazima, 1964). This genetic behavior can be explained in part by the discovery that *Gr* mutations are flanked by the two major clusters of chorion structural genes, *Ch1-2* and *Ch3* (Kaomini, 1993; Condon and Goldsmith, unpublished observations), which cover more than 500 kb (see Eickbush and Izzo, this volume). For example, one of the spontaneous gray egg mutations, Bird-eye (Gr^B), is a large deletion that eliminates most of the gene cluster, *Ch1-2,* including all late Hc gene pairs and many middle As and Bs (Durnin-Goodman and Iatrou, 1989; see also Eickbush and Izzo, this volume) but is proximal to the three other spontaneous *Gr* alleles (Nadel and Goldsmith, unpublished observations). Recent fine structure mapping of chorion electrophoretic variants reveals that at least one early gene is tightly linked to the locus represented by Gr^{16} and indicates that this mutation may reside within *Ch3* (Kaomini, 1993). These and other observations suggest that the large crossover values obtained in the region between the two genetically defined chorion clusters (>4 cM) may represent recombination hotspots (see Eickbush and Izzo, this volume). Thus it is possible that some and possibly all gray egg mutations simply involve changes in chorion structural genes located near a high frequency crossover site.

Apart from Gr^B, the well-studied *Gr* mutations show defects relatively early in choriogenesis. The collapsed mutant (Gr^{col}) exhibits a secretory defect for early and early-middle proteins, leading to the accumulation and subsequent degradation of normally synthesized proteins in large intracellular vacuoles and the formation of a thin eggshell that rapidly dehydrates (Nadel, Goldsmith, Goplerud, and Kafatos, 1980). In both European 16 (Gr^{16}) and Toyama's (*Gr*) gray eggs, a defect appears in the orientation of the inner lamellae during the early or early-middle stage of chorion formation so that the axis of the helicoidal array of fibrils comprising the inner chorion becomes oriented perpendicular rather than parallel to the egg surface (Gautreau, Zetlan, Mazur, and Goldsmith, 1993; Weare and Paul, personal communication; see also Regier et al., this volume); presumably, this is responsible for the light scattering effect for which the mutations were named. Restriction analysis of genomic DNA from Gr^{16} gives no evidence of gross physical changes in chorion structural genes (Gautreau et al., 1993; Alexopoulou and Goldsmith, unpublished observations). However, a pair of minor proteins that are synthesized early in choriogenesis show temporal changes in labeling patterns that are consistent with the possibility that they undergo rapid posttranslational modification, resulting in the accumulation of mutant-specific end products with different isoelectric points from wild-type (Gautreau et al., 1993). It has been suggested that these may be critical components involved in the assembly of the scaffolding or framework, although it is not yet certain whether they are bona fide chorion proteins. The lesion in the classic *Gr* mutation has not yet been studied on the molecular level.

The outer crust of the silk moth chorion, which is synthesized at the end of choriogenesis and found only in the Bombycidae, is thought to have evolved to protect the embryo during the prolonged egg diapause (see Regier et al., this volume). That this part of the chorion is encoded by Hc gene pairs was originally inferred by their relative timing of expression (see Regier et al. and Kafatos et al., this volume) and recently confirmed by ultrastructural analysis of eggs laid by females lacking Hc genes but carrying an essentially normal complement of other chorion gene family members (Alexopoulou et al., 1988). These females were homozygous for the Gr^B deletion and carried one of two independent sex-limited translocations in which a segment of chromosome 2 shown to contain middle A/B gene pairs had become attached to the W chromosome (Alexopoulou, Zetlan, Nelson, and Goldsmith, unpublished observations). Many additional sex-limited translocations as well as unstable chromosome fragments involving the proximal end of chromosome 2 (which carries visible larval markers of the *p* and *S* loci; see "Embryonic development – Early lethals and pattern formation defects") are available for future studies of the organization and functional diversification of the chorion structural genes.

Although preliminary evidence suggests that there are heritable inbred strain differences in the thickness of the outer crust (Nho, Sakaguchi, and Goldsmith, unpublished observations), interestingly, the copy number of Hc genes shows little variation among diapausing and nondiapausing tropical strains of *B. mori* (Xiong, Sakaguchi, and Eickbush, 1988). The outer crust and Hc proteins are especially prominent in *B. mandarina* (Sakaguchi et al., 1990), which produces only one generation per year and undergoes a prolonged and variable diapause. As a means of uncovering additional genetic factors associated with chorion morphogenesis and evolution, studies have begun to investigate the genetic and molecular bases of the chorion differences between wild and domesticated silk moths using intraspecific crosses (Sugawara, 1990).

Embryonic development – early lethals and pattern formation defects

Embryonic lethals are common in the silkworm, but few have been characterized beyond descriptions of their superficial appearance at the time of developmental arrest (Tazima, 1964). Embryos produced by homozygous females carrying the single well-studied maternal effect mutant kidney (*ki*) lack mesodermal structures (Miya, 1984, 1985b); this mutation and Burnt (*Bu*), which fails to undergo dorsal closure, are described by Nagy in this volume. Among embryonic lethals that produce well-defined phenotypic defects as heterozygotes, many appear to affect larval segment identity; these are listed in Table 2.2. There is also a large number of nonlethal mutations that modify the overall body plan and shape of larval segments; representative ones with well-defined phenotypes are also listed in Table 2.2 under the assumption that they are associated with abnormalities in early embryogenesis, although some may function during larval growth. The success of large-scale screens for maternal and zygotic embryonic mutations in *Drosophila* (Nüsslein-Volhard and Wieschaus, 1980) suggests that, despite the labor involved, this is an area that bears further investigation in the silkworm, especially to enhance opportunities for a comparative analysis of critical events in embryogenesis (see chapters by Nagy and by Regier et al., this volume).

The Extra leg, or *E*, alleles are arguably the most famous silkworm mutations and the only ones of this kind studied in light of a modern understanding of early development. These are generally characterized by supernumerary or deficient legs on specific segments, along with abnormalities in the position and number of the prominent paired crescent markings on abdominal segment 2 (A2), and stars on A5. Many are embryonic lethals as homozygotes and exhibit pleiotropic effects such as abnormal larval tracheal patterns and ganglionic commissures and abnormal development of adult gonads, wings, and legs (Tazima, 1964). That the *E* alleles affect larval segment identity was first recognized by Itikawa in 1952 (cited in Tazima, 1964). The *E* locus was recently identified as the silkworm homologue of the bithorax homeotic gene complex of *D. melanogaster* (Ueno, Hui, Fukuta, and Suzuki, 1992), and the neighboring *Nc* (No crescent) locus, mutations of which produce similar phenotypic effects, as the homologue of the Antennapedia complex (Nagata et al., in preparation). Discussions of the characteristics of *E* and *Nc* mutants and their molecular analysis are found in this volume in the chapters by Nagy and by Ueno, Nagata, and Suzuki. Thirty-five mutations affecting segment identity have been mapped to the *E-Nc* region of linkage group 6, and experiments are underway in the laboratory of H. Doira to complete its fine structure map (personal communication). This will provide important information about the functional organization of the silkworm homeotic gene complexes and their evolutionary relationships to those of fruit flies and other invertebrates and vertebrates (Akam, 1989; McGinnis and Krumlauf, 1992).

A few additional mutations affect larval pigmentation and appearance in ways suggesting that the genes responsible for their expression are controlled by the regulatory hierarchy specifying larval pattern and form (Table 2.3; see also "Larval mutations – biochemical genetics of the integument"). In the mutants Zebra (*Ze*), New striped (*S*), and striped alleles of plain (*p*), dark, melaninlike pigment is deposited in the anterior portion of each segment, as if under the influence of factors specifying compartment boundaries (Lawrence and Morata, 1977; Ingham and Martinez Arias, 1992). The *p* locus is of particular interest for its high degree of polymorphism; more than 10 alleles have been documented that produce varied patterns ranging from unpigmented, to localized pigment only in the stars, crescents, and eyespots (thoracic segment 2), to nearly uniform black. The resemblance of *S* and p^S mutations, their relatively close proximity, and the fact that most *S* alleles were induced by irradiation suggest that *S* may be a duplication or rearrangement of the *p* locus (Tazima, 1964). This may also be true of mildewed striped (*mi*), which is found on the same chromosome and modifies striped expression (Doira, 1992). In a similar vein, two other closely linked genes, Ursa (*U*) and Dirty (*Di*), deposit an irregular spattering of pigment throughout the body except along the dorsal vessel in the former and at the stars and crescents in the latter; further, their close linkage is consistent with the possibility that the *U-Di* chromosome interval comprises a complex locus. *p* modifies the

Table 2.2 *Mutations affecting larval body pattern and form*

Character	Gene[a]	Locus	Phenotype	Embryonic lethal	Gene or larval segment affected[b]	References[c]
Homeotic mutations or genes affecting segment identity markers	*E*, Plain extra legs	6–0.0	35 alleles known; affect number, position of larval segment markings and prolegs; also reproductive organs, nervous system	yes	A2, A4, A5; homologue of *bithorax* complex	Ueno et al., 1992
	Nc, No crescent[x,s]	6–1.4	crescents lacking; eyespots narrow, lightly pigmented	yes	T2, A2; homologue of *Antennapedia* complex	Nagata et al., in preparation
	Nl, No lunule	14–35.2	crescents, stars lacking	yes	A2, A5; chromosome deficiency	
	E-tr, Tr extra legs[x]	28–?	extra abdominal legs	yes	A4, A5	
	Sl-v, V-supernumerary legs	21–19.4	extra legs	yes	A5, A10 (homozygote)	
	L, multilunar	4–15.3	pair of large, yellowish-brown spots on dorsal side of each segment	yes	A5-8	
	ms, multistar	12–5.5	extra pairs of stars	no	A6-10	
	msn, new multistars	19–45.8	extra pairs of stars	no	A6-10	
	Dus, Duplication of star spots	10–3.9	extra pair of stars; homozygote lacks head capsule	yes	A9	

	es, extra spiracles	12–4.3	extra pair of small spiracles	no	A12
	Bo, Bamboo	11–26.6	inhibits formation of larval eyespots and crescents, dorsal scales on moth; larval abdomen compact, hard; posterior segment margins bulge	no	T2, A2
Body plan	*mal*, monster	4–30.1	body twisted	no	A8 (predominantly)
	hal, harelip	15–?	segments fused longitudinally	no	T1, T2
	e, elongate	1–36.4	elongated segments; intersegmental membranes stretched	no	A4, A5
	ge, geometrid	12–?	body slender; segment anterior, posterior weakly constricted	no	A2
	gn, gooseneck[g,S]	9–22.0	body slender; constricted between segments	no	
	nb, narrow breast	19–31.2	body spindle shaped; short thorax, stout abdomen	no	
	tub, tubby	23–6.9	body spindle shaped; short thorax, stout abdomen	no	

[a]Mutations spontaneous unless noted. X = X-ray induced; g = gamma induced; S = both spontaneous and induced alleles.
[b]A, abdominal segment; T, thoracic segment.
[c]Characteristics from Doira (1992) unless indicated.

Table 2.3. *Mutations affecting pattern of larval segment markings*

Character	Gene[a]	Locus	Phenotype[b]	Gene interactions
Pigmentation	*p*, plain[s,x]	2–0.0	over 10 alleles affecting segment-specific markers, ground color	certain alleles epistatic to *Di*, *U*
	S, New striped[s,x,ht]	2–6.1	anterior zone of segment black, narrow white posterior margin; like p^S	could be duplication or rearrangement involving *p* locus
	mi, mildewed striped	2–7.0	gray spots on dorsal side of segment (with striped marking)	modifies action of p^S, *S*
	Ze, Zebra	3–20.8	narrow anterior zone of segment black, posterior margin white; head cuticle dark brown	

	Di, Dirty	14–43.2	background of irregular dark brown lines, dots; crescents, stars unpigmented in homozygote	unpigmented with *p*
	U, Ursa	14–40.5	dark brown pigment except above dorsal vessel	action modified by specific alleles of *p*; U^{M} produces dilute Zebra striping with *p*
	les, light eyespot	14–?	lacks black triangular markings on both sides of eyespot	possible allele of *Di* or *U*
	bl, blind	15–0.0	eyespots solid black	
Dermal Protrusions	*K*, Knobbed	11–23.2	excess cell growth at eyespots, stars, crescents	action depends on expression of *p*, *Dus*, *L*, *ms*, *msn*, etc.

[a]Mutations spontaneous unless noted. X = X-ray induced; ht = heat induced; S = both spontaneous and induced.
[b]Characteristics from Doira (1992).

action of both genes, and in its presence, one allele of *U* (U^M) produces a Zebra-like banding pattern (Doira, 1992), suggesting that the latter may function in the same developmental pathway as other striping genes. Two other mutations, light eyespot (*les*) and blind (*bl*), alter parts of the eyespot patterns (*les*, not yet localized, may be allelic to *U* or *Di*). Finally, Knobbed (*K*) acts in concert with many of the genes listed in Table 2.2, producing cell growths (or "dragon horns") wherever eyespots, stars, and crescents are found.

Now that it has become relatively easy to isolate genes involved in early developmental pathways using heterologous probes from *Drosophila* and other species (see Hui and Suzuki and Ueno, Nagata, and Suzuki, this volume), these mutant collections should provide useful material for identifying the chromosomal location of cloned sequences and for studying their function in vivo. Isolation of the *p* locus has been undertaken in the laboratory of H. Fujiwara, taking advantage of the large number of available polymorphic stocks and genetic mosaics carrying fragments of chromosome 2 (personal communication; see "Molecular approaches to cytogenetics"). Analysis of this locus on the DNA level should help explain its mutability and provide a new system to study the interactions between selector genes in the developmental regulatory hierarchy and their targets (Akam, 1989).

Larval mutations – biochemical genetics of the integument

Considering the relatively long period of larval life and the large size of fifth instar larvae just prior to spinning, it should not be surprising that many larval integument or "skin" mutations have been found that affect the overall or ground color of the integument as well as localized pigmentation patterns (see Tables 2.4 and 2.5). These involve several physiological processes, including purine and pteridine metabolism, accumulation of metabolic end products for temporary storage in the epidermis (hypodermis), pigment biosynthetic pathways, and cuticle melanization and tanning. Emphasized here are mutations affecting the first two processes, of which there are especially large collections. Understanding the basis of larval skin color has important implications for understanding seasonal, ecological, and evolutionary polymorphism of caterpillars in natural populations (Ohashi, Tsusué, and Kiguchi, 1983; see Fuzeau-Braesch, 1985, for a review), for which the silkworm may provide a useful model. Silkworm mutations in pigment biosynthetic pathways affecting primarily egg, serosa, and eye color are reviewed elsewhere and are not considered here (see Kikkawa, 1953, and Ziegler, 1961).

The predominant ground color of wild-type silkworms from the second instar onward is opaque and white. This results from the accumulation of uric acid and isoxanthopterin, major end products of nitrogen metabolism, as a form of storage excretion in numerous low electron density, membrane-bound vesicles in the epidermis (Tsujita and Sakurai, 1964a, 1964b, 1966; Waku, Sumimoto, and Eguchi, 1968; Sakurai and Tsujita, 1976a; Tamura and Akai, 1990). These vesicles, which discharge their contents prior to each molt (Tsujita and Sakurai, 1967a; Tamura and Akai, 1990), exhibit many ultrastructural differences from the urate storage granules found in fat body, the predominant site of uric acid storage in many insects, including *Manduca* (Buckner, Henderson, Ehresmann, and Graf, 1990, and references therein) and other Lepidoptera (Dean, Locke, and Collins, 1985; Keeley, 1985). Yellow by-products of pteridine metabolism such as sepialumazine and riboflavin are taken up in the same or similar epidermal vesicles (Tsujita and Sakurai, 1966, 1967a, 1967b; Mazda, Tsusué, and Sakate, 1980). Additional pigments, notably ommochromes and melanins, which are deposited in the epidermis and cuticle, respectively, produce colors ranging from pinks, reds, and oranges to browns and blacks. For reviews of pigment formation, cuticle tanning, and melanization in insects, see Kayser (1985), Andersen (1985), and Sugumaran (1988).

The largest group of larval skin mutants are the "oily" or translucent mutations, of which more than 35 have been identified (Doira, 1978) and more than 20 mapped (Doira, 1992; Table 2.4). Some of these mutations are also expressed in the serosa and in ectodermally derived organs and tissues and may be lethal or sterile (Doira, 1978; listed in Table 2.4 under "Pleiotropic effects"). Most are defective in uric acid uptake (Tamura and Sakate, 1983), with more transparent mutants having less uric acid (Doira, 1978). Oily mutants that have been examined ultrastructurally show a dramatic reduction in the number, size, and shape of epidermal vesicles (Tsujita and Sakurai, 1966, 1967a, 1967b; Waku et al., 1968; Tamura and Akai, 1990). In some cases, proteins obtained from isolated vesicles have been shown to exhibit minor alterations in overall amino acid composition and, as assayed by gel filtration and on SDS gels, to be more diverse structurally than wild type (Tsujita and Sakurai, 1963; Sakurai and Tsujita, 1976b). These observations, although preliminary, suggest that some of the genetic changes in oily mutants affect structural genes encoding vesicle membrane proteins or proteins that bind or sequester uric acid (Sakurai

Table 2.4. *Oily skin mutations*

Character	Gene[a]	Locus	Phenotype	Process, product, or mechanism affected	References[b]
Normal uric acid uptake	*og*, giallo ascoli translucent	9–23.6	highly translucent; high pupal mortality; female sterile; male fertile, needs aid to copulate; other alleles, both sexes sterile	trace Xdh activity, excretes, xanthine, hypoxanthine; urate vesicles small or lacking	Tamura and Sakate, 1975, 1983; Tamura, 1977; Tamura and Akai, 1990
	oq, q-translucent	12–26.3	highly translucent; high mortality; both sexes sterile	low Xdh activity; excretes xanthine	Tamura, 1983
Reduced or unknown uric acid uptake	*oa*, aojuku translucent	14–42.2	moderately translucent; *oa2* allele mottled	reduced uric acid uptake; irregularly shaped urate vesicles; altered SDS profile of isolated "pteridine granules"	Sakurai and Tsujita, 1976b; Tamura and Sakate, 1983; Tamura and Akai, 1990
	oc, Chinese translucent	5–40.8	moderately translucent	reduced uric acid uptake; many small oval urate vesicles; altered SDS profile of isolated "pteridine granules"	Sakurai and Tsujita, 1976b; Tamura and Sakate, 1983; Tamura and Akai, 1990

od, distinct translucent	1–49.6	highly translucent; complete and mottled alleles	reduced uric acid uptake, limited to translucent areas in mottled allele; loss of urate vesicles during maturation of fifth instar leading to reduced number, size	Tsujita and Sakurai, 1964a, 1964b; 1976a, 1976b; Waku, Sumimoto and Eguchi, 1968, Tamura and Sakate, 1983; Tamura and Akai, 1990
ohi, hime-nichi translucent	16–32.8	moderately translucent		
oj, Japanese translucent	9–0.0	moderately translucent; light egg color		
ok, kinshiryu translucent	5–4.7	highly translucent; early larval mortality high	reduced uric acid uptake; no urate vesicles	Tamura and Sakate, 1983; Tamura and Akai, 1990
op, p-translucent	23–?	fairly high translucency; pupal mortality high; male sterile		
or, r-translucent[x,s]	22–8.9	moderate to high translucency	reduced uric acid uptake; few spindle-shaped or large urate vesicles	Tamura and Sakate, 1983; Tamura and Akai, 1990
os, sex-linked translucent	1–0.0	weakly translucent; complete and mottled alleles	reduced uric acid uptake; abundant irregular urate vesicles	Tamura and Sakate, 1983; Tamura and Akai, 1990

(continued)

Table 2.4 *(cont.)*

Character	Gene[a]	Locus	Phenotype	Process, product, or mechanism affected	References[b]
	ow, waxy translucent	17–36.4	moderately translucent	reduced uric acid uptake; spindle-shaped, disrupted urate vesicles	Tamura and Sakate, 1983; Tamura and Akai, 1990
	oy, y-translucent[M]	25–17.6	low translucency		
Mottled	*oal*, oal mottled translucent	2–26.7	*oa* allele mutates to mottled *oβ* form in somatic cells in combination with *mu-oal*	reduced uric acid uptake in translucent areas	Tamura and Sakate, 1983
	mu-aol, mutator of oal	13–40.4	induces somatic mosaicism of mutable allele of *oal*		Hatamura, 1939
	otm, Tanaka's mottled translucent	5–15.2	moderately translucent, many fine opaque dots		
	ov, mottled translucent of Var	20–15.2	size of opaque areas varies with allele	reduced uric acid uptake	Tamura and Sakate, 1983
	odk, translucent E-15	14–32.5	translucency varies with allele, fine opaque dots		

	oh, hoarfrost translucent	20–0.0	moderately translucent, many indistinct fine opaque dots		
	obt, mottled translucent B8	7–21.0	weakly translucent, small opaque dots	reduced uric acid uptake	Tamura and Sakate, 1983
Pleiotropic effects	*w-3*, white egg 3	10–19.6	translucency, color of serosa, compound eye, allele dependent; epistatic interactions with some other translucent genes	reduced uric acid, isoxanthopterin content; altered SDS profile of isolated pteridine granules; reduced numbers; size of urate vesicles	Tsujita and Sakurai, 1964a, 1964b; 1967a, 1967b; Sakurai and Tsujita, 1976b
	Obs, Dominant obese translucent	18–6.2	third instar distinctly translucent, fifth less conspicuous; body short, stout; difficulty ecdysing at fourth and pupal molts; male poor copulation	inhibitory epistatic interactions with genes affecting stars, crescents, e.g., p^{S}, *U*, *K*	
	oew, white egg translucent	?	highly translucent skin	reduced uric acid uptake; reduced numbers, size urate vesicles	Tamura and Sakate, 1983; Tamura and Akai, 1990

[a]Mutations spontaneous unless noted. X = X-ray induced and spontaneous alleles; M = MNU induced.
[b]Characteristics from Doira (1992) unless indicated.

Table 2.5. *Mutations affecting larval integument biochemistry*

Character	Gene[a]	Locus	Phenotype	Process, product, or mechanism affected	References[b]
Pteridine metabolism	*al*, albino	5–37.9	albino after first molt; cuticle porous, incomplete hardening, larva dehydrates	accumulates fluorescent sepiapterin precursor; enzyme deficiency upstream of sepiapterin reductase; cuticle defects secondary	Tsujita and Sakurai, 1971
	lem, lemon	3–0.0	deep yellow after first molt; in *lem*[l] homozygote cuticle remains soft	accumulates sepiapterin; lacks sepiapterin reductase activity in fat body; cuticle defect secondary	Tsujita and Sakaguchi, 1955; Tsujita, 1961; Matsubara, Tsusué, and Akino, 1963
	Sel, Sepialumazine	24–0.0	pale yellow	accumulates sepialumazine; increased sepiapterin deaminase activity in integument	Mazda, Tsusué, and Sakate, 1980; Mazda et al., 1981

	Xan, Xanthous	27–0.0	yellow	possibly introduced from *B. mandarina*	
	Ym, Yellow molting	27–?	larval skin covered with yellowish powder after first and second molts	yellow crystals secreted from Malpighian tubules	
Pigment storage vesicles	*i-lem*, inhibitor of lemon	2–29.5	faint yellow in combination with *lem*	reduced uptake of sepiapterin; altered tryptic peptides in isolated pteridine-binding proteins	Tsujita, 1963; Tsujita and Sakurai, 1963
	q, quail	7–0.0	quaillike pattern of black spots, thin lines on light reddish-purple skin	xanthommatin accumulation in epidermal vesicles associated with covalently bound protein	Sawada et al., 1990

[a]All mutations spontaneous.
[b]Characteristics from Doira (1992) unless indicated.

and Tsujita, 1976b) and argue that further study will provide insights into mechanisms of storage excretion in this insect.

Two oily mutations, Giallo Ascoli translucent (*og*) and oily q (*oq*), are candidates for the structural gene for xanthine dehydrogenase (Xdh), the last enzyme in the uric acid biosynthetic pathway. Both mutants are able to incorporate high levels of injected radioactive uric acid from the hemolymph into the skin (Tamura and Sakate, 1983), but the activity of Xdh is 20 percent or less than normal in fat body, the major site of uric acid synthesis, and alternate products to uric acid and isoxanthopterin accumulate in larval integument and feces (Tamura and Sakate, 1975; Tamura, 1977, 1983). The key role of Xdh in purine metabolism and the large body of work on the structure, function, and related genetics of the rosy locus, the structural gene for Xdh in *Drosophila* (Finnerty, 1976), provide an important foundation for additional characterization and future isolation of these two silkworm genes.

Mottled alleles that produce somatic mosaics with characteristic patterns in the size of translucent and opaque skin patches have been found for nearly half of the mapped oily mutations (see Table 2.4). In the few cases that have been examined, only the opaque regions take up exogenous uric acid (Tamura and Sakate, 1983). Of particular interest among this class of mutations is mutator of *oal (mu-oal)*, a gene that induces a somatic change in the degree and type of mottling expressed by a specific allele of a gene at a second, unlinked locus, *oal* (Hatamura, 1939; Tazima, 1964). Many mechanisms can be invoked to explain the genetic basis for mosaicism, including insertional inactivation events, position effects, systems of interacting transposable elements, and threshold levels of expression of critical gene products. Thus the oily mutants deserve attention not only for what they can reveal about important physiological functions of the integument but also for basic mechanisms of mutagenesis and gene action in the silkworm.

A small group of larval skin mutations affect pteridine metabolism, producing a grayish-white, brownish, or yellow background color (Table 2.5). Two of these, lemon (*lem*) and albino (*al*), also show abnormal cuticular melanization and hardening which appear in the second and succeeding instars and are lethal for certain alleles because the mandibles are too soft for feeding on leaves. This phenotype was the focus of early investigations, but is now considered to be a secondary consequence of enzyme deficiencies resulting in reduced or missing pteridine cofactors required for the conversion of phenylalanine to tyrosine (used in tanning of the cuticle) or other steps in the phenylalanine biosynthetic pathway (Tsujita and Sakurai, 1969, 1971; Kayser, 1985). Similarly, maternal rescue of first instar larvae homozygous for lemon and lethal lemon (*lem^{l}*) homozygotes (Tsujita, 1955, 1961; Tsujita and Sakaguchi, 1958) is likely to be an indirect effect of sequestering wild-type enzyme or products in the egg cytoplasm.

lem mutants have low to undetectable levels of sepiapterin reductase activity in the integument and fat body, the predominant site of the normal enzyme (Tsujita, 1961; Matsubara, Tsusué, and Akino, 1963), resulting in the accumulation of a bright yellow compound, sepiapterin (2-amino-4-hydroxy-6-lactyl-7, 8 dihydro-pteridine), in place of isoxanthopterin (Tsujita and Sakaguchi, 1955; Tsujita and Sakurai, 1971). Sepiapterin also accumulates in the sepia mutant of *Drosophila* (Nawa, 1960, cited in Tsujita and Sakurai, 1971). Albino larvae are postulated to be blocked further upstream in the sepiapterin pathway, producing large amounts of an unidentified light greenish-blue fluorescent substance in the epidermis; this hypothesis is supported by a finding of normal activity for downstream enzymes, including sepiapterin reductase, dihydrofolate reductase, and pterin dehydrogenase, and by the observation that double mutants (*al/al; lem^{l}lem^{l}*) express the upstream *al* phenotype (Tsujita and Sakurai, 1971). Finally, *Sel* mutants accumulate large amounts of a pale yellow sepiapterin metabolite, sepialumazine (7, 8-dihydro-6-lactyllumazine; Mazda et al., 1980). This mutation is hypomorphic, resulting in abnormally high sepiapterin deaminase activity (Mazda, Tsusué, Sakate, and Doira, 1981). Interestingly, *lem* larvae also express relatively high levels of this enzyme, but sepiapterin evidently masks the sepialumazine produced (Mazda et al., 1980). Two other yellow mutations potentially affecting pteridine biosynthetic pathways have been described but not yet studied biochemically (see Table 2.5).

Larvae homozygous for *lem* and a second mutation, inhibitor of lemon (*i-lem*), exhibit a dilute yellow phenotype, characterized by reduced levels of an unidentified yellow pigment, presumably sepiapterin (Tsujita, 1963). Sepiapterin reductase activity and isoxanthopterin accumulation in double mutants are equivalent to *lem* alone, suggesting that *i-lem* affects pteridine uptake rather than biosynthesis. Isolated epidermal storage vesicles or "chromogranules" have been shown to contain components that bind differentially to uric acid, isoxanthopterin, and in the *lem* mutant, sepiapterin (Tsujita and Sakurai, 1964a, 1964b, 1965a). Tryptic digests of these proteins isolated from mutant *lem, i-lem,* and wild-type larvae show a majority of common peptides comigrating with fluorescent pigments in high voltage paper electrophoresis; however, *lem* larvae have two extra peptides relative to wild-type, whereas *i-lem* have a single one whose mobility differs from either of the supernumerary *lem* peptides (Tsujita and Sakurai, 1963). These findings have been interpreted to mean that different but related proteins bind isoxanthopterin and sepiapterin and that *i-lem* is a structural variant of the latter. However, other interpretations are possible, and follow-up studies of these observations are warranted.

A different class of pigment-containing vesicle (or "granule") has been identified in the quail mutant (*q*), which accumulates larger than normal amounts of xanthommatin, a derivative of tryptophan via the kynurenin pathway (Sawada, Tsusué, Yamamoto, and Sakurai, 1990). These vesicles are smaller, more elliptical, and more electron dense than pteridine-containing granules and are not detectable in wild-type integument. Upon isolation they yield a 13 kD protein that appears to bind the pigment covalently, based on its retention in SDS gels and its continued binding to peptides separated by HPLC after protease digestion. Whether this mutation involves increased synthesis or uptake of xanthommatin is not yet known. In the silkworm, ommochrome accumulation is enhanced by administration of the juvenile hormone analogue methoprene, suggesting that quail may be a useful model for studying the action of this hormone on tryptophan metabolism and melanin formation (Ohashi et al., 1983).

A few larval mutations affect skin darkening or blackening and may also affect pigmentation at the pupal and/or adult stages. These include dilute black (*bd*), sooty (*so*), and melanism (*mln*), in addition to *p, S,* and *Ze,* described earlier (see Table 2.3), and are listed in Table 2.9 together with other mutations expressed later in metamorphosis (see "Pupal and adult mutations – appendage development and pigmentation"). Little biochemical work has been carried out on any of these presumed melanization mutants, apart from showing more than 40 years ago that the cuticle of striped (p^S) larva has a higher tyrosinase activity than that of wild-type (Aruga and Shigemi, 1952; Kawase, 1955) and, more recently, that there is no (or reduced, in *bd*) uptake of radioactive uric acid in regions of the larval integument that are darkly pigmented (Tamura and Sakate, 1983). Improvements in techniques as well as our greater understanding of the biochemistry of the integument and pigment formation since most of these observations were made argue strongly for their reinvestigation at the cellular and biochemical levels.

Larval mutations – biochemistry of the silk gland and the cocoon

Given the importance of the silk industry in the history of silkworm genetics, it is surprising that only a relatively small number of Mendelian mutations have been identified that affect silk production (Table 2.6). It is well known that silk yield is highly dependent on the growth conditions and nutritional status of the larva and exhibits a high response to heterozygosity; perhaps these are hints that most mutations affecting these important products will be found in polygenic or quantitative traits. The two predominant silk fiber proteins, fibroins H and L, are encoded by unlinked genes that were localized to linkage groups 25 and 14 using electrophoretic variants (reviewed in Shimura, 1988); sericins, which comprise a water-soluble gum, have been similarly mapped to two single copy genes on linkage group 11 (Gamo, 1982; Doira, personal communication). The silk genes, and an additional unmapped one of unknown function, P25, are coordinately expressed at high levels in the posterior (fibroins and P25) and middle (sericins) silk gland during larval intermolts. The spatial and temporal regulation of these genes is described in the chapter by Hui and Suzuki in this volume.

Larvae carrying the Naked pupa mutations, *Nd* and *Nd-s,* spin thin cocoons containing primarily sericins, and their posterior silk glands are abnormally small or missing (Shimura, 1988; Doira, personal communication). These loci probably coincide with the fibroin H and L structural genes, respectively, from which each is thus far inseparable by crossing over (reviewed in Shimura, 1988). Small amounts of both fibroin H and L messenger RNAs and their corresponding proteins are detectable in posterior silk gland cells of single homozygous *Nd* and *Nd-s* mutants, indicating that the primary defects are probably not transcriptional, but that only the forms of fibroin produced by the wild-type genes are secreted into the lumen. Further, the mutant proteins are unable to form covalent bonds with normal fibroin produced by heterozygotes (Takei et al., 1984, 1987; Shimura, 1988). Thus *Nd* and *Nd-s* are proposed to involve changes in the coding regions of the respective structural genes that prevent covalent bond formation between fibroin H and L chains, a process necessary for intracellular transport and secretion of the generally insoluble aggregated silk fiber (Shimura, 1988).

It is difficult to understand the relationship between the abnormalities in silk protein structural genes and the developmental defects of the posterior silk glands. Perhaps these mutations also involve upstream se-

Table 2.6. *Mutations affecting silk production*

Gene[a]	Locus	Phenotype	Process, product, or mechanism affected	References[b]
Nd, Naked pupa	25–0.0	cocoon contains only sericin; some homozygous lines lack a posterior silk gland	probable locus of fibroin H chain; RNA and protein detectable; no disulfide bond formation with fibroin L chain, blocking secretion	Takei et al., 1984, 1987; Shimura, 1988
Nd-s, Sericin cocoon[D,S]	14–19.2	cocoon contains only sericin; posterior silk gland degenerates	probable locus of fibroin L chain; RNA and protein detectable; no disulfide bond formation with fibroin H chain, blocking secretion	Takei et al., 1984, 1987; Shimura, 1988
flc, flimsy cocoon	3–49.0	cocoon contains excess sericin relative to fibroin; posterior silk gland degenerates	general secretion defect; fibroin H RNA synthesis normal but accumulation drops and histological abnormalities appear at midfifth instar	Adachi and Chikushi, 1977; Adachi-Yamashita, Sakaguchi, and Chikushi, 1980; Maekawa, Doira, and Sakaguchi, 1980

[a]Mutations spontaneous unless noted. D = DES induced; S = spontaneous and induced alleles.
[b]Characteristics from Doira (1992) unless indicated.

quences recently shown to be homologous to binding sites for transcription factors acting early in embryogenesis (see Hui and Suzuki, this volume), or perhaps secretion is intimately tied to integrity of the gland. Cloning of *Nd* and *Nd-s* and comparison of their sequences to the wild-type silk structural genes could test these ideas and provide additional insights into their mechanisms of action.

The flimsy cocoon mutant (*flc*) also spins a lightweight cocoon that contains both sericins and fibroins but is disproportionately reduced in fibroin content (Adachi and Chikushi, 1977). Histological evidence suggests that this mutation involves a general secretion defect. By contrast to the naked pupa mutations, the posterior silk glands of *flc* mutants appear normal until the end of the fifth instar, when fibroin synthesis should be maximal. At this time, the cells begin to degenerate (Adachi and Chikushi, 1977), displaying many ultrastructural cytoplasmic abnormalities in the rough endoplasmic reticulum, Golgi complex, and mitochondria, and accumulate numerous "fibroin globules" (Adachi-Yamashita, Sakaguchi, and Chikushi, 1980) suggestive of a general defect in intracellular transport. Fibroin H messenger RNA synthesis is essentially normal up to day 5 of the fifth instar but its accumulation drops markedly after day 3, concurrent with degeneration of the tissue (Maekawa, Doira, and Sakaguchi, 1980).

Formal genetic studies on cocoon color were first reported in 1902 (Coutagne, 1902), and the first mutation identified in the silkworm as a Mendelian trait was the dominant gene, Yellow (*Y*), which is responsible for a brilliant, brassy color in both cocoon and hemolymph (Toyama, 1906, cited in Tazima, 1964). The genetics of cocoon color is complex, affecting the color of the silk, its intensity, and which layer of the cocoon is pigmented; details of interactions among several genes involved are reviewed elsewhere (see Kikkawa, 1953, and Tazima, 1964). The pigments, generally water soluble, are obtained from the primary food plant, mulberry, and include carotenoids, yielding yellow, flesh-colored, and pink cocoons, and flavonoids, which may be modified biochemically by the silkworm and yield shades of green. Inheritance of green cocoon color is polygenic and has not yet been fully characterized, with two major loci localized and others now under investigation (Doira, personal communication). Different mutations and mutant combinations affect both the absorption and secretion of pigments by the midgut and the silk gland; histological studies of these phenotypes suggest that carotenoid uptake and secretion may be differentially controlled in the same regions of the middle silk gland that are spatially differentiated for expression of different sericin proteins (Gamo, Inokuchi, and Laufer, 1977; Hu, Shimada, and Kobayashi, 1992) and splicing of sericin transcripts (Couble, Michaille, Garel, Couble, and Prudhomme, 1987). Isolation of carotenoid-binding proteins from normal and mutant stocks of silkworms has begun in the laboratory of M. Wells using heterologous sequences isolated from *Manduca* as probes (Wells and El Jouni, personal communication). This marks an important first step in beginning a molecular investigation of this long-established genetic system in an evolutionary context.

Larval mutations – feeding behavior

Silkworms are nearly monophagous, eating primarily mulberry and close botanical relatives. This behavior is under the control of volatile attractants and phagostimulants in the leaves, as well as compounds that act as feeding deterrents (reviewed in Hsaio, 1985). The formulation of artificial diets for commercial silkworm rearing has gone hand-in-hand with the development of genetic strains that are well adapted both to feeding and to growing, indirectly taking into account both behavioral and nutritional requirements. These have resulted in a small group of induced and spontaneous mutations that convert monophagous feeding behavior into polyphagy (Table 2.7).

The first Nonpreference mutants, *Np* and *Nps,* were obtained by Tazima after an exhaustive, long-term search for dominants by testing irradiated larvae for their ability to eat beet leaves (reviewed in Tazima, 1989). Both mutations are accompanied by recessive lethals assumed to be closely linked deletions, but this has not been shown directly, and it is possible that at least some cases of lethality are due to neurogenic defects, reflecting the physiological importance of the nonpreference genes themselves. Additional mutations obtained after irradiation include the sex-linked gene, Beet feeder (*Bt*), and a still unmapped gene, *D5,* which also carries a recessive lethal (Tazima and Ohnuma, 1986) and appears to act via a cytoplasmic factor (Tazima and Ohnuma, 1987).

Breeding studies carried out using the *zairaishu* in the early 1970s by Yokoyama produced a famous polyphagous silkworm strain known as Sawa-j, which feeds avidly on many plants, including cabbage, plum, cherry, and persimmon, as well as on artificial diets (Tazima, 1989). Recent genetic experiments using this strain in combination with an artificial diet designed using a linear programming method to determine optimal nu-

Table 2.7. *Mutations affecting larval feeding behavior*

Gene[a]	Locus	Phenotype	Homozygous lethal	References[b]
Bt, Beet feeder[X]	1–40.8	wide feeding range; incomplete dominance	no	Tazima, 1989
Np, Nonpreference[X]	11–30.5	wide feeding range	yes	Tazima, 1989
Nps, Nonpreference Shokei[g]	3–2.2	wide feeding range; chromosome deficiency, includes *lem;* incomplete penetrance	yes	Tazima, 1989
D5[X]	?	wide feeding range; acts via maternal cytoplasmic factor	yes	Tazima and Ohnuma, 1986, 1987; Tazima, 1989
nfad, nonfeeding on artificial diet	25–?	repelled by feeding inhibitor in artificial diet	no	
pph, polyphagous	?	wide feeding range; discovered in strain Sawa-j	no	Kanda, Tamura, and Inoue, 1988; Kanda, 1992; Asaoka and Mano, 1992

[a]Mutations spontaneous unless noted. X = X-ray induced; g = gamma induced.
[b]Characteristics from Doira (1992) unless indicated.

trition at minimum cost (Horie and Watanabe, 1983; Yanagawa, Watanabe, and Nakamura, 1988) uncovered a major recessive gene controlling polyphagy (*pph*) (Kanda, 1992) whose action is affected by a number of undefined recessive modifier genes (Kanda, Tamura, and Inoue, 1988; Asaoka and Mano, 1992). Modifiers also influence the expression of the radiation-induced Nonpreference mutations (Tazima, Ohnuma, and Tanaka, 1987, 1988; Tazima, 1989), although none has been mapped. Finally, a spontaneous, recessive mutation that inhibits feeding on artificial diet (*nfad*) has been identified in inbred stocks of Chinese origin. Exploration of the underlying electrophysiological mechanisms associated with these mutations has just begun (Yazawa, Hirao, Arai, and Yagi, 1991; Asaoka, personal communication).

This group of mutations offers a model for studying the genetic basis of feeding behavior in a member of the insect order that contains the most destructive phytophagous larvae known. Even though the silkworm and its near relatives are not major agricultural pests, study of the mutants now in hand can provide tools for future investigation in lepidopteran species of greater economic impact for which there is no formal genetics. Despite the low yield of mutations (Tazima, 1989), the relative simplicity of selection schemes utilizing unusual foodstuffs, artificial diets, and stimulants or inhibitors of the feeding response (Yazawa et al., 1991) argues for continuation of large-scale mutagenesis, screening of inbred stocks, and possibly even crossbreeding with *B. mandarina* to identify additional feeding response genes. Although modifiers introduce a complicating factor for making initial rapid progress in this system, as the new approaches for analyzing polygenic traits become available in the silkworm, the genes controlling this complex behavior may also become accessible to direct genetic and molecular analysis (see "A molecular linkage map").

Pupal and adult mutations – appendage development and pigmentation

Returning briefly to themes discussed earlier, I list two classes of relatively abundant mutations whose action is manifest at later stages of metamorphosis. In Table 2.8 are listed mutations associated with defective pupal and adult appendage development, primarily the wings and eyes. Although none has been studied in detail, crayfish (*cf*), crayfish of Eguchi (*cf-e*), and chela (*cl*) are of special note for their superficial resemblance to *Drosophila* blister mutants, which are known to be abnormal in wing disc morpho-

Table 2.8. *Mutations affecting adult appendage development*

Character	Gene[a]	Locus	Phenotype[b]
Wings	*cf*, crayfish	13–20.9	wing buds inflated with hemolymph, protruding, fragile, bleed easily
	cf-e, crayfish of Eguchi	4–0.0	pupal wings swollen, protrude laterally from body; wing buds swollen, fragile, bleed easily
	cl, chela	21–?	wing buds inflated with hemolymph, protrude laterally from body; pupal wing buds fragile, bleed easily
	fl, flugellos (wingless)	10–13.0	wings lacking in pupa, moth; legs on T2, T3 poorly developed; often dies from bleeding in uncovered border between thorax and abdomen
	mp, micropterous	11–51.8	wings ~80% normal size
	mw, minute wing	22–25.2	pupal, adult wings extremely small (<*mp*)
	rw, rudimentary wing	1–22.8	wings poorly developed, variable expression
	tyw, tiny wing	24–?	wings poorly developed, variable expression
	Vg, Vestigial[X]	1–38.7	wings poorly developed, club-shaped and/or cut; hemizygous female embryonic lethal; deletion includes *od* locus (49.6)
	wri, wrinkled wing	14–0.0	wings poorly developed, do not extend fully
Eyes	*lu*, lustrous	16–0.0	facets of compound eye small, reflective, irregular
	ve, varnished eye	6–11.1	compound eye extremely small, lustrous; facets sparsely distributed
Pleiotropic effects	*ap*, apodal	3–22.3	thoracic legs, bursa copulatrix degenerate; female sterile
	gap, apterous and rudimentary gonads	5–?	lacks pupal wings; thorax extremely narrow, slender; ovaries undeveloped, sterile; retarded larval growth
	gon, undeveloped gonads	18–?	ovaries, testes poorly developed, accessory organs normal; pupae often *cf*-like; larval growth retarded; newly hatched larvae abnormally pigmented

[a]Mutations spontaneous unless noted. X = X-ray induced.
[b]T = thoracic segment. Characteristics from Doira (1992).

genesis (Gotwals and Fristrom, 1991) and may involve cell adhesion molecules or elements of the cytoskeleton (Fristrom, Wilcox, and Fristrom, 1993). Clearly, others listed in Table 2.8 are also likely to involve imaginal disc formation and/or differentiation, and so may be suitable for examination at the molecular level once similar genes are identified and isolated from *Drosophila* (Bryant, 1993) and, if conserved, can be used as heterologous probes.

Finally, Table 2.9 lists mutations that produce abnormal melanization or darkening at pupal and adult as well as at larval stages (see also "Larval mutations – biochemical genetics of the integument"). Two mutations expressed during pupal development, black-striped pupal wing (*bpw*) (Yamamoto, 1984) and black pupa (*bp*), have limited temperature-sensitive periods; preliminary studies indicate that the latter is involved with the prophenoloxidase activation cascade (Hashiguchi, Yoshitake, and Tsuchiya, 1970). Although it is not unusual to find some melanization in adults (Tazima, 1964), thus far it has been possible to map only three Mendelizing genes affecting this character, Black moth (*Bm*), Wild wing spot (*Ws*), and white-banded black wing (*wb*). Apart from their possible roles in pattern formation, the importance of these kinds of traits in natural populations of moths and butterflies (Sheppard, 1961; Nijhout, 1991) argues that studying such genes even in the relatively drab silkworm model may open up new avenues for their investigation in more colorful and diverse lepidopteran species.

Molecular genetics

In the following sections, I describe new techniques being introduced to silkworm genetics using recent advances in molecular biology and molecular genetics. Though most of these studies have just begun, benefits are already being gained from the advantages of using a well-developed genetic model.

Molecular approaches to cytogenetics

Recently, Tsuchida, Maekawa, and colleagues have established conditions for carrying out in situ hybridization with fluorescent probes on silkworm chromosome spreads (Okazaki et al., 1993; Figure 2.2). This significantly enhances the opportunities for correlating cloned genetic markers with specific chromosomes using both normal and aberration stocks, and perhaps even for localizing genes in relation to one another in extended chromosome preparations (Trask, 1991). The establishment of this technique for lepidopteran material may also facilitate the use of repeated sequences that show localized or uneven distribution within the genome to improve karyotyping, although appropriate ones have yet to be identified in *Bombyx* (see Eickbush, this volume). As molecular markers are correlated with known linkage groups (see the next section) and heterologous probes are identified in related species, this technique should find more widespread use in lepidopteran cytogenetics.

Pulsed field gel electrophoresis (PFGE) offers a promising alternative to cytogenetic methods for determining linkage and relative gene order on large, isolated DNA fragments and for constructing physical chromosome maps. This technique has been used to help establish the overall structure of the chorion locus (see Izzo, 1991; see also Eickbush and Izzo, this volume) and of the Antennapedia and bithorax complexes (see Ueno, Nagata, and Suzuki, this volume). Although the size of intact silkworm chromosomes exceeds current practical limitations for separation by PFGE and related approaches, a number of radiation-induced somatic mosaic stocks carrying unstable chromosome fragments marked with visible genetic mutations, which are maintained as trisomics (Tazima, 1964), are available for potential isolation of defined chromosome segments. Recently, H. Fujiwara and colleagues resolved DNA molecules on the order of 2.3–2.5 Mb from mosaic larvae using PFGE and verified the identity of one segregating band by its hybridization to a cloned chorion sequence, which was known to be linked to the dominant p^S marker carried on the unstable fragment (Fujiwara, Ninaki, Kobayashi, Kusuda, and Maekawa, 1991). Stocks carrying unstable chromosome fragments presumed to be of differing lengths from the average size of somatic mosaic patches are available for linkage groups 2 and 3, providing valuable starting material for large-scale chromosome mapping and, ultimately, for chromosome isolation, recently begun in Fujiwara's laboratory (personal communication).

A molecular linkage map

Determining the molecular lesion underlying a known mutation provides an important approach for studying gene function in vivo, especially in an organism for which germline transformation is not yet available. This

Table 2.9. *Mutations affecting melanization at various metamorphic stages*

Gene[a]	Locus	Stage expressed	Phenotype	References[b]
bd, dilute black	9–22.9	larva, moth, egg	larva grayish black with reduced uric acid uptake; egg lacks micropyle; male fertile but cannot copulate unaided; *bdf* fertile, normal copulation	Tamura and Sakate, 1983
mln, melanism	18–41.5	larva, moth	black larval head cuticle, anal plates, thoracic legs, spiracle sieve plate; moth black	
so, sooty	26–0.0	larva, pupa, moth	body smoky in larva and moth (less conspicuous); pupal case black, especially at ventral tip of abdomen	
bp, black pupa	11–40.3	pupa	pupal cuticle black; expression temperature sensitive for several hours before pupation; involves prophenoloxidase cascade	Hashiguchi, Yoshitake, and Tsuchiya, 1970

bpw, black-striped pupal wing	13–?	pupa	temperature sensitive; narrow black stripes on pupal wings when $\leq 20°C$ during 2 days before larval-pupal ecdysis	Yamamoto, 1984
Bm, Black moth	17–0.0	moth	black scales on body, wings	
Ws, Wild wing spot	17–14.7	moth	black spot on wing apex, less conspicuous in heterozygotes, especially females; from *B. mandarina*	
wb, white-banded black wing	5–35.8	moth	outer margin, proximal region of wings dark brown, leaving wide white band; body white	

[a]All mutations spontaneous.
[b]Characteristics from Doira (1992) unless noted.

usually requires direct isolation of the gene or its cDNA for structural comparison in mutant and wild-type forms. Many molecular strategies have been used for cloning silkworm genes, starting with, for example, tissue-specific mRNAs and cDNAs (see Hui and Suzuki, Eickbush and Izzo, and Riddiford, this volume), expression libraries (see Kafatos et al., this volume), defined oligonucleotide sequences (Adachi et al., 1989), or heterologous probes (Hui, Matsuno, Ueno, and Suzuki, 1992; see Hui and Suzuki and Ueno, Nagata, and Suzuki, this volume). Although cloned sequences have been correlated with the genetic linkage maps in the relatively small number of cases for which there are polymorphic protein variants, such as silk, chorion, and storage proteins (accounting for 13 linkage groups at this writing), in only a few cases have particular genetic mutations been correlated with molecular defects; these are deletions associated with the *Gr*^{B} chorion mutation (Durnin-Goodman and Iatrou, 1989) and the *Bombyx* homologue of the bithorax gene complex (Ueno et al., 1992; see Ueno, Nagata, and Suzuki, this volume). Thus the large collection of silkworm mutants remains to be fully exploited in these terms.

With the addition of molecular markers to the conventional genetic linkage maps, it becomes possible to use positional cloning to isolate a gene whose product is unknown. This requires that it be close to a site defined by a restriction fragment length polymorphism (RFLP) (Botstein, White, Skolnick, and Davis, 1980), a microsatellite sequence (Hearne, Ghosh, and Todd, 1992), a sequence-tagged site (Williams, Schrank, Huynh, Shownkeen, and Waterston, 1992), or any other physical chromosome marker that can serve as a starting place in a genomic library to walk or jump to the locus of interest. To this end, several laboratories have begun a cooperative effort to construct high density molecular linkage maps in the silkworm, to be coordinated with the conventional linkage maps.

Recent refinement of statistical techniques commonly used for human linkage analysis (Lander and Botstein, 1989) and first applied to the tomato as a model system (Paterson et al., 1988) has made it possible to localize major genes affecting quantitative traits or quantitative trait loci (QTLs) to specific chromosome segments bounded by physical markers in high density molecular linkage maps. QTLs of interest can then be tracked via closely linked molecular tags to assess their function or effects in different genetic backgrounds (Tanksley and Hewitt, 1988; Paterson et al., 1991) and to increase the efficiency of producing new commercial breeds by marker-assisted selection (Lande and Thompson, 1990; Hospital, Chevalet, and Mulsant, 1992). In principle, the molecular markers can also serve as entry points into the genome to isolate the QTLs directly, as with conventional genetic mutations.

Despite the large number of linkage groups, the genetics of map construction is straightforward because the lack of crossing over in females allows unambiguous linkage assignments to be made from a relatively small number of progeny arising either from F1 backcrosses, in which heterozygous males are mated to homozygous females, or from F2 crosses, in which both parents are heterozygous. An alternative genetic approach using a set of related inbred lines is also being developed and is described in the following section. To permit rapid correlation of the molecular and conventional linkage maps, multiply marked genetic strains have been constructed by recurrent backcrossing to one of the parental strains being used for the initial molecular map (Doira, personal communication).

In my laboratory, we are using anonymous RFLPs to begin construction of the molecular linkage maps because they show codominant expression in genomic hybridization patterns and thus yield maximum information about genetic relationships from a given cross, and because they can be used readily to map cloned single copy genes. In addition to RFLPs, we and other laboratories are developing approaches based on the polymerase chain reaction (PCR) to enable rapid screening of large sample sizes necessary for constructing high resolution maps, for QTL mapping, and for eventual marker-assisted selection. These include the use of primers constructed from short, random arbitrary polymorphic DNAs (RAPDs) (Welsh and McClelland, 1990; Williams, Kubelik, Livak, Rafalski, and Tingey, 1990; Shimada et al., 1992) as well as from cloned genes (Shimada et al., 1992). We are also investigating the feasibility of constructing primers derived from unique sequences flanking microsatellites, that is, di- and trinucleotide repeats that show length polymorphism in different strains or individuals and thus can serve as codominant markers after amplification (Dietrich et al., 1992; Hearne et al., 1992). The latter are abundant in *Drosophila* (Huijser, Hennig, and Dijkhof, 1987; Pardue, Lowenhaupt, Rich, and Nordheim, 1987) and have been used for linkage mapping in mosquitoes (Zheng, Collins, Kumar, and Kafatos, 1993). Among the Lepidoptera, a low copy number hypervariable $(GATA)_n$ microsatellite (*Bkm*) is present in *Ephestia* (Traut, 1987) which is inherited in simple Mendelian fashion (Traut, Epplen, Weichenhan, and Rohwedel, 1992), and preliminary evidence suggests that microsatellites are also present in the silkworm (Garel and Goldsmith, unpublished observations). Molecular linkage maps are also being constructed using interspecific hybrids of *Heliothis virescens* and *H. subflexa;* this work is aimed initially at isolating pesticide resistance genes (Heckel, Abbott, and Brown, 1988; Zraket, Barth, Heckel, and Abbott, 1990; Heckel, 1991) but will also be important for comparative investigations of lepidopteran genomes (Heckel, 1993).

Initial testing in *Bombyx* shows that many geographic strains are nearly monomorphic at the DNA level, but that there is a high degree of polymorphism between strains, as might be expected from their diverse biological character and the results of quantitative breeding studies noted earlier. We have found that approximately 70 percent of cloned random single or low copy cDNAs derived from an early follicular library can detect polymorphism in genomic DNA digested with at least one of three restriction enzymes (Goldsmith and Shi, 1993); short, random, single copy genomic probes (Yukuhiro, Kanda, and Tamura, personal communication) and RAPDs (Shimada et al., 1992) give similar results. The large number of mobile, repeated elements in the silkworm genome undoubtedly contributes to this level of DNA polymorphism (see Eickbush, this volume). Interestingly, anonymous cDNAs reveal much less polymorphism in *Heliothis,* even in interspecific crosses (Heckel et al., 1988). It is unclear at this stage whether the genomes of these moths contain fewer repeated elements or if there is a higher degree of conservation in the vicinity of expressed genes. At this writing, our initial molecular linkage map in *Bombyx* contains 60 RFLP markers covering 24 linkage groups, of which 8 correspond to identified chromosomes in the conventional linkage maps (Goldsmith and Shi, 1993).

Recombinant inbred lines

In addition to classic genetic approaches, in my laboratory we are investigating the feasibility of introducing recombinant inbred (RI) lines to the silkworm as a new genetic resource for constructing molecular linkage maps and for analysis of quantitative traits. This system was originally devised for mice (Bailey, 1971), which have a relatively large number of linkage groups ($n = 20$) and about the same generation time as silkworms (6–8 weeks). One establishes a family of RI lines by carrying out successive sib matings between offspring of individual pairs derived from a single F2 cross between two genetically divergent, inbred parents; for each line, only the progeny of a single, randomly chosen pair are reared to the next generation. These lines begin to approach homozygosity after approximately 8 inbreedings (although in practice this may take up to 20 generations) (Green, 1981; Burr and Burr, 1991), when each will contain a unique set of recombinant chromosomes carrying different combinations of polymorphic alleles that reflect the linkage relationships in the original parent stocks. Thus recombination frequencies can be determined from the patterns of association of parental alleles within and between lines and converted to genetic map distances using a function originally derived by Haldane and Waddington (Bailey, 1981). The number of RI lines per family affects the accuracy of linkage estimates (Taylor, 1978), but as few as seven lines have been used profitably in mice (Bailey, 1971).

Once homozygous, RI lines provide a continuous supply of genetically typed individuals that yield a cumulative data set for linkage testing (Bailey, 1981) and can be shared among many investigators. These properties are especially advantageous for constructing molecular linkage maps (Burr and Burr, 1991). A family of RI lines can also be used to generate large populations of genetically homogenous individuals for scoring quantitative traits such as disease and pesticide resistance, nutritional requirements, and behavior, and under favorable circumstances, may even be used for mapping QTLs (Taylor, 1976; Simpson, 1989; Burr and Burr, 1991). Use of recombinant inbreds to construct high density molecular linkage maps was recently reported for *Arabidopsis* (Reiter et al., 1992) and maize (Burr, Burr, Thompson, Albertson, and Stuber, 1988). It is not yet certain whether silkworm RI lines will escape potentially serious problems of inbreeding depression; however, the extensive inbreeding practiced in maintaining the geographic races and in fixing new traits in commercial silkworm stocks suggests that at least some strain combinations will prove viable.

Future prospects

The study of silkworm genetics is poised on the edge of a new era. The large collections of carefully maintained silkworm mutations and practical breeding stocks are valuable resources that have only begun to be exploited to their full potential. With continuing rapid progress in the construction of molecular linkage maps, it should be possible within a few years to devise strategies for isolating any mutant locus or QTL of interest from *Bombyx* and thence to embark on identification and cloning of homologous or related genes in other Lepidoptera. Probes generated from these studies will provide tools for chromosomal in situ hybridization and thus may help to improve the usefulness of cytogenetics for gene mapping and comparative genome analysis, in addition to making available new approaches for stock identification and breeding in sericulture. In a relatively short time, it may also be possible to begin the isolation and sequencing of one or more silkworm chromosomes as a means of revealing the structural basis of the unique character of these agents of lepidopteran inheritance and evolution. These kinds of studies will help provide a foundation to begin addressing such tantalizing and long-standing problems as sex determination, heterochromatization of the W chromosome, and lack of crossing over in lepidopteran females at the molecular level.

Acknowledgments

I am especially grateful to Hiroshi Doira for his unfailing patience in answering my questions about silkworm genetics and for giving me access to the genetic resources at Kyushu University and to his unpublished work; to Bungo Sakaguchi for his generous hospitality and use of his extensive reprint collection; and to both for their long-term sponsorship of my activities in silkworm genetics. I thank Yasuhiro Horie, Toshiki Tamura, Hideaki Maekawa, and the Science and Technology Agency for making possible my recent visits to Japan, which were essential to this project. Thanks also to Toru Shimada for help in finding Japanese references, to Adelaide Carpenter and Lynn Riddiford for helpful comments and discussion, to Georges Bosquet for calling my attention to the *Drosophila* blister mutants, to Gerard Chavancy and the International Sericultural Commission for hosting my stay during the final stages of preparation of the manuscript, and to Adam Wilkins for expert editorial advice. Finally, I must thank the Carolyn and Kenneth D. Brody Foundation, the Kingston girls, Sopchoppy Subtropical Research, and Martin Baxter for continuing encouragement and support.

3

Mobile elements of lepidopteran genomes

THOMAS H. EICKBUSH

Introduction

The enormous size and surprising fluidity of eukaryotic genomes have long been a fascinating, highly active area of investigation. One of the more remarkable examples of this fluidity is found in the genomes of insects, which can be as small as 5×10^7 base pairs (bp), one-sixtieth the size of the human genome, to 1.2×10^{10} bp, four times the size of the human genome (for a review, see John and Miklos, 1988). The initial steps in the molecular dissection of eukaryotic genomes began with a series of experiments that took advantage of the dependence of DNA:DNA reassociation rates on nucleotide sequence complexity (Britten and Kohne, 1968). These reassociation kinetic studies indicated that all eukaryotic genomes could be divided into repetitive sequences, present hundreds to millions of times per genome, and non-repetitive sequences, present once or at most only a few times per genome. The most highly repeated DNA sequences are usually short, from a few nucleotides to 200 nucleotides in length, arranged as large tandem clusters on the chromosomes (John and Miklos, 1979; Brutlag, 1980; Singer, 1982a). This tandemly repeated DNA is usually referred to as satellite DNA and is typically located at the heterochromatic centromeres of chromosomes. It has been argued that satellite DNA may play a simple structural role by preventing genes from being too near the centromere, or more functional roles such as assisting in chromosome pairing and disjunction during mitosis or meiosis (John and Miklos, 1988; Wu, True, and Johnson, 1989).

Perhaps the most unexpected discovery in these attempts to dissect eukaryotic genomes by DNA reassociation kinetics was the finding that in many cases that portion of the genome expected to contain the structural genes (the nonrepetitive sequences) is interspersed with a second class of highly repetitive DNA (Davidson, Hough, Amenson, and Britten, 1973; Graham, Neufeld, Davidson, and Britten, 1974). The nature of these interspersed repetitive sequences has been most extensively studied in mammalian genomes. It is now known that most of this interspersed repetitive DNA corresponds to mobile elements that replicate by the reverse transcription of RNA to DNA and integration of the new DNA copy into the genome. These mobile elements are also found in the heterochromatic regions of chromosomes. The interspersed repetitive DNAs were originally classified by Singer (1982b) into two groups based on their length: short interspersed nucleotide elements (SINEs) from 70 to 300 bp in length and long interspersed nucleotide elements (LINEs) up to 6–7 kilobase pairs (kb) in length.

Short interspersed nucleotide elements (retroposons)

The most extensively studied SINEs are the 300 bp *Alu* elements found within the human genome (reviewed in Deininger, 1989). These elements, named after an Alu I restriction endonuclease site found in most copies, are present at approximately 500,000 copies, representing 5 percent of the total genome mass (Rinehart, Ritch, Deininger, and Schmid, 1981). The major breakthrough in understanding the origin of these elements was the discovery that the *Alu* family was derived from 7SL RNA (Ullu and Tschudi, 1984), which is an essential component of the signal recognition particle responsible for cotranslational secretion of proteins into the lumen of the rough endoplasmic reticulum (Walter and Blobel, 1982). The nucleotide sequence of *Alu* elements differs from 7SL RNA by having small internal deletions followed by dimerization and the addition of a 3′ poly(A) tail. The internal RNA polymerase III promoter of the 7SL RNA gene is maintained in *Alu* elements, providing for their efficient transcription. *Alu* repeats are present in all primate species, while a somewhat different 7SL RNA–derived element, named *B1,* is present in rodents.

A second group of SINEs found in mammals (Daniels and Deininger, 1985; Sakamoto and Okada, 1985) as well as nonmammalian vertebrates (Endoh and Okada, 1986; Matsumoto, Murakami, and Okada, 1986) has been shown to be derived from transfer RNA (tRNA) genes. A common structural feature among SINEs derived from either 7SL RNA or tRNA genes is a poly(A) tail at the 3′ end, suggesting that the priming of reverse transcription occurs at this site (Rogers, 1985; Deininger and Daniels,

1986; Weiner, Deininger, and Efstratiadis, 1986). It appears that genes transcribed by RNA polymerase III can readily give rise to SINEs because their internal promoters accompany any new insertion within the genome, and they usually terminate in an oligo(U)-rich sequence that may be capable of self-priming reverse transcription at a poly(A) sequence (Jagadeeswaran, Forget, and Weissman, 1981; Van Arsdell et al., 1981). Because of their origin by reverse transcription of an RNA molecule, SINEs have also been called retroposons (Rogers, 1985). This term is used in the remainder of this chapter.

Retrotransposable elements

The second class of interspersed repetitive elements found in mammalian genomes is the long interspersed nucleotide elements, LINEs. Up to 50,000 copies of LINE sequences can be present in mammalian genomes, accounting for another 5 to 10 percent of the total genome mass. LINEs from different mammalian species have been shown to be members of the same family, termed *Line 1* (Singer and Skowronski, 1985). Because consensus nucleotide sequences of these elements from different species revealed open-reading frames (ORFs), *Line 1* sequences were originally believed to be derived from the reverse transcription of RNA polymerase II transcripts from protein structural genes. *Line 1* elements were therefore grouped with the various SINEs and termed *retroposons* (Rogers, 1985; Deininger and Daniels, 1986; Weiner et al., 1986). However, it is now known that the ORFs of *Line 1* elements contain features found in retroviruses, in particular the ability to encode their own reverse transcriptase, and thus should be classified as retrotransposable elements (Hattori, Kuhara, Takenaka, and Sakaki, 1986; Loeb et al., 1986). Analysis of *Line 1* elements has been frustrated because the vast majority of the copies in each species contain disruptions of their ORFs or truncations at their 5′ ends that prevent them from being autonomous retrotransposable elements.

Mammalian *Line 1* elements are members of a larger group of retrotransposable elements found in a variety of organisms. These elements include the *I* element of *Drosophila melanogaster* (Fawcett, Lister, Kellett, and Finnegan, 1986), *Ingi* from *Trypanosoma brucei* (Kimmel, ole-Moiyoi, and Young, 1987), *cin4* from *Zea mays* (Schwartz-Sommer, Leclercq, Gobel, and Saedler, 1987), and *R1* and *R2* elements of *Bombyx mori* (Burke, Calalang, and Eickbush, 1987; Xiong and Eickbush, 1988a). Several properties distinguish these elements from the first retrotransposable elements to be sequenced, the *17.6* and *copia* elements from *D. melanogaster* (Saigo et al., 1984; Mount and Rubin, 1985) and the *Ty1* element of yeast (Clare and Farabaugh, 1985). The *Line 1*-like elements do not contain long terminal repeats (LTRs), and their ORFs contain significantly lower levels of sequence similarity to retroviral genes than either *17.6*, *copia*, or *Ty1*. Together these *copia*-like and *Line 1*-like retrotransposable elements represent the largest and most diverse class of mobile elements present in eukaryotic genomes. Their current distribution in every major taxonomic group of eukaryotes suggests that these elements probably exist in virtually every eukaryotic species.

An entirely different class of transposable elements is known in eukaryotes that do not propagate via an RNA intermediate. This class of elements includes the *P* and *Hobo* elements of fruit flies (Blackman and Gelbart, 1989; Engels, 1989), the *Tc1* elements of nematodes (Moerman and Waterston, 1989), the *Ac1* and related elements of maize (Federoff, 1989), and the *Tam* elements of snapdragon (Coen, Robbins, Almeida, Hudson, and Carpenter, 1989). Like bacterial transposons, these elements are believed to transpose through DNA intermediates in which excision of the element from a preexisting site precedes insertion at a new site (Engels, Johnson-Schlitz, Eggleston, and Sved, 1990).

In this chapter we describe the mobile elements of Lepidoptera. Most of the data will be derived from investigations of the domesticated silk moth, *B. mori,* which has the best characterized lepidopteran genome. *B. mori* was the first insect shown to have an interspersed pattern of repetitive and nonrepetitive sequences typical of mammalian genomes (Gage, 1974a; Gage, Friedlander, and Manning, 1975). This organization was quite different from the *D. melanogaster* genome (Manning, Schmid, and Davidson, 1975; Crain, Davidson, and Britten, 1976), most of which is composed of very long single copy DNA sequences (up to 100 kb) completely free of repetitive DNA. Molecular characterization of the repetitive elements from *B. mori* has revealed the presence of transposable elements typical of the *D. melanogaster* genome, as well as retroposons (SINEs) typical of mammalian genomes.

Retroposons of *Bombyx mori*

Structure and distribution of *Bm1* and *Bm2* elements

Study of the interspersed short repetitive DNA of *B. mori* was initiated by Adams and co-workers (1986). Total genomic DNA was denatured, renatured for a short period to allow only the highly repeated DNA in the genome to reanneal, and treated with S1 nuclease to destroy all DNA that remained single-stranded. Three discrete-length repetitive DNA fractions were seen. The most abundant was approximately 260 bp in length, and two somewhat less abundant fractions of 470 bp and 130 bp were also found. The repetitive DNA was cloned into a bacterial plasmid and the nucleotide sequences of six clones were determined. These six clones contain greater than 88 percent nucleotide sequence identity and thus represent one repetitive family, called *Bm1* (*B. mori* element 1). *Bm1* elements can be divided into two classes, one 250 bp in length and the second 450 bp in length. This difference is due to an additional 200 bp at the 5′ end of the latter class. The 3′ ends of *Bm1* elements in both classes contain poly(A) tails.

To investigate their repetition frequency and genomic distribution, *Bm1* sequences were labeled and hybridized to a *B. mori* genomic lambda DNA library. Ninety-one percent of the phage in the library were found to contain DNA insertions that hybridized to *Bm1*. Based on the size of the *B. mori* genome, 4.8×10^8 bp (Gage, 1974a, 1974b) and the fraction of the lambda phage library that hybridized to *Bm1*, it was estimated that there are a minimum of 2.3×10^4 copies of *Bm1* per haploid genome (Adams et al., 1986). This is probably a significant underestimate of the actual number of copies present within the genome, because the calculation assumed that each lambda phage that hybridized to the probe contained only one *Bm1* element. In fact, if the distribution of *Bm1* within the genome is similar to that in the chorion locus (see Eickbush and Izzo, this volume), an example of which is shown in Figure 3.1, each lambda DNA phage that hybridizes to a *Bm1* probe actually contains multiple *Bm1* elements (Hibner, Burke, and Eickbush, 1990; Izzo, 1991). Thus we estimate the number of *Bm1* elements in the *B. mori* genome to be closer to 1×10^5. Based on this calculation, approximately 5 percent of the *B. mori* genome is composed of *Bm1* elements (1×10^5 copies $\times$ 250 bp/copy/4.8×10^8 bp). This figure is comparable to the 5 percent of the human genome represented by *Alu* sequences (5×10^5 copies $\times$ 300 bp/copy/3×10^9 bp) (Rinehart et al., 1981).

The distribution of *Bm1* elements in a defined chromosomal segment of *B. mori* is shown in Figure 3.1, which is a small region of the *Ch3* chorion gene complex described in detail elsewhere in this volume (see Figure 8.4). The nucleotide sequence of this entire 50 kb segment shown in Figure 3.1 has been determined, except for the short segment from 37 to 41 kb (Hibner et al., 1988, 1991; Izzo and Eickbush, in preparation).

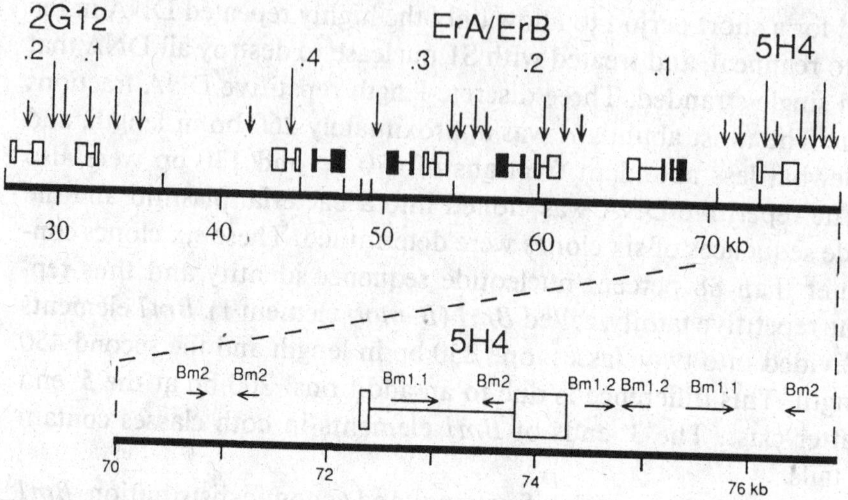

Figure 3.1. Distribution of retroposons in a segment of the early chorion gene complex of *B. mori.* Each of the 11 chorion genes is divided into a major and minor exon by one intervening sequence (open boxes, early β genes; closed boxes, early α genes). The short minor exon encodes 16–17 amino acids of the leader peptide, and thus defines the orientation of each gene. The approximate location of the *Bm1* and *Bm2* elements is indicated by a vertical line: long lines, *Bm1.1* (full length 450 bp elements); intermediate lines, *Bm1.2* (250 bp truncated elements); and short lines, 125 bp *Bm2* elements. The 7 kb region surrounding the 5H4 gene is shown on an expanded scale. The precise locations and orientation of the retroposons are indicated with horizontal arrows, the arrowhead corresponds to the location of the poly(A) tail at the 3′ end of each element.

This chromosomal segment contains 11 chorion genes, each composed of two exons. Eight of the genes are organized into four divergently transcribed ErA/ErB gene pairs. The highest density of *Bm1* elements is found in the region surrounding the 5H4 gene, which is also shown in an expanded scale. The four full-length (450 bp) *Bm1* elements are indicated by the longest vertical arrows and are referred to as the *Bm1.1* class. The 11 truncated (250 bp) *Bm1* elements are indicated by the intermediate length vertical line and are referred to as *Bm1.2*. All 15 *Bm1* elements end in poly(A) tails from 8 to 17 nucleotides in length. Many of the *Bm1* elements are surrounded by short direct repeats (2–6 bp), presumably generated during integration.

During our investigation of the repetitive DNAs surrounding the early chorion genes, we found a second class of short interspersed nucleotide elements, which we have named *Bm2*. Copies of *Bm2* are only 125 bp in length and contain poly(A) tails from 8 to 25 nucleotides in length. The consensus sequences of the 74 bp that are closest to the poly(A) tail have 70 percent sequence identity with the consensus sequence of the corresponding region from *Bm1*. No nucleotide similarity was detected between the 5′ ends of *Bm1* and *Bm2* elements. *Bm2* elements account for the third and shortest repetitive DNA length class detected in the original DNA reassociation experiments (Adams et al., 1986). The total number of *Bm2* elements in the genome of *B. mori* is approximately one-fourth that of *Bm1* (Hibner, 1989). The length and levels of sequence identity between the *Bm1.1, Bm1.2,* and *Bm2* elements are summarized in Figure 3.2.

The locations of the nine *Bm2* elements in the chromosomal region shown in Figure 3.1 are indicated by the shortest vertical lines. In total, 24 *Bm1* and *Bm2* elements are in this 50 kb segment, accounting for 11 percent of the nucleotide sequence. As has been found for the retroposons of mammalian genomes, the *Bm* elements appear to be located at random locations in and around the structural genes. Six of the *Bm1* and *Bm2* elements are located within introns (25 percent of the total number of elements). These introns account for only 12 percent of the total length of this chromosomal segment, suggesting that *Bm1* and *Bm2* insertions are at a higher density in the transcribed, but noncoding, regions of genes than in the nontranscribed regions between genes. No insertions were found within the short 5′ regions between the ErA/B gene pairs or within 200 bp of the 5′ end of the unpaired 5H4 or 2G12 genes. Because these short 5′ regions contain the control elements needed for chorion expression (see Kafatos et al., this volume), the *Bm1* and *Bm2* insertions probably have little, if any, effect on transcription of the chorion genes.

Origin of *Bm1* elements

To determine the possible origin of *Bm1,* the consensus *Bm1* sequence was used to search the GenBank DNA sequence data base (Adams et al., 1986). Significant sequence identity was found with a valine tRNA gene of *D. melanogaster* (66 percent) (Addison et al., 1982); the first 129 nucleotides of the full-length *Bm1* element could be aligned with the entire tRNA coding region of the gene and an additional 49 nucleotides located immediately downstream. Although this level of nucleotide identity is not high and involves the insertion of numerous gaps, the exact coincidence of the 5′ ends of the *Bm1* element and valine tRNA attests to the probable origin of *Bm1* elements from a tRNA gene, paralleling the situation for

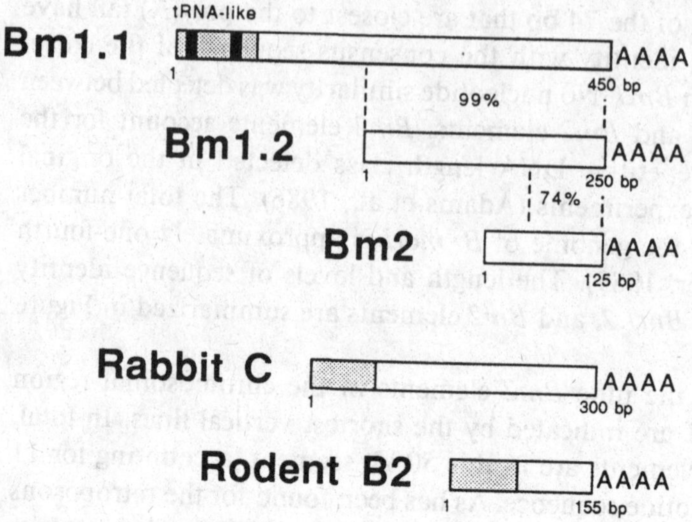

Figure 3.2. Structure of the *Bm1* and *Bm2* retroposons of *B. mori* compared to retroposons from two mammals. A variable length poly(A) tract at the 3′ end of each element is indicated. The tRNA-related sequence of the 5′ ends of certain elements is indicated by stippling. The two internal control sequences, Box A and Box B, identified in *BmX* by Wilson, Condliffe, and Sprague (1988), are indicated by the solid boxes. The nucleotide identities between the consensus sequences for the *Bm1.1, Bm1.2,* and *Bm2* elements are indicated. The rabbit *C* and rodent *B2* sequences are from Sakamoto and Okada (1985).

mammalian retroposons (Rogers, 1985; Deininger and Daniels, 1986; Weiner et al., 1986). As shown in Figure 3.2, the location of this tRNA sequence homology coincides with the tRNA homology noted for the 300 bp *C* element found in the rabbit genome and the 150 bp rodent *B2* elements (Daniels and Deininger, 1985; Sakamoto and Okada, 1985). The origin of the nucleotide sequences downstream of the tRNA-like tract is not known for any of these elements. However, it has recently been shown that the 200 nucleotides nearest the 3′ ends of the *Bm1* elements contain low levels of sequence identity to the U1 small nuclear RNAs (Herrera and Wang, 1991). U1 snRNA has been shown to be involved in pre-mRNA splicing (Steitz et al., 1988). Thus a complete *Bm1* element may correspond to the fusion of tRNA and U1 snRNA sequences.

Transfer RNA genes are known to contain internal RNA polymerase III promoters essential for their transcription (see Sprague, this volume).

Confirmation that full-length *Bm1* elements contain a functional internal polymerase III promoter within the region of tRNA homology has been obtained by Sprague and co-workers (Wilson, Condliffe, and Sprague, 1988). A comprehensive series of in vitro transcription experiments was conducted on one copy of a *Bm1* element, which they termed *BmX*. This particular *Bm1* element was selected from a collection of lambda phage clones because it exhibited high levels of polymerase III transcription. The nucleotide signals that were found to act in *cis* to control *BmX* transcription strongly resemble those that direct transcription of other silk moth polymerase III templates. In particular, deletion analysis suggested that *BmX* contains two internal promoter sequences located 50 bp apart (see solid boxes in Figure 3.2). Nucleotide sequences upstream of the transcription start site were also required, suggesting that most full-length copies of *Bm1* elements in the *B. mori* genome will not be transcribed unless they have inserted into sequences that can supply this upstream regulatory region. *BmX* itself probably does not give rise to complete *Bm1* transcripts suitable for reverse transcription because the in vitro transcripts terminate within the *BmX* sequences before reaching the poly(A) tail.

The most likely model for the origin and evolution of *Bm1* elements is similar to the one suggested for primate retroposons, in which one or a small number of founder sequences gives rise to successive waves of amplification and fixation of the repetitive family (Britten, Baron, Stout, and Davidson, 1988; Jurka and Smith, 1988). In this model, a small number of complete *Bm1* elements, the founder sequences, are in the appropriate genomic context to give rise to transcripts suitable for reverse transcription. These transcripts begin at the 5′ end of the *Bm1* sequence and extend beyond the poly(A) tail, probably ending in a U-rich region. This U-rich tail is capable of annealing to the poly(A) region, thereby serving as a primer for reverse transcription. Formation of the 5′ truncated elements, *Bm1.2*, can occur by termination of reverse transcription before reaching the 5′ end of the transcript, or by an alternative mechanism for priming the second strand of DNA synthesis. The nucleotide similarity of *Bm2* and *Bm1* elements suggests that *Bm2* may also have originated from *Bm1* sequences by establishment of a new founder element. If this is the case, then only 5′ truncated copies of *Bm2* have been found, and the sequence of a complete *Bm2* element extending to the tRNA homology at the 5′ end will be needed to confirm this hypothesis.

Transposable elements of Lepidoptera

In this section I describe the seven complete retrotransposable elements that have been reported in lepidopteran species. Spoerel and co-workers (Spoerel, Nguyen, and Kafatos, 1986) have reported another possible retrotransposable element inserted in the intron of the A.L11 chorion gene of *B. mori* (see also Kafatos et al., this volume). However, this element has not been characterized further. Most of these retrotransposable elements have been discovered in the past few years, and all but one was found in *B. mori.* Thus many additional elements will undoubtedly be found as the molecular analysis of lepidopteran genomes continues.

There have also been a number of mobile elements identified in lepidopteran species that appear to be DNA-mediated transposable elements in that they contain short inverted terminal repeats (Ueda, Mizuno, and Shimura, 1986; Carstens, 1987; Beames and Summers, 1988; Cary et al., 1989). These elements are not discussed here because in most cases only one copy of the element has been described, and in all cases it is not known whether the characterized copy is full length. Putative ORFs have been found in some of these elements (Ueda et al., 1986; Carstens, 1987; Cary et al., 1989); however, no conserved amino acid domains have been found in common with other DNA-mediated transposable elements. Thus it is not possible at present to determine which of these elements corresponds to autonomous transposable elements.

R1 and *R2* elements

DNA sequences interrupting the 28S RNA genes were first found in approximately half of the 200 ribosomal DNA (rDNA) repeats of the Oregon

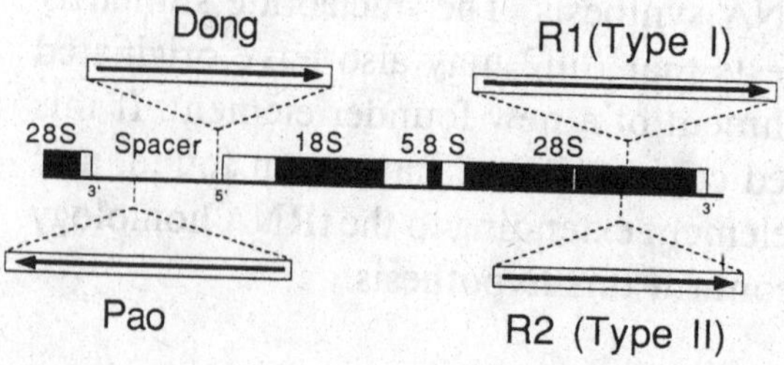

Figure 3.3. Location of the four retrotransposable elements found in the ribosomal RNA gene repeat of *B. mori* (filled bars, rRNA genes; open bars, transcribed spacer regions). A short hidden break is found near the center of the 28S gene (Fujiwara and Ishikawa, 1986). The 5′ to 3′ orientation of each retrotransposable element within the rDNA repeat is indicated by an arrow.

R strain of *D. melanogaster* (Dawid, Wellauer, and Long, 1978; Wellauer, Dawid, and Tartof, 1978). Two nonhomologous classes of elements were identified and named *Type I* and *Type II*. Each insertion is located approximately two-thirds of the distance from the 5′ end of the 28S gene (Figure 3.3), with the *Type II* insertion site located 74 bp upstream of the *Type I* insertion site (Dawid and Rebbert, 1981; Rae, 1981; Roiha, Miller, Woods, and Glover, 1981). Affected ribosomal RNA gene units (rDNA units) usually contain a single *Type I* or *Type II* insertion; however, a few rDNA units are present that contain one of each type. Although the possibility that these insertions represented transposable elements was suggested by several authors (Dover and Coen, 1981; Roiha and Glover, 1981), this idea was not pursued. Instead, considerable effort was devoted to understanding the effect of these insertions on the expression of the inserted rDNA genes. Biochemical and electron microscopic observations suggested that transcription of the entire rDNA unit containing *Type I* or *Type II* insertions was at a significantly lower level than the uninserted units (Long and Dawid, 1979; Kidd and Glover, 1981; Jamrich and Miller, 1984). The mechanism by which these elements inhibit transcription at a site over 8 kb upstream of their insertion site is not known. Given our current understanding of eukaryotic transcription, two possible mechanisms can be suggested. Either the region of the 28S gene containing the insertion sites is an enhancer that is destroyed by the insertion of *Type I* or *Type II* elements, or the elements are able to place the entire ribosomal DNA unit in a chromatin conformation that is nonconducive to transcription.

DNA insertions at the identical 28S gene location as the *Type I* and *Type II* elements of *D. melanogaster* were first found outside Diptera in *B. mori* (Fujiwara et al., 1984; Eickbush and Robins, 1985). For most strains of *B. mori* each insertion is found in approximately 10 percent of the 240 rDNA units (Xiong, Burke, Jakubczak, and Eickbush, 1988). Because the nucleotide sequence of the 5.1 kb *Type I* element and 4.2 kb *Type II* element from *B. mori* revealed many of the properties of retrotransposable elements, they have been renamed the *R1Bm* and *R2Bm* retrotransposable elements; *R* refers to its specificity for a ribosomal RNA gene and *Bm* incorporates the first letter of the genus and species name of the host genome (Burke et al., 1987; Xiong and Eickbush, 1988a). *R1Bm* and *R2Bm* lack long terminal repeats (LTRs), indicating that they are members of the *Line 1* family of elements. The absence of these terminal repeats, which are diagnostic of the *copia*-like retrotransposable elements,

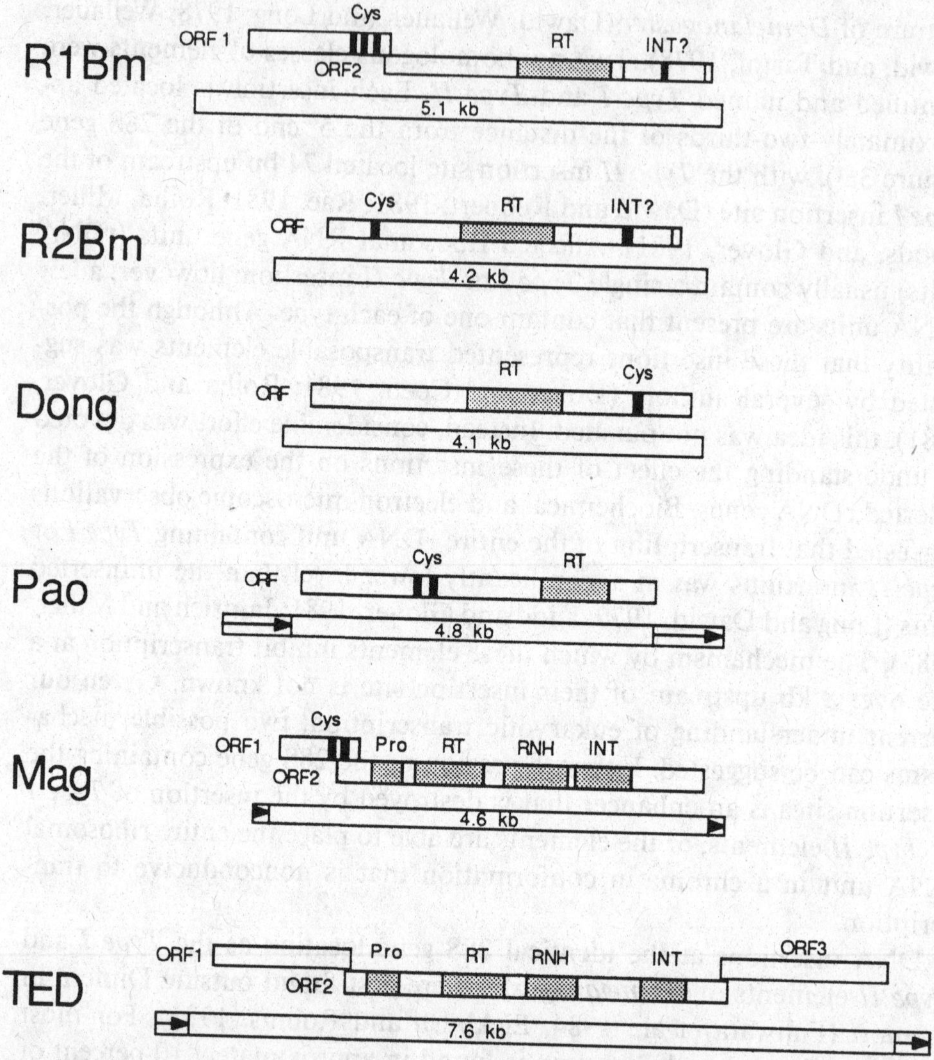

Figure 3.4. Structure of six lepidopteran retrotransposable elements. The total length of each element is shown by the lower open bar, with terminal repeats indicated by arrows. The positions of open-reading frames (ORFs) are indicated by the bars above each element. Bars at different levels are in different reading frames. Regions of those ORFs that exhibit sequence similarity to retroviral proteins are indicated with stippling. RT, reverse transcriptase; Pro, protease; RNH, RNase H; INT, integrase. Cysteine motifs (Cys) are indicated by solid bars. In *R1* and *R2* the region indicated with lighter stippling, labeled INT?, is believed to be the integrase because its sequence is conserved between *B. mori* and *D. melanogaster* elements (Jakubczak et al., 1990).

probably explains why the 28S gene insertions in dipterans went uncharacterized as transposable elements for a number of years.

Conceptual translation of the *R1Bm* nucleotide sequence reveals that the element contains two overlapping ORFs (Figure 3.4), which are similar in size, degree of overlap, and amino acid sequence to the *gag* and *pol* genes of retroviruses and *copia*-like retrotransposable elements (Xiong and Eickbush, 1988a). The *gag* gene of retroviruses encodes nucleic acid–binding proteins, which play a critical role in the packaging of the RNA intermediate in a core particle (Varmus and Brown, 1989). The *pol* gene of retroviruses encodes an aspartic protease, which processes the large polyprotein into its individual domains: the reverse transcriptase and RNAse H, which are responsible for generating a double-stranded DNA copy of the retroviral RNA genome, and the integrase, which is responsible for integrating this double-stranded DNA copy into the genome of the host cell. ORF1 of *R1Bm* exhibits two sequence features that are common in retroviral *gag* genes. First, it contains an unusually high concentration of proline residues (24 percent) near the amino terminus. Second, it contains three cysteine motifs (also called metal- or zinc-binding domains) near the carboxyl terminus. These cysteine motifs have a similar spacing of cysteine and histidine residues to those cysteine motifs of retroviral *gag* genes. ORF2 of *R1Bm* appears to be the *pol* gene, because a central region contains sequence similarity to reverse transcriptase (labeled RT in Figure 3.4). Sequence similarity to the three other retroviral *pol* gene domains (protease, RNase H, and integrase) cannot be detected in ORF2 of *R1Bm*.

The nucleotide sequence of *R2Bm* reveals that it encodes a single ORF of 1,151 amino acids (Burke et al., 1987). As shown in Figure 3.4, a central region of this ORF has similarity to reverse transcriptase but, as in the case of *R1Bm,* no sequence similarity to the protease, RNase H, and integrase of retroviral *pol* genes. The presence of only one ORF is somewhat unusual for a retrotransposable element, although single ORFs have been found in *copia* and *1731* of *D. melanogaster* (Mount and Rubin, 1985; Fourcade-Peronnet, d'Auriol, Becker, Balibert, and Best-Belpomme, 1988), *Ta1* of *Arabidopsis thaliana* (Voytas and Ausubel, 1988), and *Tnt1* of *Nicotiana tabacum* (Grandbastien, Spielmann, and Caboche, 1989).

No large regions of nucleotide or amino acid sequence similarity were detected between the *R1Bm* and *R2Bm* elements except in the reverse transcriptase domain. Even here, *R1* and *R2* have greater amino acid similarity to other retrotransposable elements than to each other. This lack of sequence similarity suggests that *R1* and *R2* were derived from two different retrotransposable elements that have independently become specialized for insertion into ribosomal genes.

The presence of *R1* and *R2* elements in both *D. melanogaster* and *B. mori,* which are estimated to be separated by more than 240 million years of evolution, provided a unique opportunity to determine what sequences at either the nucleotide or amino acid level are important for their propagation. Complete copies of *R1* and *R2* elements from *D. melanogaster* were sequenced to look for common features with the elements from *B. mori* (Jakubczak, Xiong, and Eickbush, 1990). Two overlapping ORFs are present in *R1Dm* and one ORF in *R2Dm,* nearly identical in length to those in *R1Bm* and *R2Bm.* However, neither the 5′ nor 3′ untranslated region of the R1 and R2 elements was found to be conserved between species in length or in sequence. Comparison of the ORFs revealed a cysteine motif conserved near the amino terminal end of the *R2* ORF (see Figure 3.4), which had not been noticed in the analysis of only *R2Bm.* This cysteine motif suggests that the *R2* elements may have a nucleic acid–binding component of similar function, even though they do not contain a separate *gag*-like ORF. Finally, for both elements the amino acid sequence of the entire region carboxyl terminal to the reverse transcriptase (labeled "INT?" in Figure 3.4) is conserved between the dipteran and lepidopteran species. This region also contains a conserved cysteine motif, suggesting that it may encode the integrase function responsible for integrating the element into the host chromosome. The spacing of cysteine and histidine residues in these motifs is completely different from the motif found in retroviral integrases (Doolittle, Feng, Johnson, and McClure, 1989), suggesting that the mechanism of integration by *R1* and *R2* elements differs from that of retroviruses.

Mechanism of specificity. The first direct evidence that the unusual locations of *R1* and *R2* in the genome are the result of highly sequence-specific integrase functions was obtained from the characterization of *R1* and *R2* elements in *B. mori* that had inserted into sequences outside the rDNA units (Xiong et al., 1988). A non-rDNA copy of *R1* was found to be flanked by a target site duplication of 14 bp, as are all copies inserted in 28S genes, suggesting that its mechanism of integration is the same as that used for rDNA copies. Equally important, this non-rDNA target site contained sequence similarity (11 of 16 bp) to the 28S target site. A non-rDNA copy of *R2* was also found that had inserted into a sequence with remarkable similarity to the 28S gene target site (12 of 15 bp). Insertion of this non-rDNA copy failed to generate a target site duplication, just as is found for all *R2* elements inserted within 28S genes. Clearly, the mechanism for *R1* and *R2* insertion is not dependent on any unusual feature of the 28S gene apart from the actual target sequence itself. Only a few non-rDNA copies of *R1* and *R2* exist within the genome of *B. mori.* They cannot represent the "active" elements generating intermediates for insertion into the 28S genes, because all non-rDNA inserts are substantially divergent in sequence from *R1* and *R2* elements found within rDNA units, and all non-rDNA units contain segmental mutations that destroy their ORFs (Xiong et al., 1988).

It was reasonable to assume that a critical step in the integration of *R1* and *R2* elements is a sequence-specific cleavage of the 28S insertion site. The approach used to prove that such a sequence-specific endonuclease is encoded by *R1* and *R2* was similar to that developed to demonstrate specific cleavage by the yeast HO gene product (Kostriken and Heffron, 1984), the r1 intron ORF of the yeast mitochondrial 21S gene (Colleaux et al., 1986), and the *Agrobacterium tumefaciens* virD gene (Yanofsky et al., 1986). This involved placement of the appropriate ORF from either *R1Bm* or *R2Bm* in an *Escherichia coli* expression plasmid along with the target sequence, in this case a 28S rDNA fragment containing the *R1* and *R2* insertion sites. After induction for expression of the ORF, specific cleavage of the target sequence in vivo was assayed by isolation of the plasmid DNA. A variety of expression constructs were attempted with segments of the single ORF of *R2Bm* and ORF2 of *R1Bm*. The expression construct that eventually yielded specific cleavage contained the entire *R2Bm* ORF (Xiong and Eickbush, 1988b). Similar experiments with the *R1Bm* ORF2 have not been successful.

The *R2*-encoded endonuclease cleaves the 28S gene precisely at the *R2* insertion site resulting in four nucleotide 5′ staggered ends. This activity can be detected in vivo with the target site on the same plasmid as the expression construct or in vitro using a 120 kilodalton purified protein derived from the *E. coli* expression construct and exogenous target sequences (Xiong and Eickbush, 1988b; Luan and Eickbush, unpublished data). The 28S sequences that are required for *R2* endonuclease cleavage extend from 29 bp 5′ of the cleavage site to only 3 bp 3′ of the target site. Recently, reverse transcriptase activity has also been detected with this 120 kilodalton protein (Luan, Jakubczak, and Eickbush, in preparation). *R2Bm* represents the first retrotransposable element in which it will be possible to analyze biochemically its various reverse transcription and integration functions.

Species distribution. The discovery of *R1* and *R2* in both lepidopteran and dipteran species suggested that these elements might have a wide distribution in insects. Taking advantage of the remarkable insertion specificity of *R1* and *R2* for the 28S genes, we developed a simple genomic blot assay to assay their presence and abundance in a large number of species (Jakubczak, Burke, and Eickbush, 1991). Forty-three of 47 species from nine orders of insects were found to contain specific 28S gene insertions, ranging from only a few percent to more than half of the rDNA units. The four species without insertions were from different insect orders, indicating that species without detectable insertions do not represent a particular taxonomic group. To determine whether the variant bands seen in the genomic blots corresponded to insertions in precisely the *R1* and/or *R2* sites, eight species were selected for sequence analysis (Jakubczak et al., 1991). Insertions in the *R1* site were found in all eight species and insertions in the *R2* site were found in three species. Thus *R1* and *R2* have the widest species distribution of any known transposable element.

Unlike *B. mori* or *D. melanogaster*, where all *R1* or *R2* elements within a species are nearly identical in sequence, many insect species harbor highly divergent copies of *R1* and *R2* (Jakubczak et al., 1991). For example, a parasitic wasp (*Nasonia vitripennis*) contains at least four families of *R1* elements; the Japanese beetle (*Popillia japonica*) contains at least five families of *R2* elements. The average level of sequence identity between the various *R1* families from *N. vitripennis* and *R2* families from *P. japonica* is less than 50 percent. The presence of divergent families of *R1* and *R2* elements in the same species is difficult to explain. Because a limited number of 28S gene insertion sites can be occupied, families of *R1* or *R2* within a species using the same insertion mechanism should be in competition with each other. If these families have coexisted in the same species for the time required for the extensive divergence we have observed, then each family must have been able to regulate its copy number in such a way as to prevent its own elimination and at the same time fail to displace the other competing families. Alternatively, *R1* and *R2* may be transferred between species, with the result that the number of families within a given species reflects the present state of a dynamic process of family introduction and elimination by competition or stochastic loss.

We have begun a series of experiments to address the question of the horizontal spread of these elements by determining whether the phylogeny of the *R1* and *R2* elements in different insects can be correlated with the phylogeny of the insects they inhabit (Eickbush and Burke, unpublished data). Our efforts have concentrated on species harboring multiple families of *R1* elements. Using degenerate polynucleotide primers to highly conserved regions of the *R1* elements, we have used the polymerase chain reaction (Saiki et al., 1988) to isolate segments encoding portions of the ORF2 from different species. Preliminary results are quite revealing. The gypsy moth (*Lymantria dispar*) contains two families of *R1* elements. A segment of ORF2 encoded by one of these families, *R1Ld.a,* has 56 percent amino acid identity to that encoded by the *R1* element from *B. mori* (*R1Bm*). The second element from the gypsy moth, *R1Ld.b,* has a somewhat lower level of sequence identity (43 percent) to both *R1Bm* and *R1Ld.a.* Are these two gypsy moth *R1* elements variants that have diverged in the gypsy moth lineage? The alternative explanation, that the presence of one of these two *R1* elements is the result of a horizontal transfer from another species, appears more likely. An even more striking example of this situation has come from the analysis of the *R1* elements from *N. vitripennis.* The ORF2 from one element, *R1Nv.a,* has 44 percent amino acid similarity to the *D. melanogaster R1* element, *R1Dm,* but only 29 percent amino acid identity to a second *R1* element in the same species, *R1Nv.b.* As we collect more *R1* sequences from different insects, we believe support will accumulate for the horizontal transfer of *R1* elements between insect species.

Dong

A third DNA insertion has been identified in the rDNA repeat of *B. mori,* in this case near the promoter for the rDNA transcript (see Figure 3.3) (Xiong and Eickbush, in preparation). The spacer region between consecutive rDNA transcription units of *B. mori* is composed of a number of tandem arrays of short nucleotide repeats (Fujiwara and Ishikawa, 1987). The 4.1 kb insertion is located in one such array of TAA repeats. As shown in Figure 3.4, the element contains a 1,233 amino acid ORF, whose central region contains sequence similarity to reverse transcriptase (Xiong and Eickbush, 1990). This element has been named *Dong,* which is the Chinese word for “moving.” Like *R1, R2,* and *Line 1* elements of mammals, *Dong* contains no direct or inverted terminal repeats. *Dong* elements have a nucleic acid – binding motif located carboxyl terminal to the reverse transcriptase domain similar to the motif found in other retrotransposable elements that lack terminal repeats. The spacing of cysteine and histidine residues in this motif has greatest similarity to motifs in the *I* element of *D. melanogaster* (Fawcett et al., 1986) and the *Tx1* element of *Xenopus laevis* (Garrett, Knutzon, and Carroll, 1989).

Using internal segments of *Dong* as hybridization probes to screen genomic DNA blots, we found only 7–10 additional copies, most located outside the rDNA units. Therefore, despite its initial detection in an rDNA repeat, *Dong* usually inserts in multiple places throughout the genome, typical of the distribution of most retrotransposable elements. We have cloned and sequenced one of the *Dong* elements located outside the rDNA unit. This non-rDNA copy also has inserted into a short tandem array of TAA nucleotide repeats, suggesting that, similar to a number of other retrotransposable elements (Ikenaga and Saigo, 1982; Freund and Meselson, 1984; Inouye, Yuki, and Saigo, 1984), *Dong* preferentially inserts into AT-rich nucleotide sequences.

Pao

A second retrotransposable element found in the spacer region of the *B. mori* rDNA repeat (see Figure 3.3) has been named *Pao,* the Chinese word for "running" (Xiong and Eickbush, 1990). Two lambda phage clones containing the 5′ and 3′ ends of a *Pao* element were part of the original collection of rDNA units characterized as containing aberrant rDNA repeats (Lecanidou, Eickbush, and Kafatos, 1984). Most of the 30–50 copies of the *Pao* element in *B. mori* are located outside the rDNA repeats. As shown in Figure 3.4, the 4.8 kb element contains 639 bp LTRs, with the 3.5 kb central region of the element encoding a 1,126 amino acid ORF (Xiong, Burke, and Eickbush, 1993). The reverse transcriptase domain of *Pao* is located close to the carboxyl terminal end of the ORF. This position for the reverse transcriptase domain is similar to that of *copia* and *Ty1,* whose integrase domains are located amino terminal of the reverse transcriptase domain (Mount and Rubin, 1985). A similar location of the integrase domain is expected of *Pao,* because the only nucleic acid–binding motifs detected in the ORF are centrally located.

The ability of retroviruses and *copia*-like retrotransposable elements to produce a full-length double-stranded DNA intermediate from an RNA transcript and the integration of this intermediate into the genome have been extensively described (for reviews, see Boeke, 1989, and Varmus and Brown, 1989). *Pao* elements contain three structural features, suggesting that their mechanisms of reverse transcription and integration are similar to those of retroviruses. First, *Pao* LTRs begin with the nucleotide sequence TG and end with the sequence CA. In all retroviruses the integrase removes two nucleotides from one strand at each end of the DNA intermediate, exposing these TG-CA sequences to serve as donors for integration into the genome. Second, located immediately downstream of *Pao*'s 5′ LTR is a sequence complementary to the 18 nucleotides at the 3′ end of tyrosine tRNA. Like retroviral reverse transcription, a tRNA molecule could prime synthesis of the first strand of the *Pao* DNA intermediate. Finally, located immediately upstream of *Pao*'s 3′ LTR is an 11 bp polypurine tract, which can prime second strand synthesis, as has been found for retroviruses.

One property of *Pao*'s LTRs is different from that of the retroviruses and *copia*-like retrotransposable elements. The LTRs of *Pao* are significantly longer than the LTRs of previously described elements and show considerable variation in length due to a central region that is composed of tandem repeats of a 26 bp sequence. Though different copies of *Pao* contain different length LTRs, the 5′ and 3′ LTRs of any individual *Pao* element are identical in length, confirming that these LTRs are generated during each replication cycle. It is interesting to note that two retrotransposable elements, *TAS* of *Ascaris lumbricoides* and *PAT* of *Panagrellus redivivus,* also contain unusually long LTRs with tandemly repeated central domains (Felder and de Chastonay, personal communication). Thus *Pao, TAS,* and *PAT* represent a new class of retrotransposable elements whose precise mechanism of replication or integration differs in some uniform property from that of retroviruses and *copia*-like retrotransposable elements.

Mag

During an investigation of different alleles of the sericin-2 (Ser2) gene of *B. mori,* Michaille and co-workers discovered a 4.6 kb fragment inserted within a large intron of the gene (Michaille, Garel, and Prudhomme, 1990b). The element was named *Mag,* short for *Magnan,* which is the name for silkworms in the south of France where they are reared. *Mag* elements encode two overlapping ORFs flanked by 77 bp direct repeats (Michaille et al., 1990). Different strains of *B. mori* contain from 7 to 14 dispersed copies of *Mag*. As shown in Figure 3.4, the two ORFs of *Mag* exhibit remarkable similarity to the *gag* and *pol* genes of retroviruses. ORF1 contains two nucleic acid–binding motifs with the same spacing of cysteine and histidine residues as *gag* genes. ORF2 contains regions of sequence similarity to the four enzymatic activities found in the retroviral *pol* gene: aspartic protease (Pro), reverse transcriptase, RNaseH (RNH), and integrase (INT). Surprisingly, the terminal repeats of *Mag* show little similarity to retroviral LTRs. The 77 bp flanking repeats found in *Mag* are considerably shorter than the LTRs of retroviruses and they do not end with CA sequences. However, a complementary region to a *D. melanogaster* serine tRNA is found downstream of the 5′ terminal repeat and a polypurine tract is located upstream of the 3′ flanking repeat. Furthermore, two putative promotor sites analogous to the ones found in retrotransposons such as *gypsy* or *R1Bm* and a polyadenylation site are present within the repeat.

Screening of several different copies of *Mag* from a *B. mori* genomic library indicates the constancy, hence the significance, of all these characteristics (Garel, personal communication). Therefore, despite the considerable amino acid similarity of the *Mag* ORFs to retroviral genes, the structure of the terminal repeats of *Mag* suggests that the mechanisms used for integration may be different from those of retroviruses.

TED

Another lepidopteran retrotransposable element was discovered during serial passage of a nuclear polyhedrosis virus (NPV) from the alfalfa looper, *Autographa californica,* in tissue cells of the cabbage looper, *Trichoplusia ni* (Miller and Miller, 1982). A mutant viral genome was recovered that contained a 7.6 kb insertion disrupting one of the early viral genes (Friesen and Miller, 1987). The insertion was found to be an active transposable element present in approximately 50 copies in the host *T. ni* genome. This is the only example of a fully functional eukaryotic transposable element associated with the genome of a virus.

Nucleotide sequence analysis of the element, named *TED* for *T*ransposable *E*lement in mutant *D* (Friesen and Nissen, 1990), indicates that it is a retrotransposable element that is closely related in sequence to retroviruses. *TED* contains 273 bp LTRs, a size similar to that found in most retroviruses. Transcription studies have indicated that the 5′ LTR contains the promoter for a 7.1 kb RNA transcript. Immediately downstream of the 5′ LTR is an 18 bp tRNA primer binding site for the initiation of first strand DNA synthesis, and located immediately upstream of the 3′ LTR is an 11 bp polypurine tract for the initiation of second strand DNA synthesis. The central region of *TED* encodes three overlapping ORFs. ORF1 is similar in size and location to retroviral *gag* genes. ORF2 contains regions of similarity to the four enzymatic activities that are found in retroviral *pol* genes. Finally, unlike any of the previously defined lepidopteran elements, *TED* contains a third ORF that is similar in length and position to retroviral *env* genes. The *env* gene of retroviruses encodes a membrane envelope protein that is required for the extracellular step in the retroviral life cycle. Although no significant amino acid sequence similarity was detected between ORF3 of *TED* and any retroviral *env* protein, it does encode a segment of uncharged hydrophobic residues near the carboxyl terminus that may serve as a transmembrane domain similar to that of many retroviral *env* proteins. Both the *gypsy* and *17.6* elements of *D. melanogaster,* which contain the highest amino acid similarity to *TED,* also contain a third ORF (Saigo et al., 1984; Marlor, Parkhurst, and Corces, 1986). There is no evidence, however, to indicate that any of these retrotransposable elements have an extracellular step in their life cycles.

BMC1

The final lepidopteran retrotransposable element to be described was discovered by H. Maekawa and co-workers during the analysis of an abundant repetitive DNA family in *B. mori* (Ogura et al., in press). The element, *BMC1,* has many similarities to the *Line 1* elements in mammalian species. The full-length repeat is 5 kb, although most copies are truncated to varying extents at their 5′ ends. There are approximately 2,000 copies of *BMC1* per haploid complement, representing nearly 2 percent of the *B. mori* genome. This remarkable abundance is comparable to *Line 1* elements in mammalian species (reviewed in Hutchison, Hardies, Loeb, Shehee, and Edgell, 1989). Also similar to the *Line 1* elements of some mammalian species (Hutchison et al., 1989) are the presence of short tandem repeats at the 5′ ends of full-length *BMC1* elements and a short poly(A) tail at their 3′ ends. Sequence analysis of internal regions from several cloned *BMC1* elements reveal blocks of sequence similarity to the reverse transcriptase domain, clearly confirming that it is a retrotransposable element. However, all sequenced copies contain numerous insertions and deletions, which disrupt the ORFs. Thus it is not possible to indicate the entire coding potential of a fully functional *BMC1* element in Figure 3.4. As has been done for the mammalian *Line 1* elements, it is thus necessary to sequence a number of different copies of *BMC1* in order to generate a consensus nucleotide sequence that can be used to predict the coding potential of a functional element.

Phylogenetic relationship of the lepidopteran elements to other mobile elements

The structural features of the seven lepidopteran retrotransposable elements described in the previous section are quite diverse. However, most of these features are similar to transposable elements found in other insect or noninsect species. In this section we examine the phylogenetic relationship of the lepidopteran elements to other retrotransposable elements as well as to retroviruses. This analysis is based on the amino acid sequence of the reverse transcriptase domain, as it is the only domain common to all elements (see Figure 3.4). The phylogenetic analysis also includes a number of genetic elements that are not retroviruses or retrotransposable elements but have been shown to contain ORFs encoding proteins that are similar in sequence to reverse transcriptases. These additional elements, termed *retroelements* (Temin, 1989), include the DNA viruses, hepadnaviruses of animals and caulimoviruses of plants (Toh, Hayashida, and Miyata, 1983), a group of fungal group II mitochondrial introns and a mitochondrial plasmid (Michel and Lang, 1985), and bacterial insertion elements producing multicopy single-stranded DNA (msDNA) (Inouye, Eagle, and Inouye, 1989; Lim and Maas, 1989).

The level of amino acid identity between these diverse reverse transcriptase domains can be quite low (less than 20 percent). As a consequence, the alignment of these sequences is not routine, and differences have been reported in the literature (Yuki, Ishimaru, Inouye, and Saigo, 1986; Xiong and Eickbush, 1988c, 1990; Doolittle et al., 1989). In our analyses we use an alignment that is based on seven groups of conserved amino acid residues that have been identified in all reverse transcriptase sequences (Xiong and Eickbush, 1988c, 1990). This set of residues is a modification of that originally identified by Toh and co-workers as being conserved in retroviruses, hepadnaviruses, and caulimoviruses (Toh et al., 1983). The seven peptide domains are from 11 to 35 residues in length and total 178 residues. The number of identical residues scored in these

178 positions for all pairwise comparisons of sequences was used to generate a phylogenetic tree by the Neighbor-Joining (NJ) method (Xiong and Eickbush, 1988c, 1990). This method has been shown to be reliable for determining the correct phylogenetic relationship, even for sequences with different rates of divergence (Saitou and Nei, 1987). Different rates of evolution are expected for the various genetic elements in this comparison (viruses, transposable elements, introns), given their different turnover rates and mechanisms of propagation. The phylogenetic tree of reverse transcriptase sequences derived by the NJ method was rooted by using RNA-directed RNA polymerase sequences from various prokaryotic and eukaryotic RNA viruses as an outgroup (Poch, Sauvaget, Delarue, and Tordo, 1989; Xiong and Eickbush 1990).

A simplified version of the phylogenetic tree generated from these reverse transcriptase sequences is presented in Figure 3.5. In the figure only the individual branches of the retrotransposable elements are shown; all other retroelements that are of similar structure and are localized on the same main branch are indicated with a box. A complete version of this tree showing the individual branch points for all 97 retroelements used in the analysis as well as the sequence alignments themselves can be found in Xiong and Eickbush (1990, Figure 3).

Based on reverse transcriptase sequence similarity, all retroelements can be divided into two primary branches. One branch contains the msDNA producing elements of bacteria, the group II introns, and the *Line 1*-like retrotransposable elements. All elements of this branch lack any type of direct or inverted terminal repeats. Located on the second major branch of the tree are all retroelements whose replication intermediates contain terminal repeats. This includes all retrotransposable elements that contain LTRs and three classes of viruses: retroviruses, caulimoviruses, and hepadnaviruses. Thus the presence or absence of terminal repeats is an independent structural feature that agrees with this basic classification of retroelements based exclusively on amino acid similarity.

Given this fundamental importance of terminal repeats, the *Line 1*-like elements are more appropriately termed the non-LTR retrotransposable elements. All retrotransposable elements that lack LTR clearly fall on this branch of the reverse transcriptase tree, indicating they have a common origin. No characterized non-LTR element represents an LTR-containing retrotransposable element that has simply lost its terminal repeats. The non-LTR retrotransposons are widely distributed in eukaryotes, having been found in insects, vertebrates, plants, and protozoans. *R1*, *R2*, and

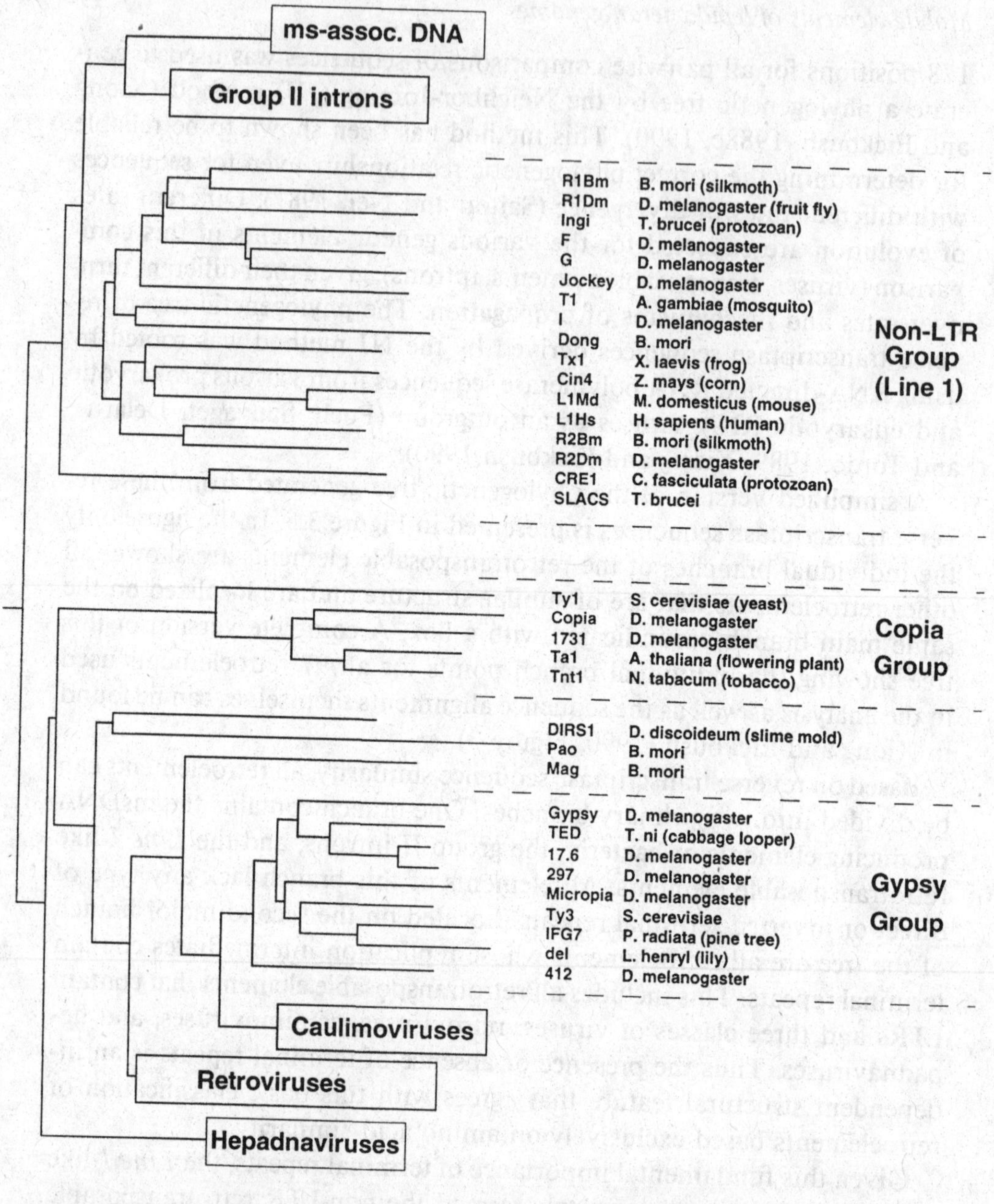

Figure 3.5. Phylogenetic tree of various mobile elements and viruses based on the comparison of their reverse transcriptase sequences. The abbreviated name and host genome are given for each retrotransposable element. To simplify visual comparison of these retrotransposable elements, other groups of genetic elements are indicated as boxes. The length of each box corresponds to the most divergent element within that box. The complete reverse transcriptase tree, including the actual sequence comparisons, references for all elements, and further discussion of the rooting of the tree, can be found in Xiong and Eickbush (1990).

Dong as well as several other elements found in *D. melanogaster* or *Anopheles gambiae* are widely divergent members of this group; thus there is no distinct insect group of non-LTR retrotransposable element. Based on conserved sequences motifs of the *BMC1* reverse transcriptase domain, this element is clearly a member of the non-LTR retrotransposable element branch. However, because the full reverse transcriptase sequence is not known, its precise position on the tree cannot be determined. The most divergent members of the non-LTR group are *SLACS* and *CRE1*. These two recently discovered elements are found exclusively in the spliced leader (miniexon) genes of trypanosomatids (Aksoy, Williams, Chang, and Richards, 1990; Gabriel et al., 1990). The basic structure of these two elements clearly identifies them as typical retrotransposable elements. Surprisingly, however, they have less amino acid sequence similarity to other non-LTR elements than do the group II introns found in the mitochondria of fungi or that shown by the reverse transcriptase genes found associated with msDNA in bacteria. Group II introns and ms-associated elements appear as a branch of the non-LTR retrotransposable elements.

In most of the literature on retrotransposable elements and throughout this chapter, retrotransposable elements with LTRs have been called *copia*-like elements. However, as shown in Figure 3.5 these LTR-containing elements should be divided into several groups based on their reverse transcriptase homology. *Copia* itself, as well as *Ty1* and three other elements, are the most divergent group, which can retain the *copia* group terminology. Most of the remaining retrotransposable elements fall within another branch called the *gypsy* group, after the best characterized element. The lepidopteran element *TED* is located within the *gypsy* group. Both the *copia* and *gypsy* groups of elements are found in insects, plants, and yeast and contain LTRs that are similar to those of retroviruses. Based on their amino acid sequences, only three LTR-containing retrotransposable elements fall outside the *copia* and *gypsy* groups. Once again the unique position of these elements on the reverse transcriptase tree is associated with the differences in their terminal sequences. The three elements are *Pao*, with its unusually long LTRs containing a central tandemly repeated segment; *Mag*, with a terminal repeat of only 77 bp; and *DIRS*, found in *Dictyostelium discoideum* (Cappello, Handelsman, and Lodish, 1985), which contains 330 bp inverted terminal repeats. The position of these elements as relatively divergent branches on the reverse transcriptase tree suggests that they are not recently derived from elements within the *copia* or *gypsy* group. Thus *Pao, Mag*, and *DIRS* represent relatively old classes of retrotransposable elements whose terminal repeats are adapted for somewhat different mechanisms of replication.

Based on this reverse transcriptase tree, a simple scenario for the origin of retrotransposable elements and viruses can be presented. Because retrotransposable elements are the only elements common to both major branches of the tree, we have suggested that they are the most likely progenitors of all current retroelements (Xiong and Eickbush, 1990). The coding capacity of this ancestral retrotransposable element, based on the presence of features in both the LTR and non-LTR branches, is a *gag* gene and a *pol* gene, either as two separate or one longer ORF. There is no evidence for LTRs in the RNA viruses used to root the tree (that is, using the RNA-directed RNA polymerase sequences they encode to root the tree) or in the hepadnaviruses, the most divergent (oldest) member of the LTR branch. We further suggested (Xiong and Eickbush, 1990) that these progenitor retrotransposable elements did not contain LTRs. The diversity of non-LTR retrotransposable elements, revealed by the very deep branch lengths of this group on the tree and their wider distribution in animals, plants, and primitive protozoans than any other class of retroelements, supports the suggestion that the non-LTR retrotransposable elements are the oldest group of retroelements.

Horizontal spread of retrotransposable elements

One of the most striking features of the reverse transcriptase tree in Figure 3.5 is that retrotransposable elements from very diverse organisms are sometimes similar in sequence. *Copia* and *1731,* both elements from *D. melanogaster,* are most related to the *Ta1* and *Tnt1* elements of plants. *Micropia* of *D. melanogaster* is most closely related to the *Ty3* element of yeast and *IFG7* and *del* elements of plants. The *R1* elements of *B. mori* and *D. melanogaster* are most closely related to the *Ingi* element found in a protozoan. The presence of related retrotransposable elements in very different taxa indicates either that the retrotransposons have spread horizontally or that most of the branch points on the reverse transcriptase tree predate the evolution of metazoans. This latter possibility seems unlikely, especially given the capacity for rapid evolution inherent in these elements due to their reliance on reverse transcription.

Unlike retrotransposable elements, the three types of viruses on the reverse transcriptase tree have a restricted distribution. Indeed, each viral type is localized to a particular taxonomic group; retroviruses and hepadnaviruses are found only in vertebrates and caulimoviruses only in plants. Although the ability of these elements to jump between species within the taxonomic group they occupy is well established, they have not jumped beyond these taxa. It seems unlikely that they have not had sufficient time, because each of these viral groups appears to predate the emergence of the *gypsy* group of elements, which are found in animals, plants, and fungi. It may be that the adaptation for extracellular mobility so effective in promoting passage between organisms of the same species has restricted the range of these highly evolved mobile elements.

The current distribution of retroelements can be explained if one assumes that retrotransposable elements have been horizontally transferred between major taxonomic groups of organisms. Once functional retrotransposable elements were within a new taxa, new types of viruses could also evolve, either from the capture of reverse transcriptase sequences from these retrotransposable elements by preexisting viruses or by these retrotransposable elements acquiring additional genes and becoming a virus. For instance, the addition of the *env* gene to an LTR-containing retrotransposable element in a vertebrate may have given rise to the retroviruses. The capture of a segment of the *pol* gene, including the reverse transcriptase, from a plant retrotransposable element by a preexisting plant virus may have led to the caulimoviruses. Additional support for this model of the evolution and spread of retroelements can only be obtained by the analysis of mobile elements from an increasingly broad range of plant, animal, and protozoan species.

Future directions

The integration of mobile elements into new genomic sites creates novel combinations of DNA sequences that undoubtedly have played a significant role in reshaping eukaryotic genomes. Because the insertion of these elements into structural genes can have an immediate detrimental effect on the host, they are usually thought of as selfish DNA (Doolittle and Sapienza, 1980; Orgel and Crick, 1980) – that is, molecular parasites that exist in the genome because of their ability to make new copies, not because they confer a selective advantage to the host. On the other hand, by keeping the genome in a constant state of flux, it can be argued, these

Table 3.1. *Comparison of mobile elements in mammalian and insect species*

	Mammalian[a]	*B. mori*	*D. melanogaster*
Retrotransposable elements			
Non-LTR elements	*Line 1*	Multiple	Multiple
LTR elements	—	Multiple	Multiple
Retroviruses	Multiple	—	—
Sines (retroposons)	*Alu*[b]	*Bm1* + *Bm2*	—
DNA-mediated elements	—	Multiple	Multiple

[a]Data derived from human and *Mus domesticus* genomes.
[b]*Alu*-like elements in rodent genomes are called *B1*.

elements give an advantage to the species (Rogers, 1985; Weiner et al., 1986).

Clearly an analysis of the distribution and dynamics of these mobile sequences in a number of different eukaryotic genomes is needed to resolve the positive and negative effects of these interesting genetic elements. As shown in Table 3.1, the *B. mori* genome provides the potential for just such an analysis because it harbors examples of many types of these genetic elements. The *B. mori* genome, like that of mammalian genomes and unlike that of the fruit fly genome, contains abundant retroposons (SINEs), representing 5 to 10 percent of the total genomic mass. Unlike the mammalian genomes but like fruit fly genomes, the *B. mori* genome also contains a wide variety of retrotransposable elements. Some of these elements contain LTRs and thus appear to replicate and integrate using mechanisms similar to those of retroviruses, whereas others lack LTRs and thus must utilize novel mechanisms yet to be described. Though not well described, the *B. mori* genome is also similar to the *D. melanogaster* genome in containing DNA-mediated mobile elements (Ueda et al., 1986).

The analysis of the distribution of mobile elements described in this chapter has already raised several interesting questions. Do most transposable elements have the wide distribution of *R1* and *R2?* Why do mammals contain only one type of retrotransposable element, *Line 1,* but abundant retroviruses, whereas silk moths and fruit flies have abundant retrotransposable elements but apparently no retroviruses? Have the retrotransposable elements been displaced by the retroviruses once they evolved in vertebrates? Why does *D. melanogaster* contain no retroposons?

What should we expect to find in the genomes of other lepidopteran species, and insects in general? In the case of retroposons it is impossible to predict. A 195 bp retroposon has been directly demonstrated in an orthopteran, *Locusta migratoria* (Bradfield, Locke, and Wyatt, 1985). Based on DNA reassociation kinetics, another silk moth, *Antheraea pernyi,* and the house fly, *Musca domestica,* probably contain retroposons, whereas the honeybee, *Apis mellifera,* probably does not (Efstradiatis et al., 1976; Crain et al., 1976). On the other hand, the presence of retrotransposable elements can be predicted with more certainty. *R1* and *R2* elements have already been demonstrated in the 28S genes of most insects (Jakubczak et al., 1991). Given the wide distribution of both LTR-containing and non-LTR elements in most animal taxa, as well as in protozoans and plants, their presence at some level in all insect species is virtually assured. The presence of the fully active retrotransposable element *TED* in a virus further suggests that these elements can be easily transferred between related species.

Finally, virtually all mobile elements have been found by their insertion in cloned genes. Where can these elements be found in species with little available genetics and no molecular probes? Perhaps the easiest place to look is in the spacer region of the rDNA repeats. The rDNA repeats are easy to clone, and their conserved organization enables rapid characterization of genomic clones. The analysis of variant rDNA units will likely reveal retrotransposable elements as well as DNA-mediated elements. These new elements, particularly from those species that are not well characterized molecularly, are critical to our analysis of the origins, spread, and genetic consequences of these interesting mobile DNAs.

Acknowledgments

I gratefully acknowledge B. Burke, D. Eickbush, and J. Jakubzcak for their comments on the manuscript. Work from our laboratory on the *B. mori* mobile elements has been supported by the National Institutes of Health Grant GM 31867 and the American Cancer Society Grant NP-691.

4

Lepidopteran phylogeny and applications to comparative studies of development

JEROME C. REGIER, TIMOTHY FRIEDLANDER, ROBERT F. LECLERC, CHARLES MITTER, and BRIAN M. WIEGMANN

Comparative development

Comparative and experimental methodologies offer two complementary approaches to deducing both ontogenetic and phylogenetic mechanisms. Both kinds of studies analyze the effects of perturbations in the generation of these mechanisms, but in comparative studies the perturbation was introduced by the evolutionary process itself. At the methodological level, however, these approaches have quite separate traditions and tools. In large part, this is because the evolutionary experiment was completed long ago and many modifications have subsequently occurred. Furthermore, the unmodified "control" organism is typically missing – that is, has become extinct or subsequently modified independently of the "experimental" organism – and can only be inferred from analysis of multiple contemporary organisms. Thus, whereas the experimentalist may, for example, be satisfied to compare a single genetic mutant with its parental type, the comparative biologist frequently must deal with multiple species that differ in many loci, most of which have nothing to do with the phenotypic change of interest. Although this approach may seem indirect to an experimentalist, there really is no alternative when studying evolutionary processes.

Before the evolution of developmental processes can be inferred, it is necessary to place the species under study within its phylogenetic context, in order to establish the actual evolutionary sequence of change in ontogeny. This is analogous to the requirement that ontogenetic stages be temporally ordered before developmental mechanisms can be deduced. Of course, there are potential difficulties, if, for example, the developmental processes have evolved faster than speciation events have occurred or if species extinction has eliminated transitional forms. But, without phylogenetics, comparative biology becomes reduced to developmental taxonomy. For example, neotenic species display juvenilized phenotypes relative to their immediate ancestors. In the case of neotenic amphibia, textbook photographs are often shown of unrelated species with varying degrees of neoteny, as if it were a single evolutionary phenomenon. In fact, as the neotenic condition evolved independently multiple times within Amphibia, specific mechanisms of developmental changes must be studied within subsets of contemporary amphibian species that are derived from a single nonneotenic ancestor. Independent origins of neoteny reflect different underlying genetic changes.

Another example that illustrates the significance of applying phylogeny to developmental studies is the evolution of insect embryogenesis. This example further illustrates the utility of studying Lepidoptera. It is often overlooked that embryogenesis in *Drosophila* is rather unusual relative to other arthropods (Anderson, 1972; Sander, 1988; French, 1990); comparison with more "typical" insects may provide improved judgment as to which aspects of embryogenesis are widespread and which are phylogenetically restricted. For example, *Drosophila* has a long germ band type egg in which the germ anlage covers much of the blastoderm and develops directly into the fully segmented germ band without major change in proportions. In short germ band eggs, the germ anlage consists only of rudiments of the protocephalon that will differentiate into preantennal and antennal segments plus a small posterior region that gradually elongates and within which segment boundaries are demarcated in an anteroposterior progression (see also Nagy, this volume).

Within the orders of neopterous insects (winged insects excluding Odonata [dragonflies and damselflies] and Ephemeroptera [mayflies]), those with short germ bands are typically more primitive and develop much more slowly, for example, 13 days for embryogenesis in *Locusta* (Orthoptera) versus 24 hours in *Drosophila* (Schwalm, 1988). There is no single cause to explain this difference in timing, but one significant factor is that many long germ eggs typically have a meroistic ovary within which there are multiple highly polyploid, nutritive sister cells (called nurse cells) attached to the oocyte, whereas short germ eggs often have panoistic ovaries lacking nurse cells, so that the oocyte alone must make many of the essential components such as ribosomes (Schwalm, 1988). Another possibility is that holometabolous insects (including Lepidoptera and *Drosophila*) have a much more advanced embryonic presumptive area within the blastoderm than the more primitive Hemimetabola and can continue development without having to recruit material from the yolky ooplasm.

It is often assumed that long germ band eggs evolved directly from short germ eggs. However, embryogenesis in most winged insects, including the Lepidoptera, paleopterous Odonata, and some Orthoptera, appears to be intermediate between these two extremes. Anterior segments are carved out of an intermediate sized germ band, not so many as for short germ eggs but fewer than for long germ eggs. The remainder of the segments are added by subterminal addition. It seems likely that the extreme long germ types of eggs evolved multiple times and possibly by different mechanisms. For example, apocritan Hymenoptera (e.g., honeybee), cyclorrhaphan Diptera (e.g., *Drosophila*), and some Homoptera (e.g., *Notonecta*) (Schwann, 1988) have long germ eggs, even though other members of their respective orders, which are basal at least for Hymenoptera and Diptera, are clearly intermediate, retaining a posterior growth zone from which additional segments are formed (Anderson, 1972). The same may also be true for the evolution of intermediate germ band insects. Thus, if one is interested in understanding the mechanistic change that resulted in the evolution of one type of development from another, it is first necessary to know the systematic relationships among the groups being compared. It is insufficient simply to compare any short germ embryo with any other long germ embryo.

The argument that the primary interest in embryogenesis is in "fundamental" (= universal or widespread) processes is sometimes used to justify dichotomous comparisons of distantly related organisms that have not been placed in an appropriate systematic context. But embryogenesis is much too complex to divide into "fundamental" and "not fundamental" categories or to justify choosing a very small number of taxa as "models" to explain the observed diversity. What may appear as a tidy categorization of development may, like germ band length, represent a nonevolutionary assemblage. This is not to say that some processes of, for example, segmentation are not conserved across major groups, but only that the mechanisms that bring about segmentation may have diversified so greatly that what little remains in common is insufficient for an in-depth understanding of the process of interest. This diversity of mechanisms may be true even though homologous genes are involved in both processes.

In sum, there is no substitute for comparing seemingly homologous processes in multiple organisms and for placing those comparisons within a phylogenetic context. Lepidoptera, together with Trichoptera, Mecoptera, Siphonaptera, and Diptera, constitute the Panorpida. Separated by more than 200 million years from the Diptera, Lepidoptera provide an important window on certain evolutionary innovations relative to *Drosophila* that would not be observed by making more distant comparisons.

The study of phylogenetic relationships is a branch of systematics, a field that in this century has overlapped little with, indeed, has been eclipsed by, experimental biology. Of late, however, phylogeny has been reemerging as a unifying theme in biology. Striking progress has been made in the theory and calculational methodology for inferring phylogenies. In addition, new data from molecular biology show promise for helping to solve many historically recalcitrant questions in phylogenetics. These advances in phylogeny reconstruction per se have, in turn, spurred rigorous reexamination of, and new methodologies for, the application of comparative (= phylogenetic) approaches in fields ranging from molecular biology to ecology.

The study of Lepidoptera seems particularly ripe for exploiting new phylogenetic approaches. Lepidoptera is one of the four most diverse insect orders, approximately 150,000 named species, and has served as model systems for major studies of plant-insect interactions, mimicry and other adaptations, hormone and pheromone action, cuticle and chorion structure, and classic and molecular genetics (see discussion that follows and other chapters in this volume). With centuries of interest in their systematics, Lepidoptera is as well characterized as any other group of insects at the level of family and below. However, relationships among families and superfamilies are very poorly understood, with the exception of more primitive groups. In part, this reflects the morphological homogeneity of the group, that is, the limited number of useful morphological characters (Kristensen, 1984; Nielsen, 1989). To circumvent this problem in other taxonomic groups, many systematists have turned to molecular data in search of the abundance of characters needed to make robust phylogenetic trees. We review here early attempts at molecular systematics of Lepidoptera.

In the second half of this essay, we describe one area of current research in Lepidoptera where comparative data have been generated and that illustrates the potential of this approach for achieving a unified understanding of phylogenetic mechanism, the ultimate goal of comparative biology. Part of the problem is that "phylogenetic mechanism" means different things to workers at different levels of organization. For instance, to a molecular biologist, the interactions between macromolecules (or cells, tissues, etc.) that lead, in a causally connected manner, to phylogenetic changes constitute its underlying mechanisms. For example, certain relatively short DNA sequences appear to promote and to undergo gene conversion within the context of a multigene family (Xiong et al., 1988; see Eickbush and Izzo, this volume). Gene conversion thus constitutes one phylogenetic mechanism for gene family diversification.

From a developmentalist's perspective, phylogenesis can be seen as a succession of ontogenies. These ontogenies are replicated and diversified as speciation occurs. Here the focus is on the specific changes in parts of the ontogeny that result in diversification; whether these involve gene family diversification, conversion, or some other genetic mechanism may be secondary. To an evolutionary ecologist seeking to understand the role of natural selection in evolution, developmental changes of a very general type may constitute phylogenetic mechanisms. An example is heterochrony, a change in the timing of an ontogenetic process in a descendant lineage relative to its ancestor. Neoteny (truncated development due to retarded somatogenesis) in salamanders and pedogenesis (truncated development due to accelerated gametogenesis) in aphids are classic examples (Gould, 1977; Leclerc and Regier, 1990). Heterochrony is considered a phylogenetic mechanism because interaction of the heterochronic organism with its environment can lead to new life history strategies. Some molecular biologists would consider heterochrony a term of classification rather than a mechanism – a consequence or result of underlying mechanisms and not a cause – but clearly ecologists do not use it in that sense. Conversely, organismically oriented researchers may find gene conversion (and most other macromolecular-level processes) without explanatory power, as these processes reveal little about how the phenotype will fare in the environment. This difference of perspective arises because heterochrony is differently considered at distinct levels of complexity. The evolution of morphogenesis of the lepidopteran chorion, which we review in some detail, provides an example in which all of these necessary and complementary levels of explanation can be sought simultaneously.

Fundamentals of phylogenetic analysis

Organic diversity is thought to arise through the processes of cladogenesis, the splitting of ancestral reproductive communities (species) into two or more descendant species, and anagenesis, genetic change within species. The history of cladogenesis by which a given set of taxa (species or groups thereof) diverged from a common ancestor can be represented by a branching diagram called a cladogram or phylogeny (see Figure 4.1). Each line segment or branch in such a figure represents the succession of generations within a single species, and each branch point or node represents the common ancestor of the lineages it subtends, at the moment of its divergence into daughter species. Groups consisting of all and only the descendants of a single ancestor, for example, the group Glossata in Figure 4.1, are called *monophyletic* (Hennig, 1966). Any two (or more) monophyletic groups resulting immediately from splitting of the same ancestor, thereby forming a more inclusive group, are called sister groups – for example, the Exoporia consisting of the sister groups Hepialoidea and Mnesarchaeoidea in Figure 4.1. As phylogenetic relatedness is defined in terms of recency of common ancestry, members of the same monophyletic group are more closely related to each other than to any species outside that group.

Phylogenies cannot be observed directly but are inferred from analysis of independently evolving, heritable characters, generally morphological or molecular, observed in extant or fossil species. The most widely accepted logic for the inference of phylogeny was outlined by the German entomologist W. Hennig (1966). If a novel character condition arises in some ancestral species, departing from the presumed "ground plan" or primitive condition for the entire study group, it will tend to be passed on to just the descendants of that ancestor. Therefore sharing of a novel or derived feature, termed a *synapomorphy,* can be taken as evidence for common ancestry – monophyly – of the subset of species possessing it. By contrast, neither symplesiomorphies (characters inherited unmodified from the ground plan) nor autapomorphies (derived conditions unique to single species) offer evidence on relationships. If an ancestral condition becomes modified in the same way more than once, the resulting parallel or convergent similarity can obscure relationships. Systematists rely on analysis of multiple characters to overcome this problem.

Hennig's principle of grouping by synapomorphy is equivalent to choosing phylogenies so as to ascribe as many similarities as possible to common ancestry, thereby minimizing the importance of convergence. The potential limitations of this "parsimony" criterion have been strongly debated, but there is declining support for its chief alternative, grouping by overall similarity (including symplesiomorphies), as this, in effect, assumes con-

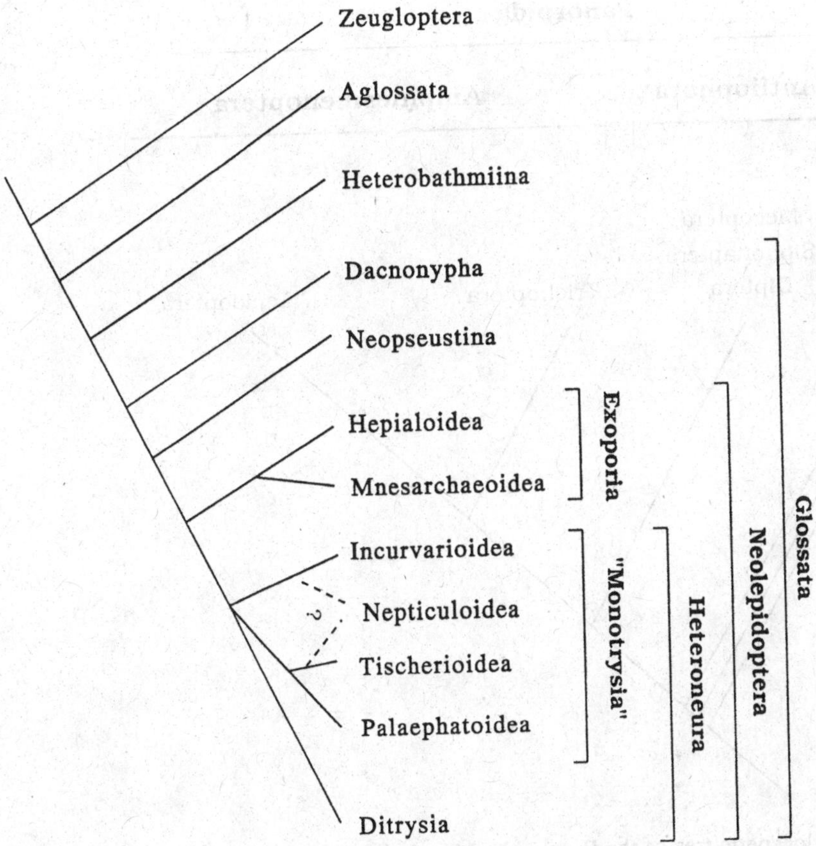

Figure 4.1. Phylogenetic tree of the Lepidoptera based on morphological analysis. (Modified from Kristensen, 1984, and Davis, 1986.)

stancy of evolutionary rates (Ridley, 1986). The parsimony criterion can also be used to estimate the history of character change (anagenesis) once the phylogeny is established (Farris, 1970; Fitch, 1971). In effect, features shared only by descendants of a given ancestor are taken to have arisen in that ancestor.

Phylogeny of Lepidoptera and allied groups

Interpretation of available morphological evidence clusters Lepidoptera and four other orders of modern insects into a monophyletic group called the Panorpida (Figure 4.2; Boudreaux, 1981; Kristensen, 1989). Two monophyletic sister superorders are recognized within the Panorpida: the Am-

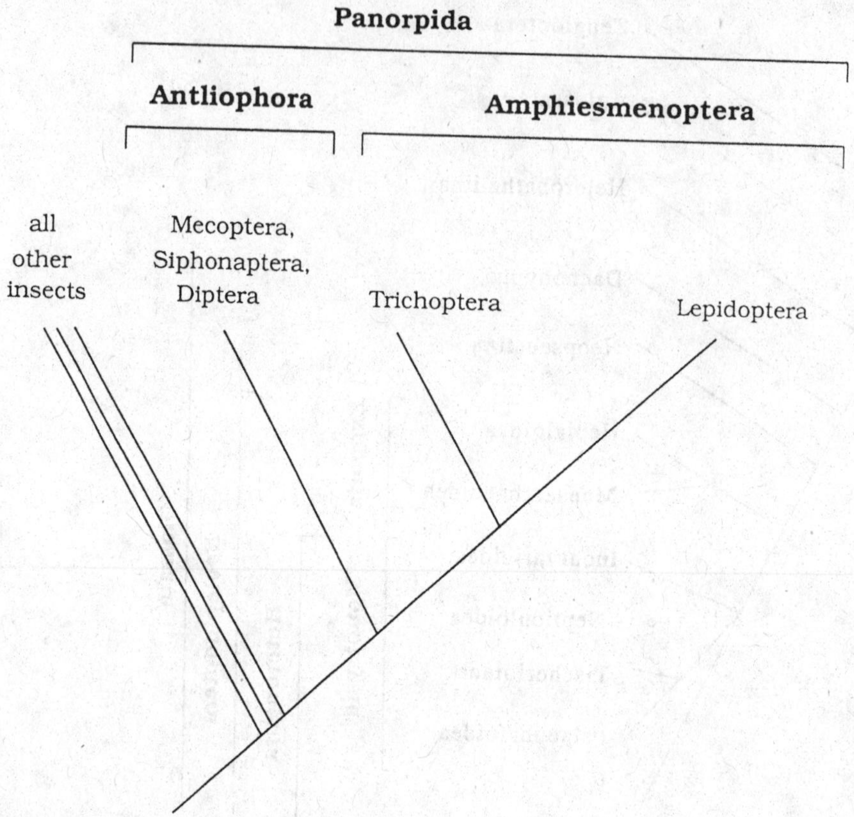

Figure 4.2. Phylogenetic tree of the Panorpida. Morphological analysis provides strong evidence for monophyly of major groups within the Panorpida. (Modified from Kristensen, 1989.)

phiesmenoptera (Lepidoptera and Trichoptera) and the Antliophora (Mecoptera, Siphonaptera, and Diptera). The sister order relationship between Lepidoptera and Trichoptera (caddis flies) is very strongly supported. Lepidoptera and Diptera are by far the most species-rich (150,000–250,000 named species versus 400–7,000 for the other three), whereas Mecoptera (scorpion flies) are the most primitive. All of these orders, except Siphonaptera (fleas), are known from Mesozoic fossils, and it is probable that the most recent common ancestor of the Panorpida dates to the Permian (~250 my Before Present [BP]) (Hennig, 1981).

Primitive Lepidoptera and Diptera are little changed in morphology from their Mesozoic ancestors, and several lineages are represented in Cretaceous fossils (~100 my BP) (Whalley, 1986; Shields, 1988). As yet, no fossil butterflies or more advanced moth families are known from that period, but they are known from the early Tertiary (roughly 50 my BP). The vast majority of Lepidoptera are of the more advanced type, called Ditrysia, whereas most of the morphological diversity is distributed among the relatively few, more primitive moths. As a consequence, the phylogeny of Lepidoptera is better known for these primitive moths and that of the Ditrysia is largely unexplored.

Intensive morphological study over the past two decades has firmly established many aspects of lepidopteran phylogeny at the most basal, that is, sub- and infraordinal levels, (see Figure 4.1). Four basal monophyletic groups are now generally recognized: Zeugloptera, Aglossata, Heterobathmiina, and Glossata (Kristensen, 1984; Kobayashi and Ando, 1988). Zeugloptera now appear to be the sister group of all other Lepidoptera; nonzeuglopterans are held together by 15 synapomorphies, and 28 synapomorphies place Zeugloptera with other Lepidoptera relative to its sister order Trichoptera. Relationships among the remaining suborders are less certain.

Within the suborder Glossata, which comprises the vast majority of extant Lepidoptera, Kristensen (1984) recognized four monophyletic subgroups: Dacnonypha, Neopseustina, Exoporia, and Heteroneura. A large number of synapomorphies link Exoporia and Heteroneura into a monophyletic group, called the Neolepidoptera. The monophyly of Heteroneura and its relationship to the Exoporia are still debated. Dugdale (1974) has placed Exoporia as sister group to Ditrysia. Minet (1986) has placed Nepticuloidea as sister group of Exoporia plus all remaining Heteroneura. However, a recent analysis of embryonic characters supports the monophyly of all Heteroneura, with Exoporia as its sister group (Kobayashi and Ando, 1988).

Despite this progress, lepidopteran phylogeny must still be regarded as only little clarified. In particular, about 98 percent of extant lepidopteran species, including all that are commonly used as laboratory tools (Figure 4.3), belong to one subgroup of the infraorder Heteroneura, the Ditrysia, within which relationships are mostly obscure (Figure 4.4). Tineoidea, Yponomeutoidea, and Gelechioidea are widely recognized as basal groups within Ditrysia because of their many symplesiomorphic characters (Nielsen, 1989). Although many higher families and superfamilies are recognized as monophyletic, the relationship of families within superfamilies and the relationship of superfamilies to each other are largely unknown. The families within the Noctuoidea are perhaps the best characterized.

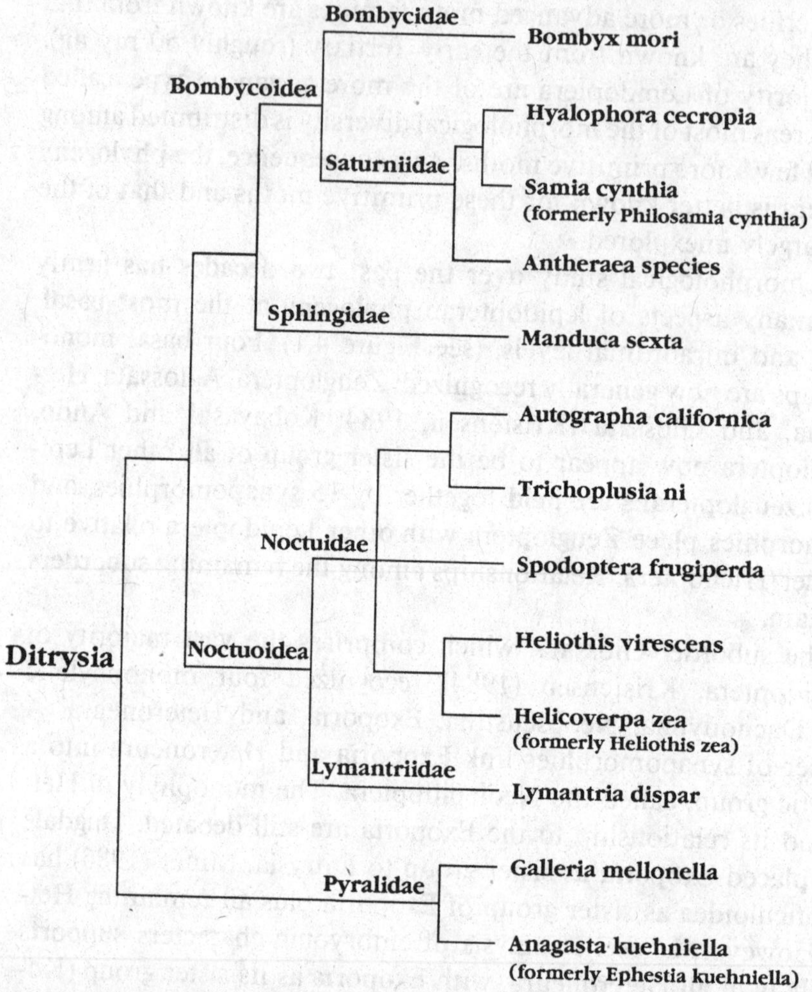

Figure 4.3. Generally accepted phylogenetic relationships among moths frequently used in experimental studies. Sphingidae are frequently included in the Bombycoidea.

Molecular studies of the higher phylogeny of Lepidoptera

The paucity of major morphological differences within the ditrysian Lepidoptera should encourage the application of molecular systematic methods. We are aware of only one study of higher level lepidopteran phylogeny in which molecular characters have been analyzed (Martin and Pashley, 1992; Weller, Friedlander, Martin, and Pashley, 1992). This preliminary study was based on partial sequences directly derived from 18S and 28S

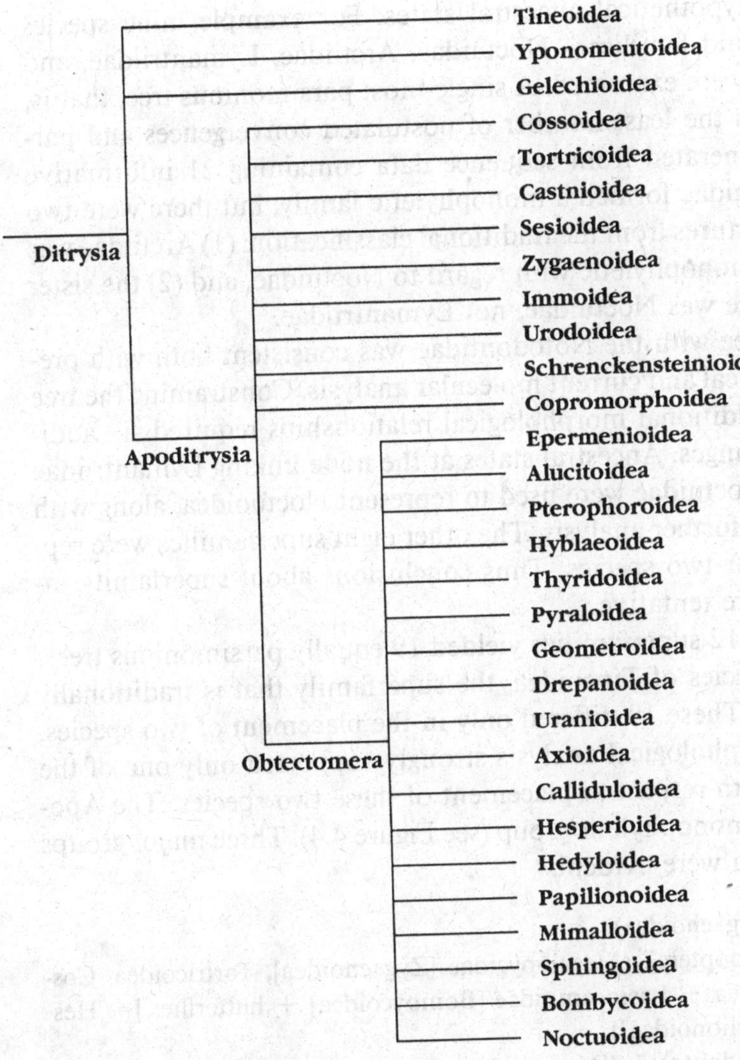

Figure 4.4. Phylogenetic tree of the Ditrysia. Only three groups are resolved based on available morphological evidence. (Modified from Nielsen, 1989.)

ribosomal RNAs, molecules that have been widely utilized for systematic studies. Forty-two informative characters – that is, presumed synapomorphies or shared derived characters – were identified from analysis of 28 species within 12 superfamilies. As a way of simplifying the analysis, species within each of the three relatively well studied groups – butterflies (Papilionoidea and Hesperioidea), Noctuoidea, and Bombycoidea – were represented as hypothetical ancestral states. For example, nine species from four noctuoid families – Noctuidae, Arctiidae, Lymantriidae, and Notodontidae – were examined. A single most parsimonious tree, that is, one that requires the least number of postulated convergences and parallelisms, was generated from sequence data containing 21 informative characters. Noctuidae formed a monophyletic family, but there were two noteworthy departures from the traditional classification: (1) Arctiidae was found not to be monophyletic with regard to Noctuidae, and (2) the sister group to Arctiidae was Noctuidae, not Lymantriidae.

Rooting the tree with the Notodontidae was consistent both with previous morphological and current molecular analysis. Constraining the tree to conform to traditional morphological relationships required six additional parallel changes. Ancestral states at the node linking Lymantriidae to Arctiidae + Noctuidae were used to represent Noctuoidea, along with Notodontidae, in further analysis. The other eight superfamilies were represented by one or two species. Thus conclusions about superfamily interrelationships are tentative.

Analysis of the 12 superfamilies yielded 19 equally parsimonious trees, rooted on one species of Tineoidea, the superfamily that is traditionally considered basal. These 19 differed only in the placement of two species, and previous morphological analysis strongly supported only one of the 19 possibilities with respect to placement of these two species. The Apoditrysia formed a monophyletic group (see Figure 4.4). Three major groups within Apoditrysia were evident:

1. Limacodidae (Zygaenoidea)
2. Most "microlepidoptera" (Megalopygidae [Zygaenoidea], Tortricoidea, Cossoidea, Castnioidea + Lasiocampidae [Bombycoidea] + butterflies [= Hesperioidea + Papilionoidea])
3. Most "macrolepidoptera" ([Noctuoidea, Geometroidea, Bombycoidea] + Pyraloidea).

The analysis strongly suggests that Lasiocampidae does not belong with other Bombycoidea. Constraining them to Bombycoidea added six steps to the overall tree length and broke up one of the strongest monophyletic groups recovered in the analysis. Lack of monophyly of Zygaenoidea is puzzling at present; it would require seven additional steps to group the two zygenoids together. It is obvious from inspection of the weaker internal branches of the tree that resolution of relationships among superfamilies of Ditrysia awaits further work, but the overall pattern is encouraging. In the future, Pashley and co-workers at Louisiana State University will continue to sequence ribosomal RNAs and begin to examine mitochondrial DNA sequences as well.

A related project on the primitive groups of Lepidoptera has begun in our laboratories at the University of Maryland, in collaboration with David Wagner at the University of Connecticut and E. S. Nielsen at CSIRO in Australia. One goal of the project is to test the utility of rRNA as a higher level phylogenetic indicator in Lepidoptera by comparing our results with those of Kristensen (1984), Davis (1986), and Nielsen (1989) that are based on morphological analysis. At the same time, we plan to look for individual and population polymorphisms within the rDNA locus, including variation in gene copy numbers. If our early results are concordant with the well-established relationships based on morphological analysis, then we will extend this analysis to other species and ultimately to the Ditrysia.

We have also begun a search for additional molecules that will be useful in discerning higher level relationships among the Lepidoptera – and undoubtedly other groups of insects. For insects generally, only ribosomal RNAs (including 5S RNA) have been explored. The results have been informative but not without controversy. The need for additional molecules is compelling, as morphological characters and the limited molecular studies to date have proved inadequate for resolving many major questions of insect phylogeny.

The proper choice of molecules is difficult in the absence of extensive comparative structural and functional studies. Nevertheless, there are criteria that should be expected of all molecules useful in higher level phylogenetic reconstruction. The following are those we have used to screen GenBank and other sources for candidate sequences:

- The molecule must be sufficiently conserved to make accurate alignments and to minimize multiple hits but variable enough to be informative at the superfamily to phylum level. This assessment will be based on comparison of homologous sequences present in the computer data bases.
- Only coding sequences will be analyzed to ensure sufficient conservation.
- Conservation should not be restricted to small domains, so that most of the molecule can be potentially informative in deducing relationships.
- Ideally, genes of interest should be single copy or, less acceptably, present in low copy number. Otherwise, identification of orthologs, those homologous genes that are related only by descent and not by gene duplication, may be difficult. Only orthologous relationships are informative of species descent.

- A clone of an orthologous gene should be available from an arthropod, preferably a holometabolous insect, so that heterologous hybridization will specifically select the sequence of interest in the species of interest.
- The sequence should be sufficiently long in order to encode sufficient information. We have restricted our search to those >500 nucleotides, as shorter sequences are unlikely to be able to resolve large numbers of species.
- The sequence should contain no simple, internal repeats in order to avoid alignment problems.
- There should be no major base biases, such as AT-rich stretches, as this complicates character analysis.
- The coding region of interest should not be interrupted by multiple and/or long introns in order to minimize the sequencing of uninformative DNA.

Based on an extensive computer and library search, we have identified 25 genes that fulfill many of these criteria. Twenty of the genes have been cloned and sequenced in *Drosophila,* and the remaining five have been cloned and sequenced in ditrysian species. In most cases, not all criteria could be tested, and thus additional tests are needed before committing to a large-scale sequencing project. Thus we have tested 20 of these sequences as heterotypic probes in genomic Southern analysis of primitive Lepidoptera. Five gave particularly simple Southern hybridization patterns. We will clone these five and determine their sequences in approximately 12 primitive lepidopteran species of known phylogenetic relationship. Final choice of genes will be from those that yield phylogenies concordant with each other and with morphological analysis.

The lepidopteran chorion

Although phylogenetic trees are of intrinsic interest, they are also crucial for addressing a multitude of evolutionary questions. Reconstruction of evolutionary pathways for diversification of the lepidopteran eggshell is one such set of questions that is of central interest in our laboratory. Based on comparative studies both within and outside the Lepidoptera, the eggshell has evolved in major ways, both as a functional whole and as a complex set of interacting macromolecules. These studies are summarized as an example of the power and utility of applied systematics in deducing phylogenetic mechanisms.

Morphology

At the end of oogenesis, follicular epithelial cells synthesize and secrete onto the oocyte's surface a complex set of proteins (Regier and Kafatos,

1985). These proteins assemble to form the tripartite structure of the eggshell, consisting of an inner vitelline membrane, a chorion, and a very thin outer sieve layer (Kafatos et al., 1977). The vitelline membrane is typically <2 μm in thickness, whereas the chorion can vary, depending on the species, from <1 μm to >50 μm (Barbier and Chauvin, 1974b; Regier and Vlahos, 1988). Two radial zones of the chorion are easily recognized, an inner trabecular layer and a typically much thicker, overlying lamellar chorion (Figure 4.5). Lamellae vary in number, from 1 to >100, with thicker chorions generally having more.

Despite their apparent discreteness when visualized in electron micrographs, lamellae actually constitute an ultrastructural continuum. The most basic lamellar component is the fiber. In turn, fibers assemble into helicoids, also referred to as "laminated plywood-like" structures (Neville, 1988) and "biological analogs of cholesteric liquid crystals" (Mazur et al., 1982). In electron micrographs, a helicoidal lamella corresponds to a 180° rotation of adjoining planes of parallel-oriented fibers.

The various functions of the insect eggshell have been extensively reviewed and its possible significance for the evolution of terrestriality noted (Zeh, Zeh, and Smith, 1989). Hinton (1981) has emphasized the importance of density-independent selective forces in molding eggshell architecture, such as the requirements for oxygen–carbon dioxide exchange, for resistance both to desiccation and flooding, and for sperm entry and hatching. However, these requirements are common to most insect eggs, whereas helicoidal lamellar chorions are restricted to Lepidoptera. We are not aware of any studies that specifically address the function of helicoidal lamellae in chorion, but consideration of other helicoidal systems in biology suggests an important role in providing mechanical strength. A few examples include cuticle of arthropods (Neville, 1975), vertebrate basement membranes (Weiss and Ferris, 1956), cell walls of plants (Neville, 1988), scales of fish (Giraud, Castanet, Meunier, and Bouligand, 1978), oothecae of mantids (Neville and Luke, 1971), eggshells of nematodes (Wharton, 1978), and steriocilia within the inner ear of vertebrates (DeRosier, Tilney, and Egelman, 1980). Fibrous systems are known to provide very high tensile strength (relative to amorphous solids), particularly in the direction of the fiber axis. The advantage of a helicoidal arrangement of fibers is that tensile strength becomes independent of direction (Barth, 1973).

Why might the strengthening of lepidopteran eggshells have occurred through the introduction of helicoidal lamellae? One reasonable hypoth-

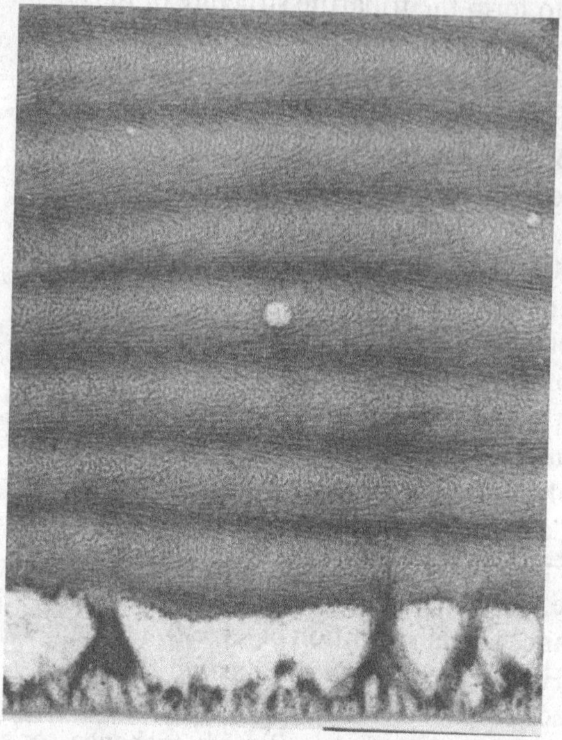

Figure 4.5. Transmission electron micrograph of an immature chorion from the silk moth *Hyalophora cecropia*. The trabecular layer is at the bottom of the micrograph with a portion (~6% of total) of the helicoidal lamellae above. (Scale marker = 1 μm.)

esis is that the lamellar chorion may resist biting or crushing by predators' chewing mouthparts, as one of many egg defense mechanisms (Hinton, 1981). As support, chickadees, ants, spiders, and mites are observed predators of gypsy moth eggs (Smith and Lautenschlager, 1978), and earwigs have been observed to kill codling moth eggs (Glen, 1982). Unfortunately, these reports have not been followed up with controlled laboratory studies.

Morphogenesis

Lamellogenesis in the silk moth *Antheraea polyphemus* occurs in four temporally distinct modes of growth, following construction of the trabecular layer (Regier, Mazur, Kafatos, and Paul, 1982; Mazur, Regier, and Kafatos, 1989):

1) The first mode is *framework formation,* in which a small number of helicoidally arranged fiber planes are laid down, largely by apposition, to form thin lamellae. The near final number of lamellae is reached when less than 20 percent of the final dry weight of the chorion has been deposited.

2) Permeation by new chorion proteins then leads to *expansion* of this early framework, with the insertion of new sheets of fibers that thicken the individual lamellae throughout the chorion. Lamellar number does not increase as a result of expansion.
3) *Densification* of the entire structure follows, apparently by thickening of individual fibers without additional lamellar expansion or an increase in lamellar number. Eventually, the structure of individual fibers is completely obscured, although lamellar banding remains obvious.
4) Finally, *surface sculpturing* occurs as a few additional thin lamellae are deposited and sculpted into surface structures called aeropyle crowns (Figure 4.6; Mazur et al., 1980).

Phylogenetics of lamellogenesis – framework formation

The available evidence is consistent with the hypothesis that the helicoidal lamellar chorion evolved after the lepidopteran lineage diverged from all other extant orders of insects. Thus a helicoidal lamellar chorion is absent from Trichoptera, the sister order of Lepidoptera and from the sister superorder Antliophora that includes Diptera and Mecoptera (see Table 4.1). In addition, a helicoidal lamellar chorion has not been found in the more distantly related orders (Furneaux and MacKay, 1972; Hinton, 1981). This seems surprising given the widespread distribution of helicoidal lamellae in other nonchorion structures, for example, the serosal cuticle within two psychid moth eggshells (Chauvin and Barbier, 1974). Many intrinsic and extrinsic factors must be analyzed in order to understand this distribution.

The taxonomic distribution of helicoidal lamellae within chorions and their modes of modification have been summarized in Table 4.1 for 60 species of Lepidoptera that represent 26 families. Although widespread, the occurrence of a helicoidal lamellar chorion is not universal. In particular, it is missing from two Zeugloptera and one Heterobathmiina but is abundantly represented in the Glossata, suggesting its origin within that group (see Figure 4.1).

As previously noted, Kristensen (1984) recognizes four glossatan infraorders – Dacnonypha, Neopseustina, Exoporia, and Heteroneura (see Figure 4.1). A large number of synapomorphies link Exoporia and Heteroneura into a monophyletic group, called the Neolepidoptera. The chorions of seven exoporians have been examined and shown to lack lamellae. Within the nonditrysian Heteroneura, two species in two distinct families have been shown to lack a lamellar chorion. Within the Ditrysia, helicoidal

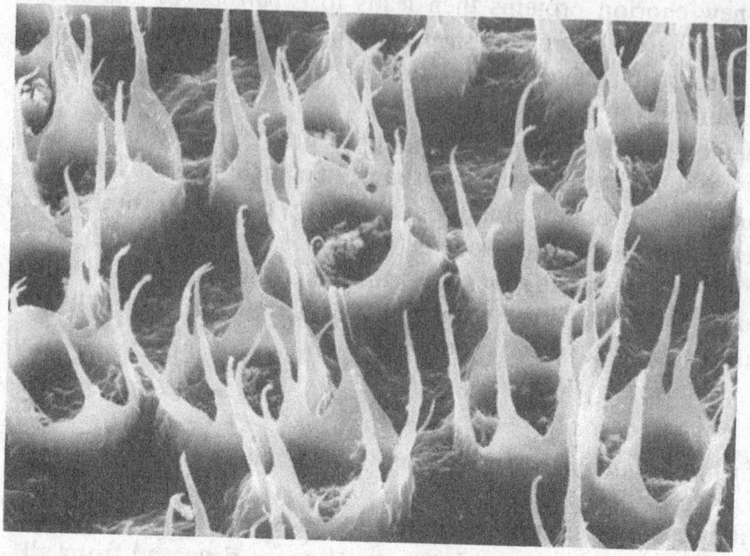

Figure 4.6. Scanning electron micrograph of aeropyle crowns on the surface of the eggshell from *Antheraea polyphemus*. (Scale marker = 10 μm.)

lamellae are uniformly present in 11 out of 12 superfamilies (see Table 4.1). Interestingly, the one superfamily, Tineoidea, whose members collectively reveal both presence and absence of helicoidal lamellae, is widely recognized to have many features of the ditrysian ground plan (Robinson, 1988). They may well represent the sister group to all other Ditrysia (see Figure 4.4).

The most reasonable hypothesis is that the helicoidal lamellar chorion evolved once very early in ditrysian evolution, perhaps while families of Tineoidea were still diversifying. As the preceding statement implies, Tineoidea may not constitute a monophyletic group, as only symplesiomorphies define the group. The interrelationships of tineoid families are currently unclear. This prevents one from drawing firm conclusions about the hypothesis of secondary losses of lamellae within the Psychidae and one species of Tineidae, *Trichophaga tapetzella* (see Table 4.1).

Yponomeutoidea and Gelechioidea are also widely recognized as primitive ditrysians and, in fact, were phenetically grouped within Tineoidea until 1915 (described in Kyrki, 1984). These three superfamilies are still considered to be more primitive than the remaining ditrysian superfamilies (Nielsen, 1989). It is unlikely that any other ditrysian superfamily will prove to be more basal on the phylogenetic tree. The occurrence of lamellar chorions within two yponomeutoids and two gelechioids (as well as all other ditrysian superfamilies) provides support for the hypothesis that the lamellar chorion evolved near the time of the initial ditrysian diversification.

Phylogenetics of lamellogenesis – expansion and densification

In Bombycoidea, deposition of the helicoidal lamellar framework constitutes an early step in choriogenesis. Subsequently, this framework is modified by expansion and densification. Although data are few, helicoidal lamellar modifications clearly vary across taxa (see Table 4.1). For example, five superfamilies within the Obtectomera undergo densification (Hesperioidea, Pyraloidea, Noctuoidea, Sphingoidea, Bombycoidea), whereas two superfamilies within the nonobtectomeran Apoditrysia (Tortricoidea, Zygaenoidea) do not. These observations are consistent with the hypothesis that densification evolved significantly later than the lamellar framework, possibly in the ancestor of the Obtectomera.

The data for lamellar expansion are confined to three superfamilies, but already an interesting observation can be made. All four bombycoids and one noctuoid examined undergo expansion, whereas the one sphingid does not. In light of the widespread recognition that Bombycoidea and Sphingoidea are closely related, perhaps sister groups, and are both more distantly related to Noctuoidea than to each other, the patterns of expansion strongly suggest either multiple origins (in Noctuoidea and Bombycoidea) or secondary loss (in Sphingoidea).

Structure and organization of chorion genes

Evolutionary changes in lamellogenesis presumably reflect genetic changes in chorion gene sequence, organization, copy number, and expression. Gene families evolve, like species, by duplication and subsequent modification. Therefore phylogenetic methods can be used to reconstruct their histories as well, although the situation may be somewhat complicated by additional processes such as gene conversion.

Comparative chorion sequence analysis of moths with differing modes of lamellogenesis, all placed within a phylogenetic context, offers an un-

Table 4.1. *Phylogenetic distribution of modes of helicoidal lamellogenesis within insect chorion*

Higher Level Classification[a]	Species	F[b]	E[b]	D[b]	References
	Lepidoptera				
Zeugloptera					
Micropterigoidea/Micropterigidae	*Neomicropteryx nipponensis*	—			Kobayashi and Ando, 1982
	Micropteryx calthella	—			Chauvin and Chauvin, 1980; Fehrenbach, in press
Heterobathmiina					
Heterobathmiidae	*Heterobathmia* sp.	—			Fehrenbach, in press
Aglossata					
Glossata/Exoporia					
Dacnonypha/Eriocraniidae	*Eriocrania subpurpurella*	—			Fehrenbach, in press
Hepialoidea/Hepialidae	*Hepialus humuli*	—			Kobayashi and Ando, 1987
	Endoclita excrescens	—			Kobayashi and Ando, 1987
	Korscheltellus lupulinus	—			Chauvin and Barbier, 1979
	Hepialus hecta	—			Fehrenbach, 1989
	Wiseana umbraculata	—			Fehrenbach, 1989
Mnesarchaeiodea/Mnesarchaeidae	*Mnesarchaea fusilella*	—			Fehrenbach, 1989
	Mnesarchaea acuta	—			Fehrenbach, 1989
Glossata/Heteroneura (Monotrysia)					
Incurvarioidea/Incurvariidae	*Adela metallica*	—			Fehrenbach, in press
Tischerioidea/Tischeriidae	*Tischeria ekebladella*	—			Fehrenbach, in press

Glossata/Heteroneura (Ditrysia)				
Tineoidea/Psychidae	*Fumea casta*	–		Chauvin and Barbier, 1974
	Luffia ferchaultella	–		Chauvin and Barbier, 1974
	Dissoctena granigerella	–		Fehrenbach, in press
	Dahlica lichenella	–		Fehrenbach, in press
Tineoidea/Tineidae/Tineinae	*Trichophaga tapetzella*	–		Chauvin and Barbier, 1972a
	Monopis rusticella	+		Chauvin and Barbier, 1972a
	Monopis crocicapitella	+		Chauvin, 1977
	Tineola bisselliella	+		Chauvin, 1977
	Tinea pellionella	+		Chauvin and Barbier, 1972b
Yponomeutoidea/Acrolepiidae	*Acrolepia assectella*	+		Chauvin, Rahn, and Barbier, 1974
Yponomeutoidea/Plutellidae	*Plutella maculipennis*	+		Chauvin, Rahn, and Barbier, 1974
Gelechioidea/Oecophoridae	*Endrosis sarcitrella*	+		Arbogast et al., 1983
	Hofmannophila pseudospretella	+		Strong, 1984
Tortricoidea/Tortricidae	*Cydia pomonella*	+	–	Fehrenbach, Dittrich, and Zissler, 1987
Zygaenoidea/Zygaenidae	*Zygaena trifolii*	+		Fehrenbach, in press
Zygaenoidea/Dalceridae	*Acraga ochracea complex*	+	–	Regier and Epstein, unpublished observations
Hesperioidea/Hesperiidae	*Calpodes ethlius*	+	+	Griffith and Lai-Fook, 1986

(continued)

Table 4.1. *(Cont.)*

Lepidoptera					
Higher Level Classification[a]	Species	F[b]	E[b]	D[b]	References
Papilionoidea/Pieridae	*Pieris brassicae*	+			Furneaux and MacKay, 1972
	Pieris rapae	+			Kim et al., 1983
Pyraloidea/Pyralidae	*Chilo suppressalis*	+			Furneaux and MacKay, 1972
	Galleria mellonella	+		+	Barbier and Chauvin, 1974a
	Anagasta kuhniella	+			Cruickshank, 1972
	Nymphula nymphaeata	+			Barbier and Chauvin, 1974b
Geometroidea/Geometridae	*Cingilia catenaria*	+			Salkeld, 1973
Noctuoidea/Notodontidae	*Phalera bucephala*	+			Chauvin, Rahn, and Barbier, 1974
	Lophodonta angulosa	+			Hinton, 1981
	Cerura modesta	+			Hinton, 1981
	Cerura vinula	+			Hinton, 1981
	Pheosia rimosa	+			Hinton, 1981
	Pheosia tremula	+			Hinton, 1981
	Pheosia gnoma	+			Hinton, 1981
	Nadata gibbosa	+			Hinton, 1981
	Notodonta dromedarius	+			Hinton, 1981
Noctuoidea/Noctuidae	*Heliothis virescens*	+		+	Fehrenbach, Dittrich, and Zissler, 1987
	Spodoptera littoralis	+		+	Fehrenbach, Dittrich, and Zissler, 1987
	Amathes c-nigrum	+		+	Salkeld, 1973
	Euxoa altera	+			Salkeld, 1975

Noctuoidea/Lymantriidae	*Lymantria dispar*	+	+	+	Regier, unpublished observations
	Orgyia antiqua	+			Hinton, 1981
Sphingoidea/Sphingidae	*Manduca sexta*	+	–	+	Regier and Vlahos, 1988
Bombycoidea/Lasiocampidae	*Lasiocampa quercus*	+			Hinton, 1981
Bombycoidea/Bombycidae	*Bombyx mori*	+	+	+	Papanikolaou, Margaritis, and Hamodrakas, 1985
Bombycoidea/Saturniidae	*Antheraea pernyi*	+	+	+	Regier and Hatzopoulos, 1988
	Antheraea polyphemus	+	+	+	Mazur, Regier, and Kafatos, 1989
	Antheraea yamamai	+			Hinton, 1981
	Hyalophora cecropia	+	+	+	Regier and Vlahos, 1988
	Rhodia fugax	+			Hinton, 1981
	Panorpida[c]				
Mecoptera	Panorpa germanica	–			Furneaux and MacKay, 1972
Diptera					
Drosophilidae	*Drosophila melanogaster*	–			Margaritis, Kafatos, and Petri, 1980
Trypetidae	*Dacus oleae*	–			Margaritis, 1985
	Ceratitis capitata	–			Margaritis, 1985

(continued)

Table 4.1. *(Cont.)*

Higher Level Classification[a]	Species	F[b]	E[b]	D[b]	References
Panorpida[c]					
Sepsidae	*Sepsis* sp.	–			Furneaux and MacKay, 1972
Culicidae	*Aedes aegypti*	–			Furneaux and MacKay, 1972
Muscidae	*Musca domestica*	–			Furneaux and MacKay, 1972
Trichoptera					
Integrepalpia/Leptoceridae	*Mystacides azurea*	–			Furneaux and MacKay, 1972
Annulipalpia/Stenopsychidae	*Parastenopsyche sauteri*	–			Matsuzaki, 1972

[a] Classification according to Kristensen (1984), and Hodges et al. (1983).

[b] F = helicoidal lamellar framework formation; E = lamellar expansion; D = lamellar densification; + = presence of a particular mode; – = absence of a particular mode; "blank" = no data available.

[c] Chorion ultrastructure has been observed in many other groups of nonpanorpid insects (see Furneaux and MacKay, 1972, and Hinton, 1981, for examples). A helicoidal lamellar chorion appears to be missing in these groups as well.

usual opportunity to seek the connection between molecular and morphogenetic levels of evolutionary explanation (Figure 4.7). Over the last 20 years, there has been a concerted effort to characterize the chorion locus in bombycoids, based on classic genetics and analysis of protein and cloned nucleic acid sequences. Six chorion gene families are recognized (Lecanidou, Rodakis, Eickbush, and Kafatos, 1986). The first, and most abundantly expressed, gene families were named A and B. They are present in both Saturniidae and Bombycidae (families within Bombycoidea). In the Bombycidae, two additional gene families, called HcA and HcB, encode chorion proteins of very high cysteine content that assemble to form the thin outer lamellae mentioned earlier. The most recently discovered, and still only partially characterized, gene families are called CA and CB. They appear to be present throughout the Bombycoidea. A reasonable estimate for the total number of genes within these six gene families is 100, based on two-dimensional protein gel analysis and molecular cloning (Regier and Kafatos, 1985).

In all six chorion families, a central domain of the protein sequence is highly conserved relative to the "arm" sequences. This conservation is also reflected at the level of protein secondary structure, which may be necessary for the highly regular chorion ultrastructure (Mazur et al., 1982; Hamodrakas, Etmektzoglou, and Kafatos, 1985). Extensive central domain sequence similarities (symplesiomorphies + synapomorphies) group together the A, HcA, and CA families and separately the B, HcB, and CB families. Similarities between these two groupings are more distant but still indicate homology. Thus, at its deepest level the chorion superfamily consists of two branches, α and β, each containing three gene families.

The six families of the α and β branches encode proteins that assemble to form the predominant helicoidal lamellar chorion substructure. Two additional substructures, filler and the trabecular layer, each contributes no more than 5 percent to the chorion's mass and contains unique subsets of proteins (Regier, Mazur, and Kafatos, 1980; Regier et al., 1982). The two filler-forming proteins, called E1 and E2, are homologous to each other but not to the lamellar-forming components (Regier, 1986). No sequences are yet available for trabecular layer components.

Even though greater than 50 of the approximately 100 chorion genes have been at least partially sequenced in *B. mori* and *A. polyphemus*, our knowledge of the evolutionary history of chorion genes may be incomplete and limited to relatively recent events within a single lineage. To remedy this situation, sequence analysis must be extended to additional taxonomic

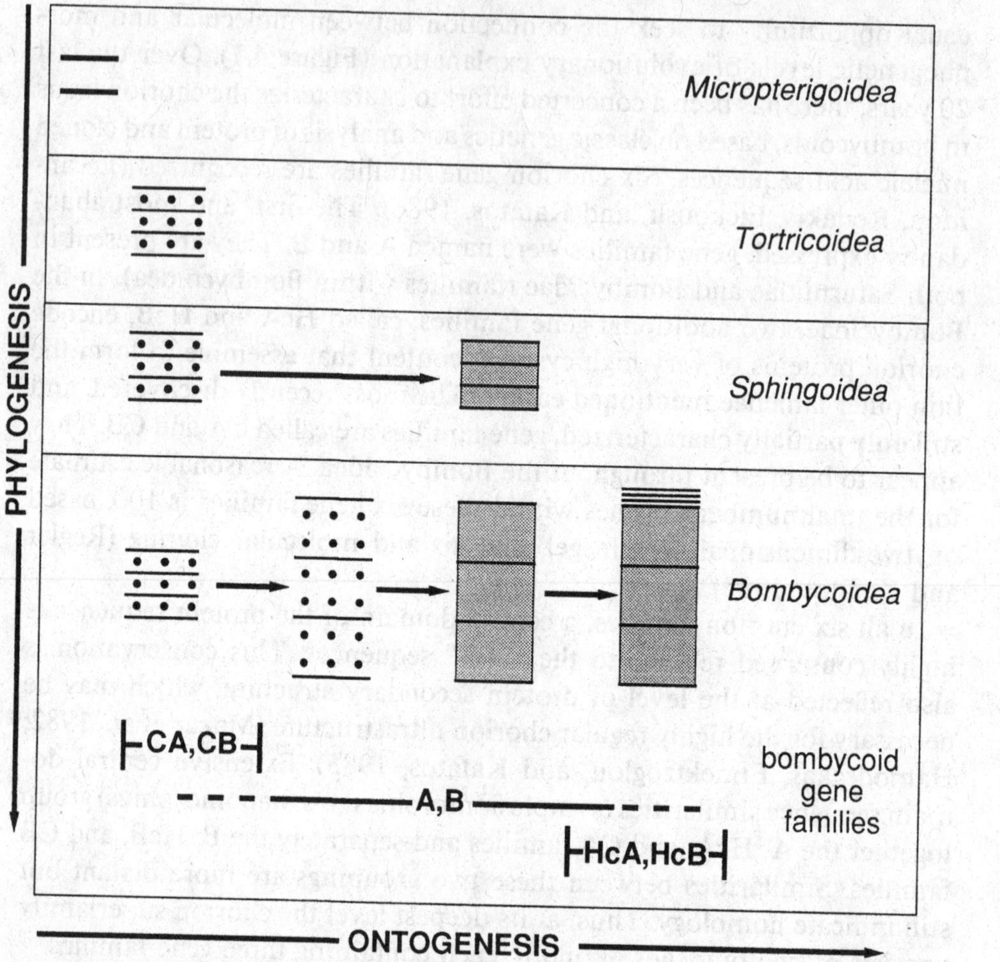

Figure 4.7. Schematic representation of helicoidal lamellogenesis in four superfamilies of moths, placed within a reasonable, but speculative, evolutionary framework. The first column represents a chorion that lacks helicoidal lamellae, as occurs in nonditrysian Lepidoptera such as Micropterigoidea; the second column, presence of a helicoidal lamellar framework, as occurs throughout almost all Ditrysia; the third, occurrence of lamellar expansion; the fourth, occurrence of lamellar densification; the fifth, assembly of distinct surface structures. For Bombycoidea (*Bombyx mori*, in this case), helicoidal lamellar framework formation to surface structure assembly forms a linear ontogenetic sequence that correlates with the temporally restricted expression of specific chorion gene families (identified as CA, CB, A, B, HcA, and HcB at the bottom of the bottom panel).

groups. The effort to obtain sufficient information to assess accurately the multigenic nature of even a single species is not trivial. But, to extend generalizations to the entire order of 112 families is indeed daunting. As a start, any serious attempt at placing major evolutionary events in chorion gene diversification within a phylogenetic context will require characterization of perhaps 10 well-chosen species at a minimum.

Toward this goal we have begun characterization of chorion sequences in the gypsy moth (Noctuoidea, Lymantriidae). Our first step was to generate a cDNA library to total follicular mRNA. Clones specific to choriogenic stages were identified by differential screening. Subsequently, partial sequence analysis was performed on approximately 50 clones. Then these clones were grouped by similarity, and representatives are now being completely sequenced. The conceptual translation for the middle portion (64 percent of total) of one clone (Ld3) that has been completely sequenced is shown in Figure 4.8.

Ld3 can be easily arranged into an orderly set of short repeating peptides, mostly hexapeptides. Within this repeating region, alignment of its central domain (Hamodrakas et al. 1985) with those from the α branch of bombycoids reveals striking similarity in sequence and in length (see Figure 4.8). For example, pc18 and pc292 (both A family sequences from *A. polyphemus*) show 78 and 76 percent sequence identity, respectively. Other members of the bombycoid A family, as well as of the HcA and CA families, have between 63 and 73 percent identity. Ld3 is clearly within the α branch of chorion sequences. Nevertheless, assignment of Ld3 to one of the existing gene families (based on previous analysis of bombycoid sequences) does not appear reasonable at present. Further analysis of Ld3 and other gypsy moth chorion sequences will be necessary using an explicitly phylogenetic approach.

The central domain of bombycoid A family sequences shows strong sixfold periodicities for various residues (see Figure 4.8; Hamodrakas et al., 1985). These have been interpreted to support a secondary structure model in which short β-sheet strands alternate with β-turns to form an antiparallel β-sheet (Hamodrakas et al., 1985). Very similar distributions of residues are present in Ld3, suggesting a similar secondary structure. This is consistent with the similar ultrastructure and morphogenesis of gypsy moth and bombycoid chorions.

Nonchorion sequences

Similarities between bombycoid chorion sequences and nonchorion sequences have been documented in the past (Pau, Weaver, and Edwards-Jones, 1986; Willis, 1989), but these similarities do not necessarily indicate homology. Similar sequences tend to be relatively short and contain repeats with simple amino acid compositions. However, a recent report of

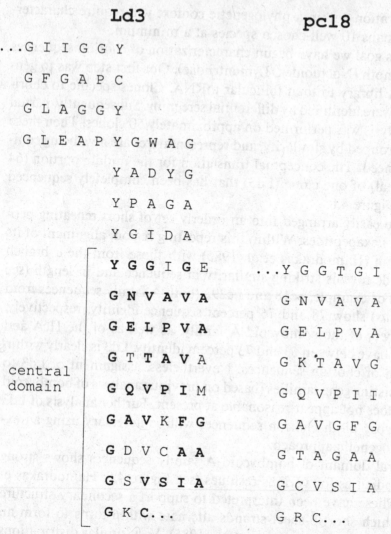

	Ld3	pc18
	...G I I G G Y	
	G F G A P C	
	G L A G G Y	
	G L E A P Y G W A G	
	Y A D Y G	
	Y P A G A	
	Y G I D A	
central domain	**Y G G** I **G** E	...Y G G T G I
	G N V A V A	G N V A V A
	G E L P V A	G E L P V A
	G T **T A V** A	G K T A V G
	G Q V P I M	G Q V P I I
	G A V K **F G**	G A V G F G
	G D V C **A A**	G T A G A A
	G S **V S I A**	G C V S I A
	G K **C**...	G R C...

Figure 4.8. Internally repeating peptide sequences within chorion proteins from two moth superfamilies (see text). The central domains of Ld3 and pc18 (Hamodrakas et al., 1985) are displayed side by side with identical residues at equivalent positions in boldface for Ld3.

an oocyte-specific sequence (called Ec20) from the silk moth *Hyalophora cecropia* shows striking similarity with the B gene family over the entire central domain (Kastern, Watson, and Berry, 1990). Fifty-six out of 61 residues (92 percent) are identical, with no gaps needed for alignment. This degree of similarity meets the minimal criterion for inclusion in the B subfamily that includes the 10a gene sequence. Although this high degree of similarity makes it obvious that we are dealing with homology and not convergence, it could be interpreted as implying that the ancestral Ec20 gene diverged from chorion genes after appearance of the B family. This would imply that there exist nonchorion sequences that are more closely related to some chorion sequences than are other chorion sequences, so that chorion genes do not constitute a monophyletic assemblage. An alternative explanation for these observations is that chorion genes have transferred part of their sequence to nonchorion genes by the process of gene conversion. This explanation may not be unreasonable, as convincing evidence has been presented for large-scale conversion events between distantly related chorion genes in the α and β branches (Iatrou, Tsitilou, and Kafatos, 1984; see also Eickbush, this volume). Clearly, there is a need for explicit phylogenetic analysis of these sequences.

Genes and morphogenesis

Production of the chorion is the terminal stage in the developmental history of follicle cells (Kafatos et al., 1977). In *A. polyphemus,* choriogenesis has been divided into a small number of successive time frames or periods. Each period corresponds to the formation of a unique chorion substructure or to a distinct growth mode. Furthermore, unique subsets of chorion proteins are maximally synthesized during each of these periods and, by correlation, may be considered responsible for the corresponding substructures or growth modes (see bottom panel of Figure 4.7). For example, in *Bombyx mori,* C protein synthesis, which includes products of the CA and CB gene families, occurs only during the early period, when framework formation occurs. Similarly, HcA and HcB protein synthesis occur exclusively at the end of choriogenesis, when the thin outer lamellae are forming. A and B proteins are synthesized predominantly at intermediate periods, when lamellar expansion and densification occur.

Within the Bombycidae and Saturniidae, there is considerable variation in surface structure morphology that correlates with major changes in patterns of protein synthesis. For example, in *A. polyphemus,* completion of lamellar densification is followed by chorion surface sculpturing to form "aeropyle crowns" (see Figure 4.6). These crowns consist of a few additional thin lamellae molded into a characteristic shape by the action of filler, which consists of E1 and E2 proteins. We have generated nucleic acid clones that encode aeropyle crown–specific lamellar and filler-forming proteins. These clones were used as probes to isolate the most similar cloned genes from *H. cecropia,* a related silk moth lacking aeropyle crowns (Hatzopoulos and Regier, 1987). Using these cloned sequences, we have quantified time-dependent and cell-specific changes in specific mRNA populations. E1 and E2 RNAs are present in *A. polyphemus* at levels that are 35–40 times that in *H. cecropia,* even though the genes are expressed at similar times and are present in similar copy numbers. Elevated levels of E RNAs have resulted in elevated levels of the corresponding proteins and of the filler substructure and, ultimately, in the ability to form aeropyle crowns. By contrast, the aeropyle crown lamellar–forming gene that we have isolated from *A. polyphemus* is expressed significantly later in that species than the corresponding gene in *H. cecropia* (Regier, Cole, and Leclerc, in preparation). Expression of lamellar genes at the end of choriogenesis is essential for aeropyle crown formation, as only lamellae in the process of assembly can be molded by filler. These two examples illustrate the importance of regulatory alterations in gene expression in generating morphological evolution of the chorion.

Two major morphogenetic events distinguish choriogenesis in *Manduca sexta* (Sphingoidea) from that in *B. mori* (Bombycoidea), in addition to the terminal events (Regier and Vlahos, 1988). The first represents a heterochronic change, namely, that new lamellar formation occurs throughout choriogenesis rather than largely at the beginning. This morphogenetic difference correlates with the extended period of C-size protein synthesis in *M. sexta.* The second morphogenetic difference is that lamellae do not expand, although they do densify, correlating with the underproduction of both A- and B-size proteins in this species. As a consequence, the chorion of *M. sexta* is thinner than the bombycoid chorions we have studied, measuring 9 μm in *M. sexta* versus 27 μm in *A. polyphemus.*

It is a reasonable hypothesis that the diversity in modes of chorion morphogenesis indicated in Table 4.1 and Figure 4.7 results, at least in part, from structural and regulatory changes in the expression of chorion genes such as have been described. Further analyses of other groups are needed to strengthen this hypothesis.

Conclusion

Over the last few decades, the revolutions that have occurred in systematic theory and in molecular biology have encompassed an increasing number of disciplines. Specifically, comparative biology has much to gain from the rigorous evolutionary framework of phylogenetic systematics applied to developmentally significant genes. Within the Lepidoptera, as for all other insect orders, progress is currently limited by insufficient knowledge of phylogenetic relationships. Gene sequences offer an extensive evolutionary record that, when properly interpreted, should enable robust species tree construction. Given such a tree, developmental pathways, such as early embryogenesis or choriogenesis, can be mapped in multiple species onto an established phylogeny. Subsequently, systematic methods can be used to infer pathways and mechanisms of evolutionary change.

Acknowledgments

We thank Dr. Dorothy Pashley for permitting us to cite studies prior to publication and Dr. Marc Epstein for drawing Figure 4.7. The writing of this essay and some of the research cited herein have been supported by the U.S. Department of Agriculture and the Center for Agricultural Biotechnology, University of Maryland Biotechnology Institute.

5

A summary of lepidopteran embryogenesis and experimental embryology

LISA M. NAGY

Lepidopterans have been the object of both genetic and experimental analysis since the turn of the century (see Table 5.1). In spite of this long history of research, very little is known about the establishment of the embryonic axes during early lepidopteran development. Recent progress in the field of insect embryology has been dominated by work on the dipteran *Drosophila melanogaster,* due to the success of combining experimental embryological manipulations with developmental genetics and molecular biology (see Sander, Gutzeit, and Jäckle, 1985, Akam, 1987, and Ingham, 1988, for reviews). Consequently, more is known about the establishment of embryonic axes in *D. melanogaster* than in any other organism. Many of the genes involved in the early determination events of *D. melanogaster* have been cloned and their putative homologues have subsequently been isolated from a wide range of other metazoans, from vertebrates to cnidarians (see Kessel and Gruss, 1990; Murtha, Leckman, and Ruddle, 1991; McGinnis and Krumlauf, 1992). Thus the early development of the Lepidoptera and other insects formerly intractable to molecular analysis can now be studied by taking advantage of this widespread sequence homology.

In this chapter I review the literature on lepidopteran development, with an emphasis on integrating the recent molecular data with the older descriptive and experimental work. I begin with a summary of the morphological features of lepidopteran embryos as compared to other insects. I then compare what is known about development in *D. melanogaster* to what is known in the Lepidoptera, with an eye toward understanding the conservation of the mechanisms of segmentation and embryonic axis formation.

Table 5.1. *Literature on early lepidopteran development*

Species	Duration of embryogenesis hr (temp.)	Egg size l × w × d	Nuclear migration[a]	Reference
Zeugloptera				
Micropterygidae				
Neomicropteryx nipponensi	336 (23°)	0.32 × 0.35	s	Kobayashi and Ando, 1982
Dacnonypha				
Eriocraniidae				
Eriocrania sp.	168 (15°–20°)	0.48 × 0.23	s	Kobayashi and Ando, 1987
Monotrysia				
Hepialidae				
Endoclyta signifer	288 (25°–30°)	0.62 × 0.55	a	Ando and Tanaka, 1976, 1980
Endoclyta excrescens		0.75 × 0.65	a	Ando and Tanaka, 1976, 1980
Ditrysia				
Tineidae				
Tineola bisselliella	180			Lüscher, 1944
Psychidae				
Solenobia triquetrella				Lautenschlager, 1932
Gelechiidae				
Pectinophora gossypiella				Berg and Gassner, 1978

Tortricidae				
Carpocaspa pomonella				Weismann, 1935
Choristoneura fumiferana	150	0.7 × 0.5 × 0.2		Stairs, 1960 Reed and Day, 1966; Anderson and Wood, 1968
Epiphyas postvittana		1.0 × 0.8		Huie, 1918
Eudemis naevana				Richardson et al., 1982
Laspeyresia pomonella				Bassand, 1965
Zeiraphera griseana				
Limacodidae				
Cochlidion limacodes				Christensen, 1943
Pyralidae				
Ancylolomia japonica				Tanaka, 1970
Anagasta (Ephestia) kuehniella	144		a	Sehl, 1931; Maschlanka, 1938
Chilo suppresalis	150	1.0 × 0.60		Okada, 1960
Galleria mellonella				Beck, 1960
Papilionidae				
Acraea horta				Balinsky, 1986
Byasa (Atrophaneua) alcinous				Tanaka, 1987
Leuhdorfia japonica				Tanaka, 1987
Parnassius glacialis				Tanaka, 1987
Nymphalidae				
Euvanessa antiopa				Woodworth, 1889
Pieridae				
Pieris rapae				Eastham, 1927

Table 5.1. *(cont.)*

Species	Duration of embryogenesis hr (temp.)	Egg size l × w × d	Nuclear migration[a]	Reference
Geometridae				
Chesias legatella		0.83 × 0.51		Wall, 1973
Chesias rufata	216 (20°)	0.62 × 0.40		Wall, 1973
Operophtera brumata				Gaumont, 1950
Lasiocampidae				
Lasiocampa sp.			e	Schwangart, 1905
Bombycidae				
Bombyx mori	10 days	1.2 × 0.80	a	[b]Toyama, 1902; Krause and Krause, 1964, 1965
Saturnidae				
Antheraea pernyi	240	0.35 × 3.0	a	Saito, 1934, 1937
Hyalophora cecropia	252			Riddiford, 1970
Noctuidae				
Catocala nupta				Hirschler, 1905
Heliothis zeae	84	0.6 spherical		Presser and Rutschky, 1956
Mamestra configurata	130	0.45 spherical	b	Rempel, 1951
Prodenia eridania		0.45 × 0.55	v	Gross and Howland 1940

Sphingidae				
Manduca sexta	100		a	Dorn et al., 1987; Dow et al., 1988; Broadie et al., 1990
Arctiidae				
Amata fortunei	136	0.75 × 0.7	a	Tanaka, 1985
Diacrisia virginica		0.75 spherical	b	Johannsen, 1929
Orgyia antigua				Christensen, 1943

Classification based mainly on Borror et al. (1981). l × w × d = length × width × depth in mm.

[a]Nuclear migration describes the manner in which the cleavage energids migrate to the periphery: s, simultaneous arrival of the energids at the periphery; a, anterior arrival; b, both poles; p, posterior arrival; e, equatorial arrival; v, ventral arrival. Hatching time is at 25°.

[b]Papers devoted to development of the silk moth are too numerous to list here; see Johannsen and Butt (1941), Tazima (1964), and Miya (1985a) for recent reviews. See also Johannsen and Butt (1941) for an extensive bibliography of lepidopteran embryology prior to 1940.

Morphological features

A common approach to comparing the way in which different insects pattern their early embryos is encapsulated in the categories of long, intermediate, and short germ embryos. Only a portion of the cells comprising the insect cellular blastoderm are destined to form the insect embryo; these cells are called the germ anlage (also sometimes referred to as the germ rudiment, or germ disc). The remaining cells of the blastoderm will form the extraembryonic tissues. The size of the germ anlage in relation to the length of the egg differs in different types of insects. Insects with a very small anlage are termed short germ insects, those with anlage that occupy a third to one-half the length of the egg are called intermediate germ, and insects with anlage that extend over the entire length of the egg are termed long germ (Figure 5.1; Krause, 1939; Sander, 1976).

The germ anlage develops into the germ band, a stage that is conserved throughout all insects and is similar in some ways to the vertebrate gastrula (Sander, 1983). Unlike the germ anlage, which is a single layer of visually indistinguishable cells, the germ band consists of mesodermal and ectodermal layers that have differentiated into a head and three thoracic and eight to eleven abdominal segments. The morphological transition from the germ anlage to the germ band is ostensibly different in short, intermediate, and long germ insects. In long germ insects, all of the future segments are already represented in the fate map of the blastoderm, and segmentation occurs by a nearly simultaneous subdivision of the germ anlage. In short germ insects, the future abdomen is represented by a disproportionately small region of the fate map of the blastoderm and undergoes substantial cell division. This cell division is thought to occur in a posterior zone of proliferation that is held to be responsible for the sequential addition of more posterior segments and the anteroposterior extension of the germ anlage that results in a fully segmented germ band (Sander, 1976, 1983), although for most insects the actual mechanics by which growth and elongation occur is not well described.

Long germ development also correlates with meroistic oogenesis, where the oocyte remains connected with its sister cells (nurse cells). Through a connection maintained at the anterior of the oocyte, the nurse cells supply the oocyte with cellular components and large amounts of RNA, including the maternal coordinate gene products utilized at the top of the segmentation gene hierarchy (discussed later), thereby providing the oocyte with its initial anteroposterior polarity (Telfer, 1975; King and Büning, 1985; St. Johnston and Nüsslein-Volhard, 1992). In contrast, most (but not all)

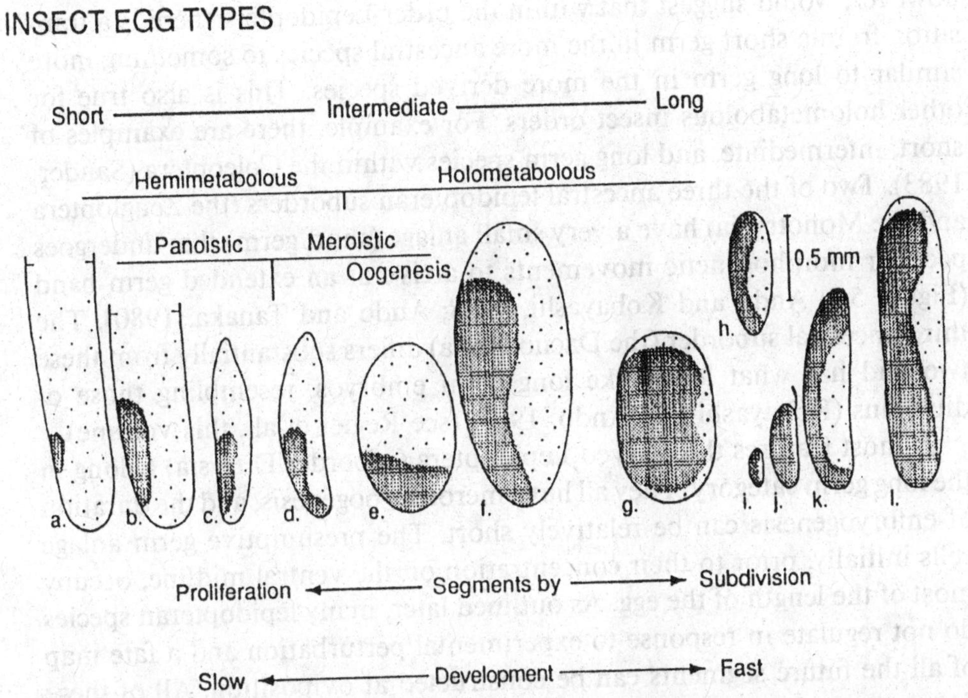

Figure 5.1. Summary table of representative categories of insect egg types and correlated characters (adapted from Sander, 1983). The insect eggs are drawn to scale from a lateral view, with anterior to the top. The germ anlage are shaded. Orthoptera: a. *Oecanthus pellucens;* B. *Acheta domesticus.* Odonata: c. *Platycnemis pennipes.* Homoptera: d. *Euscelis plebejus.* Coleoptera: e. *Atrachya menetriesi;* f. *Leptinotarsa decemineata;* h. *Bruchidius obtectus.* Lepidoptera: g. *Bombyx mori.* Diptera: i. *Smittia* sp.; j. *Drosophila melanogaster;* k. *Calliphora erythrocephala.* Hymenoptera: l. *Apis mellifera.*

short germ embryos have panoistic oogenesis, where each oocyte develops independently in the absence of any nurse cells (King and Büning, 1985; Stys and Bilinski, 1990).

Long germ insects also develop much more rapidly than short germ insects and are unable to regulate in response to experimental perturbation. For example, if the embryos of the short germ beetle *Atrachya menetriesi* are ligated into two separate egg fragments at 35 percent egg length during the blastoderm stage, two complete embryos, one in each fragment, can form (Miya and Kobayashi, 1974). On the other hand, if long germ embryos are ligated at the blastoderm stage, only partial embryos form in each fragment (Sander, 1976, 1983).

Historically, the Lepidoptera have been grouped with the intermediate germ insects or professed "unclassifiable" (Krause and Krause, 1964; Sander, 1983). The published accounts of lepidopteran embryogenesis, however, would suggest that within the order Lepidoptera there is a transition from a short germ in the more ancestral species to something more similar to long germ in the more derived species. This is also true for other holometabolous insect orders. For example, there are examples of short, intermediate, and long germ species within the Coleoptera (Sander, 1983). Two of the three ancestral lepidopteran suborders (the Zeugloptera and the Monotrysia) have a very small anlage (short germ) that undergoes peculiar morphogenetic movements to arrive at an extended germ band (Figure 5.2; Ando and Kobayashi, 1978; Ando and Tanaka, 1980). The third ancestral suborder (the Dacnonypha) differs substantially from these two and has what looks like long germ embryos, resembling those of dipterans (Kobayashi and Ando, 1987) (see Regier et al., this volume).

In most features the derived Lepidoptera (suborder Ditrysia) belong in the long germ category. They all have meroistic oogenesis, and the duration of embryogenesis can be relatively short. The presumptive germ anlage cells initially, prior to their concentration on the ventral midline, occupy most of the length of the egg. As outlined later, many lepidopteran species do not regulate in response to experimental perturbation and a fate map of all the future segments can be constructed at oviposition. All of these features correlate with long germ embryogenesis. Yet the morphological steps through which the germ anlage passes to form a fully segmented embryo look more like what is expected for a short or intermediate germ embryo (Figure 5.3). The shape of the germ anlage at the onset of gastrulation resembles a short germ insect: two head lobes followed by a non-segmented terminus, from which the thorax and abdomen will develop. During the ensuing shape changes the gnathal and thoracic segments are the first segments visibly delineated posterior to the head lobes, followed by a sequential appearance of the abdominal segments. These shape changes are deceiving, however. The early germ anlage is very wide (occupying roughly 75 percent of the egg diameter) before the onset of gastrulation (Krause and Krause, 1964; see Figure 5.3). Although the abdomen undergoes a process of extension that appears analogous to that in short or intermediate germ insects, there is no visible posterior zone of proliferation: Mitotic figures are equally abundant in both the anterior and posterior of the abdomen (*Bombyx mori,* Krause and Krause, 1964; *Manduca sexta,* Nagy, personal observation). In fact, the abdomen can extend in the presence of inhibitors of cell division (*M. sexta,* Dorn, personal communication). However, elongation is halted in the presence of cytochalasin B, an inhibitor of actin polymerization, implicating cell rear-

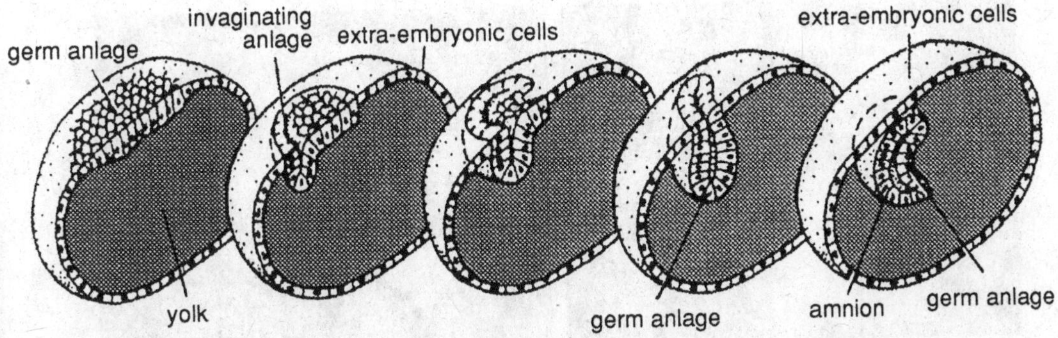

Figure 5.2. Scheme of the unusual early embryogenesis of a primitive lepidopteran, *Neomicropteryx nipponensis.* Ventroposterior view of a half egg, chorion omitted. Notice that the presumptive germ anlage completely invaginates into the yolk and separates from the remaining blastoderm cells prior to elongation and segmentation. (From Kobayashi and Ando, 1981.)

rangements in the extension of the germ band. The overall shape change of the *B. mori* embryo looks remarkably like the overall change in shape of the dorsal axial mesoderm of the *Xenopus laevis* embryo, which elongates by convergent extension, whereby more lateral cells interdigitate, or converge with more dorsal cells (Keller, Danilchik, Gimlich, and Shih, 1985a, 1985b; Wilson and Keller, 1991). It is easy to understand why the Lepidoptera have been difficult to categorize as long, intermediate, or short germ insects.

It is possible that long germ development evolved more than once within the insects. Both the Diptera and members of the Lepidoptera have evolved toward faster embryogenesis, changing many of the slower, proliferative features of ancestral insects into the early establishment of pattern. In *D. melanogaster* this resulted in a cascade of transcriptional regulatory interactions that occur prior to gastrulation, as prompted by maternally supplied products (Figure 5.4; reviewed in Anderson, 1989; St. Johnston and Nüsslein-Volhard, 1992). It would be useful to know if this cascade of molecular events is conserved in the Lepidoptera. In the remainder of this chapter I go through the major categories of the *D. melanogaster* segmentation genes and compare them to what is known about pattern formation in the Lepidoptera.

Briefly, in *D. melanogaster* both the anteroposterior and dorsoventral axes are organized through the coordinated activity of four groups of maternal genes: the anterior, posterior, terminal, and dorsoventral group genes (reviewed in Nüsslein-Volhard, Frohnhöfer, and Lehman, 1987; St.

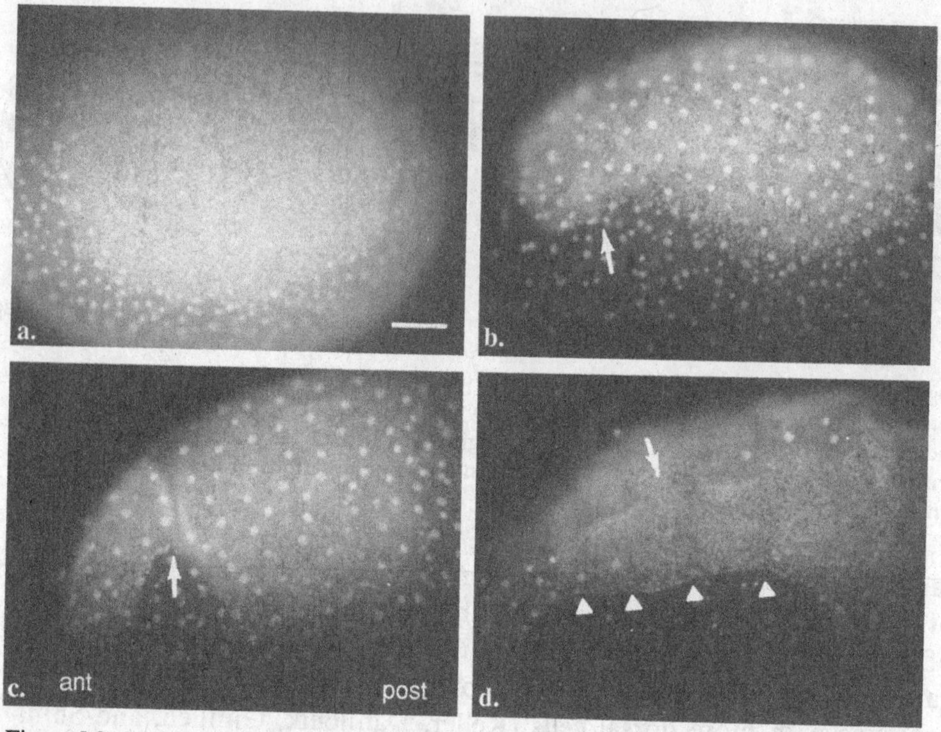

Figure 5.3. Morphological transition from germ anlage to extended germ band in *Bombyx mori*. *B. mori* embryos from the nondiapausing strain *pnd* stained with the nuclear specific dye Hoechst 33258 (Nagy, Riddiford, and Kiguchi, in press). Anterior is to the left; ventral is up. (a) 20 hr after egg laying (AEL); the germ anlage is nearly condensed on the ventral surface and has not yet begun gastrulation. The large, polyploid nuclei visible on the anterior, posterior, and dorsal surfaces are the nuclei of the serosal cells, which have just begun to migrate over the germ anlage. (b) 24 hr AEL; the germ anlage has begun to elongate in the anteroposterior direction, and the head lobes are forming at the anterior end (*arrow*); the serosa cells now completely envelope the embryo. (c) 27 hr AEL; the head lobes are separated from the presumptive posterior segments by a visible furrow (*arrow*). (d) 36 hr AEL; the serosa cells have been partially removed to reveal the extending germ band beneath them. The developing head segments have now migrated over the anterior pole and are no longer visible; the posteriormost abdominal segments are still forming over the posterior pole, the ventral furrow is visible down the midline (*arrow*). Visible furrows separate the thoracic and anterior abdominal segments (*arrowheads*). (ant = anterior; post = posterior; scale bar = 150 μm.)

Johnston and Nüsslein-Volhard, 1992). The anteroposterior axis is specified by the largely independent action of the anterior, posterior, and terminal group genes (see Figure 5.4), whereas the dorsoventral group genes act separately. Mutations in the anterior group result in loss or reduction of head and thoracic structures, mutations in the posterior group cause abdominal defects, and terminal class mutations affect the unsegmented ends of the embryo, the acron, and telson. Mutations in the dorsoventral group genes lead to dorsalized or ventralized embryos. The activities of these four groups of genes provide the egg with a global and initially crude pattern of maternal positional information. This maternal information lies at the top of a hierarchy of genes that specifies the character of individual segments along both axes.

Along the anteroposterior axis the hierarchy consists of a series of interactions among transcription factors that occurs during the early syncytial stage of embryogenesis. The maternal coordinate genes regulate the gap genes, which in turn regulate the pair-rule genes. Pair-rule genes then modulate expression of segment-polarity genes and establish the segment-specific patterns of expression of the homeotic genes (reviewed in Akam, 1987, and Ingham, 1988). Along the dorsoventral axis, the activity of the maternal dorsoventral genes results in the establishment of a nuclear gradient of the *dorsal* protein. The *dorsal* protein gradient is thought to provide positional information through the differential activation of the zygotic genes in the ventral and dorsal cells of the blastoderm (reviewed in Govind and Steward, 1991).

Experimental and molecular features

Morphogenetic centers at the egg poles and the independence of the terminalia

The cascade of transcriptional regulatory events that establishes the anteroposterior axis in *D. melanogaster* begins with the activity of maternally supplied morphogens, localized to the anterior and posterior poles. The presence of localized polar morphogens was predicted from the results of ligation and cytoplasmic removal experiments and subsequently confirmed by genetic and molecular approaches (Frohnhöfer, Lehmann, and Nüsslein-Volhard, 1986; Driever and Nüsslein-Volhard, 1988a, 1988b; Lehmann and Nüsslein-Volhard, 1991). The activity of morphogens at either pole has been difficult to demonstrate in lepidopterans. Experiments designed to disturb the determination of the anteroposterior axis have failed, thus leading to the hypothesis that determination occurs very early in the lepidopteran egg (Kuwana and Takami, 1968). As will be seen in subsequent discussion, however, this view is not completely compelling.

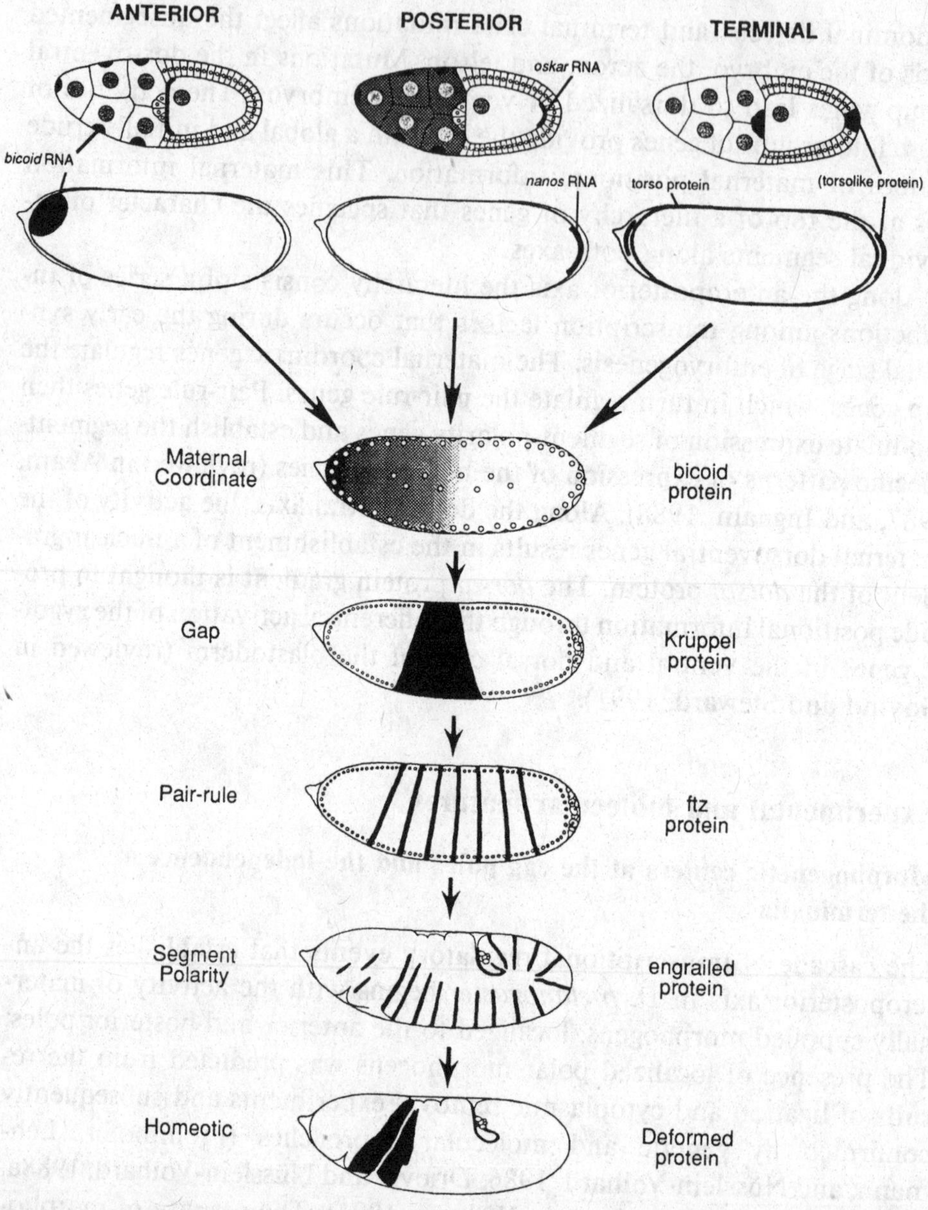

Figure 5.4. Summary of the segmentation gene hierarchy in *Drosophila melanogaster*. The three independently acting systems, the anterior, posterior, and terminal group genes, are depicted at the top. The activity of these three groups of genes results in the localized deposition of products of the maternal coordinate genes in the newly laid egg, exemplified here by the graded distribution of the *bicoid* protein. The maternal coordinate genes regulate the transcription of the next tier of the hierarchy, resulting in localized expression of the gap genes, shown here by the distribution of the *Krüppel* protein. The gap genes

Centrifugation of the early embryo in several dipteran species results in double abdomens, in which anterior segments are replaced by abdominal segments with reversed polarity or, more rarely, in double heads, which show head duplications in the absence of posterior segments (see Kalthoff, 1983). Centrifugation does not disturb the anteroposterior axis in *B. mori* (Sakaguchi, 1952; Miya, 1973). Segments nearest the centripetal pole are often disturbed, perhaps due to mechanical damage, but there is no transfer of positional information, whereby a head or a tail develops in an unexpected site. However, caution must be taken in interpreting these results as evidence for the absence of anterior or posterior morphogens in lepidopterans. Despite ample evidence in *D. melanogaster* for the presence of anterior and posterior morphogens, double tails and especially double heads are difficult to generate by centrifugation (Sander, 1988). Double abdomens are created only if the embryo is centrifuged within 15 min after deposition and if the orientation of the embryo in the centrifuge is reversed once or twice. Even under these conditions, a double cephalon formed in only a single case (Sander, 1983).

The results of cauterization and irradiation experiments in several species of lepidopterans could also support the argument for an earlier determination of segmental specification in the Lepidoptera than in the Diptera. No visible defects were seen after cauterizing 20 percent of the posterior of the egg in *Ephestia kuehniella* (Maschlanka, 1938), suggesting that at this time the posterior pole does not harbor a morphogen that can be destroyed by cauterization. In *B. mori* the posterior pole and the regions immediately anterior to it fate map to amnioserosa (Katsuki, Murakami, and Watanabe, 1980; Myohara and Kiguchi, 1990). If the fate map of *E. kuehniella* is similar and the posterior pole is already destined for an extraembryonic fate, cauterizing the posterior pole should not result in any segmental defects. Similarly, by irradiating small spots on the egg surface, Myohara and Kiguchi (1990) were able to construct a segment-specific fate map for *B. mori* within 2 hr of egg deposition (prior to the

---

are expressed in broad domains that include several segment primordia and begin the subdivision of the embryo. The gap genes regulate the transcription of one another, as well as members of all three of the subsequent tiers, the pair-rule, segment-polarity, and homeotic genes, represented here by the products of the *fushi-tarazu* (*ftz*), *engrailed,* and *Deformed* genes. The pair-rule genes are activated in alternating parasegments and transcriptionally regulate one another, as well as genes downstream in the segmentation hierarchy. The segment-polarity genes are expressed in a segmentally repeating pattern and organize and maintain the pattern and polarity of the segments. And finally, the homeotic genes function to specify segment identity.

first embryonic cleavage). Lüscher (1944) also induced disruption of individual segments by UV-irradiating transverse strips of the lepidopteran *Tineola bisselliella* egg prior to the blastoderm stage.

The ability to construct a fate map including all the abdominal segments just after oviposition is consistent with viewing *B. mori* as a long or intermediate germ insect. In short germ insects the abdomen occupies a relatively small portion of the fate map, and the fate of each individual abdominal segment cannot be localized to a single transverse strip by irradiation until germ band elongation (Sander, 1976). In fact, the capacity to map segment-specific fates with irradiation at oviposition is unusual. A similar segment-specific fate map cannot be generated in *D. melanogaster* and other insects (both long and short germ) until the stage when the nuclei have reached the surface (nuclear division cycle 10) (Sander, 1976; Lohs-Schardin, Cremer, and Nüsslein-Volhard, 1979). One explanation for these results is that a UV-sensitive product necessary for segment specification is localized to the egg cortex during oogenesis, providing an even earlier determination of segment-specific fate in *B. mori* than in the Diptera. However, the methods employed are unreliable for determining the exact timing of the restriction of cell fates in embryogenesis. The early UV irradiations could be damaging the cortex and preventing cellularization at the injured site (Myohara, personal communication) or disrupting the capacity for internuclear or intercellular communication (Sander, 1976). Thus, although the UV irradiation was inflicted early, the developmental process affected by the irradiation could be occurring later.

Histological evidence has also been used to support the hypothesis of an early cortical localization of determinants in *B. mori*. Pyronin-positive granules, which correspond to large concentric whorls of rough endoplasmic reticulum (rER) (Miya, 1978), are localized to the lateral and ventral surfaces of the *B. mori* egg at oviposition (the presumptive germ anlage region) and are absent from the anterior, posterior, and dorsal surfaces (the presumptive extraembryonic region) (Kobayashi and Miya, 1987). Cytological differences along the embryonic axes, other than the specialized germinal plasm, have not been observed in other insect eggs (see Nüsslein-Volhard et al., 1987). This unequal distribution of rER could function in establishing a differential rate or type of protein synthesis between embryonic and extraembryonic regions of the cortex. Similarly, it is known that the cytoarchitecture of the cortex of the newly laid *Hy-*

alophora cecropia egg harbors poly(A)+ mRNA, which is thought to be potentially active in the programming of the invading nuclei to region- or segment-specific fates (Jarnot et al., 1988). However, regional differences in the cortex were not assayed for. Unless regions of the cortex can be transplanted elsewhere and shown to develop autonomously, these observations are not revealing about the timing of determination of segment character.

Two pieces of evidence suggest that the anteroposterior axis is not fixed at oviposition, but undergoes subsequent modification. When large regions of the *B. mori* egg near the posterior pole were cauterized, the number of segments affected decreased with time (Takami, 1942). This result could indicate the disruption of the final stages of activity of a posterior morphogen acting on more anterior regions. Alternatively, it could also reflect the contraction of the germ anlage toward the ventral midline. Immediately following cellularization, the *B. mori* germ anlage extends almost from the anterior to the posterior pole. Over the next several hours it contracts toward the ventral midline (see Figure 5.5; Krause and Krause, 1964). Thus an early posterior cauterization would damage more anterior cells. Later irradiations would affect mostly the amnioserosa, as is thought to occur in the *E. kuehniella* experiments. The second piece of evidence for a possible interaction between the posterior pole and more anterior structures comes from cytoplasmic removal experiments in *B. mori.* If cytoplasm was removed from the posterior pole of the *B. mori* egg shortly after oviposition, in a small percentage of the survivors the terminal segment or telson was preserved, whereas anterior segments, distant from the site of removal, were missing or altered (Nagy, Sugiyama, and Myohara, personal observations). If the egg just after oviposition were a mosaic of localized determinants, removal of posterior cytoplasm would be expected to affect only the most terminal features of the fate map. The preservation of posterior structures in these embryos also suggests that the determination of the terminalia is independent from the other abdominal segments, as seen in *D. melanogaster* (reviewed in Nüsslein-Volhard et al., 1987).

Gap genes and the gap phenomenon

The gap phenomenon, first witnessed in ligation experiments (whereby a gap is seen in the pattern of segments formed between the resultant halves

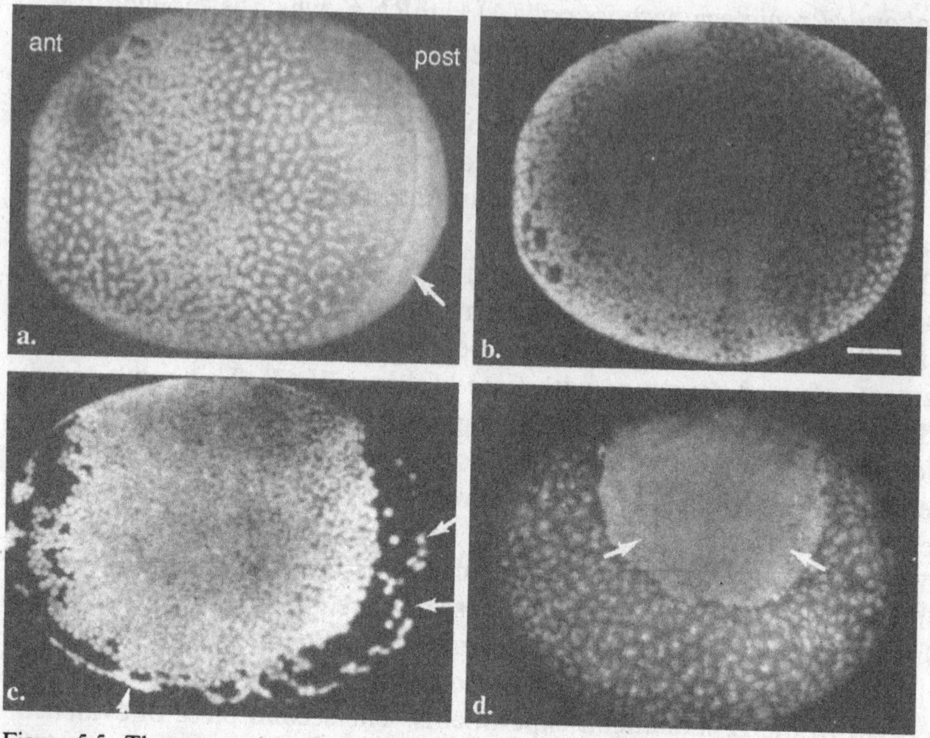

Figure 5.5. The contraction of the *Bombyx mori* germ anlage toward the ventral midline. *B. mori* embryos from the nondiapausing strain *pnd* stained with the nuclear specific dye chromomycin (Nagy, Riddiford, and Kiguchi, in press). a. 12 hr AEL; blastoderm formation; the nuclei in the posteriormost region of the embryo have not yet reached the surface (*arrow*). b. 13 hr AEL; completion of the blastoderm. c. 17 hr AEL; the presumptive anlage cells have aggregated to form the germ anlage; the larger cells remaining on the periphery will form the serosa or extra-embryonic membrane (*arrows*). d. 22 hr AEL; the germ anlage has contracted toward the ventral midline; the serosal cells are beginning to migrate over the germ anlage (*arrows*). (ant = anterior; post = posterior; scale bar = 170 μm.)

of the ligation), showed the need for interactions between the opposite egg poles for the complete specification of segments (Sander, 1960). The gap phenomenon is found in virtually every holometabolous insect tested (Sander, 1976). Unfortunately, ligation experiments have not been successful in lepidopterans due to the formidable rigidity of the chorion. A similar phenomenon, however, the failure of several adjacent segments to form, was found after the irradiation of transverse strips of preblastoderm *E. kuehniella* embryos by Maschlanka (1938). Either all or none of the three gnathal segments formed, suggesting the disruption of some gaplike determinant responsible for all the gnathal segments. Sander (1976), however, suggested that these results could be explained if the gnathal segments occupied a very small portion of the fate map at the time of irradiation, so that they could not be irradiated individually. The all or nothing response of the gnathal segments was not found in similar irradiation of *T. bisselliella* (Lüscher, 1944) and *B. mori* (Myohara, personal communication).

The ligation results, at least in *D. melanogaster,* can now be partially explained by the disruption of the activation of the gap genes by the maternally provided anterior and posterior morphogens. In spite of the absence of evidence for either anterior or posterior morphogens or any experimental evidence for a gap phenomenon in the Lepidoptera, sequences with significant homology to the DNA-binding regions of the *D. melanogaster* gap genes *hunchback* and *Krüppel* have recently been isolated from *M. sexta* (Kraft and Jäckle, in press). As gap genes in *D. melanogaster,* both *hunchback* and *Krüppel* encode putative transcription factors, are activated in response to maternal cues during the late syncytial cleavage stages, and are expressed in large, slightly overlapping domains in the early syncytial blastoderm (see Figure 5.4; Hülskamp and Tautz, 1991). These domains are thought to be refined by transcriptional regulatory feedback resulting from the localized diffusion of the gap gene products between nuclei in the syncytial blastoderm stage. The final gap gene expression pattern is crucial for establishing the domains of expression of the genes downstream in the segmentation hierarchy, the pair-rule, segment-polarity, and homeotic genes (see Hülskamp and Tautz, 1991). Interestingly, some lepidopteran embryos do not have a syncytial blastoderm stage, where all the nuclei reside at the periphery for several cell divisions prior to cellularization. In *B. mori,* for instance, cleavage energids arrive at the anterior egg surface much sooner than they arrive at the posterior pole (Keino and Takesue, 1982; Kobayashi and Miya, 1987). The energids at the anterior initiate cellularization shortly after their arrival at the cortex, whereas the posterior energids are still in the central yolk mass. If the *M. sexta* gap gene homologues are expressed similarly to their *D. melanogaster* counterparts, it will be interesting to analyze how their initial expression patterns become refined in the absence of a syncytial blastoderm.

Pair-rule genes

Downstream from the gap genes in the *D. melanogaster* segmentation gene cascade are the pair-rule genes, which are expressed in every other segment primordium on the *D. melanogaster* blastoderm (see Akam, 1987, and Ingham, 1988, for review). The domains of expression of the pair-rule genes do not correspond to any morphological feature of the *D. melanogaster* embryo. That the blastoderm would be subdivided into double-segment iterations prior to segmental iterations was first suggested on the basis of the phenotypes of developmental mutants in which every other segment primordium was deleted (Nüsslein-Volhard and Wieschaus, 1980; Jürgens, Wieschaus, Nüsslein-Volhard, and Kluding, 1984; Nüsslein-Volhard, Wieschaus, and Kluding, 1984).

To date, homologues to the *D. melanogaster* pair-rule genes *hairy, runt,* and *sloppy-paired* have been isolated from *Manduca sexta,* but their expression patterns have not yet been analyzed (Kraft and Jäckle, in press). There is, however, circumstantial evidence for the existence of pair-rule genes in *M. sexta:* A cell surface antigen *(TN-1)* is expressed in a pair-rule pattern during gastrulation (Carr and Taghert, 1989). It is unlikely that the *M. sexta TN-1* antigen is the product of a pair-rule gene itself, as many genes in *D. melanogaster* are expressed initially in a pair-rule pattern but are not themselves pair-rule genes in the segmentation hierarchy. In most instances, this type of pair-rule expression is thought to reflect genes regulated by pair-rule genes (Gauger et al., 1987; Doe, Chu-LaGraff, Wright, and Scott, 1991). In addition, the *Manduca TN-1* antigen is expressed exclusively in mesodermal cells (Carr and Taghert, 1989).

The onset of expression of the *Manduca TN-1* antigen is particularly interesting and suggests that the *M. sexta* embryo undergoes a progressive subdivision of the germ band analogous to the segmentation process that occurs in the *D. melanogaster* embryo. The elongating germ band is first split up into units of four-segment periodicity; then these are broken into two-segment units; finally expression is detected in a portion of every segment (Carr and Taghert, 1989). This pattern of refinement of expression is not common to the *D. melanogaster* pair-rule genes or the later neurogenic genes thought to be controlled by the pair-rule genes, which in most cases are initially expressed ubiquitously, then repressed in the interstripe regions (Carroll and Vavra, 1989). The onset of the *M. sexta TN-1* expression pattern is, however, more similar to the onset of expression of the segment-polarity gene *engrailed* in *D. melanogaster* (Carr and Taghert, 1989).

Interestingly, homologues to the homeobox-containing pair-rule gene *fushi tarazu* (*ftz*) have not yet been uncovered outside the Diptera in extensive homeobox homology searches (see Akam and Dawes, 1992). In

D. melanogaster, the *ftz* transcription unit lies between the *Antennapedia* (*Antp*) and *Sex combs reduced* (*Scr*) genes in the Antennapedia complex (Weiner, Scott, and Kaufman, 1984). A *Tribolium castaneum* homeobox-containing cDNA maps to a similar location between the *T. castaneum* homologues of the *Antp* and *Scr* genes. Based on its sequence, this *T. castaneum* cDNA is not an obvious homologue of the *D. melanogaster* *ftz* gene, yet its expression pattern is somewhat similar to the latter, with both segmental and neurogenic expression (Brown, Hilgenfeld, and Denell, personal communication). Surprisingly, a deletion of the *T. castaneum* homeotic complex does not result in a pair-rule lethal phenotype; the resultant embryos contain the proper number of segments (Stuart, Brown, Beeman, and Denell, 1991). *Bombyx mori* could easily provide a third point of comparison on the chromosomal positioning and function of the *ftz* gene. The *B. mori Antp* and *Scr* homologues also map closely to one another within the *B. mori* homeotic gene complex (see Ueno, Nagata, and Suzuki, this volume). Hence this region could be scanned for additional homeobox-containing genes.

Whether pair-rule gene function is an evolutionarily conserved feature of the segmentation gene cascade is uncertain. Homologues to the pair-rule gene *eve* have been cloned from *Schistocerca americana* (Patel, Ball, and Goodman, 1992), *T. castaneum* (Parrish, Brown, and Denell, personal communication), *Xenopus laevis* (Ruiz i Altaba and Melton, 1989), and the mouse (Bastian and Gruss, 1990; Dush and Martin, 1992). The *T. castaneum eve* expression pattern displays a pair-rule function (Denell and Patel, personal communication), whereas the remaining *eve* homologues do not. Thus the function of the pair-rule gene *eve* does not appear to be evolutionarily conserved as a member of the segmentation gene hierarchy, even within the insects.

Parasegments and compartments

From a pattern of double-segment iterations demarcated by the pair-rule genes, the *D. melanogaster* blastoderm is further subdivided into single-segment domains; these are subsequently resolved into anterior and posterior compartments (see Akam, 1987, and Ingham, 1988, for review). The single-segment iterations in the embryo are out of register by one-half segment with the visible morphological boundaries separating segments in the larvae; hence they are termed *parasegments.* The first indications of parasegments arose from analysis of the expression patterns of the bithorax genes (Morata and Kerridge, 1981) and the detailed analysis of the genetic effects of the bithorax complex (Hayes, Sato, and Denell, 1984; Struhl, 1984). The domains of gene activity that were labeled parasegments have anatomical correlates in the grooves that appear in the extended germ band of *D. melanogaster* (Martinez-Arias and Lawrence, 1985).

Parasegments as domains of gene expression are conserved in the lepidopteran species *Manduca sexta:* The anterior border of the embryonic expression of the *abdominal-A* homologue bisects the first abdominal segment (Nagy, Booker, and Riddiford, 1991), as is most clearly seen in the first abdominal ganglia. Whether parasegments will be visible as anatomical features of early lepidopteran development remains to be seen.

Compartments, which measure half-segment iterations in *D. melanogaster,* were first established on the basis of lineage restrictions of clones induced by mitotic recombination (Garcia-Bellido, Ripoll, and Morata, 1973). The idea of compartments was further solidified by the finding that the segment-polarity gene *engrailed* modifies the fate of and is expressed in half-segment domains, which are thought to correspond directly to the lineage compartments (Morata and Lawrence, 1975; Kornberg, Siden, O'Farrell, and Simon, 1985). Homologues to segment-polarity genes *engrailed/invected* and *wingless* have now been isolated from *M. sexta, B. mori,* and *Precis coenia* (Kraft and Jäckle, in press [*M. sexta*]; Hui et al., 1992; Suzuki, personal communication [*B. mori*]; and Keys, Seleque, Williams, and Carroll, personal communication [*P. coenia*]). The expression of patterns of the *Bombyx engrailed* gene has only been analyzed in the larval silk gland to date. The silk gland consists of three morphologically distinct regions: the anterior, middle, and posterior silk glands, which have unique patterns of gene expression (see Hui and Suzuki, this volume). The *engrailed* gene is expressed in the middle, but not the posterior silk gland, a compartmentalized expression that may be instrumental in establishing the restricted expression of glue and silk proteins (Suzuki et al., 1990; Hui, et al., 1992). Whether these parasegmental or compartmentalized patterns of gene expression correspond to lineage compartments is unknown.

Homeosis and the homeotic genes

The homeotic genes comprise the final step in the *D. melanogaster* segmentation gene cascade. Homeotic mutations, which result in "variations which consist in the assumption by one member of a meristic series of the form or characters proper to other members of the series" (Bateson, 1894), were first described for *D. melanogaster* (Bridges and Morgan, 1923) and the silk moth *B. mori* (Hashimoto, 1941; Itikawa, 1943; reviewed in Tazima, 1964). Twenty-four known mutations in *B. mori* that modify segment character, as indicated by altered pigmentation patterns and the loss or addition of appendages and gonads, have been mapped to the E complex, at the 0.0 locus on the sixth linkage group. A similar homeotic phenotype results from the *No crescent* (*Nc*) mutation, which is located 1.4 map units away (see Itikawa, 1943; Tazima 1964; Doira, 1978). Several features of the phenotypes caused by mutant *E* alleles resemble the homeotic mutant phenotypes of *D. melanogaster:* (1) Individual segments are transformed into the character of another segment along the anteroposterior axis; (2) the transformations affect either dorsal or ventral anlage or both, but do not transform dorsal features into ventral features; and (3) the positions of the genes on the chromosome are colinear with the body region affected by a mutation (Lewis, 1963). Interestingly, the thorax is the most anterior body region affected by either the *E* or the *Nc* alleles. Mutations in the *D. melanogaster* Antennapedia complex affect the segments anterior to the mesothoracic segment; analogous mutations have been described in *B. mori,* but homologues to the Antennapedia complex genes responsible for these mutations have only been recently described in *B. mori* (see Ueno, Nagata, and Suzuki, this volume).

There are several striking differences between the phenotypes of the *B. mori* and *D. melanogaster* mutants. In most cases in *D. melanogaster,* a dominant gain of function mutation results in the transformation of an anterior segment into a more posterior segment; loss of function alleles result in the transformation of posterior segments to ones of more anterior character (see McGinnis and Krumlauf, 1992, for review). Moreover, if both the dorsal and ventral surfaces are transformed, they are both transformed in the same direction (Garcia-Bellido, 1977). In *B. mori,* however, many of the mutations cause bidirectional transformations of two types. In one type, the dorsal half of the larva is modified in one direction and the ventral half is changed in the other direction (Tazima, 1964). In the other type, more anterior segments are transformed in one direction and posterior segments are transformed in the other direction. This is elegantly demonstrated by the analysis of dorsal fusion in larvae mosaic for the *E* alleles (Shimada, Ebinuma, and Kobayashi, 1986). This bidirectionality is also seen in some of the *T. castaneum* homeotic mutant phenotypes (Beeman, 1987; Beeman, Stuart, Haas, and Denell, 1989). Finally, in *D.*

melanogaster, the segments transformed by any given single homeotic mutation (with the exception of *Scr*) mimic only one type of segment. In *B. mori,* what is thought to be a single mutation can cause multiple segment transformations, mimicking more than one type of segment. For instance, a heterozygous mutation at the E^{Ca} locus causes the dorsal transformation of the third abdominal segment (A3) to the second abdominal segment (A2), A5 to A4, and A6 and A7 to A5, as though the entire body axis had been shifted (see Tazima, 1964).

Superficially, these differences suggest that the E complex in *B. mori* is in some ways structured or regulated differently from the *D. melanogaster* homeotic gene complexes. However, more extensive genetic and molecular studies are needed to confirm this idea. Unfortunately, many of the known *E* alleles are insufficiently mapped to ascertain whether they are single point mutations, deletions, or rearrangements. As the *E* alleles are closely linked on the chromosome, it has not yet been resolved whether an individual mutant phenotype results from the perturbation of one or more genes or regulatory regions.

Our knowledge is likely to expand rapidly within the next few years. Sequences similar to the homeobox regions of the *Antennapedia (Antp), Deformed* (*Dfd*), *Ultrabithorax* (*Ubx*), *abdominal-A* (*abd-A*), and *Abdominal-B* (*Abd-B*) genes have been isolated from *M. sexta* (Nagy et al., 1991; R. Booker, personal communication). Sequences with significant similarity to the *D. melanogaster* homeotic genes *Antp, Scr, Ubx, abd-A,* and *Abd-B* have also been uncovered in *B. mori* (Ueno et al., 1992; Y. Suzuki, personal communication). All of these genes are included within the *D. melanogaster* Antennapedia and bithorax complexes. By analyzing the relative levels of hybridization signals on Southern blots of genomic DNA from wild-type and homeotic mutant strains probed with the *Ubx, abd-A,* and *Abd-B* sequence homologues, Ueno et al. (1992) were able to confirm that some of the *E* alleles correspond to deletions of these homeotic gene sequences. *Ubx* and *abd-A* sequences appear to be deleted from the E^N mutant, whereas *abd-A* is missing from the E^{Ca} mutant. This localization of the homeotic gene sequences to the *E* locus verifies that the *D. melanogaster* bithorax gene homologues *Ubx, abd-A,* and *Abd-B* are clustered on the *B. mori* chromosome and should facilitate further analysis of the regulation of the *B. mori* homeotic genes.

Comparisons of the expression patterns of the *M. sexta* and *D. melanogaster abd-A* genes did not clarify any of the variations in the lepidopteran and dipteran homeotic mutations. However, details of the *abd-A* mRNA expression domains are different between *D. melanogaster* and *M. sexta,* suggesting that the gene homologues are under different regulatory control in each organism. The anterior boundary of *abd-A* mRNA expression is conserved between *M. sexta* and *D. melanogaster,* whereas the posterior boundary is not conserved and extends three segments more posterior in *M. sexta.* The difference in expression domains suggests that some of the regulatory interactions that delimit the initial domain of *abd-A* expression have been altered between these two species (Nagy et al., 1991).

The pattern of *abd-A* mRNA expression also bears on the issue of classification of *M. sexta* as a long, intermediate, or short germ insect. The onset of *abd-A* mRNA expression in *M. sexta* is more like that of *abd-A* in *D. melanogaster,* a long germ insect, than that of *Schistocerca gregaria,* a short germ insect. The *M. sexta abd-A* mRNA expression is initially seen in a broad region that is thought to correspond to the presumptive abdomen (Nagy et al., 1991), whereas in *S. gregaria, abd-A* protein expression appears sequentially, as each segment makes its morphological appearance (Tear, Akam, and Martinez-Arias, 1990).

Pole cells and the primordial germ cells

In *D. melanogaster,* as in many other holometabolous insects, the primordial germ cells are first visible in the posterior pole of the embryo (Anderson, 1972). Their location is anticipated by the presence of a specialized cytoplasm called the posterior pole plasm (Mahowald, 1971). In *D. melanogaster* the posterior localization of the germline determinants is intimately related to the posterior localization of the axis determinants. The *nanos* gene product is thought to be the primary posterior morphogen in the establishment of the anteroposterior embryonic axis and its posterior localization is controlled by genetic factors that are also responsible for generating the specialized posterior germinal plasm (Ephrussi, Dickinson, and Lehmann, 1991; Lehmann and Nüsslein-Volhard, 1991).

In the Lepidoptera, the presumptive germ cells are thought to ingress from the germ anlage on the ventral midline. In many species the germ cells first appear in a group and later disperse to several segments (*B. mori* [Miya, 1958]; *Parnassius glacialis, Luehdorfia japonica, Byasa alcinous* [Tanaka, 1987]; and *M. sexta* [Nardi, 1993]), whereas in others they are already scattered in several segments when they ingress (*Antheraea pernyi* [Saito, 1937]; *Diacrisia virginica* [Johannsen, 1929]; *Pieris rapae* [Eastham, 1927]; *Neomicropteryx nipponensis* [Kobayashi and Ando, 1981;

Kobayashi, 1983]). In the absence of a specialized cytoplasm with visible polar granules (Miya, 1958) the site of origin for the *B. mori* germ cells was suggested by the fact that cauterizing the midventral region of the germ anlage results in hatchlings with gonads devoid of germ cells (Miya, 1958). Because the presumptive germ cells in the Lepidoptera occupy a midventral position, it is unlikely that the genes involved in setting up the posterior embryonic axis will interact with and be dependent on the genes involved in the localization of germline determinants, as occurs in *D. melanogaster.*

Dorsoventral patterning

The early lepidopteran germ anlage shows a remarkable degree of regenerative capacity along the dorsoventral axis. In an elegant set of experiments, Krause and Krause (1965) were able to bisect the early *B. mori* germ anlage down the ventral midline into two lateral halves; each half was capable of forming an entire germ band. Whether *D. melanogaster* has a similar capacity to regulate along the dorsoventral axis has not been reported, although when cells are transplanted from the cellular blastoderm to new positions in another embryo, they will adjust their fate to their new position along the dorsoventral, but not the anteroposterior, axis (Simcox and Sang, 1983).

Although the timing of a particular cell's irreversible commitment to a dorsoventral fate in *D. melanogaster* has not been established, a great deal is known about the molecules that function to establish the dorsoventral polarity of the eggshell and embryo. The dorsoventral axis is established using a separate cascade of events from those that determine the anteroposterior axis. A pathway involving 18 maternal-effect loci results in a graded distribution of the dorsal-ventral morphogen *dorsal* in the cellular blastoderm (reviewed in Anderson, 1987, and Govind and Steward, 1991). The *dorsal* morphogen differentially activates the zygotic genes that specify the dorsoventral pattern of the embryo. The earliest acting maternal-effect genes, such as *fs(1)K10* and *torpedo,* affect the dorsoventral polarity of both eggshell and embryo, whereas the shape of the egg is not altered in the remaining 12 dorsal group mutations (see Govind and Steward, 1991).

In *B. mori* there is one known maternal-effect mutation, *kidney-shaped egg* (*ki*), which is thought to affect the determination of the dorsoventral axis. All embryos from *ki/ki* mothers develop without mesodermal tissues (Sakaguchi, 1982; Miya, 1984, 1985b). The follicle cells on the ventral side of the oocytes from *ki/ki* mothers are elongate and more similar to the cells on the dorsal side of the wild-type oocyte (Sakaguchi et al., 1982). Phenomenologically, the phenotype appears similar to that caused by the maternal-effect dorsalizing mutations in *D. melanogaster,* which result in ventral positions of both the follicle cells and embryo assuming more dorsal fates. An additional morphological feature of the embryos from *ki/ki* mutant mothers corroborates the idea of ventral positions adopting more dorsal fates: Miya (1985a) observed that during the contraction of the germ anlage to form the germ band, cytoplasmic masses are extruded from the most lateral edges of the anlage. In the embryos from homozygous *ki* mothers these masses are extruded across the entire width of the anlage and are not restricted to the lateral edges (Miya, 1985b).

Embryos from *ki/ki* mutant mothers do not develop a ventral nervous system (Miya, 1985a). The absence of a nervous system could result from an extreme dorsalizing effect, whereby not only the mesoderm but also the neurectoderm is converted to more lateral ectoderm. The presumptive germ cells, however, ingress normally in the embryos from *ki/ki* mutant mothers. Miya (1984) suggests that the primary defect is not in the dorsal-ventral patterning, but reflects a disruption of the midline ingression of mesoderm. Alternatively, the germ cells could be localized to the ventral midline by a mechanism independent of the information used by the germ anlage to establish the dorsal-ventral axis.

The isolation of additional alleles of this gene, which is linked to the E complex, could reveal the extremes of lateralization possible. It would also be very simple to determine if these mutants can be rescued with wild-type cytoplasm, as can some of the *D. melanogaster* dorsalizing mutants (Santamaria and Nüsslein-Volhard, 1983; Anderson and Nüsslein-Volhard, 1984; Anderson, Bokla, and Nüsslein-Volhard, 1985).

Summary

The derived Lepidoptera can best be described as long germ insects, with several distinctive characteristics of their own. If one compares lepidopteran embryogenesis to the early patterning processes known to occur in *D. melanogaster,* there are both similarities and differences. Homologues to the maternal coordinate genes have not yet been isolated. Sequence homologues to the *D. melanogaster* gap and pair-rule genes have been isolated, but their expression patterns have not yet been analyzed; whether or not they function in lepidopteran segmentation remains unknown. There are also sequence homologues to at least some of the segment-polarity genes and circumstantial evidence for the existence of parasegments, both of which divide the developing germ band into segment length iterations. The strongest evidence for similarity lies in the homeotic genes. These genes are clearly conserved in the Lepidoptera, based on homeobox sequence comparisons and domains of mRNA activity. Nonetheless, both the mutant phenotypes and the limits of mRNA expression suggest that the *B. mori* homeotics may be regulated differently from their *D. melanogaster* counterparts during development.

As mentioned, both the Diptera and members of the Lepidoptera have lost many of the slower, proliferative features of other classes of insects and have evolved toward a more rapid, early establishment of pattern. Although the Lepidoptera may have chosen a generally similar path toward the early determination of the anteroposterior axis, lepidopteran embryogenesis nonetheless differs in detail from the *D. melanogaster* plan. Exactly how these deviations will correlate with what is known in *D. melanogaster* remains to be seen.

Acknowledgments

The author greatly appreciates the innumerable contributions made by Dr. Lynn Riddiford in the preparation of this chapter, the help and information provided by Drs. K. Kiguchi, M. Myohara, Y. Suzuki, K. Miya, and C.-c. Hui, the editorial efforts of Drs. C. Berg, S. Blair, B. Wakimoto, and R. Warren, and the assistance provided by Leanne Olds and S. Blair with the figures.

6

Roles of homeotic genes in the *Bombyx* body plan

KOHJI UENO, TOSHIFUMI NAGATA, and
YOSHIAKI SUZUKI

Introduction

The body of an insect is constructed in segmental fashion, with segmented subdivisions of the head, a set of mouthparts, the thorax, the abdomen, and the terminal region (Snodgrass, 1935; Anderson, 1973). In early embryogenesis, after determination of the anteroposterior and dorsoventral axes of the embryo, the segments of the mouthparts, the thorax, and the abdomen are specified and formed, acquiring their individual identities (Sander, 1976).

The specific identities of body segments are known to be determined by homeotic genes in many kinds of insects (Ouweneel, 1976). In the silkworm, *Bombyx mori,* for instance, a homeotic gene complex (the E complex) that specifies the identities of body segments has long been known (Hashimoto, 1941). In other insects, notably *Drosophila melanogaster* and the flour beetle *Tribolium castaneum,* homeotic genes are also clustered. In *Drosophila* the Antennapedia complex (ANT-C) (Wakimoto and Kaufman, 1981) and the bithorax complex (BX-C) (Lewis, 1978), which are located on the right arm of the third chromosome, specify identities of the body segments. The ANT-C determines identities from the head to the mesothoracic segment, and the BX-C determines identities of the metathoracic segment and all the abdominal segments. In *Tribolium,* similar homeotic genes are known to determine identities of the body segments, where Beeman demonstrated six clustered homeotic genes in the second linkage group (Beeman, 1987). These loci include elements with apparent homology to the homeotic genes in the ANT-C and BX-C in *Drosophila.*

We are studying the molecular structure and function of the homeotic genes in *Bombyx*. These genes specify identities of the segments during early embryogenesis. The silkworm is unique among the lepidoptera because its developmental processes can be studied by genetic analysis. As already noted, the E complex was found to be a cluster of homeotic genes that play important roles in the specification of the larval abdominal segments (Hashimoto, 1941). Recently we have identified the homeobox genes, *Bombyx Ultrabithorax (Bm Ubx), Bombyx abdominal-A (Bm abd-A),* and *Bombyx Abdominal-B (Bm Abd-B),* which constitute the E complex (Ueno et al., 1992). We have also analyzed the molecular structure of the *Nc* locus, which contains a homeotic gene specifying the thoracic segments in early embryogenesis and which we now know corresponds to *Bombyx Antennapedia (Bm Antp)* (Nagata et al., in preparation).

In this chapter we review the embryonic development and the molecular structure and function of homeotic genes of *Bombyx mori.* We also discuss the unity and diversity of the structure and function of the homeotic genes between *Bombyx* and *Drosophila.*

The embryonic development of *Bombyx*

Bombyx embryos develop and hatch in about 10 days after fertilization at 25°C. Figure 6.1 shows a schematic drawing of the stages of embryogenesis.

After fertilization, the nucleus in silkworm eggs divides mitotically and the cleavage nuclei migrate toward the surface of the egg (stage 2), as observed in other insects. At stage 3, the nuclei and associated cytoplasm become surrounded by cell membranes and the blastoderm is formed. At this stage the regions where the germ band or organ primordia will be formed are thought to be predetermined in the embryo (see Nagy, this volume). About 16 hr after deposition, the blastoderm thickens and the germ band is formed (stage 4). After gradual thickening (stage 5), the germ band undergoes gastrulation and segmentation (stage 7). The stomodeum becomes manifest and the proctodeum begins to be formed at the caudal end, and the thoracic appendages appear in the gnathal region of the head (stage 17). Three days after deposition the embryo gradually becomes shorter, and the appendages of the abdominal segments begin to develop (stage 20).

The *Bombyx* embryo develops such that the ventral side faces toward

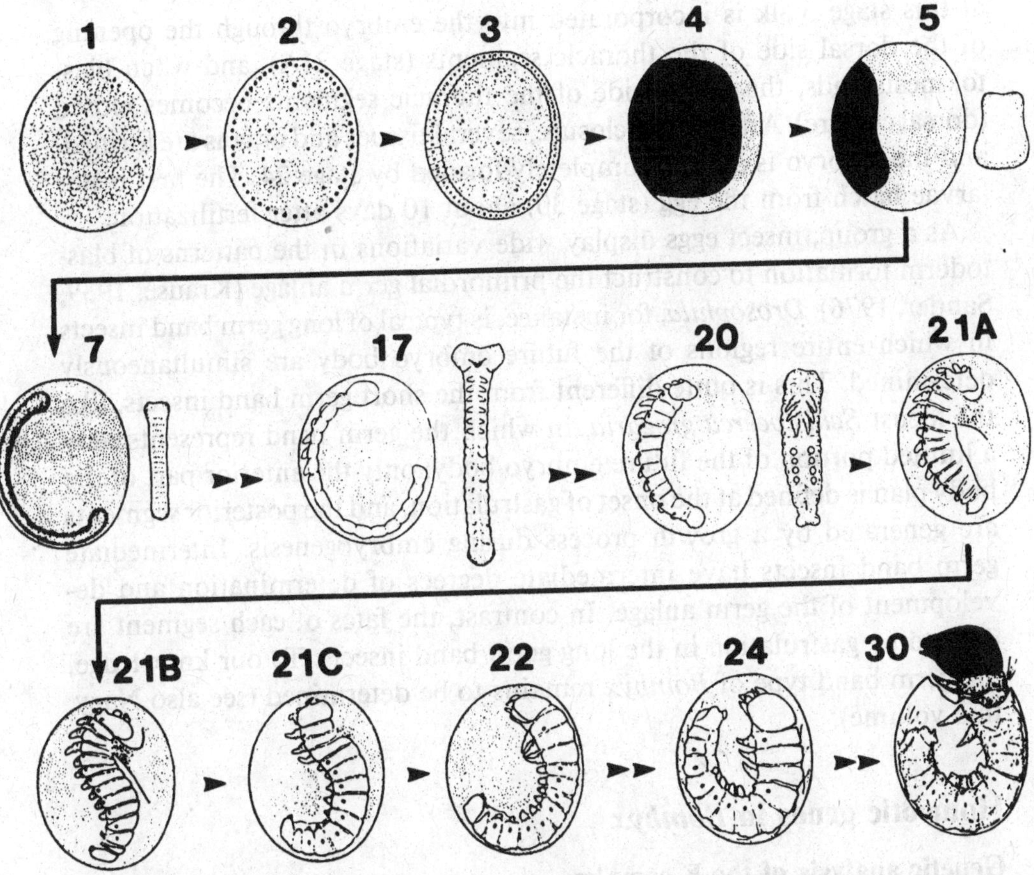

Figure 6.1. Schematic diagrams of *Bombyx* embryogenesis. The developmental stages are modified from Takami and Kitazawa (1960). Stage 1, fertilization, 2 hr after oviposition; stage 2, late cleavage; stage 3, formation of blastoderm; stage 4, formation of the germ band, 15 hr after oviposition; stage 5, germ band, 24 hr after oviposition; stage 7, gastrulation; stage 17, formation of stomodeum and proctodeum, 2 days after oviposition; stage 20, segmentation of head and thorax, 3 days after oviposition; stage 21A, early blastokinesis; stage 21B, blastokinesis, 4 days after oviposition; stage 21C, late blastokinesis; stage 22, completion of blastokinesis; stage 24, completion of embryo, 8–10 days after oviposition; stage 30, first instar larva hatching from the egg.

the outside of the egg until stage 20 (see Figure 6.1). From stages 21A to 21C the embryo turns around and the dorsal side comes to face toward the outside of the egg (see Figure 6.1). These stages of *Bombyx* embryogenesis are called periods of body position reversal or blastokinesis; blastokinesis is commonly observed in Lepidoptera and many types of insects. The inversion movement is an important step for embryogenesis because, at this stage, yolk is incorporated into the embryo through the opening of the dorsal side of the thoracic segments (stage 21A), and when blastokinesis ends, the dorsal side of the thoracic segments becomes closed (dorsal closure). After dorsal closure, several tissues and organs are formed, and the embryo is almost completely formed by stage 24. The first instar larvae hatch from the egg (stage 30) about 10 days after fertilization.

As a group, insect eggs display wide variations in the patterns of blastoderm formation to construct the primordial germ anlage (Krause, 1939; Sander, 1976). *Drosophila,* for instance, is typical of long germ band insects in which entire regions of the future embryo body are simultaneously determined. This is quite different from the short germ band insects, like the locust *Schistocerca gregaria,* in which the germ band represents only a limited portion of the future embryo body; only the anterior part of the body plan is defined at the onset of gastrulation, and the posterior segments are generated by a growth process during embryogenesis. Intermediate germ band insects have intermediate degrees of determination and development of the germ anlage. In contrast, the fates of each segment are defined by gastrulation in the long germ band insects. To our knowledge, the germ band type of *Bombyx* remains to be determined (see also Nagy, this volume).

Homeotic genes in *Bombyx*

Genetic analysis of the E complex

The E complex contains homeotic genes that specify the identities of the larval abdominal segments and is located at the proximal end (0.0) of the sixth linkage group in *B. mori,* as shown in Figure 6.2 (Hashimoto, 1941; Tazima, 1964). So far, more than thirty types of mutations have been reported in the E complex and all of them are dominant. *E* mutants not only display extra abdominal markings on the larval epidermis and supernumerary thoracic or abdominal legs on the abdominal segments but

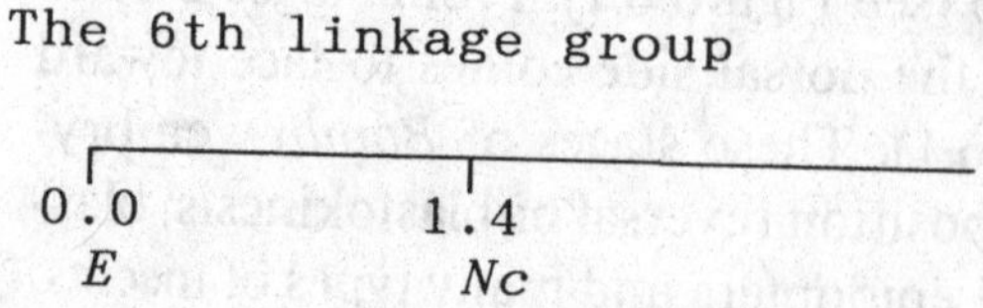

Figure 6.2. The *E* locus and the *Nc* locus in the sixth linkage map of *Bombyx mori.* The *E* (E-pseudo-allelic group) locus is located at 0.0 cM and the *Nc* (No crescent) locus is located at 1.4 cM.

also abnormalities in the formation of the reproductive organs in the abdomen (Sakata, 1938), the network pattern of the trachea, and the pattern of the nerve commissures (Itikawa, 1944). The nomenclature of the E complex was derived from the first letter of "extra" with reference to the leg and marking phenotype.

Most of the *E* mutations cause a particular directional shift, anterior to posterior, or vice versa, on both dorsal and ventral sides of the larva. However, these mutations reveal one interesting aspect that is different from those in the homeotic gene complexes of *Drosophila:* In some of the E mutants a specific mutation causes shifts in one direction on the dorsal side and in the opposite direction on the ventral side (Itikawa, 1943; Tazima, 1964). Figure 6.3 shows phenotypes of these two categories of *E* mutants. In E^N (New additional crescent), a fifth instar larva heterozygous for the mutation expresses extra markings on the dorsal side of the third abdominal segment (A3) and small supernumerary thoracic-type legs on the A1 segment (Itikawa, 1943). The phenotype of the E^N mutation is typical, with the expression of the extra crescents and legs caused by the abnormal specification of segmental identity resulting in a conversion of abdominal segments toward thoracic ones. Both dorsal and ventral sides show this shift. Therefore the overall effect of the E^N mutation is to create transformations of posterior segments toward anterior ones on both the dorsal and ventral sides of the larva (Hashimoto, 1941).

A contrasting phenotype is shown by fifth instar larvae homozygous for E^{Kp} (Kp supernumerary legs) (see Figure 6.3), which causes different shifts on the dorsal and ventral sides. In this case the mutant larva reveals extra crescents on the A3 segment (anterior transformation) and small supernumerary abdominal-type legs on the A2 segment (posterior transformation). The E^D (Double crescents) mutation reveals a similar dorsoventral difference, but the dorsal and ventral shifts are reversed in orientation so that larvae heterozygous for E^D express both extra crescent markings and supernumerary thoracic-type legs on the A1 segment. This phenotype is understood as a duplication of the A2 segment, causing the expression of extra crescents on the dorsal side of the A1 segment, whereas on the ventral side, the metathoracic segment is duplicated. Therefore the effect of the E^D mutation causes a posterior-type transformation on the dorsal side and an anterior-type transformation on the ventral side.

In *Drosophila,* when a loss of function mutation occurs in the BX-C, anteriorward transformation is observed on both the dorsal and ventral

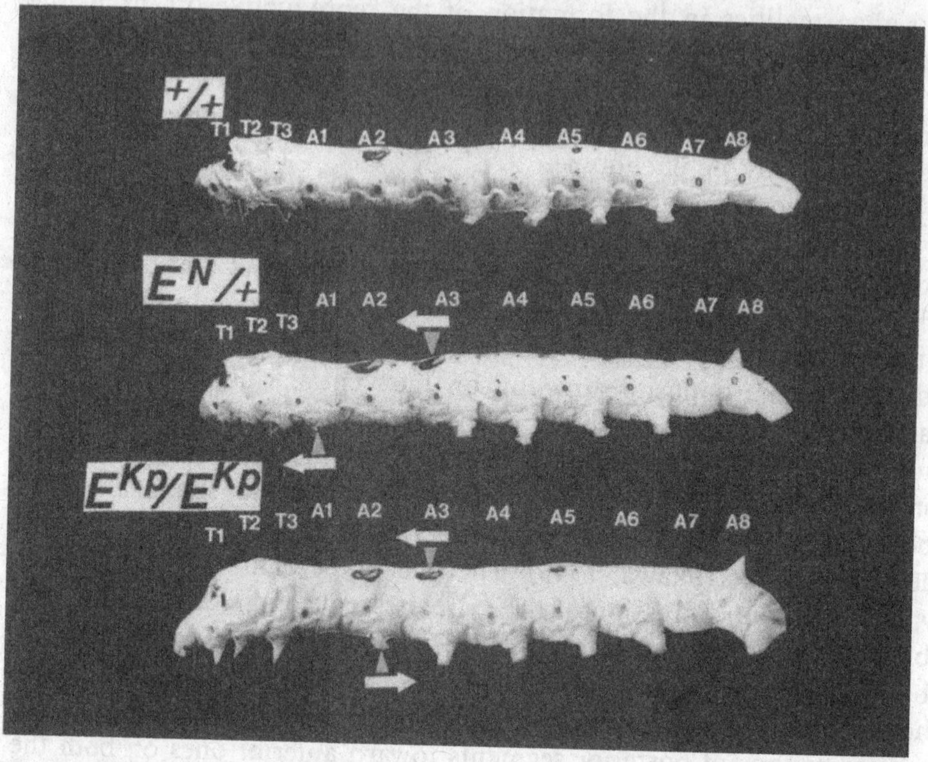

Figure 6.3. The phenotypes of *Bombyx mori* (A) wild-type, (B) heterozygous $E^{N}/+$, and (C) homozygous E^{Kp}/E^{kp} larvae. Thoracic segments are numbered T1–T3 and abdominal segments are numbered A1–A8. Extra markings and supernumerary legs are indicated by vertical arrowheads and mutational shifts are indicated by horizontal arrows.

sides, so that posteriorward segments become more anteriorlike. And when a gain of function mutation occurs in the BX-C, a posteriorward transformation takes place on both the dorsal and ventral sides (anterior segments become more posteriorlike) (Duncan, 1987). Because the independent determination of the dorsal and ventral sides observed in *Bombyx* larvae has not been observed in *Drosophila,* there must be differences in the regulatory mechanisms that specify abdominal segments between these two insects. Genetic studies had earlier revealed that the E complex is a pseudo-allelic gene complex, similar to the BX-C in *Drosophila.* Itikawa (1951) first found and analyzed a recombinant between E^{Gd} (Deformed gonad) and E^{Nc} (No crescent, supernumerary legs), which was distinct from either of the heterozygotes; larvae heterozygous for E^{Gd} express extra crescents on the A1 segment and reveal an abnormality in the gonads

(Sakata, 1938), whereas larvae heterozygous for E^{Nc} express supernumerary abdominal-type legs and lack crescents on the A2 segment (Sasaki, 1940). Tsujita (1955) demonstrated the occurrence of recombination between genes E^{H} (H extra crescents) and E^{Kp}; heterozygotes for the E^{H} mutation express extra crescents on the A1 segment (Hashimoto, 1941). Additional evidence indicates that many genes of the E complex are located in the same region but at separate sites in close proximity to each other, suggesting that the E region constitutes several subunits partly shared by different alleles. Detailed genetic and molecular analyses of the E complex are urgently required.

Molecular analysis of the E complex

Homozygotes for the E^{N} and E^{Ca} (Additional crescent) mutations produce dramatic phenotypic transformations (Figure 6.4). Itikawa (1943) reported that embryos homozygous for E^{N} express many thoracic-type appendages in segments that must otherwise be regarded as abdominal. As shown in Figure 6.4B, homozygous E^{N}/E^{N} embryos reveal thoracic-type legs from the first to the seventh abdominal segments and intermediate thoracic/abdominal-type legs in the A8 segment. These embryos cannot develop beyond stage 20, and fail to invert the body sides at blastokinesis. Itikawa also showed that the nerve commissures and tracheae in the first to the eighth abdominal segments in embryos homozygous for E^{N} show patterns similar to those in the thoracic segments of normal embryos (Itikawa, 1943). From these observations, Itikawa proposed that the first to the seventh or eighth abdominal segments in homozygous E^{N}/E^{N} embryos are transformed to thoracic-type segments.

Molecular studies of the structure of the ANT-C and BX-C in *Drosophila* revealed that these complexes consist of several homeobox genes (Gehring and Hiromi, 1986). In the BX-C, three homeobox genes, *Ultrabithorax (Ubx), abdominal-A (abd-A),* and *Abdominal-B (Abd-B),* are involved in the determination of the abdominal segments (Duncan, 1987). Lewis (1978) reported that the functional deficiency of *Ubx* and *abd-A (DfUbx109)* causes transformation of most abdominal segments to thoracic-type segments in *Drosophila.* Because the *Ubx* gene represses the expression of *Antennapedia (Antp),* which specifies the thoracic segments at early embryogenesis, the deficiency of the function of *Ubx* is thought to cause transformation of the abdominal segments to thoracic segments by allowing *Antp* expression in those segments. From the similarity of the

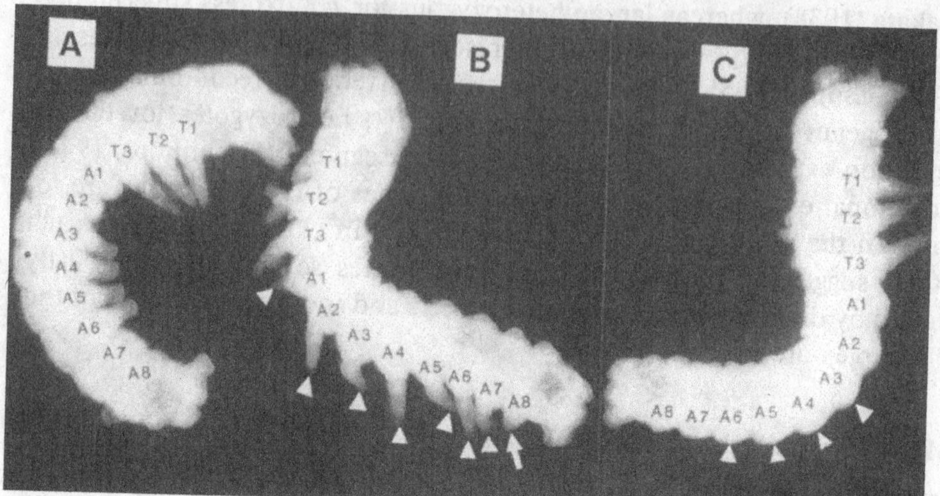

Figure 6.4. Phenotypes of *Bombyx mori* (A) wild-type, (B) homozygous E^N/E^N, and (C) homozygous E^{Ca}/E^{Ca} embryos. The wild-type embryo has three pairs of thoracic legs from the first to the third thoracic segments (T1–T3) and four pairs of abdominal legs from segments A3 to A6. Embryos homozygous for E^N express thoracic-type legs from segments A1 to A7 (arrowheads) and intermediate thoracic/abdominal-type legs on segment A8 (arrow). The homozygous E^{Ca} embryo does not express any abdominal legs (arrowheads). Embryos are approximately 2 mm long.

phenotypes between the homozygous E^N embryo and *DfUbx109* embryo, we assumed that the E complex has a similar structure and function to the BX-C in *Drosophila.*

Itikawa (1943) also showed that embryos homozygous for E^{Ca} express three pairs of normal thoracic legs in the thoracic segments, but no legs in the abdominal segments (see Figure 6.4C). These embryos cannot develop beyond embryonic stage 20 and fail to invert, as observed in the homozygous E^N/E^N embryos. We speculated that the A3–A6 segments, which normally express abdominal legs, might have transformed to other abdominal segments that have no legs (Ueno et al., 1992). Although the precise transformation in segments A3–A6 in embryos homozygous for E^{Ca} was not understood, it was known that some bithorax complex mutations in *Drosophila* cause a transformation of some abdominal segments to other more posterior ones (Lewis, 1978). Therefore we speculated that the E^{Ca} mutation is also causing a defect in a homeotic gene complex that functions like the BX-C in *Drosophila.*

To test these assumptions, we first isolated the *Bombyx* homologues of the *Ubx, abd-A,* and *Abd-B* genes in the BX-C (Ueno et al., 1992) and

```
(a) Ubx                                                                  3'
             5'
Bombyx         RRRGRQTYTRYQTLELEKEFHTNHYLTRRRRIEMAHALCLTERQIKIWFQNRRMKLKKEI
               ************************************************************
Drosophila     RRRGRQTYTRYQTLELEKEFHTNHYLTRRRRIEMAHALCLTERQIKIWFQNRRMKLKKEI

(b) abd-A                                                                3'
             5'
Bombyx         RRRGRQTYTRFQTLELEKEFHFNHYLTRRRRIEIAHALCLTERQIKIWFQNRRMKLKKEL
               ************************************************************
Drosophila     RRRGRQTYTRFQTLELEKEFHFNHYLTRRRRIEIAHALCLTERQIKIWFQNRRMKLKKEL

(c) Abd-B                                                                3'
             5'
Bombyx         VRKKRKPYSKFQTLELEKEFLFNAYVSKQKRWELARNLNLTERQVKIWFQNRRMKNKKNS
               ************************************** *********************
Drosophila     VRKKRKPYSKFQTLELEKEFLFNAYVSKQKRWELARNLQLTERQVKIWFQNRRMKNKKNS
```

Figure 6.5. Comparison of the homeodomain amino acid sequences encoded by (a) *Ubx*, (b) *abd-A*, and (c) *Abd-B* genes from *Bombyx* (upper) and *Drosophila* (lower). The amino acid sequences of the homeodomain protein homologues from *Bombyx* are predicted from the nucleotide sequences of genomic DNAs we have cloned. Sequences encoded by *Drosophila Ubx, abd-A,* and *Abd-B* are from McGinnis et al. (1984), Karch et al. (1990), and Zavortink and Sakonju (1989), respectively. Asterisks denote matches between *Bombyx* and *Drosophila*.

compared the homeodomain amino acid sequences with those encoded by the three homeobox genes in the *Drosophila* BX-C (Figure 6.5). The sixty amino acids encoded by *Ubx* homeodomains (see Figure 6.5a) or by *abd-A* homeodomains (see Figure 6.5b) are identical between *Bombyx* and *Drosophila;* only a single amino acid in the homeodomain of the *Abd-B* genes is different, but involves similar amino acids, an asparagine residue in *Bombyx* substituted for a glutamine residue in *Drosophila* (see Figure 6.5c). The conservation of homeodomain amino acid sequences between *Bombyx* and *Drosophila* suggests that these silkworm homologues may have functions similar to their *Drosophila* counterparts. Therefore we have named these genes *Bm Ubx, Bm abd-A,* and *Bm Abd-B,* respectively. Analyses of the chromosomal defects in the E^{N} and E^{Ca} mutations supported the idea that these encompass genes with homologous function to the *Drosophila* BX-C.

The homeodomain regions of the *Bm Ubx* and *abd-A* genes are deleted in the E^{N} chromosome, as shown in Figure 6.6 (Ueno et al., 1992). Because the homeodomain is essential for the function of the homeoprotein, we speculate that complete functional deficiency of the *Bm Ubx* and *abd-A* genes may have caused drepression of the *Bm Antp* gene and transfor-

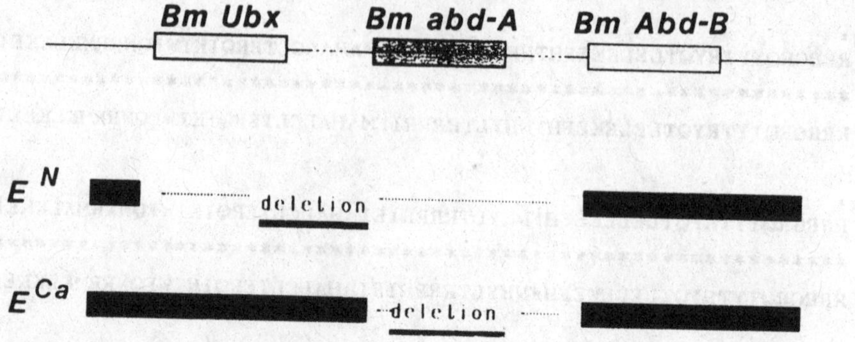

Figure 6.6. Schematic structure of the E complex and the deletions in the E^N and E^{Ca} chromosomes. The order of the *Bm Ubx, abd-A,* and *Abd-B* genes is hypothetical and based on the gene structure of the bithorax complex in *Drosophila.*

mation of the A1–A7 segments to the thoracic type segments in the homozygous E^N embryos (see Figure 6.4B), with intermediate thoracic/abdominal-type legs expressed in the A8 segment. This latter observation suggests that the *Bm Abd-B* gene is intact and functional in homozygous E^N embryos.

In *Drosophila,* deletion mutations of the *Ubx* and *abd-A* genes such as *DfUbx109* cause a transformation of the A1–A7 segments to thoracic-type segments (Lewis, 1978), with the A8 segment expressing an abdominal phenotype. The comparable phenotypes of the homozygous E^N/E^N embryos and the *DfUbx109* embryos suggest that the *Bm Ubx, abd-A,* and *Abd-B* genes have similar functions to their *Drosophila* homologues.

We have also shown that the E^{Ca} chromosome has a deletion around the homeobox region of the *Bm abd-A* gene, as shown in Figure 6.6 (Ueno et al., 1992). In *Drosophila,* functional deficiency of the *abd-A* gene is thought to cause transformation of the A2–A7 segments to A1 type segments (Lewis, 1978). By analogy, we infer that a complete deficiency of *abd-A* function in homozygous E^{Ca}/E^{Ca} embryos may have caused transformation of the A2–A7 segments to A1 type segments, which have no abdominal legs (see Figure 6.4C).

The analyses of E^N and E^{Ca} chromosomes suggested that the deletions of *Bm Ubx* and/or *abd-A* genes are responsible for the phenotypes of these E mutants and functions of these genes are similar to those of homologues in the BX-C (see Figure 6.6). Although no chromosomal abnormality was detected in the homeobox region of the *Bm Abd-B* gene in either the E^N or E^{Ca} chromosomes, nevertheless, we speculate that the E complex also contains the *Bm Abd-B* gene, for the following reasons. Because the E^{Ds} (Double stars) mutation causes expression of extra markings flanking the A5 segment (Takasaki, 1947), we think that this mutation reveals an abnormal specification of the segments flanking the A5 segment. This phenotype seems to resemble the phenotype associated with *Abd-B* mutations in *Drosophila* (Duncan, 1987). Therefore the E complex in *Bombyx* probably consists of *Ubx, abd-A,* and *Abd-B* genes and the functions of these homeobox genes resemble those of the *Drosophila* BX-C. Lewis (1978) has suggested the hypothesis that "leg-suppressing" genes like the bithorax complex are responsible for preventing leg development from abdominal segments of millipedelike ancestors during arthropod evolution. The E^N/E^N embryo, displaying many thoracic-type legs (see Figure 6.4B), might be expressing atavistic characteristics through the deletion of some "leg-suppressing" genes.

Phenotypes of the *Nc* mutations

The *Nc* (No crescent) locus is located 1.4 cM from the E complex, as shown in Figure 6.2. Itikawa (1943) suggested that this locus resembles the E complex in several respects. As shown in Figure 6.7, larvae heterozygous for the *Nc* mutation lose the crescent markings on the A2 segment and the star spots on the A5 segment, and reveal thorax-specific eyespots of reduced size on the thoracic segment. Itikawa reported that the homozygotes for this mutation are lethal and die at later embryonic stages, lacking one of the three thoracic segments.

From Itikawa's observations, the *Nc* locus seemed to accommodate a homeotic gene specifying thoracic segments during *Bombyx* embryogenesis; however, the effects of the *Nc* mutation were not extensively characterized by him. Therefore we first carefully analyzed the phenotypes of embryos and larvae carrying the *Nc* chromosome. Embryos homozygous for *Nc* terminate their development at embryonic stage 24, and the prothoracic and mesothoracic segments are fused (Nagata et al., in preparation). The prothoracic legs of the embryos homozygous for *Nc* express two brown stripes and two filaments, which are antennal characteristics, and the mesothoracic legs in the mutant embryos express filaments that are antennal and maxillary features. These phenotypes suggest that the prothoracic and mesothoracic segments might have transformed to more anterior segments such as the maxillae or some parts of the head (Nagata et al., in preparation). Furthermore, the silk glands in the *Nc/Nc* embryos

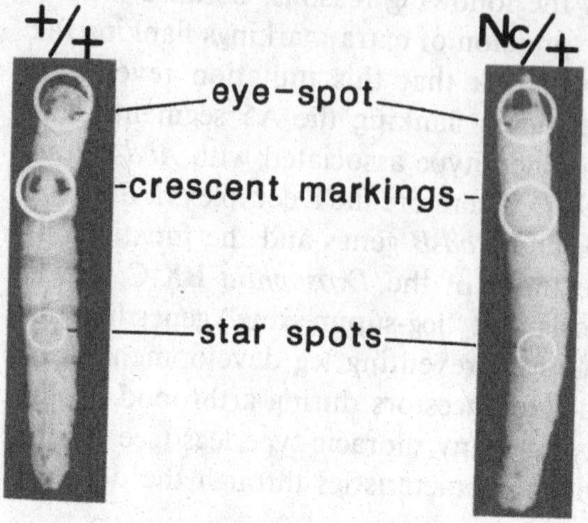

Figure 6.7. Phenotypes of *Bombyx mori* wild-type +/+ and heterozygous *Nc*/+ fifth instar larvae. Heterozygous larvae for the *Nc* mutation have lost the crescent markings on the A2 segment and the star spots on the A5 segment, and express eyespots of reduced size on the thoracic segments.

cannot develop efficiently and remain dwarf when compared with those in normal embryos. Larvae and adults heterozygous for *Nc* are viable (see Figure 6.7), and the forewings of the *Nc*/+ adults are shrunken and sometimes completely absent. From these observations we speculated that the *Nc* locus corresponds to a gene that is required for the determination of the thoracic segments, the development of the silk gland, which originates from the labial segment, and the development of other parts of the head in *Bombyx.*

Molecular analysis of the *Nc* locus

In *Drosophila,* the ANT-C accommodates the homeotic genes that, when mutated, alter identities of the head and thoracic segments (Denell, Hummels, Wakimoto, and Kaufman, 1981). *Antp* specifies identities of the labial to metathoracic segments, and *Sex combs reduced (Scr)* specifies identities of the maxillae, labial, and prothoracic segments. *Deformed (Dfd), proboscipedia (pb),* and *labial (lab)* are involved in determining the head structures. Molecular analysis revealed that all of these genes contain homeobox regions (Gehring and Hiromi, 1986).

The ANT-C and BX-C are located in 84E and 89B of the third chromosome, respectively, in *Drosophila* (Kaufman, 1981). Because the *Nc* locus is located near the E complex, which is analogous to the *Drosophila* BX-C, we hypothesized that the *Nc* locus might accommodate a homeotic gene homologous to that in the *Drosophila* ANT-C. We have isolated the

Bm Antp gene, which contains the same homeodomain amino acid sequence as that of *Drosophila* (Hara and Suzuki, in preparation). We found that the transcript of the *Bm Antp* gene accumulates in the middle silk gland (as described by Hui and Suzuki in this volume), so we analyzed the transcription of the *Bm Antp* gene in the middle silk gland of fifth instar larvae heterozygous for *Nc*. In mutant larvae we detected two transcripts, 4.8 kb and 1.8 kb, whereas in wild-type larvae we found only the 4.8 kb transcript. Cloning and sequencing of the cDNAs of these transcripts indicated that the 1.8 kb transcript lacks the homeobox region of the *Bm Antp* gene, as shown in Figure 6.8 (Nagata et al., in preparation). Upon analyzing the chromosomal abnormality of the *Nc* mutation by Southern blot analysis, we found that the homeodomain of the *Bm Antp* gene is deleted from the *Nc* chromosome (Nagata et al., in preparation). We have also isolated a homologue of the *Drosophila Scr* gene from *Bombyx* (Ueno, unpublished observations) which contains the same homeodomain amino acid sequence as that of *Drosophila*. By Southern blot analysis, the *Nc* chromosome was found to carry an intact homeodomain for the *Bm Scr* gene.

From these results we conclude that *Bm Antp* maps to the *Nc* locus. Loss of function of *Bm Antp* might cause transformation of the prothoracic and mesothoracic segments to more anterior segments in embryos homozygous for *Nc*. This phenotype resembles that of the null mutation of *Antp* in *Drosophila* (Struhl, 1981), which strongly supports the idea that the region around *Nc* may accommodate a counterpart to the *Drosophila* ANT-C.

Perspectives

We have described the roles of the E complex and *Nc* locus on the body plan during *Bombyx* embryogenesis. From our analyses we propose that the E complex is analogous to the *Drosophila* BX-C. However, the entire structure of the E complex remains to be analyzed at the molecular level. So far we have isolated only the homeoboxes and their flanking regions from the *Bm Ubx*, *abd-A*, and *Abd-B* genes and speculated about the possible structure of the E complex from analyses of mutant chromosomes. Because the *Drosophila* BX-C extends over approximately 300 kb of DNA (Bender et al., 1983), we estimate that the *Bombyx* E complex might also extend over a long DNA region. Upon isolation of the whole E complex

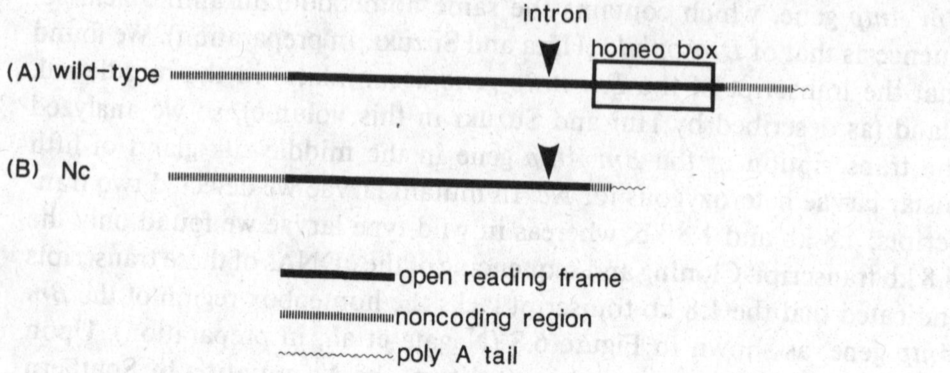

Figure 6.8. Schematic structures of the transcripts expressed from the *Bm Antp* gene, (A) wild-type and (B) *Nc* mutant. The transcript of *Bm Antp* from the *Nc* chromosome lacks the homeobox region.

and identification of the transcriptional units of the three homeobox genes, we will be able to compare the similarities and differences between the silkworm and fruit fly counterparts of this large complex at the molecular level. In the *Drosophila* BX-C, the *Ubx, abd-A,* and *Abd-B* genes are in the same orientation and regulate the determination of the metathoracic and eight abdominal segments (Duncan, 1987). We are particularly interested in comparing the structures of the three *Bombyx* homologues, especially in regions outside the homeoboxes, and comparing the *cis*-acting elements between the two insect species, for the following reasons.

As already mentioned, the *E* mutant phenotypes reveal an important difference from those of the BX-C. Some *E* mutants, like E^{Kp}/E^{Kp} (Hashimoto, 1941) and $E^{D}/+$ (Hashimoto, 1930), cause shifts in one direction on the dorsal side and in the opposite direction on the ventral side (Tazima, 1964). Because such an independent determination of the two sides is not seen in *Drosophila,* there must be differences between the *Bombyx* E complex and the BX-C of *Drosophila* in the regulatory mechanisms that specify abdominal segments. In *Drosophila,* it is well known that the *cis*-regulatory regions of the *Ubx, abd-A,* and *Abd-B* genes are crucial in the determination of each segment, and that the *cis*-regulatory regions are arranged in the same order as the body axis (Duncan, 1987). We speculate that the homeobox genes in the E complex may have similar features, but that the structures and functions of the *cis*-regulatory regions and the coding regions outside the homeodomains may be different from those of *Drosophila.* This makes it necessary to isolate the regions corresponding to complete open reading frames and the putative *cis*-regulatory regions and compare the structures and functions between *Bombyx* and *Drosophila* at the molecular level. We already know that several regions outside the homeodomains in *Bombyx engrailed (en), invected (in),* and *Antp* are different from those in the corresponding *Drosophila* genes (Hui et al., 1992; Nagata et al., in preparation).

Based on our finding that the *Nc* locus corresponds to the *Bm Antp* gene, we speculate that *Bombyx* may have a gene complex analogous to the *Drosophila* ANT-C in the *Nc* locus, as shown in Figure 6.9. One reason for this speculation is the close proximity of the *Nc* and *E* loci in the sixth chromosome of *Bombyx* (see Figure 6.2), a situation that resembles that of the ANT-C and BX-C in *Drosophila.* Another reason is that the distance between the homeoboxes of the *Bm Antp* and *Scr* genes is within approximately 100 kb, which we have shown by Southern blot analysis after pulse field electrophoresis of *Sfi*I digested genomic DNA (Ueno, unpublished observations). In *Drosophila,* the homeobox of the *Scr* gene is positioned only about 70 kb from the homeobox of the *Antp* gene (LeMotte, Kuroiwa, Fessler, and Gehring, 1989). Moreover, C.-c. Hui has isolated genes that contain sequences similar to part of the homeodomains of the *Drosophila Dfd, pb,* and *lab* genes by PCR cloning (see Hui and Suzuki, this volume). From our results, we speculate that the *Bm Antp, Scr,* and probably *Dfd, pb,* and *lab* genes constitute a gene complex like the *Drosophila* ANT-C. As there is no genetic information about the homeotic genes responsible for the determination of the thoracic segments except for the *Nc* mutants, it is necessary to determine at the molecular level whether the *Bombyx Antp* and the other genes are clustered in a single gene complex.

We can now study the molecular mechanisms of the *Bombyx* developmental processes with these homeobox genes as probes. It is necessary to analyze in what stages the *Bm Ubx, abd-A, Abd-B,* and *Antp* genes begin to be expressed during embryogenesis and where in embryos the transcripts of these genes are distributed. We can also localize the homeodomain proteins encoded by the *Bm Ubx, abd-A, Abd-B,* and *Antp* genes by immunohistochemical methods. It is important to find out what factors regulate the temporal and spatial expression of these homeotic genes, and how the homeodomain proteins regulate the specification of the body segments in controlling their target genes. For the study of these problems, the use of a transgenic system in which we can transfer modified DNA into *Bombyx* would be a powerful approach, but such a transgenic system

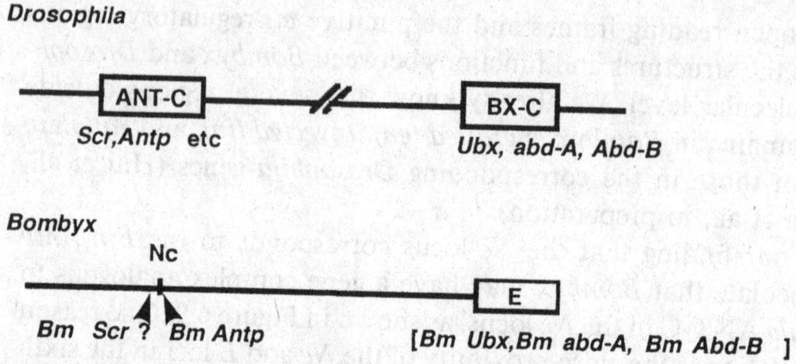

Figure 6.9. Similarities of the structures of the homeotic gene complexes between *Drosophila* and *Bombyx.* We speculate that the E locus accommodates the *Bm Ubx, abd-A,* and *Abd-B* genes and the *Nc* locus corresponds to *Bm Antp* and is flanked by the *Bm Scr* gene. These homeotic gene complexes are speculated to be homologous to the BX-C and ANT-C gene complexes in *Drosophila.* The brackets indicate that the gene order is not yet established.

is not yet available (see Iatrou, this volume). Using the transgenic system in *Drosophila,* we will be able to obtain limited information about the basic functions of the homeobox genes and their flanking regions and about the compatibility between the two insect species.

The homeodomain proteins encoded by homeotic genes are proposed to act as transcriptional regulators on target genes that are necessary to specify the identities of the segments. To determine the molecular mechanisms of determination of the segments, these target genes must be identified. A few target genes have been tentatively identified in *Drosophila* (Gould, Brookman, Strutt, and White, 1990). Because a transcript of the *Bm Antp* gene is detected in the middle silk gland and the development of the silk gland is affected by the *Nc* mutation, we expect that the homeodomain protein encoded by *Bm Antp* may have a role in regulating the development of the silk gland as well as the expression of the silk protein genes.

Acknowledgments

We thank Dr. H. Doira for the gifts of eggs and larvae of *E* group mutants and the members of our laboratory for permitting us to cite their unpublished data. The work in our laboratory was partly supported by Grants-in-Aid for Research of Priority Areas and a research grant from the TERUMO Life Science Foundation.

7

Chorion genes: an overview of their structure, function, and transcriptional regulation

FOTIS C. KAFATOS, GEORGE TZERTZINIS, NIKOLAUS A. SPOEREL, and HANH T. NGUYEN

Introduction

The silk moth eggshell (chorion) has been studied intensively since the first biochemical report on its composition and synthesis was published twenty years ago (Kawasaki, Sato, and Suzuki, 1971). To date, the work has made significant contributions to two distinct but interrelated fields: molecular evolution and developmental biology. This chapter will serve as a general introduction to the biology of the chorion, as well as a review of our current understanding of the mechanisms for developmentally regulated chorion gene expression. Other aspects of the system are treated in greater detail in chapters by Eickbush and Izzo, Regier et al., and Goldsmith.

Two features have made eggshell formation (choriogenesis) an excellent model system for the study of development: a high degree of tissue, spatial and temporal regulation of the structural genes encoding chorion proteins, and informative similarities and differences in choriogenesis among different insects. Indeed, the comparative approach has been an important hallmark of chorion research from the outset. Thus, although our primary interest here is choriogenesis in Lepidoptera, it is important to place the information in the context of what is known about eggshell formation in the other well-studied group of insects, the Diptera. It should be noted that these two orders had their last common ancestor approximately 250 million years ago, and thus have been separated in evolution approximately as long as mammals from birds.

In both Lepidoptera and Diptera, the eggs are formed in follicles (often called egg chambers in *Drosophila*), which consist of three cell types. The first is a single oocyte. The second type is 7 or 15 nurse cells, respectively, which are sister cells of the oocyte and have been formed by three or four sequential and terminal divisions of an oogonium; these divisions are incomplete, leaving intercellular bridges through which materials produced in the nurse cells are conveyed to the growing oocyte. Third is a monolayer of 1,000 to 10,000 follicular epithelial cells (follicular cells for short), which are of somatic mesodermal origin, surround the oocyte-nurse cell complex, and, when the nurse cells degenerate near the end of oogenesis, alone envelop the oocyte. At this late stage, after the end of vitellogenesis (the uptake of yolk by the oocyte), the epithelial cells embark on the process of choriogenesis: They synthesize chorion proteins and secrete them into the space between themselves and the oocyte, where they assemble to form the eggshell. The chorion genes are exclusively active in the follicular cells, that is, they are extremely tissue specific. Once choriogenesis is complete, the follicular epithelium sloughs off (ovulation) and the oocyte, surrounded by the proteinaceous chorion, is ready to enter the oviduct, as a prelude to fertilization and subsequent embryonic development.

Spatial and temporal regulation in choriogenesis

In both moths and *Drosophila,* the eggs develop in assembly-line fashion: In the ovary, progressively more mature contiguous follicles are interconnected in strings (ovarioles), each string representing successive developmental stages, from the earliest in the germarium at the anterior end to the most mature near the oviduct at the posterior end (Figure 7.1). Thus developmental stages can be recovered merely by dissecting successive follicles of the ovariole. In *Drosophila,* stages are easily distinguished morphologically and are numbered 1 to 14 (King, 1970), but in silk moths the morphological landmarks are limited and the follicles are usually staged by position within the ovariole, relative to a size discontinuity that results from water uptake into the oocyte in a terminal growth phase at the end of vitellogenesis, just before choriogenesis.

Paralleling the developmental polarity of the ovariole as a whole is a clear-cut anterior-posterior polarity within each follicle (see Figure. 7.1). The nurse cells are always found at the anterior end, where the passageway for sperm (micropyle) subsequently forms, traversing the chorion to permit fertilization, and where the anterior, head structures of the embryo develop later. Many species have additional regional differentiations of the follicular cells, which are clearly manifested in the chorion that they

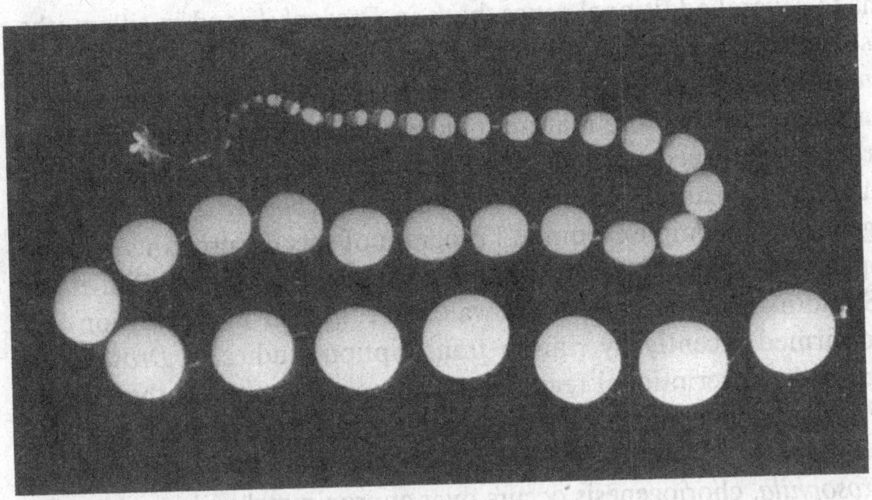

Figure 7.1. Ovariole dissected from a developing *A. polyphemus* moth. Anterior is at upper left, posterior at bottom right. Note the anterior cap of gray nurse cells in the top row of young follicles, and the abrupt increase in size of follicles in terminal growth (between the second and third rows), just before choriogenesis. (From Paul et al., 1972.)

produce. For example, in the flattened ellipsoid egg of the saturniid silk moth, *Antheraea polyphemus,* three distinct regions of the chorion are evident, in addition to the micropyle: a disordered strip along the longest equator, two flanking equatorial strips decorated by air-capturing protrusions (aeropyle crowns), and two lateral, flattened regions that bear simple polygonal imprints of follicular cells (Kafatos et al., 1977). In *Drosophila,* differentiations along both the anterior-posterior and the dorsal-ventral axes of the chorion indicate the existence of 10 recognizably distinct subpopulations of the follicular epithelium (Margaritis et al., 1980). For example, whereas cells at the extreme anterior pole of the chorion construct the micropyle, cells of the posterior pole form an aeropyle protrusion; in the anterior-dorsal region the epithelium produces a complex operculum (a trapdoor specialization for the hatching of the larva) and two long dorsal appendages, which are respiratory in function. Genetic evidence indicates that the regional differentiations of the epithelium are related to the polarity of the underlying oocyte (Schüpbach, 1987). For present purposes it is important to note that the regional differentiations are reflected in the spatial regulation of at least some of the chorion genes. Spatial regulation has been studied in detail for the *Drosophila s36* gene, using P-element transformation procedures (Tolias and Kafatos, 1990), and also pertains to several additional genes, both in *Drosophila* and in silk moths (see also subsequent discussion).

Even more dramatic is the temporal regulation of chorion gene expression. In progressively more mature follicles, the synthesis of specific chorion proteins is turned on and then off, paralleling the appearance and subsequent disappearance of the corresponding mRNAs (Figure 7.2). Thus choriogenesis is a complex temporal program of specific protein synthesis that appears to be regulated largely at the transcriptional level (Sim et al., 1979; Spradling, Waring, and Mahowald, 1979). That interpretation has been confirmed recently by run-on transcription studies in *Drosophila,* although posttranscriptional regulation also appears to affect the expression of at least two chorion genes (Romano, Martínez-Cruzado, and Kafatos, 1991).

In *Drosophila,* choriogenesis occurs over approximately 6 hr and entails the programmed expression of approximately 20 genes. Seven of these, named *s38, s37, s36, s19, s18, s16,* and *s15,* according to the approximate size of the encoded proteins in kD, are found in two tight clusters, the first three on the X chromosome (at the 7F1 cytogenetic location) but the rest on 3L (at 66D11-15). A minor gene is found at yet a third location, at 7C (Bauer and Waring, 1987). Conventionally we consider *s38, s37,* and *s36* early genes, *s19* and *s16* of middle temporal specificity, and *s18* and *s15* late genes. In reality, however, each of these genes, except for *s36* and *s38,* is distinguishable in terms of timing of expression. In *B. mori* and other silk moths, choriogenesis lasts approximately 10 times longer than in *Drosophila* and entails expression of approximately 10 times as many genes.

In *Bombyx* it has been shown that the chorion structural genes are clustered on a single chromosome (reviewed in Eickbush and Izzo, this volume). Here, again, one distinguishes conventional developmental classes (early, middle, late), but in reality there is greater temporal heterogeneity (Bock, Campo, and Goldsmith, 1986). These genes are related in evolutionary terms and are members of an ancient superfamily with two major branches, α and β. Both α and β components are included in each temporal class, and these components are classified in families according to their sequence similarities. Two families of α and β early proteins are now designated ErA and ErB, respectively (including those previously identified as CA and CB). The middle proteins belong to two other families, A and B, which have been subdivided into early-middle, middle I, and middle II subclasses (Spoerel, Nguyen, Eickbush, and Kafatos, 1989).

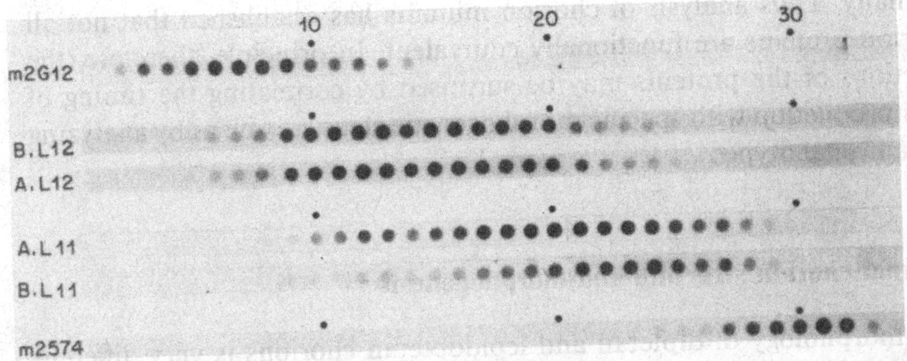

Figure 7.2. Temporal regulation of chorion gene transcripts in *B. mori.* Total RNA was isolated from sequential, individual follicles, which are numbered in the abscissa, beginning at the start of the choriogenesis; positions 30 to 34 represent pools of four follicles each, thereby compressing the expression profile of the late genes. The RNA dotblots were hybridized to specific ^{32}P-labeled single-stranded probes from the genes identified in the ordinate. For each probe the exposure time was adjusted to yield approximately similar intensities during the peak of accumulation. Note the completely coordinate appearance and disappearance of transcripts from the A.L12 and B.L12-like genes. The A.L11 probe detects two subgroups of gene pairs, only one of which is detected by the B.L11, late-middle probe. (From Spoerel et al., 1986.)

The cysteine-rich late proteins are members of the HcA and HcB families (see Eickbush and Izzo, this volume). In both Diptera and Lepidoptera, the complex developmental program of chorion transcription/translation is presumably necessitated by the elaborate morphogenesis of the chorion. In the homozygous *Gr^{col}* mutant of *B. mori,* a gross deficiency in early proteins results in failure of the normally produced later proteins to assemble properly, ultimately leading to the collapse of the chorion (Nadel, Goldsmith, Goplerud, and Kafatos, 1980). The homozygous *Gr^{B}* mutant, which lacks some middle and all of the late proteins (while over-producing some of the late-middle components), also produces a collapsing egg. Even the *Gr^{B}* heterozygote, where the affected proteins are present in half dose (Nadel et al., 1980), has abnormal chorion structure in some genetic backgrounds (Takasaki, 1962). Similarly, a set of nested translocations that replace only middle components fails to rescue all of the ultrastructural defects in the *Gr^{B}* homozygote (Alexopoulou, Nelson, and Goldsmith, personal communication; see also Goldsmith, this volume). In *D. melanogaster,* the absence of proteins encoded by either one of two early genes (Digan et al., 1979; Bauer and Waring, 1987) results in major structural abnormalities, even though the rest of the proteins are produced normally. Thus analysis of chorion mutants has established that not all chorion proteins are functionally equivalent. In principle, therefore, the functions of the proteins may be surmised by correlating the timing of their production with specific morphogenetic steps, as well as by analyzing mutant phenotypes.

Internal chorion structure and morphogenesis

The morphology of dipteran and lepidopteran chorions is very different (Figure 7.3; Kafatos et al., 1977; Margaritis et al., 1980). The *Drosophila* chorion is a thin structure, consisting largely of the so-called endochorion, where a fenestrated floor, closest to the oocyte, is separated by air spaces from a more substantial, outer domed roof; these two layers are connected by vertical pillars. The chorion is deposited sequentially, in the order floor, pillars, roof. Although rotary shadowing reveals granules and interconnecting fibrils, no fibrous substructure is evident by ordinary electron microscopy.

In contrast, the silk moth chorion consists of numerous layers of fibers, oriented largely parallel to the surface and forming prominent helicoidal lamellae (Mazur et al., 1989; see also Regier et al., this volume). It is constructed in four sequential morphogenetic modes. The earliest one is assembly of a framework consisting of thin, low density lamellae. By temporal correlation, we presume that this initial framework is largely made up of ErA and ErB proteins. The phenotype of Gr^{col} homozygotes indicates that this framework is crucial in morphogenetic terms. During the middle period of choriogenesis, the lamellae of the framework do not change in number, but undergo profound modification through two morphogenetic modes: expansion and densification. The framework expands in height as additional fiber sheets are inserted into preexisting lamellae. This expansion mode begins farthest from the follicular epithelial cells and progresses in reverse toward them, suggesting that the proteins intercalate into the framework after diffusing away from the secretory cells (Blau and Kafatos, 1978). Following upon expansion is densification: Individual fibers grow in thickness, presumably through accretion of newly synthesized proteins, and eventually fuse. This process, which also begins in the most distant lamellae and proceeds in reverse, results in an approximately twofold increase in overall chorion density without further lamellar expansion. These two modes have been studied most carefully in *A. polyphemus,* but

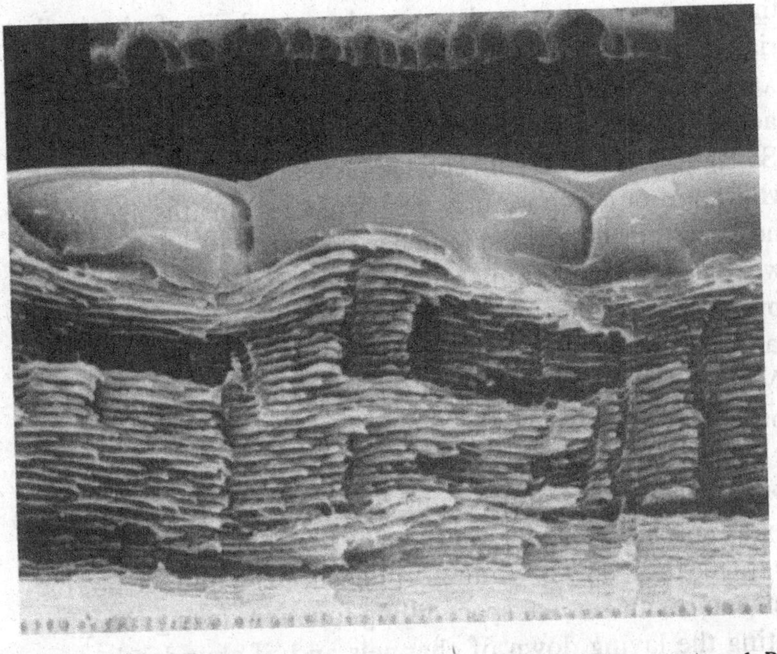

Figure 7.3. Comparison of chorions from *D. melanogaster* (top) and *B. mori* (bottom). Scanning electron micrographs of transverse rips through mature eggshells are shown. In both cases the oocyte would be toward the bottom of the picture. Note that the fly chorion at most resembles the innermost (trabecular) layer of the moth chorion; both structures largely consist of air spaces separated by multiple vertical pillars. However, the bulk of the moth chorion is made up of multiple horizontal lamellae, which are not evident in the fly chorion. These fibrous, helicoidal lamellae consist of proteins of the moth chorion superfamily (discussed in the text); the composition of the trabecular layer is as yet unknown. Bar = 5 μm. (From Kafatos et al., 1985.)

they also occur in *B. mori*. Presumably the various types of middle A and B proteins are responsible for expansion and densification.

The fourth and final morphogenetic mode is a recapitulation in miniature of lamellogenesis: At the chorion surface a few additional lamellae are assembled and molded into prominent surface layers and decorations. In *B. mori*, the late lamellae are very compact and are deposited throughout the surface of the chorion, forming a thick, electron dense outermost layer. This layer consists of HcA and HcB proteins, as indicated by temporal correlation between its appearance and the changing protein synthetic pattern, by the phenotypes of the Gr^B mutant and of the earlier mentioned chromosome translocations that lack both the high cysteine proteins and the outermost chorion layer, and by analysis of physically isolated material. Similarly, the outer chorion layer of *B. mandarina*, which forms exaggerated mountainlike peaks, is also composed of densely packed lamellae, scrapings of which consist almost entirely of cysteine-rich proteins (Sakaguchi et al., 1990). The Hc proteins are absent from *A. polyphemus*, where the outer chorion is very different. In this species it appears that the very thin outermost lamellae are made of late A and B proteins. In two equatorial strips mentioned earlier these lamellae are molded into prominent surface decorations, the chimneylike aeropyle crowns that lead into cross-chorion aeropyle channels (see Regier et al., this volume). Molding of the crowns is achieved by locally abundant secretion of a loose "filler" material consisting of late proteins known as E1 and E2. These proteins are produced at low levels in other parts of the follicle, where aeropyle crowns are missing and only aeropyle channels are present. In summary, E1 and E2 are thought to be architectural components (to use the terminology of viral morphogenesis) that have no role in the mature structure but serve a space-filling function during morphogenesis, permitting the laying down of channels and, if abundantly produced, the elaboration of open protuberances.

The structural and morphogenetic differences of the outer chorion in the two species, *B. mori* and *A. polyphemus*, correlate with important physiological differences. The small *B. mori* eggs usually undergo diapause, as the *B. mandarina* eggs invariably do. In these species, the eggs must remain viable and resist desiccation for months; sealing the entire surface with Hc proteins presumably aids survival of the embryo. In contrast, the large and nondiapausing *A. polyphemus* eggs possess a thick chorion and must have as their major challenge the efficient exchange of respiratory gases during embryogenesis. The inner chorion of this species has an extensive network of air spaces interconnected with the aeropyle channels.

The densification and especially the expansion modes suggest considerable fluidity, and together with the geometrical arrangement of the chorion fibers are consistent with the interpretation that the developing chorion is a cholesteric liquid crystalline structure. This interpretation is also consistent with deviations from the ideal helicoidal fibrous array, which are well known from studies on liquid crystals (Figure 7.4). Continuous deformations of the fibrous array, called *distortions* and *defects*, resulting from insertion or excision of material from the array, arise after the initial deposition of the framework (Mazur et al., 1989). In addition to being informative about the presumed liquid crystalline nature of the developing

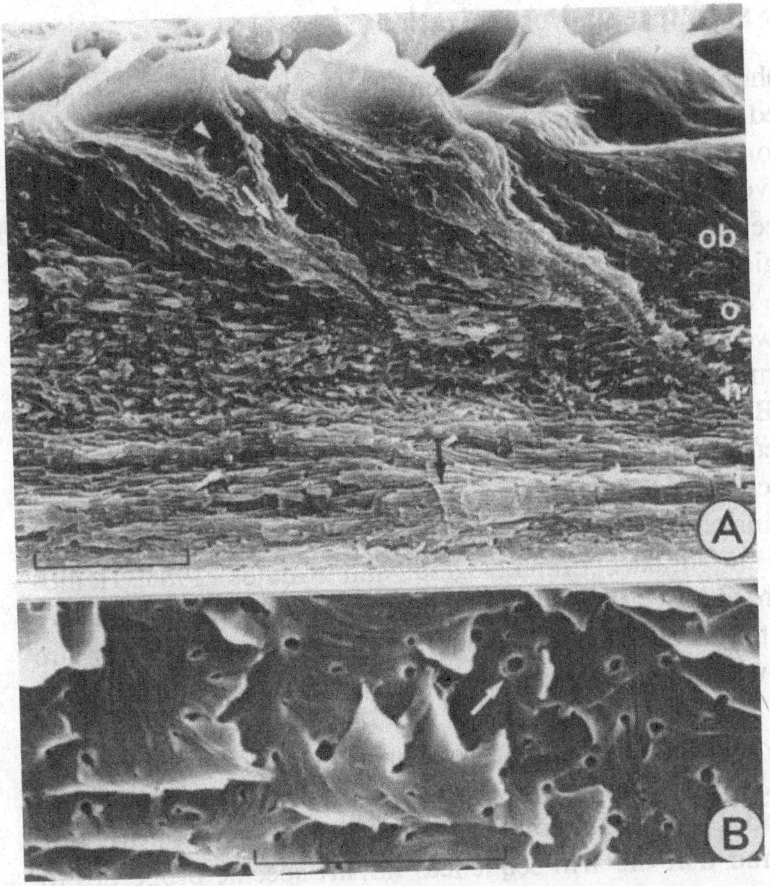

Figure 7.4. Chorion lamellae and screw dislocation defects in *A. polyphemus. Panel A:* Low power view of a transverse rip through the aeropyle region. The aeropyle openings (white arrowhead) lead to internal channels (white arrow) which clearly penetrate the oblique (ob), outer (o), and holey (h) lamellar zones. In the inner lamellar zone (i), frequent screw defects (black arrow) are seen as vertical breaks in the lamellae. Scale marker = 10 μm. *Panel B:* Detail of an almost horizontal rip through the inner lamellar zone. Abundant screw defects appear as round holes (arrow), and the surrounding lamellae often show a slight helical twist. Scale marker = 4 μm. (From Mazur et al., 1989.)

chorion, various defects and distortions may be important physiologically. For example, a type of defect called screw dislocation creates channels through the chorion, probably serving for respiratory gas exchange and possibly also permitting hardening of the mature chorion through an ultrastructurally evident late-permeating material.

Chorion gene structure and organization

The vast morphological difference between silk moth and *Drosophila* chorions is reflected in major differences in the sequence as well as the number and organization of the corresponding structural genes (see also Eickbush and Izzo, this volume).

With the exception of the E1 and E2 filler proteins, which are unrelated to the rest (Regier, 1986), all the characterized silk moth chorion proteins belong to the α and β branches of the chorion superfamily. What distinguishes these two branches is the sequence of the central domain. This is very similar among all α families (ErA, A, HcA), as it is among all β families (ErB, B, HcB); the α and β central domains are clearly distinct, although limited similarities indicate that they share distant homology (Figure 7.5). Secondary structure predictions indicate that both the α and β central domains consist of short (6 residue) β-sheet strands, alternating with β-turns (Hamodrakas, Etmektzoglou, and Kafatos, 1985; Hamodrakas, Bosshard, and Carlson, 1988). Flanking the central domain are protein segments (N- or left and/or right arms, respectively), which are highly variable within and between families. These arms frequently contain peptide repeats, such as Cys.Gly, Cys.Gly.Gly, and Gly.Tyr.Gly.Gly.Leu. The variability of the arms helps define subfamilies and gene types within each family. A possible model is that the central domains of α and β proteins interact to form a core unit for constructing chorion fibers, and that the arm sequences impart specific properties to the fibers (concerning assembly, cross-linking, etc.). This hypothetical model provides a rationale for the paired organization of moth chorion genes (see subsequent discussion).

The sequences of chorion proteins in *Drosophila* are quite different from all those that are known in moths. No conserved central domain is apparent, and peptide repeats (Gly.Tyr.Gly.Gly, polyalanine) are distributed throughout each protein rather than being sequestered near the ends. The similar, albeit variably organized, peptide repeats suggest that different *Drosophila* proteins are distantly homologous to each other (see Figure 7.5). However, the evidence for homology is only marginal: Alignment scores between two *Drosophila* chorion proteins, *s18* and *s15*, are only 2 to 3 standard deviations above the mean for 50 randomly scrambled sequences of the same composition (Kafatos et al., 1987). Similarly, the most extensive similarity that we have detected between a *Drosophila* and a *Bombyx* chorion protein only marginally suggests homology: The initial

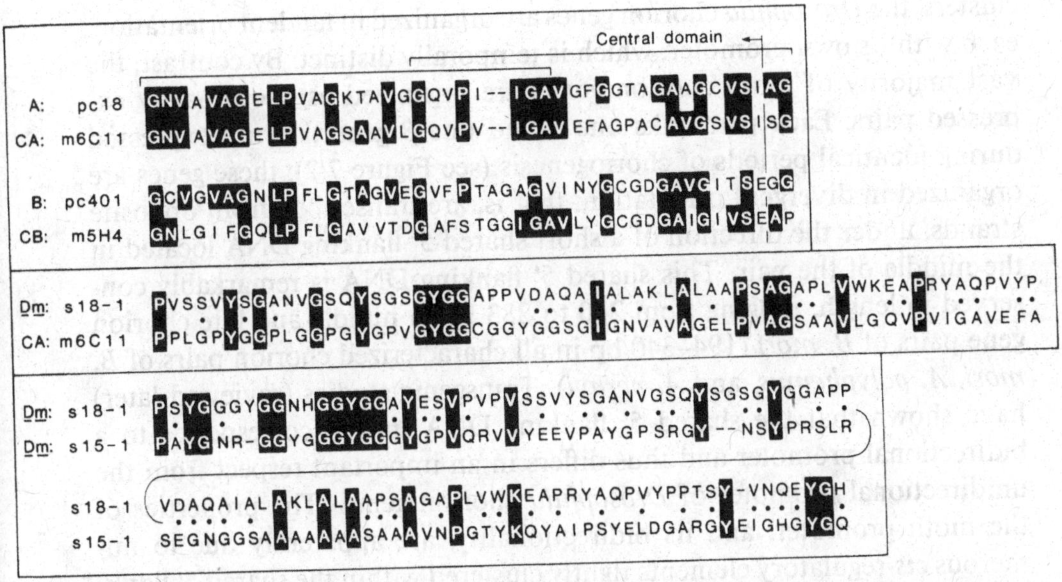

Figure 7.5. Distant chorion protein similarities. *Top panel:* Homology of the α and β branches of the moth chorion superfamily. The typical α sequences pc18 (A family) and m6C11 (ErA family) are compared with the typical β sequences pc401 (B family) and m5H4 (ErB family). Amino acid residues present in one or both of the α as well as β sequences are shown in white against black background. The regions compared encompass the COOH-terminal part of the central domains and the first two residues of the right arms. Sequence identities, neglecting conservative replacements, are 40% for pc18 × pc401, 40% for pc18 × m5H4, 40% for m6C11 × pc401, and 33% for m6C11 × m5H4. Good alternative alignments offset by 18 residues are also possible (horizontal bars), suggesting a subrepeat structure of the central domain (cf. Hamodrakas et al., 1985, 1988). *Middle panel:* Sequence similarity of the *D. melanogaster* chorion protein *s18* (approximately the middle third of the sequence) with the *B. mori* m6C11 sequence (ErA family; left arm except for the first 10 residues and central domain except for the last 11 residues). *Bottom panel:* Sequence similarity of *D. melanogaster s18* (approximately the middle half of the sequence) and *s15* (all but the first 15 and the last 4 residues of the large exon). The matches of the middle and bottom panels were detected by the Lipman-Pearson FASTP computer program and also take into account conservative replacements (dots). Respectively, they showed 70 initial and 73 optimized alignment score values (3.8 and 2.0 standard deviations above the mean of scores for 50 randomly scrambled sequences) and 54 initial and 74 optimized alignment score values (2.3 and 2.4 standard deviations above the mean for 50 randomly scrambled sequences). (From Kafatos et al., 1987.)

and optimized alignment scores between *s18* and the ErA protein, m6C11, are 3.8 and 2.0 standard deviations above the mean for 50 randomly scrambled sequences (Kafatos et al., 1987; see Figure 7.5).

The chorion genes of flies and moths are very different, not only in sequence but also in organization. In both the X-linked and the autosomal clusters, the *Drosophila* chorion genes are organized in tandem orientation, each with its own promoter, which is temporally distinct. By contrast, the vast majority of moth chorion genes are organized as coordinately expressed pairs. Each pair links one α and one β gene that are expressed during identical periods of choriogenesis (see Figure 7.2); these genes are organized in divergent orientation; that is, are transcribed from opposite strands, under the direction of a short shared 5′ flanking DNA located in the middle of the pair. This shared 5′ flanking DNA is remarkably conserved in length, ranging from 270 to 283 bp in middle and late chorion gene pairs of *B. mori* (194–340 bp in all characterized chorion pairs of *B. mori, A. polyphemus,* and *A. pernyi*). Transgenic studies (reviewed later) have shown that the shared 5′ flanking DNA indeed corresponds to a bidirectional promoter and thus differs in an important respect from the unidirectional promoter of *Drosophila* chorion genes. The properties of the moth promoter, and its bidirectionality, are apparently due to numerous *cis*-regulatory elements tightly clustered within the shared 5′ flanking DNA.

Cis-regulation of chorion gene expression

Expression of chorion genes in nuclear polyhedrosis viral vectors

The standard procedure for identifying *cis*-acting DNA elements that are important for developmentally regulated gene expression is to modify the DNA by site-directed in vitro mutagenesis and assay its ability to direct gene expression in vivo. Because of the convenience of P-element mediated transformation this approach has been used extensively to identify the regulatory sequences of a wide variety of *Drosophila* genes. Unfortunately, no such germline transformation procedure is available as yet for Lepidoptera. However, two methods have been used to express cloned moth chorion genes in vivo: transduction using nuclear polyhedrosis virus vectors and heterologous transformation in *Drosophila* using P-element vectors.

Transduction of Bm5 tissue culture cells with polyhedrin promoter-directed Hc genes revealed that these chimeric genes are actively transcribed and that a significant proportion of the correctly initiated transcripts are also spliced correctly (Iatrou, Meidinger, and Goldsmith 1989; see also Iatrou, this volume). As a result, authentic Hc proteins can be produced in infected cells, and the coding potential of these genes can be identified with specific protein spots in two-dimensional gels. More recently, an entire Hc gene pair, HcA/HcB.12, with its own bidirectional promoter was incorporated within the intron of a promoter-inactivated polyhedrin gene and used to infect developing adults of the *Gr*^{B} mutant, which totally lacks Hc genes (Iatrou and Meidinger, 1990). Aberrant HcA/HcB.12 transcripts were seen in a variety of tissues and in Bm5 cells, but correctly initiated transcripts were exclusively detected in infected follicular cells, thus establishing that the chorion promoter is correctly regulated in these transduced constructs. The aberrant transcripts may possibly be related to the fact that some of the endogenous chorion locus DNA is transcribed in the follicles in the antisense direction, in both *Bombyx* and *Drosophila* (Skeiky and Iatrou, 1990; Romano et al., 1991), sometimes outside the boundaries of the canonical transcription units. The significance of these unexpected antisense transcripts is still unknown.

Heterologous in vivo expression in *Drosophila*

Despite the major differences between flies and moths in chorion gene organization and sequence, we reasoned that the regulatory mechanisms may be similar enough to permit expression of moth chorion promoters introduced into *Drosophila* within P-element vectors. Indeed, we were able to confirm this conjecture (Mitsialis and Kafatos, 1985). Regulated expression, essentially limited to choriogenic follicular cells, was documented with DNA from four different chorion gene pairs (A/B.L12 of *B. mori*, Po18/401 and Po292/10 of *A. polyphemus*, and Pe18/401 of *A. pernyi*). In every case, the 5′ flanking DNA of the chorion gene pair was sufficient for imparting chorion-specific regulation on a reporter bacterial gene, encoding chloramphenicol acetyltransferase (CAT) (Mitsialis, Spoerel, Leviten, and Kafatos, 1987; Mitsialis, Veletza, and Kafatos, 1989) or β-galactosidase (Fenerjian, 1991). Expression in follicles at prechoriogenic stages was detectable but extremely low. Ectopic expression elsewhere in the body or at different stages of the life cycle was either extremely low or undetectable. The reporter gene could be attached to either end of the 5′ flanking DNA, replacing either the α or the β gene, with similar results. Thus these experiments showed that the short 5′ flanking DNA of moth chorion gene pairs is indeed a bidirectional promoter, which can be appropriately recognized across species lines by the *trans*-regulatory factors present in the follicular cells of transgenic *Drosophila*.

Because of chromosomal position effects, the level of expression observed varies widely (by up to three orders of magnitude) in independent transformant lines bearing the same construct. Nevertheless, when a number of independent lines sufficient for statistical evaluation is analyzed, it can be seen that some of the moth chorion promoters (A/B.L12 and Pe18/401) are expressed at high levels, up to the levels of the strong *Drosophila* chorion promoter, *s15*. Others are expressed at generally low levels (Po18/401 and Po292/10) (Mitsialis, Veletza, and Kafatos, 1989). In general, the average levels of expression in the two possible orientations of each promoter differ by no more than tenfold.

The broad temporal specificity of moth promoters is preserved in transgenic *Drosophila,* in the sense that expression is largely limited to choriogenic stages. The differences between fly and moth choriogenesis are so extensive that we cannot ask directly whether the detailed temporal specificity is conserved. However, the question has been answered in the negative indirectly by comparing the relative heterologous expression patterns of promoters that are known to differ in endogenous temporal specificity. In the moth, the endogenous 18/401 genes of both *A. polyphemus* and *A. pernyi* are expressed relatively late in development (Sim et al., 1979; Moschonas, Thireos, and Kafatos, 1988), whereas A/B.L12 and Po292/10 clearly are expressed earlier (early to middle stages) (Sim et al., 1979; Spoerel, Nguyen, and Kafatos, 1986). Nevertheless, in transgenic *Drosophila* the late Pe18/401 promoter begins its expression relatively early, at approximately stage 12, whereas its late *A. polyphemus* homologue, Po18/401, as well as the early moth promoters, is expressed very late, at stage 14, in a manner resembling the late *Drosophila* gene *s15.* We conclude that not all *cis-trans* interactions that define the precise temporal specificity are conserved between flies and moths. However, in every case heterologous temporal regulation is identical in both orientations, again emphasizing the bidirectionality of the moth promoters.

At least some of the *Drosophila* chorion genes are spatially regulated, being expressed uniquely, earlier, or more intensely in some of the follicular cell subpopulations, such as the anterior dorsal cells that construct the operculum and the dorsal appendages (Parks and Spradling, 1987; Tolias and Kafatos, 1990). This is also observed with a β-galactosidase gene under the control of the Pe18/401 promoter, suggesting that at least some elements of spatial regulation may also be conserved between flies and moths (Fenerjian, 1991).

Localization of *cis*-regulatory elements

To date, the most detailed analysis of moth chorion promoters has been performed on the *B. mori* A/B.L12 promoter, which was assayed after extensive deletion and linker-substitution mutagenesis (Mitsialis et al., 1987; Spoerel, Nguyen, Towne, and Kafatos, 1993). Figure 7.6 summarizes the results of this analysis, which took into account sequence comparisons between this promoter and the proximal 5′ flanking DNA region of the *s15* gene in four *Drosophila* species (Figure 7.7; cf. Martínez-Cruzado et al., 1988). The latter region in *D. melanogaster* is sufficient for imparting the characteristic late follicular expression on reporter genes (Mariani, Lingappa, and Kafatos, 1988, and unpublished observations). In other *Drosophila* species the *s15* gene is expressed in an indistinguishable manner (Fenerjian, Martínez-Cruzado, Swimmer, King, and Kafatos, 1989), but this DNA region shows very uneven sequence divergence, with certain strictly conserved motifs embedded in more or less extensively diverged DNA.

Consideration of the conserved sequences identified a single motif, GTCACGT, which is shared by *s15* in all *Drosophila* species, as well as A/B.L12. Elimination of this motif totally inactivated the A/B.L12 promoter in both orientations (Mitsialis et al., 1987). Furthermore, this motif was shown to be essential for expression of the *Drosophila s15* gene in the homologous transformation system (Mariani et al., 1988). Point mutations changing either of the two Cs in this motif to T similarly inactivate the Pe18/401 bidirectional promoter (Fenerjian, 1991).

Indeed, TCACGT is likely to be an important element in almost all chorion genes of Lepidoptera as well as Diptera. It has been found just upstream of all four autosomal chorion genes in all four *Drosophila* species tested (Levine and Spradling, 1985; Wong, Pustell, Spoerel, and Kafatos, 1985; Martínez-Cruzado et al., 1988; Fenerjian et al., 1989; Swimmer, Fenerjian, Martínez-Cruzado, and Kafatos, 1990) and in a similar location on two X-linked chorion genes in both *Drosophila* (Spradling et al., 1987) and the medfly, *Ceratitis capitata* (Konsolaki et al., 1990). It also occurs in most moth chorion 5′ flanking DNA regions (Spoerel et al., 1986), including the unrelated "filler" E1 and E2 promoters (Regier, Hatzopoulos, and Durot, 1986), but not in the promoters of ErA and ErB genes (Hibner et al., 1988). A less precisely conserved 17-nucleotide consensus sequence, which encompasses TCACGT, is represented with variations in all moth

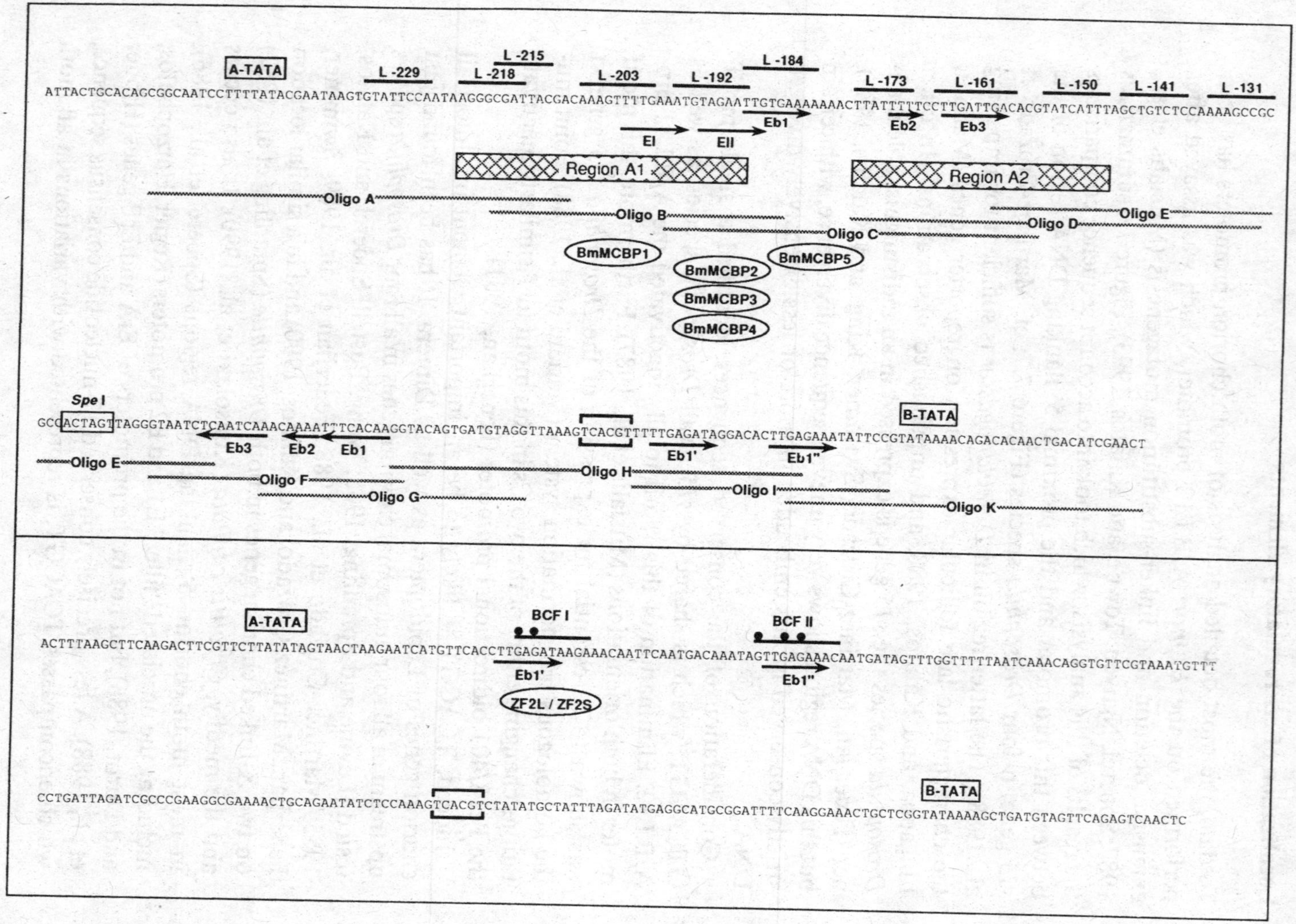
A-TATA
L -229
L -218
L -215
L -203
L -192
L -184
L -173
L -161
L -150
L -141
L -131
ATTACTGCACAGCGGCAATCCTTTTATACGAATAAGTGTATTCCAATAAGGGCGATTACGACAAAGTTTTGAAATGTAGAATTGTGAAAAAAACTTATTTTTCCTTGATTGACACGTATCATTTAGCTGTCTCCAAAAGCCGC
Eb1
Eb2
Eb3
EI
EII
Region A1
Region A2
Oligo A
Oligo B
Oligo C
Oligo D
Oligo E
BmMCBP1
BmMCBP2
BmMCBP3
BmMCBP4
BmMCBP5
Spe I
B-TATA
GCCACTAGTTAGGGTAATCTCAATCAAACAAAATTTCACAAGGTACAGTGATGTAGGTTAAAGTCACGTTTTTGAGATAGGACACTTGAGAAATATTCCGTATAAAACAGACACAACTGACATCGAACT
Eb3
Eb2
Eb1
Eb1'
Eb1"
Oligo E
Oligo F
Oligo G
Oligo H
Oligo I
Oligo K
A-TATA
BCF I
BCF II
ACTTTAAGCTTCAAGACTTCGTTCTTATATAGTAACTAAGAATCATGTTCACCTTGAGATAAGAAACAATTCAATGACAAATAGTTGAGAAACAATGATAGTTTGGTTTTTAATCAAACAGGTGTTCGTAAATGTTT
Eb1'
Eb1"
ZF2L / ZF2S
B-TATA
CCTGATTAGATCGCCCGAAGGCGAAAACTGCAGAATATCTCCAAAGTCACGTCTATATGCTATTTAGATATGAGGCATGCGGATTTTCAAGGAAACTGCTCGGTATAAAAGCTGATGTAGTTCAGAGTCAACTC

Figure 7.6. Sequences and features of the bidirectional promoter regions of two *B. mori* chorion gene pairs. *Top panel:* The A/B.L12 promoter. The TATA boxes of the A and B genes are indicated, and the TCACGT hexamer is boxed. Elements EI, EII, Eb1, Eb2, and Eb3 are also indicated (cf. Figure 7.7). L-131 to L-229 are linker-scanning mutations of the A-half of the promoter (numbered relative to the transcription start site) that identified Region A1 as important for quantitative and temporal regulation (cf. Figure 7.8) and Region A2 as quantitatively important by testing in transgenic *Drosophila.* Oligonucleotides A–K were used to detect specific DNA-binding proteins (see Figure 7.11). Oligonucleotides B and/or C were used to clone five sequence specific DNA-binding proteins, BmMCBP1 to BmMCBP5, as indicated. Additional binding sites for these factors may be present in the A/B.L12 promoter (Spoerel et al., 1993 and unpublished data). *Bottom panel:* The HcA/HcB.12 promoter. The TATA boxes, Eb1 elements, and TCACGT hexamer are indicated as in the top panel. Binding sites of factors BCFI and BCFII are shown, together with contacted bases as determined by methylation interference analysis. (From Skeiky and Iatrou, 1991.)

Figure 7.7. Recurrent motifs in chorion promoter sequences. *Right panel:* The bidirectional moth promoters of four divergent A and B chorion gene pairs, which have been shown to be regulated in transgenic *Drosophila:* the *B. mori* A/B.L12 promoter (MoL12), the *A. pernyi* 18/401 promoter (Pe18/401), and the *A. polyphemus* 18/401 and 292/10 promoters (Po 18/ 401 and Po 292/10, respectively). For comparison, the *Drosophila s15* promoter (Dm15) is also shown. Small open triangles mark the position and orientation of TATA elements (aligned on the B side); terminal filled triangles indicate the orientations in which the frag- ments were tested in vivo. The TCACGT hexamer is indicated by H and filled arrows. Open arrows indicate the recurrent elements I and II, and thin arrows below the promoters (a to g) indicate other motifs that occur symmetrically (cf. Figure 7.6 for a more detailed view of the A/B.L12 promoter). Below the promoter diagrams is shown the transformation vector pCarCat-1 (Mitsialis et al., 1987), which has been used in in vivo promoter analysis. *Left panel:* Sequences of the hexamer-containing motif (H) and elements I and II in various chorion promoters. (From Mitsialis et al., 1989.)

Symbol	Promoter	Motif
H	Mo*L12*	TTAAaGTCACGTtTTTG
	Pe*18/401*.RC	TTTtGGTCACGTcTTTc
	Po*18/401*	GgcttcTCACGTATTTG
	Po*18/401*.RC	aAAActTCACGTAAATt
	Po*292/10*.RC	TTAAacaCACtTtTTaA
	Dm *s15*	TATAGGTCACGTAAATG
	Ds *s15*	TATGGGTCACGTAAATA
	Dv *s15*	GcAGtGTCACGTAAATA
	Dg *s15*	GcAGGGTCACGTAAgTA

```
Consensus   TAAA         AA A
                GGTCACGTA  T
            GTTG         TT G

            5354         54 4
Matches         468999897  8
            3333         44 3
```

Symbol	Promoter	Motif
I	Mo*L12*	TtTGAAAT
	Po*18/401*	TtTGAAAT
	Po*292/10*	TATGAAAT
	Dm, Ds, Dv, Dg *s15*	TATGAAAT
II	Mo*L12*	gTAGAATTG
	Pe*18/401*	aTAGAATTG
	Po*18/401*	aTAGAATTG
	Dm *s15*	gTAGAATaG

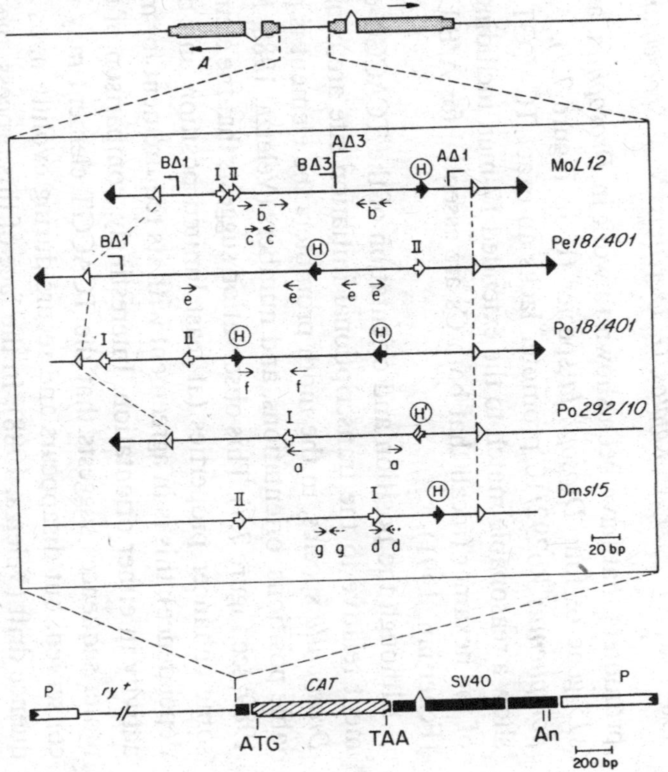

promoters that have been shown to work in *Drosophila*, as well as in the *s15* gene of four *Drosophila* species (H in Figure 7.7). Although the *A. polyphemus* Po292/10 promoter lacks an exact TCACGT motif, it does show a reasonable match to the extended 17-mer, including the CAC core of the hexamer (recall that both Cs are essential for A/B.L12 expression) (Fenerjian, 1991).

Although the position and orientation of the TCACGT-containing element relative to the transcriptional initiation site are conserved in four *Drosophila* species, in the moth promoters the element is present in variable positions, orientations, and numbers (Veletza, 1988; Mitsialis et al., 1989; see Figure 7.7). This observation suggests that the element may have some enhancer properties (at least limited position and orientation independence); this is in agreement with its requirement for moth promoter activity in either orientation. Interestingly, comparison of the moth promoter sequences suggests that the TCACGT element may not be strictly conserved, but disappears and recurs during evolution, as a result of sequence drift (Veletza, 1988). In the face of this process, the occurrence of two or more copies of TCACGT in some promoters (such as Po18/401) might stabilize them against random inactivation.

Although the extended TCACGT-containing element, which is present in the B gene half of the A/B.L12 promoter, appears to be necessary for *s15*-like expression, it is certainly not sufficient. A 40 bp terminal deletion from the A-end of the promoter reduces expression from the B-end by approximately one order of magnitude, and a more extensive deletion, to a central *Spe* I site, inactivates the promoter. The distance between these two breakpoints, 112 bp, defines a segment of the A-half of the promoter that clearly contains essential *cis*-regulatory elements. This segment was extensively mutagenized (see Figure 7.6), and 11 different linker substitutions were tested in vivo in the A orientation; representative mutations were also tested in the B orientation (Spoerel et al., 1993). To overcome random quantitative variations due to chromosomal position effects, approximately 15 independent transformant lines were tested for each mutation or orientation, 263 transformants in all. Each and every mutation acted in an identical manner in both orientations (Figure 7.8), once again emphasizing the bidirectionality of the promoter. Three linker mutations, spanning a region designated as A2 (−179 to −150 relative to the B transcription start site), showed activation in the normal late period (stage 14B), but at an average level less than 1 percent of the wild type; activity at all earlier stages was proportionally affected. A more complex phenotype

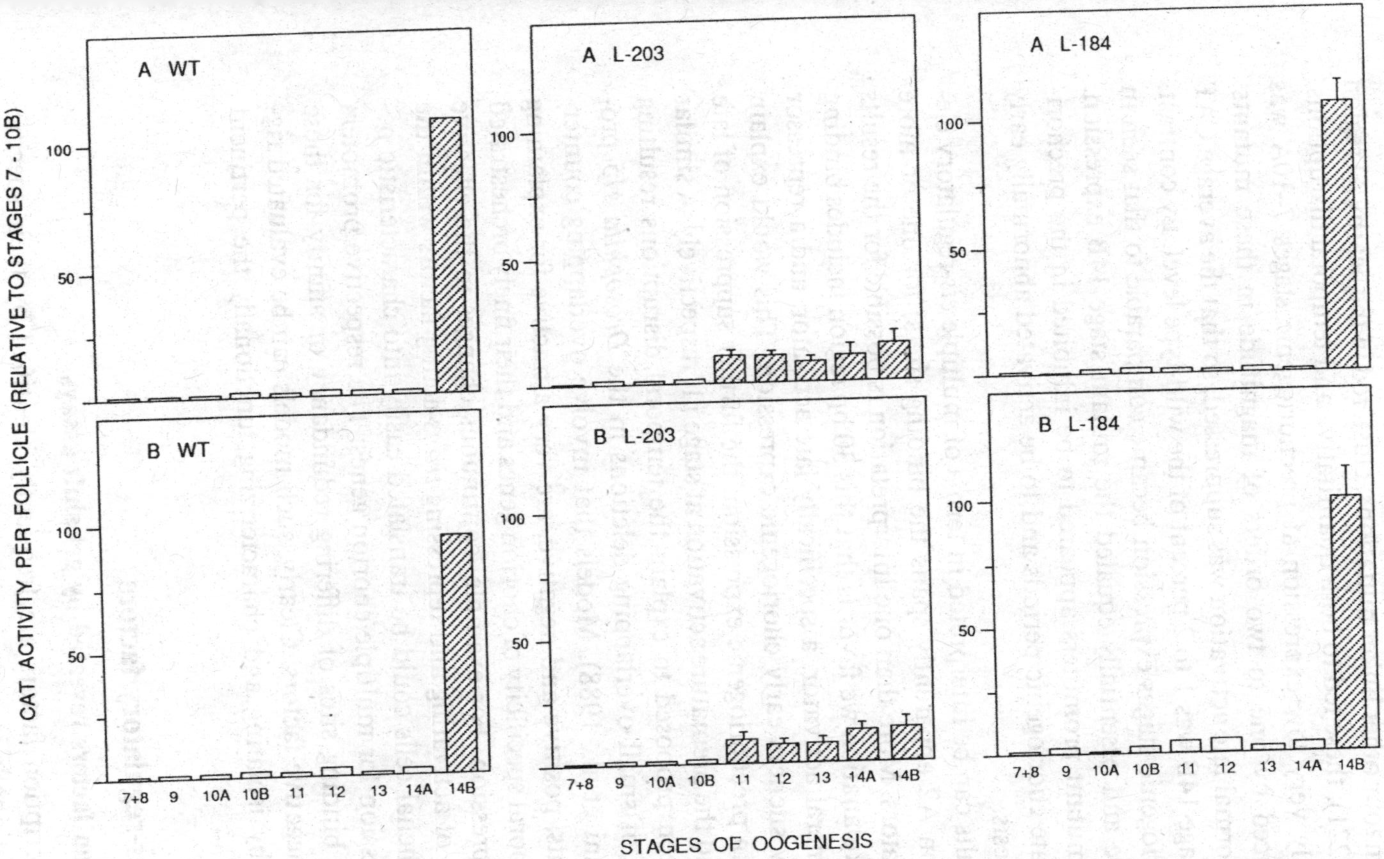

Figure 7.8. Temporal specificity of CAT activity levels in follicles of wild-type (WT) and mutant transformant flies, bearing the CAT reporter gene under the control of the bidirectional moth A/B.L12 promoter, either in the A or the B orientation, as indicated. For the location of the −184 linker-scanning mutations, see Figure 7.6. For each construct, heterozygous flies were assayed from three different single-insert transformant lines. Extracts from equal numbers of follicles from each indicated stage of oogenesis were assayed. For each line, the results of choriogenic stages are plotted as the activation ratio relative to the prechoriogenic stages, 7–10A. The histogram shows this ratio, averaged for the three lines. Note that the temporally and quantitatively disrupting L-203 mutation acts bidirectionally. (From Spoerel et al., 1993.)

resulted from four other linker mutations, clustered in the nearby Al region (−192 to −221); these led to both quantitative and temporal disruptions. The normally very low expression at prechoriogenic stages 7–10A was further reduced by one to two orders of magnitude in these mutants; moreover, normal late activation was suppressed, so that the average CAT activity at stage 14B was 1 to 2 percent of the wild-type level. By contrast, at early choriogenic stages expression became comparable to that seen in the wild type and essentially equaled the mutant stage 14B expression. Thus these mutant promoters appeared to be inhibited in the prechoriogenic and late choriogenic periods and to be activated abnormally early in choriogenesis.

These results can be interpreted in terms of multiple *cis*-regulatory elements. Region A2 apparently spans the binding site(s) for one or more general activators. More than one interpretation is possible for the results of A1. The explanation we favor is that this 30 bp region includes binding sites for a general activator, a specifically late activator, and a repressor that normally silences early choriogenic expression. This would explain the decrease in prechoriogenic expression, the further suppression of late activation, and the premature activation at stage 11, respectively. A similar model has been proposed to explain the temporal disruptions resulting from a series of small overlapping deletions in the *Drosophila s15* promoter (Mariani et al., 1988). Models that involve overlapping counteractive elements, positive and negative, are very attractive for explaining the strict temporal specificity of chorion genes and their finely orchestrated sequential expression. For example, continuous changes in the effective concentration of activating and repressing *trans*-acting factors within the follicular epithelial cells could be translated easily into characteristic periods of expression for multiple chorion genes: The respective promoters could include binding sites of differing redundancy or affinity for these positive and negative factors. Clearly, such models can be evaluated rigorously only by isolating and characterizing functionally the pertinent factors.

Chorion *trans*-regulatory factors

Putative chorion factors revealed by gel shift assays

Putative transcription factors are frequently identified by their specific binding to functionally important *cis*-regulatory DNA elements. To assay binding, the mobility shift procedure (band shift, gel retardation) (Fried and Crothers, 1981) is widely used. In this assay, nuclear proteins are allowed to bind to regulatory ("promoter") DNA fragments, and the specificity and site of binding are established by competition experiments and the use of mutated fragments. The exact DNA sites where the proteins bind are determined by DNase I footprinting (Galas and Schmitz, 1978) and methylation interference analysis (Brunelle and Schleif, 1987). Once the binding site of a putative factor on the DNA is defined, it can be used as a tool to clone the corresponding protein: Expression libraries are screened by binding of oligonucleotide probes corresponding to the DNA target site of the factor (Vinson, LaMarco, Johnson, Landschulz, and McKnight, 1988; Singh, LeBowitz, Baldwin, and Sharp, 1988). Alternatively, if the factor of interest is thought to be homologous to a previously isolated factor from another species, its cloning can be achieved by cross-hybridization or by PCR amplification using primers designed according to the sequence of the known factor. In the last two years, these techniques have been used to clone 15 putative chorion transcription factors, 7 from *Drosophila* and 8 from *Bombyx* (Figure 7.9).

The first published identification of putative chorion factors used the mobility shift procedure to assay binding to the *Drosophila s15* chorion gene (Shea, King, Conboy, Mariani, and Kafatos, 1990). The proximal regulatory region, where deletion analysis in vivo had indicated the existence of *cis*-regulatory elements, was mutagenized by linker substitution and used to detect components of follicular nuclear extracts capable of specifically recognizing important promoter elements. Three major factors or factor assemblies were thus detected and mapped by comparing their interactions with wild type and mutated probes (Figure 7.10). Complex I binds to the TCACGT hexamer and immediately upstream; complex II has an overlapping target site, just upstream of the hexamer; and the major complex, X, binds even further upstream in the broad region that behaves as a negative element in vivo. Interestingly, complex II is highly enriched in late follicles, in agreement with its binding to a positive, late activating DNA element. Complexes I and X are generated by both early and late follicular extracts, as well as extracts from embryos, but not KC_0 and S3S tissue culture cells (Shea, personal communication). Two factors, CF1 and CF2 (see Figure 7.9), which correspond to complexes I and II, have been cloned and analyzed in some detail (Shea et al., 1990; see later discussion). The factor(s) responsible for complex X has not been cloned as yet.

***Drosophila melanogaster* factors:**

name	homology group	promoter containing binding site	isolated by
CF1	Nuclear receptors	***s15*** -80 -42	M. Shea
CF2	Zinc finger	***s15*** -80 -42	"
CF3	Zinc finger	***s15*** -120 -76	"
CF4	?	***s36*** -93 -49	P. Tollias
CF5	HMG-box	***s15*** -66 -32	T. Hsu, C. LaBonne
CF6	?	***s15*** -66 -32	"
CF7	?	***s15*** -80 -32	A. Khoury, D. Thanos

***Bombyx mori* factors:**

name	homology group	promoter containing binding site	isolated by
BmCF1	Nuclear receptor	A/B.L12 -1 -1	G. Tzertzinis
ZF2L / ZF2S	Zinc finger	HcA/B.12 -1 -1	K. Iatrou
BmMCBP1	?	A/B.L12 -1 -1	E. Rhee & N.Spoerel
BmMCBP2	trithorax HMG-I motifs	A/B.L12 -1 -1	"
BmMCBP3	HMG-I motif	A/B.L12 -1 -1	"
BmMCBP4	SNF2	A/B.L12 -1 -1	"
BmMCBP5	HMG-I	A/B.L12 -1 -1	"

Figure 7.9. Fifteen putative chorion transcription factors cloned by cross-hybridization or PCR amplification.

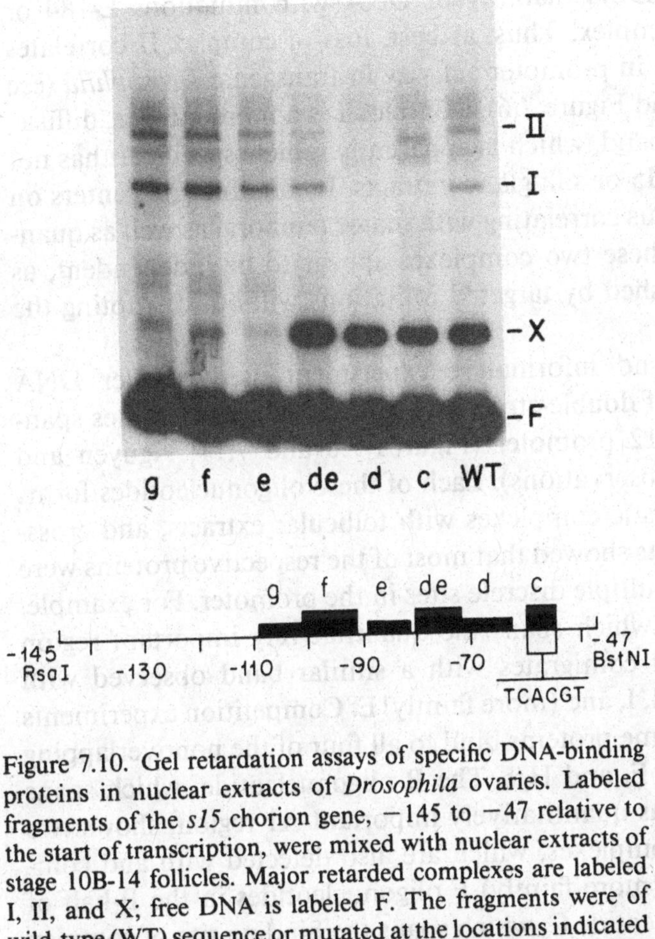

Figure 7.10. Gel retardation assays of specific DNA-binding proteins in nuclear extracts of *Drosophila* ovaries. Labeled fragments of the *s15* chorion gene, −145 to −47 relative to the start of transcription, were mixed with nuclear extracts of stage 10B-14 follicles. Major retarded complexes are labeled I, II, and X; free DNA is labeled F. The fragments were of wild-type (WT) sequence or mutated at the locations indicated by the filled boxes (c to g). Note that mutation c eliminates the TCACGT hexamer (open bar) and interferes with formation of complex I, as does mutation d just upstream. Complex II is only disrupted by mutation d, whereas complex X is disrupted by mutations e, f, and g. (From Shea et al., 1990.)

Specifically binding factors have also been detected in *B. mori* follicular extracts, using fragments of both middle A/B and late HcA/HcB bidirectional promoters. Two major complexes, I and II, were observed reproducibly with a relatively large DNA fragment, spanning the A-half of the A/B.L12 promoter (Nguyen, unpublished observations). Complex II is a sharp band that appears not to be follicle specific (it is also observed with extracts of the ovarian cell line Bm5 and, at a much lower level, with silk gland extracts). Its binding site spans the −192 to −173 region, as shown by the complete or partial inability of DNA with mutations L-184 or L-173 to form this complex. Thus, at best, loss of complex II correlates with a minor decrease in promoter activity in transgenic *Drosophila* (see previous discussion and Figure 7.6). Complex I is a more intense, diffuse, and slower migrating band, which is apparently follicle specific; it has not been detected with BM5 or silk gland extracts. Its binding site centers on the L-203 mutation, thus correlating with major temporal as well as quantitative disruptions. These two complexes appear to be independent, as each one can be abolished by targeted mutations without disrupting the other.

A more sensitive and informative experiment used smaller DNA probes, namely, a set of double-stranded synthetic oligonucleotides spanning the entire A/B.L12 promoter (Figures 7.6 and 7.11; Nguyen and Spoerel, unpublished observations). Each of these oligonucleotides forms one or more characteristic complexes with follicular extracts, and cross-competition experiments showed that most of the respective proteins were capable of binding to multiple discrete sites in the promoter. For example, the D oligonucleotide, which spans the quantitatively important region A2, forms a band that comigrates with a similar band observed with oligonucleotides A, F, H, I, and (more faintly) E. Competition experiments demonstrate that the same proteins bind to all four of the nonoverlapping oligonucleotides, A, D, F, and H/I. The B oligonucleotide, which spans the temporally as well as quantitatively important A1 region, shows two closely spaced major complexes, which are also detected with and competed by the G, I, and (more faintly) F oligonucleotides in the B-half of the promoter. Oligonucleotide C, which spans the binding site of complex II, forms a unique complex in addition to two major complexes that are competed by oligonucleotides B, G, and I. Finally, oligonucleotide A forms a complex that comigrates closely with the unique complex of oligonucleotide C but is competed only by oligonucleotides E, H, and (weakly) K. Thus numerous factors bind to the A/B.L12 promoter, frequently at multiple sites. The bidirectional effects of mutations at individual sites in vivo suggest that a multicomponent assembly is responsible for coordinate transcription in both directions, instead of each half of the promoter operating independently.

Four major complexes, A1 to A4, were detected with a 170 bp fragment of the HcA.12 promoter (Skeiky and Iatrou, 1991). In this case complex formation appears to be synergistic and hierarchical. Formation of complex A1 requires a factor named BCFI, A2 requires both BCFI and a second

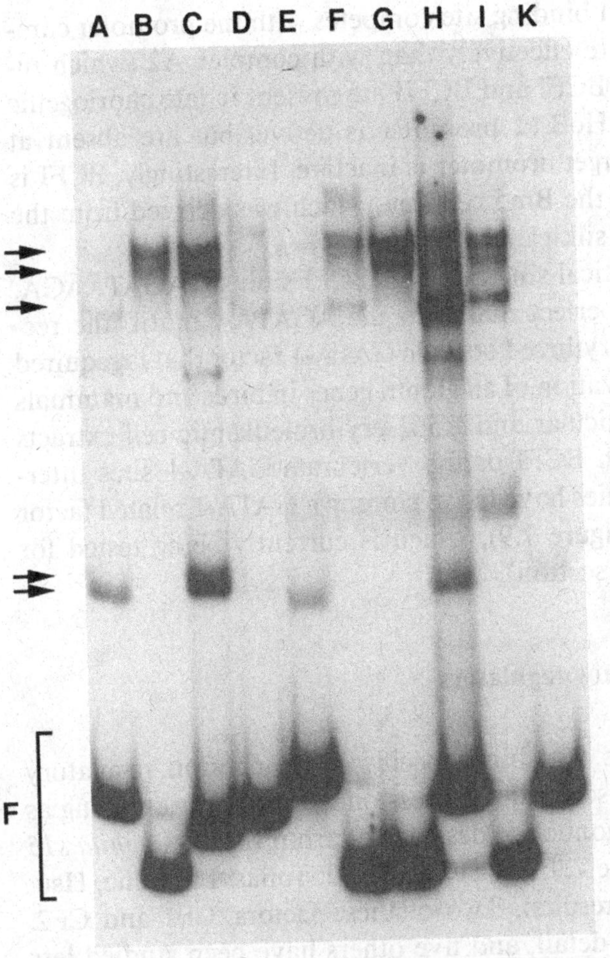

Figure 7.11. Gel retardation assays of specific DNA-binding proteins in nuclear extracts of choriogenic *B. mori* follicles. Labeled partially overlapping double-stranded oligonucleotides (see Figure 7.6, top) were mixed with nuclear extracts, incubated for 30 min at 4°C, and analyzed on a 5% polyacrylamide gel. Positions of major retarded complexes are marked by arrows; free DNA is labeled F. Note the identical migration of several of these complexes (see text).

factor, BCFII, and A3 and A4 are higher order complexes that require these two factors as well as additional ones that have not been studied in detail. The binding sites of BCFI and BCFII were defined by DNase I footprinting and methylation interference assays (see Figure 7.6, bottom panel). Oligonucleotides were used to confirm the binding sites and to demonstrate the interactions of these factors through competition experiments. An oligonucleotide that encompasses the BCFI binding site is sufficient for forming a stable complex with follicular extracts. The oligonucleotide corresponding to BCFII is effective in competing with the promoter for BCFII binding, but does not by itself yield a complex that is stable under the conditions of the mobility shift assay. This suggests that stable BCFII binding requires prior or simultaneous binding of BCFI. Conversely, BCFII appears to stabilize BCFI binding: The oligonucleotide corresponding to the BCFI binding site competes with the promoter complex A1 (BCFI alone) more effectively than with complex A2 (which includes both factors). Both BCFI and BCFII are present in late choriogenic follicles, where the HcA/HcB.12 promoter is active, but are absent at earlier stages, when the target promoter is inactive. Interestingly, BCFI is also present in extracts of the Bm5 cell line, which was derived from the ovary, but is absent from silk gland extracts.

The footprint of the critical sine qua non BCFI site is GAGATAAGA, and thus encompasses a perfect match to the AGATAA motif, the recognition sequence for the erythroid-specific GATA-1 factor that is required for the transcriptional activation of all globin genes in birds and mammals (Orkin, 1990). Indeed, follicular and R562 erythroleukemic cell extracts bind DNAs with the moth BCFI or the vertebrate GATA-1 sites interchangeably. These similarities have led to cloning a GATA-1 related factor from *B. mori*, ZF2 (see Figure 7.9), which is currently being tested for identity to BCFI (see next section).

Cloned putative chorion *trans* regulators

***Drosophila* factors.** Factors that bind specifically to chorion regulatory DNA have been cloned by screening expression cDNA libraries, using as a binding target labeled oligonucleotides derived from the *Drosophila s15* or *s36* promoters (see Figure 7.9; Shea et al., 1990; Tolias, LaBonne, Hsu, and Khoury, unpublished results). Two of these factors, CF1 and CF2, have been characterized in detail, and five others have been studied less extensively. Of the latter five, CF3 is a member of the C_2H_2 zinc finger family (Shea et al., in preparation), and CF4 has a basic DNA binding domain that does not show strong similarities to any other protein in the data bases (Tolias, personal communication). CF5 contains a domain homologous to the nuclear high-mobility group proteins HMG1 and HMG2. Several other transcription factors with this structure have been recently described, including the nucleolar transcription factor UBF (Jantzen, Admon, Bell, and Tjian, 1990) and SRY, a gene from the human sex-determining region (Sinclair et al., 1990). CF6 and CF7 have not been characterized as yet.

CF1 is a member of the nuclear hormone receptor superfamily, an apparent homologue of the mammalian receptors H-2RIIBP (mouse) (Hamada et al., 1989) and hRXRα (human) (Oro, McKeown, and Evans,

1990). CF1 has two zinc fingers of the C_2C_2 type, like other members of the superfamily, and shares with them strong sequence similarity throughout the DNA binding domain. Indeed, this similarity was the basis of its independent cloning by two other groups (Henrich, Sliter, Lubahn, MacIntyre, and Gilbert, 1990; Oro et al., 1990), in parallel with the cloning based on specific binding to the *s15* promoter (Shea et al., 1990). CF1 also has significant similarities to the carboxy terminal ligand binding domains of known hormone receptors, although its ligand has not been identified. As shown by DNase I footprinting and methylation interference analysis, the target site of CF1 on the *s15* promoter encompasses the beginning of the hexamer and immediate upstream nucleotides, GGTCAC. This is the first half of a hyphenated palindrome, GGTCACGTAAATGTCC, spaced differently but otherwise very similar to typical vertebrate hormone receptor recognition elements, such as the estrogen response element, (A/T)GGTCAC_n(G/T)TG(A/T)CCT. Many hormone receptors are known to bind as heterodimers (Glass, Lipkin, Devary, and Rosenfeld, 1989). It is not yet known whether the second half-palindrome also binds CF1, perhaps at lower efficiency, or an as yet unknown factor.

Identification of a putative chorion *trans* regulator as a member of the nuclear hormone receptor superfamily suggests the possibility that choriogenesis may be hormonally regulated in *Drosophila*. This possibility is further supported by the observation that a mutation that disrupts formation of the chorionic appendages, *fs(1)de12* (Orr, Galanopoulos, Romano, and Kafatos, 1989), is allelic to the *Broad-Complex* locus, which is implicated in the mediation of steroid hormone action in *Drosophila* (Belyaeva et al., 1981; Meyerowitz et al., 1985). CF1 is highly enriched in follicles, including the nuclei of follicular cells, but is also present in numerous other tissues at lower levels, suggesting a wider role in hormone action. In fact, the gene encoding this factor has been identified with the *ultraspiracle (usp)* locus, because it rescues embryonic and early larval lethal *usp* mutations (Oro et al., 1990). These results necessitate a genetic clonal analysis involving the creation of *usp*-deficient clones in the follicle cells, for testing rigorously whether CF1 indeed is a chorion regulator, as suggested by its binding properties.

CF2 is a member of another superfamily of transcription factors, the well-known C_2H_2 zinc finger superfamily exemplified by the vertebrate TFIIIA and Sp1 factors, as well as by many *Drosophila* pattern regulators, such as the product of the *Krüppel* gene (Shea et al., 1990). It shows two distinct zinc finger domains, one at the amino and the other at the carboxy terminal region, and alternative splicing within the latter domain (Hsu, Gogos, Kirsh, and Kafatos, 1992). A major binding site for this factor on *s15* is just upstream of CF1, although in the presence of zinc, additional minor sites are evident (Shea et al., 1990). CF2 apparently has some latitude in target site recognition (Gogos, Tzertzinis, and Kafatos, 1991), and its sequence preferences are modulated by the participation of different zinc fingers from the differently spliced forms of the protein (Gogos, Hsu, Bolton, and Kafatos, 1992). The CF2 transcript maps at 25A5-8 on polytene chromosomes, in a region that has been subjected to extensive genetic analysis, and thus the functional significance of this factor should also be amenable to genetic analysis. Like CF1, CF2 is highly enriched in follicles relative to other tissues.

A highly encouraging observation is that CF1 and CF2 bind specifically not only to the promoter used in their isolation but also to other chorion promoters, including moth promoters that are expressed in transgenic *Drosophila* (Gogos and Mitsialis, unpublished observations). On the A/B.L12 promoter, CF1 binds not only to the region of TCACGT in the B-half of the promoter but also to an imperfect match, CACGT, in the A-half. Similarly, CF2 binds to multiple sites in both halves of the promoter. Finally, the uncharacterized CF7 factor has been shown to bind to both *s15* and A/B.L12 promoter fragments (Khoury, personal communication).

***Bombyx mori* factors.** Both direct binding of their products to DNA and homology to cloned factors are being employed to isolate *Bombyx* cDNAs encoding factors that regulate chorion gene expression. With the assumption that the key domains of the proteins will be sufficiently conserved between flies and moths, the latter method was used successfully to clone BmCF1, the homologue of the *D. melanogaster* chorion factor CF1. This was achieved by PCR amplification of *B. mori* DNA, using as primers regions that are specifically conserved between CF1 and its vertebrate homologues; the amplified fragment was then used to recover cDNA clones, which were sequenced in their entirety (Figure 7.12; Tzertzinis, Malecki, and Kafatos, in press). Interestingly, the DNA binding domains of BmCF1 and CF1 are identical except for three amino acid replacements, two of which are within a motif of the second zinc finger that is thought to be involved in DNA recognition specificity through protein-protein interactions (D-box) (Umesono and Evans, 1989). The ligand binding/dimerization domains show 51 percent identical (71 percent similar) amino acid residues, whereas the central hinge segments and the probable

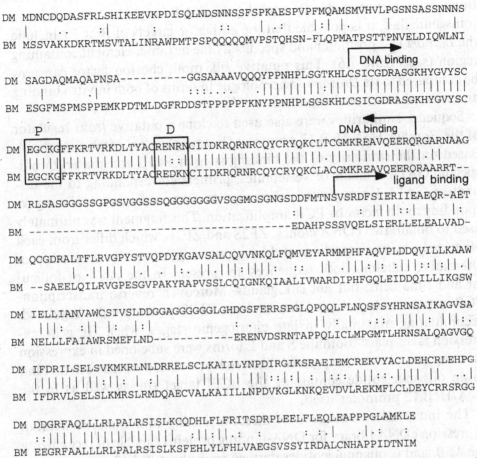

```
DM MDNCDQDASFRLSHIKEEVKPDISQLNDSNNSSFSPKAESPVPFMQAMSMVHVLPGSNSASSNNNS
BM MSSVAKKDKRTMSVTALINRAWPMTPSPQQQQMVPSTQHSN-FLQPMATPSTTPNVELDIQWLNI

                                                         DNA binding
DM SAGDAQMAQAPNSA----------GGSAAAAVQQQYPPNHPLSGTKHLCSICGDRASGKHYGVYSC
BM ESGFMSPMSPPEMKPDTMLDGFRDDSTPPPPPFKNYPPNHPLSGSKHLCSICGDRASGKHYGVYSC

   P              D                                   DNA binding
DM EGCKGFFKRTVRKDLTYACRENRNCIIDKRQRNRCQYCRYQKCLTCGMKREAVQEERQRGARNAAG
BM EGCKGFFKRTVRKDLTYACREDKNCIIDKRQRNRCQYCRYQKCLACGMKREAVQEERQRAARRT--
                                                      ligand binding

DM RLSASGGGSSGPGSVGGSSSQGGGGGGVSGGMGSGNGSDDFMTNSVSRDFSIERIIEAEQR-AET
BM ------------------------------------------EDAHPSSSVQELSIERLLELEALVAD-

DM QCGDRALTFLRVGPYSTVQPDYKGAVSALCQVVNKQLFQMVEYARMMPHFAQVPLDDQVILLKAAW
BM --SAEELQILRVGPESGVPAKYRAPVSSLCQIGNKQIAALIVWARDIPHFGQLEIDDQILLIKGSW

DM IELLIANVAWCSIVSLDDGGAGGGGGGLGHDGSFERRSPGLQPQQLFLNQSFSYHRNSAIKAGVSA
BM NELLLFAIAWRSMEFLND------------ERENVDSRNTAPPQLICLMPGMTLHRNSALQAGVGQ

DM IFDRILSELSVKMKRLNLDRRELSCLKAIILYNPDIRGIKSRAEIEMCREKVYACLDEHCRLEHPG
BM IFDRVLSELSLKMRSLRMDQAECVALKAIILLNPDVKGLKNKQEVDVLREKMFLCLDEYCRRSRGG

DM DDGRFAQLLLRLPALRSISLKCQDHLFLFRITSDRPLEELFLEQLEAPPPGLAMKLE
BM EEGRFAALLLRLPALRSISLKSFEHLYLFHLVAEGSVSSYIRDALCNHAPPIDTNIM
```

Figure 7.12. Sequence alignment of the *Drosophila* and *Bombyx* CF1 proteins (DM and BM, respectively). Identical amino acid residues are marked with vertical tick marks, strongly conservative replacements with two dots and conservative replacements with single dots. Deletions are indicated by dashes. Note the high degree of conservation in and immediately around the DNA binding domain, contrasting with the high divergence of the preceding putative activation domain and the ensuing short hinge region. The extensive, C-terminal ligand binding domain is moderately conserved. Note that of the two boxes that are thought to be important in DNA recognition, the P box is identical between the two species, but the D box contains two of the three replacements of the DNA binding domain. (From Tzertzinis et al., in press.)

transactivation domains at the amino terminus of BmCF1 and CF1 are not conserved (see Figure 7.12). Bacterially expressed fragments of BmCF1 containing the DNA binding domain fused to glutathione-S-transferase bind specifically the hexamer (TCACGT) containing region of the *Drosophila s15* promoter (Tzertzinis, unpublished results). Given the functional equivalence of the *Drosophila s15* and *Bombyx* L12 promoters in transgenic flies, it is possible that the authentic targets of BmCF1 include the *B. mori* L12 promoter and specifically the hexamer element containing region (see Figure 7.6). This putative silk moth chorion factor can now be compared to its *Drosophila* homologue in terms of both in vitro binding and in vivo functional significance.

Sequence similarities were also used to clone a putative *trans*-regulator of the *Bombyx* HcA/HcB.12 promoter (Iatrou, personal communication). Based on the similarity of the binding sites of BCFI and the erythroid-specific GATA-1 factor of vertebrates, primers corresponding to the distinctive GATA-1 zinc fingers were used to clone a fragment of a corresponding moth factor by PCR amplification. This fragment was ultimately used to obtain two cDNA clones, ZF2S and ZF2L, which differ from each other only by the insertion of 14 codons, suggesting the possibility of alternative splicing. Both sequences are present in the RNA of follicular and Bm5 cells, but not silk glands. Moreover, reverse transcription-PCR experiments indicate that at least ZF2S is temporally regulated, accumulating maximally in late choriogenic stages, when Hc gene expression takes place. Both the S and L forms were subcloned in expression vectors, and the overexpressed bacterial proteins were shown to bind to an oligonucleotide containing the BCFI target site as well as to the HcA/HcB.12 promoter itself.

The independent approach of direct screening of a *Bombyx* ovarian expression cDNA library for DNA-binding proteins capable of recognizing the A, B, and C oligonucleotides derived from the A/B.L12 promoter (see discussion and Figure 7.6) has also been highly successful for isolating clones corresponding to putative chorion regulatory factors (Spoerel et al., in preparation). Five different classes of cDNA clones have been isolated, each corresponding to a single-copy silk moth gene, and each encoding a product with distinct binding specificity. None of these clones corresponds to the *Drosophila* CF factors.

The BmMCBP5 factor is a silk moth homologue of mammalian HMG-I and HMG-Y proteins. It is most similar in length to HMG-I (115 vs. 107 amino acid residues), and contains several oligopeptide repeats of alternating arginines and glycines with the core consensus sequence K-(R)-P/G-R-G-R-P. The HMG-I proteins have three such repeats, which are thought to form β-sheets and to mediate binding to AT-rich sequences via the minor groove of DNA, (Kolodrubetz, 1990); the BmMCBP5 protein possesses four repeats (Figure 7.13). Like HMG-I, the BmMCBP5 protein has a highly acidic C-terminus, similar to that of several transcriptional activators. When overexpressed as a fusion product with maltose-binding protein, BmMCBP5 has been shown to bind to the −184 to −192 region of the A/B.L12 promoter (see Figure 7.6), where it also forms higher order structures. Its transcript is developmentally regulated, being present in ovarioles, embryos, and second instar larvae, but is absent from adult males. It has recently been demonstrated that HMG I(Y) stimulates the binding of NFκB to the human interferon β promoter (Thanos and Maniatis, 1992), suggesting that BmMCBP5, which was cloned by binding to fragments of the A/B.L12 promoter, may well be functionally significant.

The DNA binding region of the BmMCBP2 protein contains two motifs that are similar to the K-(R)-P/G-R-G-R-P DNA binding motifs of the HMG-I proteins and the BmMCBP5 factor, whereas BmMCBP3 contains one such motif. The presence of these motifs suggests that DNA binding of these two proteins may at least partially be mediated by an HMG-I-like DNA-binding domain. Thus proteins that interact with little specificity with DNA (HMG-I) seem to share features of their DNA binding domain with sequence-specific DNA-binding proteins. Similarly, the DNA binding domain of the HMG1/2 proteins is also found in a growing number of sequence-specific DNA-binding proteins, including the *Drosophila* factor CF5 (see earlier discussion). BmMCBP2 binds to the quantitatively and temporally important A1 region of the A/B.L12 promoter (see Figure 7.6). An intriguing observation is that this factor is encoded by two forms of cDNA, one of which has an insertion of three codons just in the middle of one of the HMG-I like motifs. This insertion changes the regular spacing of alternating arginines and thereby substantially decreases the potential of approximately 20 adjacent amino acid residues to form β-sheet. Differential expression of the two proteins could have important implications for chorion gene regulation. Most interestingly, the BmMCBP2 protein shows elsewhere in its sequence a cysteine-rich region (see Figure 7.13) with repeated Cys, His motifs, which resembles a Zn binding domain in the product of the *Drosophila* pattern formation gene, *trithorax* (Mazo, Huang, Mozer, and Dawid, 1990), but is distinct from consensus Zn finger motifs. This region could either function as a second DNA binding domain or in protein–protein interactions.

The BmMCBP4 factor is a large protein (1,813 amino acid residues). The central third of this protein is clearly homologous to the yeast transcriptional activator, SNF2 (33 percent identity and 78 percent similarity in 677 residues), suggesting a highly conserved function. In contrast, the

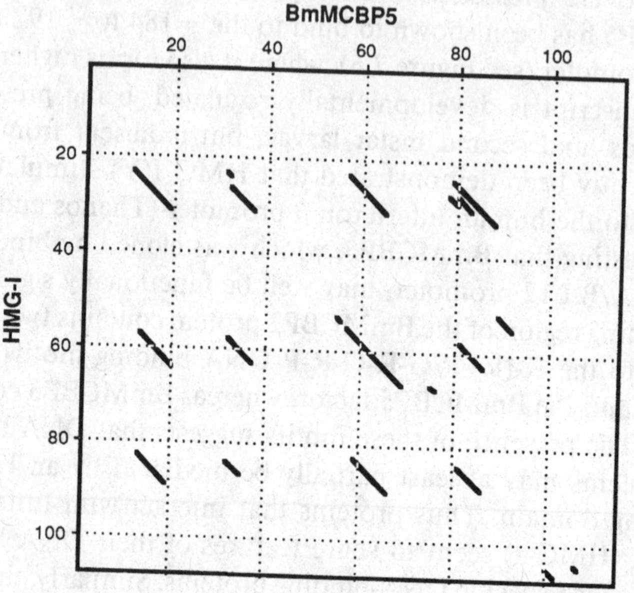

A)

546 PAEVEDTCRACKLRLEGNRKHTHERFLVCATCNAKLHPGCVELGP 590
591 DTIRKCREYPWQCAECKTCGQC-----SRPADDDKMLFCDLCDRGFHSYCVGLPEVP 642
643 TGRWHCVECAICKSCGARSPAGAAAAGGAAGDAAP 677
678 EWHHQTKRGPGGHKVYSHSLCTPC 701
702 ARAYRIGRYCPLCDRS-----FVGPKGTMQLVICKLCDRQLHQECVRT 744
745 TVSVLKVLDYTCGECRRGGITSR 767

B)

1251 CQTGFGLIVTETVAQRALCFLC-------GSTGLDPLIFCACCCEPYHQYCVQDEYNL 1301
1326 PSSSLNQLTQRLNWLCPRCTVCYTC-------NMSSGSKVKCQKCQKNYHSTCIGTSKRL 1378
1379 LGADRPLICVNCLKCKSC 1396
1397 STTKVSKFVGNLPMCTGC 1414
1415 FKLRKKGNFCPICQRC----YDDNDFDLKMMECGDCGQWVHSKCEGL 1457
1458 SDEQYNLLSTLPESIEFICKKCARRNESSKIKAEEWR 1494

C-terminal half, which contains the DNA binding region, has no significant similarities to proteins that are currently represented in the data bases. Like BmMCBP2, the BmMCBP4 factor binds to the A1 region of the A/B.L12 promoter. Finally, BmMCBP1 shows no significant homology to other transcription factors or DNA-binding proteins.

Concluding remarks and future prospects

It is clear that for a mechanistic understanding of the developmental regulation of chorion gene transcription, further work is needed both on the *cis*-regulatory elements of these genes and on the *trans* factors that interact with them. It would be advisable to concentrate attention on a small number of carefully selected moth promoters: A/B.L12, about which the most is known, Pe401/18, which is expressed with different temporal specificity in transgenic *Drosophila*, a late HcA/HcB, and an early ErA/ErB promoter. At least for the first two, detailed analysis of the *cis* elements is possible in *Drosophila*, but with considerable effort. Furthermore, it is possible that not all the *cis* elements can be detected in this heterologous system. Thus development of a convenient homologous system is highly desirable: either a streamlined viral transduction and assay system or, better still, moth germline transformation. In addition, consideration should be given to the ultimate development of a follicular cell culture line, where chorion genes might be expressed and their *cis*-regulatory elements analyzed by transfection studies. Such improvements in technology will probably be necessary, for example, to sort out the multiple elements of the A/B.L12 A1 region at the level of single base substitutions.

As far as the *trans* factors are concerned, the present biochemical approach is obviously very fruitful. Undoubtedly, over the next couple of years much will be learned about the proteins that can recognize and bind

Figure 7.13. *Top:* Homology matrix of the human HMG-I protein and the BmMCBP5 protein. Repeated short diagonals demonstrate the existence of three repeated motifs in HMG-I and four similar repeats in BmMCBP5. The pam250 scoring matrix (MacVector, IBI) was used to weigh amino acid similarities (Window = 8, Hash Value = 2, Minimum score 60%). *Bottom:* Alignment of putative finger domains of the predicted BmMCBP2 (A) and *trithorax* (B) proteins. Cysteine (C) and histidine (H) residues are highlighted. This region may be folded in different ways and has been aligned to emphasize the alternating CxxCxxCxxCxxC and CxxCxxxxHxxC motifs. Other identical amino acid residues in BmMCBP2 and *trithorax* have been boxed. The incomplete BmMCBP2 sequence is numbered from an internal EcoRI site in the cDNA.

with some specificity to the chorion promoter sequences. For now, the following main points can be made:

1) We are still in the early stages of analysis. It is obviously a complex story, because of the large number of genes and their varied patterns of expression, but some similarities to other well-studied developmentally regulated systems are already apparent (see Hui and Suzuki, this volume).
2) In view of the ability of some of the moth promoters to be properly regulated in the heterologous *Drosophila* system, parallel studies on both fly and moth factors will undoubtedly reinforce each other.
3) Thus far, putative factors have been isolated that bind to promoter motifs that may be involved in both quantitative and temporal regulation, acting in either a positive or a negative manner.
4) The interactions of these factors may be complex, synergistic, and hierarchical.
5) The factors belong to (or contain motifs characteristic of) several different families of transcription factors that are known from yeast to mammals (zinc finger proteins, hormone receptors, HMG proteins, etc.). Some of the factors are as yet novel in sequence.
6) An additional component of the regulatory hierarchy is the potential for some of these factors to undergo alternative splicing.

Looking into the future, the central question is how to progress from knowledge of in vitro binding activity to a rigorous assay of the in vivo function of these factors. There are precedents that caution against presuming that in vitro binding reflects in vivo function (e.g., Bray and Kafatos, 1991). In vitro transcription systems can offer evidence on the functional properties of putative transcription factors (e.g., activation or repression), but this should not be accepted uncritically as evidence that the factor indeed is important in the appropriate tissues in vivo. Cotransfection studies may be more informative, but they also must be interpreted with caution. Genetic evidence is the most desirable and will undoubtedly be forthcoming for the *Drosophila* factors. Even though *Bombyx* is the best genetically characterized insect other than *Drosophila,* direct genetic testing of putative *Bombyx trans* factors would be enormously difficult. The heterologous assay of moth promoter sequences in *Drosophila,* in conjunction with manipulations of moth putative *trans* factors introduced in flies by transformation, could be used as a substitute genetic assay system. Once again, a homologous moth transformation procedure would be invaluable. For example, with such a procedure one could envisage assaying the function of putative moth factors on appropriate reporter constructs, by overexpressing these factors in homospecific follicle cells, expressing antisense constructs, or expressing in vitro mutated factors that might act as dominant disruptors of the interacting endogenous factor. This is yet another argument why development of a germline transformation technique would be the single most important advance for lepidopteran molecular genetics.

Acknowledgments

We thank our colleagues who participated in this work and provided unpublished information, especially M. Shea, T. Hsu, J. Gogos, S. A. Mitsialis, A. Khoury, D. Thanos, and P. Tolias. We are grateful for the excellent editorial advice of M. R. Goldsmith and A. S. Wilkins. We thank E. Fenerjian for secretarial help and B. Klumpar for help with the figures. Work from our laboratories has been supported by NSF and NIH grants (F. C. K.), a Gerondellis Foundation Fellowship (G. T.), a NIH grant (N. A. S.), and a NSF grant (H. T. N.).

8

Chorion genes: molecular models of evolution

THOMAS H. EICKBUSH and JOHN A. IZZO

Introduction

Multigene families can be defined as sets of genes with significant nucleotide similarity encoding RNA or protein products with related functions. The ubiquitous nature of multigene families has made the study of their organization, expression, and evolution essential to understanding the eukaryotic genome. In some instances multiple, nearly identical, genes appear to be needed by an organism to produce the large quantity of products that are required at specific developmental periods. The classic example of such multigene families is the hundreds of ribosomal RNA (rRNA) genes needed to produce the high levels of rRNA found in all cells. In most instances, however, the genes within a multigene family encode slightly different gene products that are required by the cell for a series of related functions. Examples of such families are the genes encoding the actin proteins in most eukaryotes, cuticle proteins in insects, and the major histocompatibility proteins found in mammals.

When one compares the same multigene family in two species, characteristic sequence features can be found that are shared by all family members in one species but are not seen in the family members of the second species. Although selection pressures can be cited to explain the maintenance of key sequence features within a family, these pressures are not sufficient to explain why apparently neutral changes can also be shared by all family members, a process termed *concerted evolution.* The phenomenon of concerted evolution implies the operation of recombinational processes that spread and fix new sequence variants in a multigene family (reviewed by Ohta, 1980; Baltimore, 1981; Dover, 1982; Arnheim, 1983). For families of tandemly repeated genes, sequence identity can be maintained by unequal crossovers. Even in the absence of selection pressures, a stochastic process of expansion and contraction in the number of genes can lead to the fixation of new variants within a family (Smith, 1976). The concerted evolution of rRNA genes by unequal crossovers has been well documented in yeast (Petes, 1980; Szostak and Wu, 1980), *Drosophila* (Coen, Strachan, and Dover, 1982), and humans (Seperack, Slatkin and Arnheim, 1988).

For most multigene families, however, the members of the family are not tandemly repeated along the chromosome. The ability of neutral characters within these families to undergo concerted evolution implies an alternative mechanism for the exchange of sequence information between genes. What is the recombinational process responsible for these sequence exchanges? In fungi, where it is possible to score directly all the individual products of recombination events by tetrad analysis, the major interaction between repeated genes on either the same or different chromosomes has been shown to involve gene conversion (Scherer and Davis, 1980; Ernst, Stewart, and Sherman, 1981; Jackson and Fink, 1981; Klein and Petes, 1981). Gene conversions were originally detected as tetrads not showing the typical 1:1 Mendelian segregation ratio in tetrad products resulting from one heterozygotic meiotic cell. The mechanism of gene conversion involves the formation of heteroduplexes between the DNA strands from two genes, followed by DNA repair processes utilizing the sequence information from one gene to "correct" the sequence of the second gene (see review by Szostak, Orr-Weaver, Rothstein, and Stahl, 1983). In higher eukaryotes, many examples have been observed of sequence transfers between members of a gene family that are similar to the gene conversion events described in fungi (Slightom, Bechl and Smithies, 1980; Ruppert, Scherer, and Schüz, 1984; Crain et al., 1987; Reynaud, Auquez, Grimal, and Weill, 1987; Geliebter and Nathenson, 1988; Parham et al., 1988). Gene conversion has been suggested as a major contributor to the concerted evolution of heat shock genes of *Drosophila* (Brown and Ish-Horowicz, 1981), transfer RNA genes of yeast (Munz, Amstutz, Kohli, and Leupold, 1982), and globin genes of humans (Powers and Smithies, 1986). However, it is not known why some gene families undergo conversion events and other gene families apparently do not.

A favorable system for studying the concerted evolution of multigene families is the series of genes encoding the chorion (eggshell) proteins of Lepidoptera. The most extensively studied species are the silk moths,

Antheraea polyphemus and *Bombyx mori* (see Kafatos et al., this volume). The eggshells in these silk moths are composed of more than 100 proteins that are encoded by a similar number of genes. These genes can be placed into several multigene families based on their nucleotide sequence similarities and periods of expression (Goldsmith and Kafatos, 1984). Based on the amino acid sequence and secondary structure of their encoded proteins, all of the chorion multigene families can in turn be placed onto either the α or β branch of a chorion protein superfamily (Lecanidou et al., 1986). Chorion gene families appear to be able to evolve rapidly, because eggshell morphology and chorion protein composition can be quite distinct for different species of silk moth (see Regier et al., this volume). For each species, however, the many chorion proteins assemble into a single macromolecular structure, suggesting that this rapid evolution is a concerted process.

The individual chorion protein families serve different functions in the formation of the eggshell. The first sets of proteins to be synthesized during choriogenesis in both *A. polyphemus* and *B. mori* are called the early or C proteins. These proteins assemble into a framework on which all later synthesized chorion proteins are deposited. Because the eggshell is regionally differentiated (Regier et al., 1980), this initial framework must itself contain significant spatial information. The second major sets of proteins synthesized during choriogenesis are called the middle or A and B proteins. This middle period is sometimes divided into two subperiods, early-middle and middle, to reflect the synthesis of the two major subgroups of A and B proteins (Bock et al., 1986; Spoerel et al., 1986). The A and B proteins expand and increase the density of the eggshell framework made by the early proteins, and are extremely abundant, accounting for up to 80 percent of the total mass of the eggshell in *A. polyphemus* (Regier and Kafatos, 1985).

A third major group of chorion proteins is synthesized at the end of choriogenesis. These late, or high cysteine A-like (HcA) and B-like (HcB), proteins form a series of additional highly cross-linked outer layers to the eggshell, in addition to cross-linking the entire structure. Hc proteins are found in *B. mori* and its putative ancestor *B. mandarina* (Sakaguchi et al., 1988, 1990), which diapause as eggs, but not in *A. polyphemus*, which, like most silk moths, diapause as pupae. The correlation of HcA and HcB families with a change in the developmental stage at which the species undergoes diapause suggests that these families have recently evolved in

Bombyx as an adaptation to prevent desiccation of the egg during the cold, dry winter months. *A. polyphemus* also produces a late set of chorion proteins, the E proteins, which are responsible for the formation of specialized breathing structures on the surface of the eggs (see Regier et al., this volume).

In previous chapters (see Regier et al. and Kafatos et al.), chorion morphogenesis, the organization of the middle and late chorion gene families into divergently transcribed A/B and HcA/HcB gene pairs, and their transcriptional regulation have been described in detail. In this chapter we describe efforts to determine the global organization of the entire set of chorion genes in *B. mori*. Based on an analysis of the sequence relationships of the individual members within each family, we then discuss what can be inferred about the recombinational mechanisms that give rise to their concerted evolution.

The chorion gene complexes of *B. mori*

Genetic analysis

The structural genes for the chorion proteins of *B. mori* have been shown by Goldsmith and co-workers to be clustered between the larval marker *p* at the proximal end of chromosome 2 and the cocoon color marker *Y* (Goldsmith and Basehoar, 1978; Goldsmith and Clermont-Rattner, 1979, 1980; Goldsmith, 1989). This chromosomal location maps close to a series of spontaneous and X-ray induced mutations, termed *Gr,* that turn the normally transparent chorion opaque (Tazima, 1964; see also Goldsmith, this volume). The strategy used by Goldsmith and co-workers for mapping the chorion structural genes involved standard Mendelian analysis, using protein electrophoretic variants as genetic markers. A standard reference strain of *B. mori* (C108) was crossed to various other strains that contained electrophoretic differences in their chorion profiles. Because interchromosomal recombination occurs only in males, heterozygous F1 males were backcrossed to homozygous females of the original reference strain, and recombinants were scored by screening mature chorion protein patterns produced by female offspring on one- and two-dimensional polyacrylamide gels. All strain-specific chorion markers were shown to exhibit codominance in the progeny; thus it was unlikely that any of the electrophoretic variants corresponded to posttranslational modifications.

As shown in Figure 8.1, the chorion genes are linked to two major clusters designated *Ch1-2* and *Ch3* separated by a map distance of 4.0 centiMorgans (cM) (Goldsmith, 1989). *Ch3* appears to contain all of the developmentally defined early chorion genes and approximately one-third of the middle genes (the A and B protein families), and *Ch1-2* appears to contain the remaining two-thirds of the A and B chorion families and all of the late HcA and HcB genes. Based on a single recombination event, the *Ch1-2* cluster can be separated into two segments, *Ch1* and *Ch2*, with genes encoding A, B, HcA, and HcB proteins present in both segments.

Cloning of the multigene families

To study the organization and evolution of the chorion genes of *B. mori*, a number of genes from each family have been cloned and analyzed in detail. Because the clustered arrangement of the genes revealed by the genetic analysis suggested that there was a level of organization above that of the individual gene, extensive efforts were made from the inception of this project to recover a large fraction of each gene complex on recombinant clones (Eickbush, Jones, and Kafatos, 1981). At present, more than 500 kilobases (kb) of chromosomal DNA have been recovered from the chorion gene clusters of *B. mori* (Eickbush and Kafatos, 1982; Hibner et al., 1988, 1991; Izzo, 1991).

As is true for many eukaryotic genomes, most of the structural genes in the *B. mori* genome are interspersed with repetitive DNA (see Eickbush, this volume). As a consequence, the approach used for recovering large contiguous regions of the chromosome was different from the standard "chromosomal walking" procedure utilized so successfully to clone the euchromatic regions of *Drosophila melanogaster*, which are essentially free of repetitive DNA (Bender, Spierer, and Hogness, 1983). The procedure developed to clone the *B. mori* chorion gene complexes was based on the assumption that most chorion genes are less than 20 kb from each other. The first step involved the isolation of a large pool of lambda phage clones, each containing one or more chorion genes on the approximately 15 kb segment of chromosomal DNA present in each recombinant clone. The goal was to encompass the entire set of *B. mori* chorion genes in this "chorion gene sublibrary." Large continuous chromosomal segments of the chorion gene complexes could then be recovered by identifying regions of overlap between the individual lambda clones.

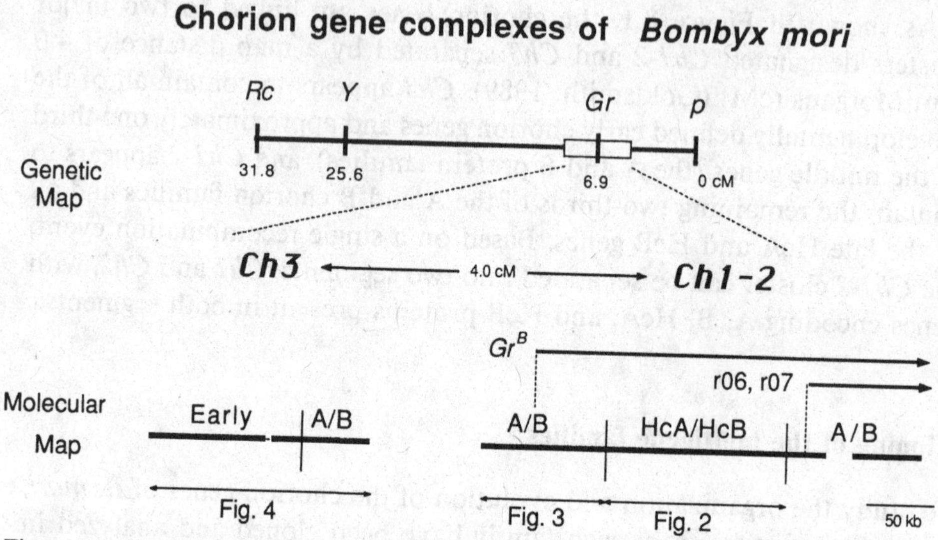

Figure 8.1. Comparison of the genetic and molecular maps of the chorion gene complexes of *B. mori*. *Upper:* A portion of the genetic map of chromosome 2 is shown with the genetic distances in centiMorgans (cM) for the following markers: *p*, plain larval markings; *Gr*, gray eggs; *Y*, Yellow hemolymph and cocoons; and *Rc*, rusty cocoon color. The structural genes for the chorion proteins have been mapped to the gene complexes *Ch1–2* and *Ch3* (Goldsmith, 1989). The proximal end of the chromosome has been placed to the right in this figure in order that the orientation of the cloned fragments in *Ch1–2* will be the same as in previous reports (Eickbush and Kafatos, 1982; Burke and Eickbush, 1986; Spoerel et al., 1989). *Lower:* The thick horizontal lines correspond to chromosomal segments from *Ch1–2* and *Ch3* that have been recovered as series of overlapping recombinant DNA clones. The major families of chorion genes (early, A, B, HcA, and HcB) present on each cloned segment are indicated. Vertical lines on these chromosomal segments correspond to the boundaries between these gene families. An additional 80 kb of chromosomal DNA (not shown in the figure) has been recovered as a series of unlinked genomic clones containing A and B genes. The orientation and distance of the cloned segments from *Ch3* relative to that of the segments from *Ch1–2* are unknown. The segments of the *Ch1–2* complex deleted by the mutation, Gr^B, or translocated in mutants r06 and r07, are shown by the arrows above the cloned segments. The location of the chromosomal segments shown in Figures 8.2–8.4 are indicated by the arrows below the chromosomal segments.

Chorion mRNAs are the most abundant messages present in choriogenic follicle cells (Thireos and Kafatos, 1980). This mRNA abundance suggested that the complete chorion gene sublibrary could be obtained by isolating from a lambda phage genomic library all clones that hybridized to a radioactive probe made by reverse transcription of total choriogenic follicular poly(A)+ mRNA. Analysis of the clones from this sublibrary resulted in the recovery of more than 400 kb of chromosomal DNA from the chorion gene complex *Ch1-2* (Eickbush and Kafatos, 1982). The two largest continuous segments of chromosomal DNA recovered were 270 kb and 50 kb in length (see Figure 8.1).

Because the early chorion proteins are synthesized for much shorter periods of time than the middle and late chorion proteins, the original chorion gene sublibrary did not contain clones with early chorion genes. A new sublibrary was therefore generated by isolating all genomic clones from the total lambda library that hybridized with a series of early chorion cDNA clones (Lecanidou, Eickbush, Rodakis, and Kafatos, 1983; Eickbush, Rodakis, Lecanidou, and Kafatos, 1985). Most of the clones within this second early chorion gene sublibrary could be assembled into two overlapping arrays (Hibner et al., 1988, 1991). All genomic clones that did not overlap with these two segments were found to be located on the 270 kb segment from *Ch1-2* (see further discussion on HcA/HcB gene families in "Organization and evolution of the gene families"). In an attempt to link the two chromosomal segments containing early chorion genes, as well as to recover any additional early chorion genes for which cDNA probes were unavailable, a cosmid library (containing on average 40 kb genomic DNA inserts) was constructed and screened with the same set of early cDNA clones used to screen the lambda library (Izzo, 1991). The final result of placing all genomic clones derived from the lambda and cosmid libraries was the recovery of 155 kb of genomic DNA as two segments of 85 kb and 70 kb (see Figure 8.1). One end of the 70 kb segment was also found to contain a series of A and B genes.

What fraction of the genes within the different chorion multigene families has been recovered? We are confident that in the case of the Hc genes all 30 members of the HcA and HcB families were cloned. All lambda clones in the original sublibrary that hybridized to probes from these families were positioned within a 130 kb region of the 270 kb chromosomal segment. In addition, reconstruction Southerns using lambda clones from this 130 kb region, performed in parallel with genomic Southerns, confirmed that all family members had been isolated (Eickbush and Kafatos, 1982).

In the case of the A and B chorion genes, the 54 genes recovered correspond to most but not all members of these families. A small fraction of the lambda clones in the original chorion gene sublibrary that contained A and B chorion genes could not be assigned to either the 270 or 50 kb chromosomal segment shown in Figure 8.1. Added together, these lambda clones contain at least 10 genes on approximately 80 kb of chromosomal

DNA but have not been further characterized. Reconstruction Southerns of the A and B multigene families suggest that most of the family members are located on the chromosomal segments shown in Figure 8.1 (Spoerel et al., 1989; Eickbush and Izzo, unpublished data). The exact number of A and B genes that remain to be mapped is difficult to estimate, due to the large numbers of genes within each family and their different levels of sequence divergence.

Finally, all early chorion genes detected on genomic Southerns using the early cDNA clones as probes have been localized to one of the chromosomal segments shown in Figure 8.1. The number of early chorion genes recovered (16) is similar to that of the early proteins detected as the first battery of proteins synthesized during choriogenesis (Bock et al., 1986). However, a direct correlation of cloned genes with specific proteins seen on polyacrylamide gels has not been conducted. It is also not known whether any of these early genes encode proteins of the trabecular layer (see Regier et al., this volume). We have made repeated attempts to isolate additional early chorion cDNA clones by probing a follicle cell cDNA library at low hybridization criteria with the characterized cDNAs or with total cDNA synthesized from early follicles. All cDNA clones that have been obtained correspond to one of the preexisting cloned early genes (Hibner and Eickbush, unpublished data). If other early chorion genes exist, they represent minor transcripts that have little sequence similarity to those previously characterized.

In summary, we believe that the 100 chorion genes recovered on the nearly 500 kb of chromosomal DNA shown in Figure 8.1 correspond to virtually all of the chorion genes of *B. mori*. The only known chorion genes that have not yet been located on the chromosomal map correspond to a small fraction of the A and B gene families.

Correlation of the genetic and physical maps

The location of the 50 kb chromosomal segment from *Ch1-2* has been determined relative to the 270 kb segment using the spontaneous deletion mutant, Gr^B (Iatrou, Tsitilou, Goldsmith, and Kafatos, 1980; Durnin-Goodman and Iatrou, 1989). As shown in Figure 8.1, the region of *Ch1-2* deleted in the Gr^B mutation includes all but approximately 45 kb of the left end of the 270 kb segment. The orientation of the 270 and 50 kb chromosomal segments has been determined relative to the genetic map using a set of translocations that involve chromosome 2, covering the 50 kb segment and only the A/B genes located to the right of the Hc gene cluster (Alexopoulou and Goldsmith, personal communication). Because these translocations also include the *p* locus and thus probably the tip of chromosome 2, the 50 kb segment appears to be proximal to the 270 kb segment.

The relative orientation and distance separating the 85 and 70 kb cloned segments of the *Ch3* gene cluster were determined by genomic Southerns using restriction enzymes that cleave infrequently (Izzo, 1991). Only 8 kb of uncloned genomic DNA separate these two segments. Because there are no rearrangements involving the *Ch3* complex, the orientation of this 163 kb segment with respect to the chromosomal genetic map is not known. Given the remarkable clustering of gene families that is expressed at the same time, in Figure 8.1 we have placed the A and B genes at one end of *Ch3* nearest to the *Ch1-2* complex.

The major unanswered question in our correlation of the genetic map of the chorion genes with the physical map is the distance separating the two major gene complexes. Based on a haploid genome size of 0.5 picogram (Gage, 1974b) and a total chromosomal length of 1,000 cM (Doira, 1992), the 4 cM between *Ch1-2* and *Ch3* would correspond to approximately 2,400 kb of DNA. A second argument for a large distance separating the two chorion complexes is that the original chorion mutant, *Gr,* is located within this region, mapping 1–2 cM from both *Ch1-2* and *Ch3* (see Goldsmith, this volume). If such a large distance does exist between the two cloned chorion gene complexes, then this region must be essentially devoid of chorion genes. As described in the previous section, all chorion genes except for a portion of the A and B gene families have already been localized to approximately 500 kb. An alternative possibility is that the chromosomal segment separating *Ch1-2* and *Ch3* contains one or more recombination hotspots that have artificially increased the apparent distance between the two chorion gene complexes. Such recombination hotspots, in which all recombination events detected within regions of the chromosome several hundred kilobases in length are localized to less than 1 kb, have been found in the major histocompatibility gene complex in the mouse (Steinmetz, Stephan, and Lindahl, 1986). Attempts to link the two chorion gene clusters physically, using infrequently cleaving restriction enzymes and pulse-field electrophoresis to separate very large restriction fragments, have not been successful because all currently available enzymes cleave the chromosomal segments containing the A and B gene families into fragments less than 100 kb in length (Izzo, unpublished

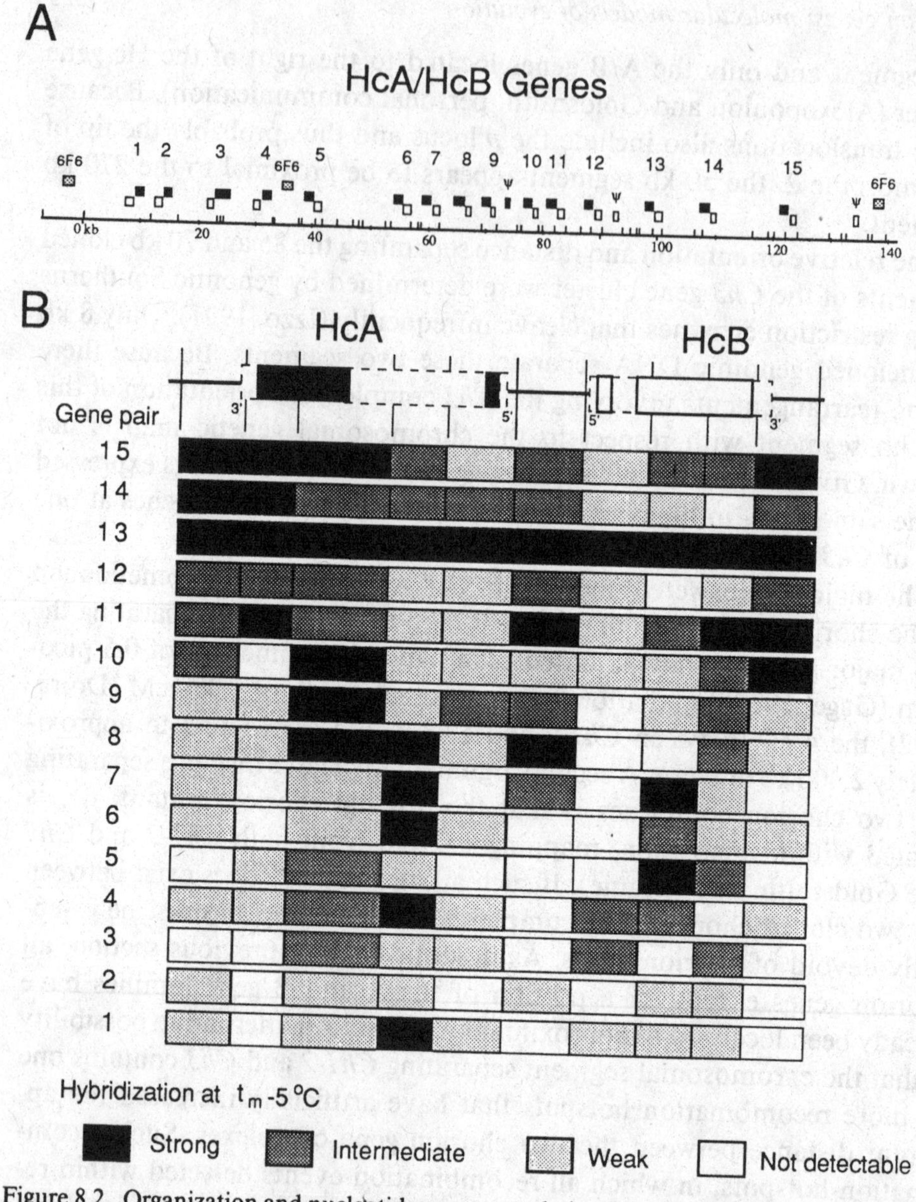

Figure 8.2. Organization and nucleotide sequence relationships of the high cysteine (Hc) chorion gene families of *B. mori*. (A) Diagram of the HcA and HcB gene cluster. The locations of EcoR1 restriction enzyme cleavage sites are indicated by short vertical lines above the horizontal line. The locations of all chorion genes and pseudogenes are indicated by boxes, shaded to represent the family membership of each gene. Filled boxes, HcA genes, open boxes, HcB genes; stippled boxes, 6F6 genes. Each HcA gene is paired with a HcB gene, as diagrammed in panel B. The 6F6 genes are not paired. (B) Summary of cross-hybridization analysis with HcA/HcB.13 gene region probes and all 15 HcA/HcB gene pairs. Top, schematic diagram representing all HcA/HcB gene pairs.

data). Direct evidence for the close proximity of the *Ch1-2* and *Ch3* complexes will require the cloning and direct physical linkage of the remaining A and B chorion genes.

Organization and evolution of the gene families

The HcA and HcB gene families

The organization of the 140 kb chromosomal region containing the HcA and HcB gene families of *B. mori* is shown in Figure 8.2A. The 15 functional genes of each family are organized into 15 divergently transcribed HcA/HcB gene pairs. Each gene pair has been numbered, starting at the left end of the complex. The detailed structure of the Hc gene pairs first determined for pairs 12 and 13 (Iatrou et al., 1984; Rodakis, Lecanidou, and Eickbush, 1984) is shown in Figure 8.2B. The general features of this structure are similar to those of the A/B gene pairs of both *B. mori* and *A. polyphemus* (Jones and Kafatos, 1980; Spoerel et al., 1986) and are more extensively discussed in Kafatos et al. (this volume). That all 15 Hc gene pairs conform to the specific structure shown in Figure 8.2B was initially determined by detailed restriction mapping and nucleic acid hybridization experiments (Eickbush and Burke, 1985) and eventually confirmed by sequence analysis (Burke and Eickbush, 1986).

The HcA/HcB gene cluster shows several features. First, the four gene pairs at the right end of the cluster (HcA/HcB.1 through HcA/HcB.4) are in an inverted orientation relative to the remainder of the gene pairs, suggesting that at least one inversion event occurred in the evolution of this complex. Second, the HcA and HcB gene pairs are not tandem repeats, as the distance separating consecutive gene pairs ranges from 2 to 15 kb. The variation in the distance separating gene pairs is a result of numerous insertions of repetitive DNA sequences (see Eickbush this volume). Third, interspersed with the 15 typical gene pairs are three pseudogenes. The HcB pseudogenes located to the right of genes pairs 12 and 15 appear to have

Filled boxes, protein-encoding regions of HcA genes; open boxes, protein-encoding regions of HcB; boxes in dotted outline, introns, 3′ and 5′ untranslated regions; horizontal lines, 3′ and 5′ flanking regions; vertical lines, restriction sites in HcA/HcB.13 used to generate short fragments for the hybridization experiments. The hybridization intensity of each gene region fragment to each HcA/HcB gene pair is indicated by the shading of the 15 bars below the gene pair diagram (see the key at the bottom). (Data from Eickbush and Burke, 1985.)

had a common origin, in that they share a series of unique sequence changes (Fotaki and Iatrou, 1988, and unpublished observations). These pseudogenes appear to be transcribed with the proper developmental kinetics but cannot produce functional chorion proteins due to a series of insertion and deletion events disrupting the open-reading frame. The third pseudogene is located to the right of gene pair 9 and corresponds to approximately the 3′ one-third of a typical HcA gene (Burke and Eickbush, 1986).

The most unexpected feature of the Hc gene cluster is the location, at either end and between Hc gene pairs 4 and 5, of three genes that are clearly not members of either the HcA or HcB gene family. These genes were identified by their strong hybridization to the cDNA clone m6F6 (Eickbush et al., 1985) and are referred to as 6F6 genes. Based on its temporal pattern of mRNA accumulation and its conceptual translation, 6F6 has been designated as encoding a β protein expressed during early choriogenesis (Eickbush et al., 1985). Analysis of the structure and expression of the 6F6 family will be of interest, because its members are the only examples of chorion genes that are flanked by genes expressed at different developmental stages (Lecanidou and Rodakis, in preparation). It has been suggested that clusters of genes expressed with the same developmental kinetics could be organized into functional chromosomal domains whose higher order structure controls accessibility of all genes within the domain to transcription (Eissenberg, Cartwright, Thomas, and Elgin, 1985). The interspersion of 6F6 and Hc genes strongly argues against the possibility that the Hc gene region is organized as a single large functional domain within the nucleus.

Patterns of sequence identity. Complete or partial nucleotide sequence determinations have been made of all 15 HcA and HcB gene pairs (Iatrou et al., 1984; Rodakis et al., 1984; Burke and Eickbush, 1986). The level of sequence identity between gene pairs was similar to that which would be predicted from the sequence constraints on different functional regions of a gene. Protein-encoding regions of the major and minor exon had the highest level of nucleotide identity, 95 to 96 percent; the 5′ flanking regions were nearly as well conserved, 93.5 percent identity; and least conserved were the intron and 3′ flanking regions, with 86 to 89 percent identity. However, detailed analysis of the nucleotide variation between the gene pairs revealed a more complicated pattern of sequence relationships between the individual genes than what would be predicted from selection pressure alone.

The complex pattern of sequence relationships for the HcA/HcB gene pairs was first demonstrated by a series of nucleic acid hybridization experiments using short (150–300 bp) DNA restriction fragments from HcA/HcB.13 (Eickbush and Burke, 1985). The results of these hybridizations are summarized in Figure 8.2B. The 10 restriction fragments from HcA/HcB.13 that were used as hybridization probes are shown by the vertical lines on the diagram of the Hc gene pair in Figure 8.2B. Each restriction fragment was hybridized to equal aliquots of DNA from all 15 gene pairs bound to nitrocellulose. The extent to which each gene hybridized to each probe is revealed in Figure 8.2B by the shading of the bar regions. The hybridizations were conducted at high stringency ($= t_m - 5°C$) to permit maximum discrimination between genes that differ only slightly in nucleotide sequence.

Each gene segment from HcA/HcB.13 exhibited a unique pattern of hybridization with the other 14 gene pairs. For example, the fragment corresponding to the 3′ terminal region of the HcA.13 major exon hybridized most intensely to gene pairs 11, 14, and 15; the fragment from the central region of the HcA.13 intron hybridized most intensely to gene pairs 1, 4, 6, 7, and 8; and the 5′ flanking regions of HcA/HcB.13 hybridized most intensely to pairs 8 and 11. We have described this pattern of nucleotide sequence similarity between the various members of the Hc gene families as a patchwork (Eickbush and Burke, 1985). Each gene pair contained a unique pattern of regions or "patches" that were highly similar in sequence to the reference gene pair, interspersed with regions that exhibited lower similarity.

The analysis of this patchwork pattern at the nucleotide level (Eickbush and Burke, 1986) revealed that the vast majority of the nucleotide differences accumulated in any particular gene pair were not new mutations but, rather, corresponded to nucleotide variants that were present within other members of the gene family. Indeed, at each nucleotide position within the HcA and HcB genes only one type of nucleotide variant was found in the entire family, suggesting that when new nucleotide substitutions arose in one member of the family by mutation, they were either fixed in all 15 genes or eliminated, before a new mutation occurred in any of the genes at that position. This extensive sharing of nucleotide variation between the members of the HcA and HcB gene families clearly suggested that these families were evolving in concert. Concerted evolution had occurred with both sequences from apparently neutral regions (e.g., introns) as well as from coding regions. The question became, therefore, which of the two recombinational mechanisms, unequal crossover or gene conversion, was responsible for this concerted evolution.

Evidence for gene conversion. In most instances the patchwork arrangement of sequence similarities extended for only a short distance into the 3′ flanking region surrounding each Hc gene pair. When DNA fragments more than a few hundred base pairs from the 3′ end of a number of different Hc genes were used as hybridization probes, they seldom showed hybridization even at low stringency to the regions flanking other Hc gene pairs (Eickbush and Burke, 1985). This restriction of high levels of sequence similarity to the region immediately surrounding each Hc gene pair was more readily explained by recombination mechanisms similar to gene conversion than to unequal crossover. However, clear evidence of occasional unequal crossovers can be detected in the Hc gene cluster as expansions and contractions in the number of gene pairs. Genomic Southern analysis of six geographic races of *B. mori* has revealed that the number of Hc gene pairs varies from 14 to 19 (Xiong, Sakaguchi, and Eickbush, 1988). Did these unequal crossovers occur often enough to account for the concerted evolution of the gene families?

The most effective method of determining the relative frequency of unequal crossover and gene conversion events within the Hc gene cluster is to look at the accumulation of sequence exchanges in the Hc gene pairs with time. If the nucleotide exchanges observed in the genes are usually associated with similar changes in the 3′ flanking regions, then crossover events expanding and contracting the size of the locus directed the evolution of the gene families. If the sequence exchanges are not associated with changes in the 3′ flanking region (i.e., the exchanges involved only the gene regions), then the probable mechanism of these exchanges was gene conversion.

To determine how the nucleotide sequence of the Hc genes changed with time, we analyzed a segment of the Hc gene complex from strain C108 (Xiong, Sakaguchi, and Eickbush, 1988). C108 is a standard Chinese strain of *B. mori* used for the genetic studies of the chorion locus (Goldsmith, 1989). Strain 703, which had been used for all the previous molecular studies, was derived from a European strain (Iatrou et al., 1980). Based on morphological characteristics of the eggshell and chorion protein profiles, the chorion genes of C108 were expected to exhibit significant differences from 703. Although all strains of *B. mori* originated in China where it was domesticated an estimated 5,000 years ago, 703 has been genetically separated from C108 since at least the time of its transfer to Europe, approximately 1,000 years ago (Tazima, 1964).

A lambda genomic library was made and a segment of the C108 HcA/HcB gene complex was cloned, corresponding to the region of the 703 complex from gene pair 11 to the unpaired pseudogene to the right of pair 12 (see Figure 8.2A). The HcA/HcB gene pair of C108 corresponding to gene pair 12 of 703 was sequenced. In order to compare nucleotide changes that had accumulated in the 3′ flanking region between gene pairs, a 1.1 kb region 3′ of the HcA.12 gene was also sequenced from the two strains.

Within the 3.1 kb of nucleotide sequence that could be compared, 37 substitutions and 3 single base pair insertions or deletions were detected. The nature and location of these nucleotide differences strongly suggest that the HcA/HcB.12 gene pair from either strain 703 and/or C108 had been involved in several gene conversion events (Xiong, Sakaguchi, and Eickbush, 1988). First, the nucleotide differences that had accumulated in the gene pairs are clustered. More than 60 percent of these differences are localized to three regions; ten differences are within a 140 bp region of the 5′ flanking region, six differences are within a 65 bp region of HcB major exon, and four differences are within a 45 bp region of the HcA intron. Second, almost all nucleotide differences detected within the gene regions correspond to nucleotide sequence variants present in other members of the Hc families. Finally, the numerous transfers of sequence information that occurred in this gene pair are not accompanied by exchanges of the 3′ flanking regions. Indeed, the number of nucleotide differences that have accumulated in the highly conserved regions of the gene pairs (exons and 5′ flanking) is significantly higher (2.3 substitutions/100 bp) than the number of differences in the 3′ flanking regions (0.6 substitution/100 bp). That is because the changes in the gene regions were sequence exchanges with other gene pairs, whereas the changes in the 3′ flanking region were mutations. This comparison of HcA/HcB.12 from two strains of *B. mori* suggests that sequence exchanges similar to gene conversions were responsible for the concerted evolution of the Hc genes. The unequal crossover events occasionally detected in different strains as expansions or contractions in the number of gene pairs are infrequent relative to the rates of gene conversionlike events between the gene pairs.

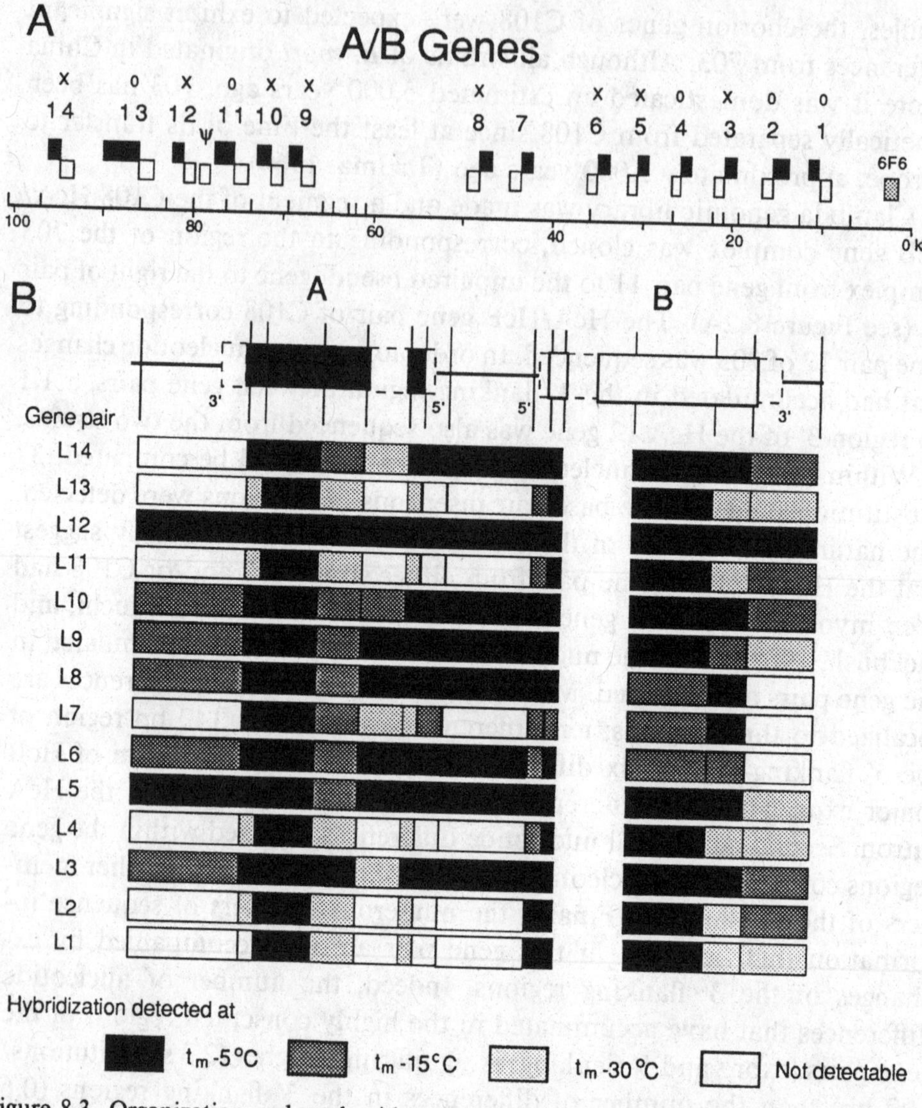

Figure 8.3. Organization and nucleotide sequence relationships of the A and B chorion genes of *B. mori*. (A) Diagram of the A/B gene cluster immediately to the left of the HcA/HcB gene cluster shown in Figure 8.2. (Marker at 0 kb for Figures 8.2 and 8.3 represent the same chromosomal EcoR1 restriction site.) The locations of all chorion genes and pseudogenes are indicated by boxes shaded to represent the family of each gene. Filled boxes, A genes; open boxes, B genes; stippled box, 6F6 gene. Each A gene is paired with a B gene, as indicated by the gene pair diagram in panel B. (B) Summary of the cross-hybridization analysis between A.B.L12 and the other A/B gene pairs. Top, schematic diagram of the A/B.L12 gene pairs. The other A/B gene pairs are similar but contain introns of different lengths. The various functional regions of each gene are diagrammed as in Figure 8.2. Vertical lines indicate the specific fragments from A/B.L12 used for the hybridization experiments. For each A/B gene pair, the highest stringency compatible with

The A and B gene families

A and B chorion genes of *B. mori* are organized into divergently transcribed gene pairs similar in most details to those just described for the HcA and HcB genes (see Kafatos et al., this volume). The most thorough characterization has been of those A/B gene pairs on the 100 kb region to the left of the Hc gene clusters (see Figures 8.1 and 8.3). The 14 A/B gene pairs on this segment have been numbered A/B.L1 to A/B.L14 (abbreviated L1 to L14). Adjacent A/B gene pairs usually have the same orientation, although pairs L2 to L8 are inverted relative to the remainder. The distances separating the gene pairs are variable in length, with the largest distance separating L8 and L9. At least one pseudogene is present within the gene complex, located to the left of L11, and corresponding to the 3′ half of an A gene (Spoerel et al., 1986).

The nucleotide sequence of the A/B gene pairs L1, L11, and L12 has been determined (Spoerel et al., 1986; Tsitilou and Kafatos, 1989). The structure of the A/B.L12 gene pair is shown in Figure 8.3B. All 14 A/B gene pairs shown in Figure 8.3A conform to this structure except for segmental variation in the lengths of their introns (Spoerel et al., 1989). Unlike the HcA and HcB families, where sequence identity between all members of the same family is high for both coding and noncoding regions, nucleotide sequence comparison of the A and B genes reveals uniformly high levels of nucleotide identity only for the protein coding regions (>90 percent). Based on the sequence of their noncoding regions, the A/B gene pairs can be divided into two major subgroups, one characterized by the L11 gene pair and one by the L12 pair. The subgroup classification of each A/B gene pair is best determined by the nucleotide sequence of its 5′ flanking region. Using 5′ flanking regions from the L11 and L12 gene pairs to probe genomic DNA blots, it was found that there are a total of 14 A/B gene pairs in the genome of *B. mori* that are members of the L11 family. Five of these L11-like gene pairs are located on the 100 kb segment shown in Figure 8.3A (indicated with O's above the gene pairs). A total of eight A/B gene pairs present in the genome of *B. mori* are members of the L12 subfamily. All eight of these gene pairs are located on the 100 kb segment shown in Figure 8.3 (indicated with X's above the gene pairs).

hybridization to the A/B.L12 probes is indicated by the shading of the bars (see the key at the bottom). Experiments were not conducted with the B gene intron regions. o denotes member of the L11 subfamily; x denotes member of the L12 subfamily. (Data redrawn from Spoerel et al., 1989.)

Sequence comparison of the 5′ flanking regions of the L11 and L12 subfamilies indicates that they are nearly identical in length (277 vs. 278 bp) but have only short regions of nucleotide identity, presumably corresponding to the promoter regions (Spoerel et al., 1989). The subgroups of A/B gene pairs defined by these 5′ flanking regions are transcribed with different but overlapping developmental kinetics. The L12 subfamily is transcribed somewhat earlier in choriogenesis than the L11 subfamily and thus is defined as early-middle genes, whereas the later transcribed L11 family is defined as middle (see Kafatos et al., this volume). Additional subfamilies of A/B gene pairs exist in *B. mori;* for example, a number of A/B gene pairs on the 270 and 50 kb cloned segments of *Ch1-2,* including L7, are not members of either the L11 or L12 subfamily. Further, as described in the next section, a series of A/B gene pairs detected in *Ch3* represents a distinct subfamily.

To characterize in greater detail the complex patterns of sequence identity between different A/B gene pairs, a series of nucleic acid hybridization experiments was conducted in a manner similar to that used for the Hc gene pairs (Spoerel et al., 1989). Short DNA restriction fragments from gene pair L12, representing the different functional regions of the genes, were used as probes to hybridize to equal aliquots of DNA from each of the 14 A/B gene pairs shown in Figure 8.3A. The length and location of the restriction fragments from A/B.L12 that were used as the hybridization probes in this study are indicated by the vertical lines on the gene pair diagram in Figure 8.3B, and the results are summarized below this diagram by the shading of the boxes. It is important to note that because of the lower levels of sequence identity, it was necessary to vary the hybridizations over a wide range of stringencies ($= t_m - 5°$ to $-30°C$). This differed from the hybridization conducted with the HcA/HcB gene pairs, where high levels of sequence identity allowed all hybridizations to be conducted at high stringency ($= t_m - 5°C$).

The eight members of the L12 subgroup (L3, L5, L6, L8, L9, L10, L12, and L14) are easily identified by their intense hybridization to the L12 5′ flanking region probe. In general the members of this subgroup exhibit the highest level of hybridization to all of the L12 probes. However, this correlation of highest sequence identity within the L12 subfamily is not absolute. Several instances were found where segments of genes that are not members of the L12 subgroup exhibit higher levels of hybridization to the L12 probes than that exhibited by members of the L12 subfamily. For example, the central region of the major exon of the B.L6 and B.L8 genes, both of the L12 subfamily, hybridizes at a lower criterion than do certain genes of the L11 family (L1, L2, L4, L11, and L13), and the 3′ end of the major exon of B.L9 or B.L6, both of the L12 subgroup, has lower sequence identity to the L12 probe than B.L7.

This patchwork pattern of sequence identity between the A/B gene pairs, like that of the Hc gene pairs, can be explained by sequence exchanges between the gene pairs similar to gene conversion events. However, unlike the pattern for the HcA/HcB gene pairs, most of the exchanges that involved sequences outside the major exons of the A and B genes appear to be between members of the same subfamily. This preferential exchange between members of the same subfamily is not related to distance, as the L11 and L12 subfamilies are interspersed along the chromosome. Nucleotide sequence comparisons of the 5′ flanking region of 18 A/B gene pairs confirm that these sequence transfers can lead to the concerted evolution of the gene families (Spoerel et al., 1989). Most of the nucleotide variation within the 5′ flanking regions of each gene are shared by other genes of the subfamily, suggesting that new variants are either fixed or eliminated from each subfamily. Limited nucleotide sequence comparisons (Rodakis, Moschonas, and Kafatos, 1982; Tsitilou and Kafatos, 1989) suggest that the concerted evolution of major exons includes all the A/B gene pairs independent of their subfamily. Additional sequence data from a larger number of A/B gene pairs are needed to determine the relative rates of sequence transfer within and between the A/B subfamilies.

The ErA and ErB gene families

The location of 23 chorion genes within the *Ch3* locus of *B. mori* is shown in Figure 8.4A. The two chromosomal regions recovered as overlapping recombinant DNA clones were shown to be only 8 kb apart by restriction mapping large fragments of genomic DNA separated by pulse-field electrophoresis (Izzo, 1991). Ten of the chorion genes correspond to five divergently transcribed gene pairs named ErA/ErB.1 to .5 (ErA = early A-like genes; ErB = early B-like genes). Three other chorion genes, 5H4, 2G12.1, and 2G12.2 (named after the cDNAs with which they were identified), are unpaired. All 13 of these chorion genes have similar transcription kinetics, with mRNA initially accumulating in follicle 1 and disappearing by follicles 5–8.

Located immediately to one side of this cluster of early genes (125–150 kb in Figure 8.4A) are 10 chorion genes expressed later in choriogenesis.

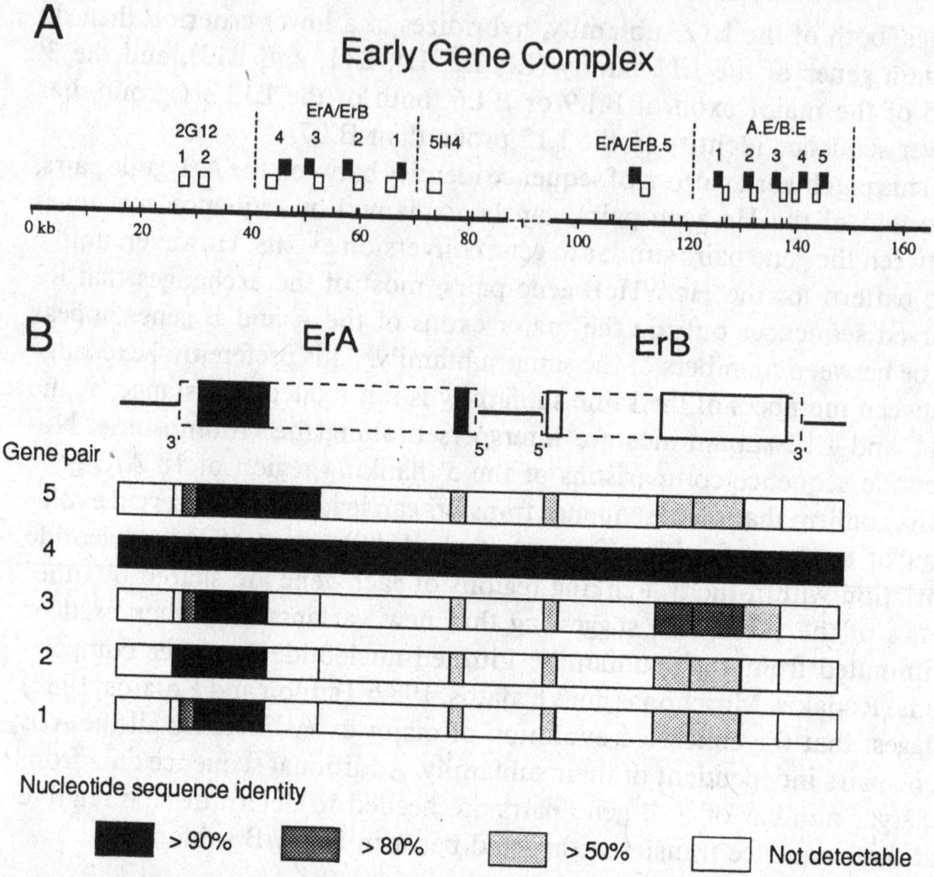

Figure 8.4. Organization of the early chorion genes and nucleotide sequence relationships of the ErA and ErB genes of *B. mori.* (A) Diagram of the early gene cluster. The dotted line at 90 kb represents an uncloned chromosomal region with distance inferred from restriction enzyme analysis. The locations of all chorion genes are indicated by boxes shaded to represent the family of each gene. Filled boxes, α genes (ErA and A.E); open boxes, β genes (ErB, B.E, 5H4, and 2G12). The 5H4 and two 2G12 genes are unpaired. (B) Nucleotide sequence relationships between the members of the ErA and ErB gene families. Top, schematic diagram of the ErA/B.4 gene pairs. All other ErA/B gene pairs contain the same basic organization but have different length introns. The various gene regions are diagrammed as in Figure 8.2. The nucleotide sequence identity of each gene region to that of ErA/ErB.4 is indicated by the shading of the bars (see the key at the bottom). (Data from Hibner et al., 1991.)

Partial nucleotide sequence analysis of these genes indicates that they are organized as five divergently oriented gene pairs (Izzo, Burke, and Eickbush, unpublished data). One gene of each pair is an A gene, and the second gene of each pair is clearly a B gene. Because the 5′ flanking regions of these A/B gene pairs are different from the L11 or L12 subfamily, they represent a third distinct subfamily, which we have named A/B.E1–.E5 (A and B genes from the early chorion complex). RNA for these genes is first detected in follicle 3 and remains present until follicles 14–16. This expression is somewhat earlier in developmental timing from that of the A/B.L12 subfamily. Further sequence analysis is necessary to determine the extent to which the A.E and B.E genes undergo sequence transfer with the L11 and L12 subfamilies.

The ErA and ErB families are the most extensively studied early genes from the *Ch3* locus. The complete nucleotide sequence of the five ErA/ErB gene pairs (Hibner et al., 1988, 1991) indicates that their structure (see Figure 8.4B) is similar to that of the HcA/HcB and A/B chorion gene pairs. However, the variation between the ErA/ErB gene pairs is even greater than that between the A/B pairs. For example, the length of the 5′ regions of the ErA/ErB pairs varies from 190 to 270 bp, whereas the 5′ flanking region of all A/B pairs is a highly conserved 277 to 278 bp in length. The ErA and ErB introns are also highly variable in length and sequence. Located within many of these introns are copies of the middle repetitive oligo-A terminated mobile elements, *Bm1* and *Bm2* (Adams et al., 1986; Hibner et al., 1991). As is discussed in greater detail by Eickbush in this volume (see Figure 3.1), the variable location of these elements and their high levels of nucleotide identity indicate that they have inserted well after the gene duplications that gave rise to the ErA and ErB families.

A summary of the sequence similarities among the ErA/ErB gene pairs is shown in Figure 8.4B. The figure shows the sequence identity of each gene pair with that of gene pair 4, and is based on the determined nucleotide sequence of the genes rather than on the DNA: DNA hybridization data used for the comparison of the HcA/HcB (see Figure 8.2) and A/B gene pairs (see Figure 8.3). Only coding regions of the ErA major exons have high levels of nucleotide sequence identity (96 percent). For comparison, nucleotide identities for the coding regions of the minor exons are 62 percent (ErA genes) and 55 percent (ErB genes) and 63 percent for the coding region of the ErB major exons. No nucleotide sequence identity can be detected between the noncoding regions, except for short regions immediately adjacent to the ErA major exons. The absence of nucleotide identity in the noncoding regions of the ErA/ErB gene pairs even includes the 5′ flanking regions. As previously described, the 5′ flanking regions are highly conserved in the HcA/HcB and A/B gene pairs and are predicted to contain the *cis*-acting DNA elements sufficient for the correct temporal and tissue-specific regulation of the genes (see Kafatos et al., this volume).

The sequence comparisons of the ErA/ErB gene pairs raise an intriguing question: Why is there a significantly higher level of sequence identity between the ErA major exons than between the ErB major exons? One explanation that could account for the higher sequence identity between the ErA major exons is a higher selection pressure to maintain the sequence of the ErA proteins. To examine this possibility, we compared the percentage of nucleotide substitutions at synonymous codon positions within the ErA and ErB genes (positions that will not result in amino acid changes). Nucleotide changes at such positions are not subject to direct selection pressure at the protein level and thus should be equal between the two gene families. We used exclusively fourfold synonymous sites, that is, the third position of codons that can be mutated to any base without changing the encoded amino acid. The average percent divergence at fourfold synonymous positions is 7.7 in the ErA genes and 57.8 in the ErB genes. These values become 8.2 percent and 110 percent, respectively, after correction for multiple substitutions at the same site (Jukes and Canter, 1969). Thus ErB genes have 13 times the level of divergence at fourfold synonymous sites found in the ErA genes. This conservation of neutral sites in the ErA genes clearly suggests that recombination between the genes is responsible for their high nucleotide similarity, not selective pressure on the amino acid sequences of the encoded proteins.

Short regions of sequence identity immediately 3′ and 5′ of the ErA major exons also suggest recombination between the ErA exons. As shown in Figure 8.4B, approximately 85 percent nucleotide identity is found for the 50 bp 3′ of the termination codons of the ErA genes. In the case of the ErA.2 and ErA.4 genes, the sequence identity of this region is higher (92 percent) and extends for a greater distance beyond the termination codon (100 bp). A similar segment of greater sequence identity is also present between ErA.1 and ErA.3 3′ of their major exons (not shown in Figure 8.4B). Finally, a 214 bp region in the intron of ErA.4 and ErA.5 immediately 5′ of the major exon contains 96 percent nucleotide similarity. We interpret these short regions of sequence identity between the noncoding regions of individual ErA genes as sequence exchanges originating in the major exon and extending for variable distances in the 3′ and 5′ directions.

In summary, the levels of sequence identity in and around the major exons of the ErA genes are best explained by recombination events giving rise to sequence exchanges between the five members of the family. Because the level of nucleotide diversity in all other regions of the gene pair indicates the complete absence of recent unequal crossovers, these sequence exchanges once again appear to be gene conversionlike events. Unlike the situation within the HcA/HcB or A/B gene pairs, these conversion events occur only between the ErA genes of the ErA/ErB gene pairs. Both coding and noncoding regions of the ErB genes are highly divergent in sequence. Indeed, ErB is the only chorion gene family we have studied whose members have diverged considerably in sequence.

Concerted evolution of the chorion gene families

The results of our sequence comparison of the chorion multigene families of *B. mori,* described earlier, suggest that the major families expressed during the early, middle, or late periods of choriogenesis have maintained a precise structural organization. Each α gene (HcA, A, and ErA) is divergently paired with a β gene (HcB, B, ErB). All but one of these families are evolving in concert, resulting in all members of the family encoding proteins that are highly similar in sequence. The predominant mechanism for this concerted evolution is recombination events similar to gene conversions. The exact pattern of these gene conversion events is somewhat different for each family. In the case of the HcA and HcB families, they extend throughout the entire 2.0 kb region of the HcA/HcB gene pairs, encompassing both the coding and noncoding segments, but not extending for more than a few hundred base pairs into the 3′ flanking region. In the case of the A and B families, there is a greater tendency for the conversions to occur between genes within the same subfamily. High levels of sequence identity are again found for all regions immediately surrounding the gene pairs of the same subfamily. Gene conversions also occur between subfamilies, but these are usually limited to the major exons. Finally, for the ErA and ErB gene families, the gene conversion events are limited exclusively to the major exon of the ErA genes. The ErB genes are diverging in sequence largely independently of one another.

In the following section we propose a model to explain the different patterns of gene conversion for the individual chorion gene families, how these patterns may have evolved, and the effects such recombinational change can have on the evolution of lepidopteran eggshells.

Initiation sites for gene conversion

Given the many gene conversion events detected within five of the six chorion gene families that have been studied, perhaps the most interesting issue to address is why the ErB genes are not undergoing such recombinations. The model we have proposed is based on the observation that the ErB genes lack putative recombination initiation sites that can be identified in the five chorion gene families that do undergo these conversions (Hibner et al., 1991). The first suggestion of these initiation sites came from a detailed comparison of the distribution of recombination events within the Hc genes (Eickbush and Burke, 1986). It was found that these events were not uniformly distributed along the genes, but were highest in the major exon near the 3′ end of each gene and lowest in the 5′ flanking region between the genes. These gradients of gene conversion are similar to the initiation of conversion events at preferred sites along a gene as previously described in fungi (Gutz, 1971; Fogel, Mortimer, Lusnak, and Travares, 1979; Whitehouse, 1982). In the case of the ARG4 gene of *Saccharomyces cerevisiae,* it has been shown that the initiation site is located near the promoter and is capable of stimulating conversions in both directions (Nicolas, Treco, Schultes, and Szostak, 1989; Schultes and Szostak, 1990).

We propose that the gradient of gene conversion in the Hc genes can be explained by the preferential initiation of recombination events leading to gene conversion in a region within the major exon of each gene and then extending to variable distances towards the 5′ and 3′ ends of the genes as diagrammed in Figure 8.5. The heteroduplexes that form between the participating genes are indicated in this figure by the horizontal arrows. The DNA synthesis that gives rise to the conversion "patches" (solid lines) initiates and terminates at random along the heteroduplex, decreasing in frequency with distance from the initiation site. The length of these conversion patches is usually short, consistent with what has been proposed for conversion events in other higher eukaryotic systems (Powers and Smithies, 1986; Reynaud et al., 1987; Geliebter and Nathenson, 1988; Parham et al., 1988).

The region of the Hc genes with the highest apparent rate of gene conversion is composed of variable numbers of repeats of the DNA sequence, TG(T/C)GGXGGX (where X = A, T, or C). This 9 bp repeat encodes the amino acid sequence cysteine-glycine-glycine. Each Hc gene has, on average, eight of these repeats, comprising the major part of the carboxyl-

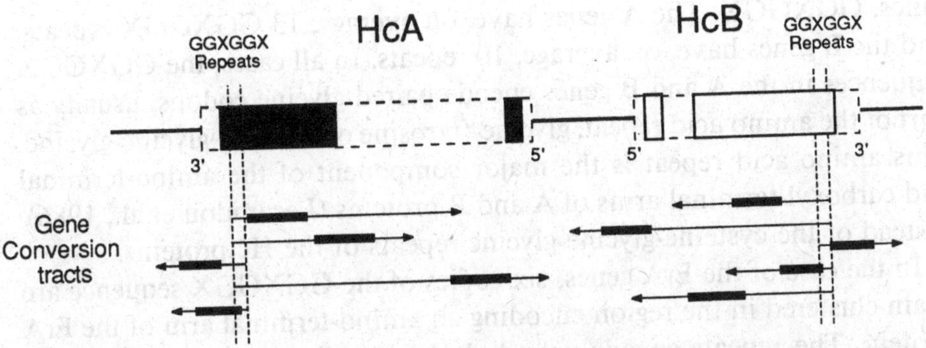

Figure 8.5. Model to explain the gradients of gene conversion between HcA/HcB gene pairs. Recombination events leading to gene conversion events can potentially initiate between any two regions of homologous DNA. Based on the patterns of shared nucleotide variants (Eickbush and Burke, 1986), we suggest that most conversion events between the HcA/HcB gene pairs initiate at GGXGGX repeats in the major exons of the genes. These repeats encode paired glycine residues in the carboxyl terminal arm of each gene. After initiation, the extent and direction of the subsequent conversion events (indicated by horizontal arrows) are dependent on the stability of the heteroduplexes that form between the two pairs. The actual regions of DNA synthesis that give rise to the conversion "patches" (thicker segments of each line) are usually short and initiate and terminate randomly along the heteroduplexes.

terminal arm of the mature Hc protein. This short DNA repeat contains nucleotide similarity to several eukaryotic recombination hotspots that have been correlated with reciprocal and nonreciprocal recombination events (Hibner et al., 1991). These include a region next to the Eβ gene in the mouse histocompatibility locus (Kobori, Strauss, Minard, and Hood, 1986; Steinmetz, et al., 1986), deletion mutants in the ADE8 gene of *S. cerevisiae* (White et al., 1988), a sequence within the phosphoglycerate kinase gene of *Trypanosoma brucei* (Le Blancq, Swinkels, Gibson, and Borst, 1988), the hypervariable human minisatellite DNA (Jeffreys, Wilson, and Thein, 1985), and the human variable number of tandem repeat (VNTR) markers used for gene mapping (Nakamura et al., 1987). These putative hotspots have in common the occurrence of paired guanine nucleotides, separated by either one or two base pairs. This sequence feature is also shared with the prokaryotic recombination signal, *chi* (Smith, Kunes, Schultz, Taylor, and Triman, 1981).

The A, B, ErA, and ErB chorion genes contain only a few copies of the complete TG(T/C)GGXGGX repeat found in the HcA and HcB gene families. However, except for ErB, these genes do contain a much larger number of copies of that portion of the repeat containing the paired guanines, GGXGGX. The A genes have, on average, 13 GGXGGX repeats, and the B genes have, on average, 10 repeats. In all cases, the GGXGGX sequences in the A and B genes encode paired glycine codons, usually as part of the amino acid repeat, glycine-(tyrosine or leucine)-glycine-glycine. This amino acid repeat is the major component of the amino-terminal and carboxyl-terminal arms of A and B proteins (Lecanidou et al., 1986), instead of the cysteine-glycine-glycine repeats of the Hc proteins.

In the case of the ErA genes, six copies of the GGXGGX sequence are again clustered in the region encoding an amino-terminal arm of the ErA protein. The repeats encode paired glycine residues as part of tyrosine-glycine-glycine or cysteine-glycine-glycine repeats. The ErB genes, on the other hand, contain only three copies of the GGXGGX nucleotide sequence. These copies are widely separated in the ErB genes, one in the region encoding the N-arm and one at either end of the region encoding the central domain. A few additional GGXGGX sequences can also be found in certain of the ErB genes; however, these copies are not conserved between the different genes.

We suggest that the five chorion gene families that appear to be undergoing gene conversion do so as a result of closely spaced copies of the nucleotide sequence, GGXGGX, which serve as initiation sites for gene conversion events. The GGXGGX sequences encode paired glycine residues within the repetitive amino-terminal and/or carboxyl-terminal domain of the mature protein. Because the only gene family that does not undergo gene conversion, ErB, contains a few dispersed copies of GGXGGX, it appears that either the density of these repeats or their precise spacing must be important for the efficient promotion of recombination events.

Two explanations could account for the absence of GGXGGX repeats in the ErB genes. First, the function of the ErB proteins in the formation of the eggshell may be incompatible with clusters of paired glycine residues in their amino- or carboxyl-terminal arms. Second, the different ErB genes may serve unique functions in the formation of the eggshell. In this case, sequence transfers between these genes would result in the loss of function and would be selected against. Thus, in order for the ErB genes to have evolved and maintained distinct functions, they must have lost (or never possessed) the gene conversion initiation sites.

We have found GGXGGX repeats in two other multigene families, both of which appear to be undergoing concerted evolution. High levels of sequence identity determined for two collagen genes in *Caenorhabditis elegans* have been attributed to gene conversion (Park and Kramer, 1990). Each collagen gene contains numerous examples of the GGXGGX sequence, and its reverse complement, CCXCCX (where X = A, T, or G). The former sequence encodes paired glycine residues, and the latter encodes paired proline residues. Similarly, the human salivary proline-rich protein genes undergo frequent intragenic recombination events (Lyons, Stein, and Smithies, 1988). These genes contain many examples of CCXCCX, encoding paired proline residues, and GGXGGX, encoding paired glycine residues. Thus we predict that any clustered multigene family that encodes proteins with frequently paired glycine or proline residues is likely to be undergoing concerted evolution promoted by gene conversion. Although it is possible that recombination hotspots adjacent to the genes of a family could promote gene conversions, such an arrangement would be evolutionarily less stable, because, as is described in the next section, random mobile DNA insertion could occasionally separate the hotspots from the gene.

Pattern of gene conversion as a function of family age

If most of the gene conversion events between chorion genes initiate as a result of GGXGGX sequences in the major exon of each gene, the next question to be explained is why there are significant differences in the extent to which these conversions extend to other regions of the chorion genes. Within the ErA genes, few of the conversion events extend beyond their putative origin in the major exon, whereas in the HcA and HcB families, the introns, minor exons, and 5′ flanking regions are also involved. Gene conversions in the A and B chorion families also extend throughout the gene, but only among genes from the same subfamily.

The difference in the extent of the gene conversion between the two extreme cases, ErA and the Hc families, cannot be explained by the conversion events simply occurring less frequently in the ErA genes. One can estimate the relative rates of gene conversion between the different chorion gene families by comparing the nucleotide divergence of the genes at four-fold synonymous sites. Because synonymous sites are probably not under selective restraint, each chorion gene should accumulate sequence changes at these sites by mutations occurring at similar rates. The rate at which these changes are fixed or eliminated from each family is a relative estimate of the conversion rate for each family. The level of nucleotide divergence between the ErA genes at four-fold synonymous sites is 7.7 percent (Hibner et al., 1991). In the case of the HcA and HcB genes, the divergence at these sites is 6.7 and 7.9 percent, respectively (Burke and Eickbush, 1986). Thus the frequency of conversion events in the major exon of the ErA genes is similar to that in the HcA and HcB families. Clearly, an explanation other than rate of exchange must account for why these conversion tracts extend outside their origin in the major exon of the Hc genes but not in the ErA genes.

It is possible to explain the differences between the extent of gene conversions in the various chorion gene families as being a result of the age of each family. The early gene families are believed to be the oldest, based on amino acid sequence considerations (Lecanidou et al., 1986) and the similar function of the early chorion proteins in making a framework in all lepidopteran eggshells (Regier and Hatzopoulos, 1988). Of the other chorion families, HcA and HcB are probably the most recent to have evolved, because they are present only in *Bombyx* species, whereas highly similar A and B gene families are present in both *B. mori* and *A. polyphemus* (Kafatos et al., 1977). Given this relative age of the different chorion gene families and a constant rate of insertion and deletion of repetitive DNA sequences in noncoding regions, the older ErA and ErB families would have undergone a greater number of insertion/deletion events. Each insertion or deletion into the introns or immediately 3′ of the gene could serve as a barrier to the passage of conversion events initiated in the major exon. Thus the youngest families, HcA and HcB, may have simply expanded too recently to have accumulated sufficient insertion/deletion differences to affect significantly the passage of conversion events.

An interesting example of a DNA insertion that has occurred in an HcA gene adds support to the hypothesis that insertions can block the passage of gene conversion events. The insertion is 0.9 kb in length and located near the middle of the HcA.3 intron (Eickbush and Burke, 1985). The segment of the HcA.3 intron on the major exon side of the insertion has undergone the same rate of gene conversion as that found in all other HcA genes. However, as judged by the level of unique mutations (i.e., nucleotide changes not found in any of the other HcA genes), the frequency of gene conversion has been significantly reduced on the minor exon side of this insertion (Eickbush and Burke, 1986). These data both support our model for the gradient of the conversion events in the HcA genes and demonstrate that DNA insertions can act as a barrier to these gene conversion events. Unless the HcA.3 insertion is eliminated (e.g., by unequal crossovers deleting all or part of the gene), the intron, minor exon, and 5′ flanking region of the HcA.3 gene will become increasingly more divergent from the other members of the family.

In fact, the introns of the HcA genes can be divided into two major sequence classes (Eickbush and Burke, 1985, 1986; Burke and Eickbush, 1986). Near the center of the 450 bp HcA intron, one class contains two 17–20 bp deletions separated by 85 bp that are missing in the second intron class. Nucleotide identity for this region of the HcA intron within each intron class averages 94 percent, whereas sequence identity between intron classes is only 74 percent. Further increase in the diversity of this intron region could serve as a barrier to the free movement of conversion events between genes of the different intron classes, from their origin in the major exon through the intron to the 5′ end of the gene. The eventual outcome of this barrier would be the evolution of two subclasses of HcA genes that share similar major exons but differ in their intron and 5′ flanking regions. This outcome is similar to that of the 14 A/B gene pairs shown in Figure 8.3. Two major subfamilies of the A/B gene pairs are present, L11 and L12, which contain similar exons but distinct intron and 5′ flanking regions.

The division of the total gene family into smaller subgroups will also be unstable in the long term. Given enough time, and the continual bombardment of insertion sequences in the intron and 3′ flanking regions, the gene conversion events will eventually be limited to the major exon where they originated. This is precisely the situation with the oldest known chorion gene family that is still undergoing gene conversion, the ErA genes.

The ErA genes reveal the remarkable level of precision that appears possible by gene conversion. The level of sequence identity of the ErA.5 major exon is similar to that of the other four ErA genes, indicating that it is undergoing sequence exchanges at a rate similar to the other ErA genes. However, ErA.5 is located nearly 40 kb distant from the 25 kb segment containing the other four ErA genes (see Figure 8.4). This would suggest that even short DNA segments (the ErA major exon is only approximately 300 bp in length) can undergo concerted evolution by gene conversion over considerable distances along a chromosome. The ability of short DNA segments to promote gene conversion over large distances has also been demonstrated for the 200 bp regions containing tRNA genes in yeast (Munz et al., 1982). However, the yeast genome has few repeated sequences, and it has been shown that any sequence duplication placed in yeast on the same or different chromosomes will undergo conversion events (Scherer and Davis, 1980; Ernst et al., 1981; Jackson and Fink, 1981; Klein and Petes, 1981). The *B. mori* genome, on the other hand, has a high density of repetitive DNA sequences that surrounds most genes (Gage, 1974a). The ErA genes are no exception, being embedded in a complex array of repeated sequences (see Eickbush, this volume, Figure 3.1). The ability of the ErA genes to undergo gene conversion implies an extraordinary level of specificity for the recombinational apparatus to find the short ErA exons in a sea of repetitive DNA.

Concluding remarks

Formation of lepidopteran eggshells provides a remarkable opportunity to dissect the relationship between genotype and phenotype of a complex developmental system at the molecular level. It also provides an opportunity to follow the evolution of the eggshell under different ecological conditions, involving the different periods of time and conditions under which eggs from related lepidopteran species can develop. For example, as identified earlier, presence of the Hc gene families of *B. mori* correlates with changes in the eggshell to prevent desiccation of the embryo during the dry winter months (Kafatos et al., 1977). The formation of specialized breathing structures on the surface of *A. polyphemus* eggs, the aeropyle crowns, has been shown to correlate with a dramatic increase in the levels of expression of the chorion gene E family (Hatzopoulos and Regier, 1987; see also Regier et al., this volume).

To understand fully the evolutionary forces exerted on a gene system, we must understand not only the selective forces at work on the expressed proteins (selection at the phenotypic level) but also the recombinational forces at work on the genes themselves, a process that has been described as molecular drive (Dover, 1982). As has been described in this chapter, a major factor contributing to the concerted evolution of the chorion gene families in *B. mori* is the presence of recombination hotspots, GGXGGX, in the major exons of most genes. Because the primary function of this sequence is to encode paired glycine residues, these genes provide a particularly clear example of how features of a DNA sequence can have important evolutionary consequences independent of their phenotypic expression. Indeed, our analysis of the concerted evolution of the chorion gene families raises the interesting question of why these paired glycine residues are present in the chorion proteins. Are they critical to the function of the proteins, with the concerted evolution of the gene families thus being only a secondary effect? Or has the recombination induced by this sequence driven the genes to encode as many GGXGGX repeats as are compatible with the function of their encoded proteins? Continued studies of the structure and evolution of the chorion proteins in different lepidopteran species should provide insight into this problem and further delineate the conditions under which gene conversion occurs in multigene families.

Acknowledgments

We gratefully acknowledge the role of William Burke in all aspects of our structural studies of the *B. mori* chorion gene families. The work described in this chapter from the laboratory of the authors was supported by the National Institutes of Health.

9

Regulation of the silk protein genes and the homeobox genes in silk gland development

CHI-CHUNG HUI and YOSHIAKI SUZUKI

Introduction

In order to decipher the mechanisms that control the utilization of genetic information for time- and space-dependent expression we have been studying the developmental regulation of the silk protein genes of *Bombyx mori* (Suzuki, 1977; Suzuki et al., 1987, 1990b). The silk protein genes, which encode the protein components of the silk cocoon, are specifically transcribed in the silk gland and are temporally and spatially regulated during embryonic and larval development. The promoters/enhancers that are responsible for the highly specific expression patterns of the silk protein genes are expected to be correspondingly simpler than those specifying more complicated patterns of gene expression, such as embryonic patterning and segmentation (see Ingham, 1988, for a review), and correspondingly more tractable experimentally. Furthermore, the silk gland itself is highly amenable to biochemical experimentation and thus represents a facile model system for studying developmental gene activation of a group of differentiation-specific genes. We anticipate that studies of such differential gene activation will enable us to identify the regulatory components that underlie the events of cell differentiation and specialization.

Genetic analysis is a complementary approach that has recently provided a great deal of information about the general mechanisms governing development in *Drosophila melanogaster* and *Caenorhabditis elegans.* A group of developmental control genes, including those for membrane-bound receptors, proteases, growth factors, and transcription factors that form a temporally ordered regulatory cascade, has been identified. In particular, many of these genes that encode transcription factors like the homeodomain-containing proteins (Scott, Tamkun, and Hartzell, 1989; Affolter, Schier, and Gehring, 1990), the zinc finger proteins (Klug and Rhodes, 1987), and the helix-loop-helix proteins (Murre, McCaw, and Baltimore, 1989), are believed to control spatial and temporal patterns of gene expression (Ingham, 1988). Currently, enormous efforts have been invested in understanding how these transcriptional regulatory genes control the expression of their as yet largely unidentified downstream genes (Biggin and Tjian, 1989; Hayashi and Scott, 1990). These two approaches, one starting from tissue-specific genes to regulatory genes and the other starting from transcriptional regulatory genes to downstream genes, have already started to provide some ideas of the regulatory networks in the development of multicellular organisms. A trial of the latter approach in *Bombyx* has been described in a separate chapter (see Ueno, Nagata, and Suzuki, this volume).

In this chapter we review the current state of knowledge concerning the developmental regulation of silk protein genes. In our recent analyses of the silk protein gene promoters, we have obtained interesting observations suggesting that the silk protein genes might be targets of some homeobox genes. We discuss here our working hypothesis that a number of homeobox genes might be involved in the control of silk gland development, and that their involvement in the transcriptional regulation of silk protein genes might be a part of the molecular mechanisms for silk gland cell differentiation and specialization. For other reviews of the silk gland system, the reader is referred to Akai (1984), Goldsmith and Kafatos (1984), Prudhomme, Couble, Garel, and Daillie, (1985), Shimura (1988), and Sprague (this volume).

Temporal and spatial gene expression in the silk gland

Specialization in the silk gland

Some authors describe the *B. mori* silk gland as a modified salivary gland, probably because of its origin from the labial segment. Indeed, though silk moth larvae also possess a pair of salivary glands that is derived from the mandibular segment (Toyama, 1909), the silk gland shares many similarities with the larval salivary gland of *Drosophila* in its function and postembryonic development (Berendes and Ashburner, 1978; Meyerowitz, Raghavan, Mathers, and Roark, 1987): Both organs grow in size without undergoing cell division, contain mostly secretory cells producing specialized products, and undergo histolysis shortly after puparium formation (see Suzuki, 1977, for references on silk gland development).

The silk gland begins as an ectodermal invagination of the basal part of the labium around stage 18 (Kanda, unpublished observations) or stage 19 (Toyama, 1909; Nunome, 1937) of embryonic development (see Figure 9.1) and its morphological development is complete by stage 25 (Nunome, 1937; Kanda, unpublished observations). No cell division takes place during the entire larval period, but the gland grows dramatically through multiple rounds of endomitoses. By the end of larval development, the silk gland, together with silk protein products, constitutes about 40 percent of the body weight. Because silk proteins constitute more than 95 percent of total protein synthesis in the gland, "silk production" can be considered a main quantitative function of the silkworm after the middle of the fifth intermolt. However, the silk gland degenerates after the silkworm enters its pupal stage.

Morphologically, a pair of silk glands is divided into three parts (Figure 9.2): the anterior silk gland (ASG), middle silk gland (MSG), and posterior silk gland (PSG), which contain about 600, 500, and 1,000 cells for the pair, respectively (Akai, 1984). These cells are all polyploid, and it has been estimated that 13, 19–20, and 18–19 endomitotic cycles occur during larval life in the cells of the ASG, MSG, and PSG, respectively (Perdrix-Gillot, 1979). The cells of the MSG and PSG are specialized for the production and secretion of silk proteins; the function of the ASG cells is still poorly known. The PSG produces fibroin, which is the main silk component. The gelatinous silk components, consisting of three layers of sericin which coat the fibroin, are secreted by different regions of the MSG. The MSG can be subdivided into four regions (see Figure 9.2): the anterior region (am) secreting the outer sericin layer, the anterior-middle region (amm) secreting the middle and outer sericin layers, the posterior-middle region (pmm) secreting the middle sericin layer, and the posterior region (pm) secreting the inner sericin layer (reviewed by Akai, 1985). Though it is not clear whether these regions are composed of different secretory cells, the expression of the sericin genes and several others is known to be regulated differently in various regions (see the section "Temporal and spatial gene expression"). There are also subtle differences in the ultrastructure of cells taken from different regions of the MSG (Akai, 1984). In contrast, the cells of the PSG appear to be uniform.

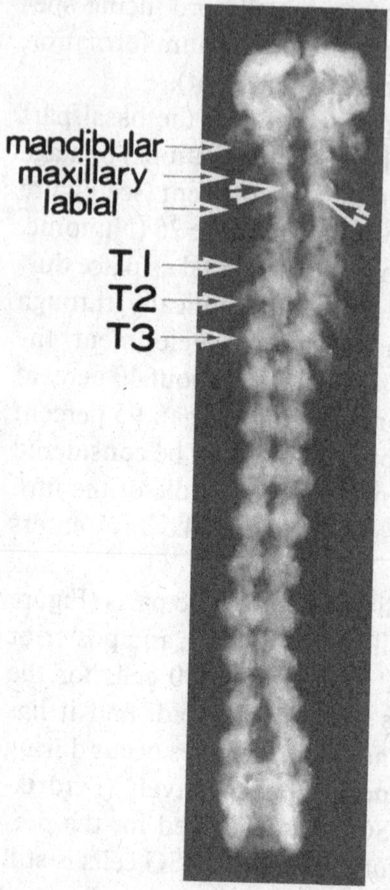

Figure 9.1. A stage 18 *Bombyx mori* embryo (generously provided by T. Kanda and T. Tamura of the National Institute of Sericulture and Entomology). Large arrows indicate the two placodes at the basal part of the labial segment that eventually invaginate to give rise to the silk glands. T1 to T3 represent the three thoracic segments.

Temporal and spatial gene expression

To date, at least five *B. mori* silk protein genes have been identified. Three of them are the structural genes of the fibroin that are expressed specifically in the PSG: the fibroin gene (Suzuki and Brown, 1972; Suzuki, Gage, and Brown, 1972; Ohshima and Suzuki, 1977; Tsujimoto and Suzuki, 1979a, 1979b), which encodes the large component of the fibroin; the P25 gene (Couble, Chevillard, Moine, Ravel-Chapuis, and Prudhomme, 1985); and the fibroin light chain gene (Yamaguchi et al., 1989; Mizuno, unpublished observations), which encode the small components of the fibroin. The other two are MSG-specific genes, the sericin-1 gene (originally called the sericin gene by Okamoto, Ishikawa, and Suzuki [1982] and renamed the sericin-1 gene by Michaille, Couble, Prudhomme, and Garel, [1986]) and the sericin-2 gene (Michaille, Garel, and Prudhomme, 1990b), which encode the sericin proteins.

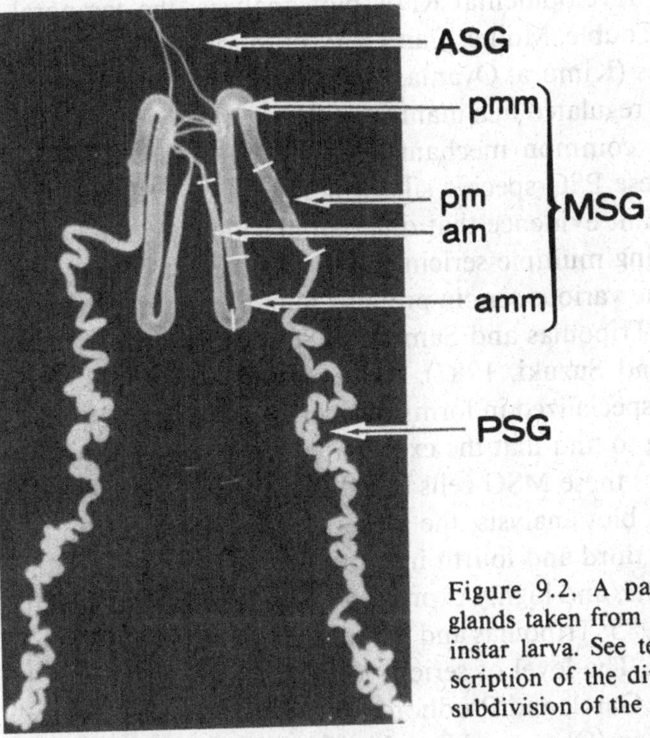

Figure 9.2. A pair of silk glands taken from a late fifth instar larva. See text for description of the division and subdivision of the silk gland.

A newly hatched larva is already capable of spinning silk. It has been shown by a ribonuclease protection assay that fibroin gene transcripts can be first detected about stages 25 and 26 of embryonic development (Ohta, Suzuki, Hara, Takiya, and Suzuki, 1988). This suggests that the transcription of the fibroin gene as well as other silk protein genes might be activated once embryonic development of the silk gland is completed. After hatching, larvae at all stages except during molting can spin small amounts of silk to facilitate molting by anchoring the old skin to the substratum. Fibroin gene transcripts can be detected in the PSG during the first, second (Suzuki, unpublished observations), third, fourth, and fifth intermolts but not during the third and the fourth molts (Suzuki and Brown, 1972; Suzuki and Suzuki, 1974; Maekawa and Suzuki, 1980). By nuclear run-on transcription assays, it has been recently documented that this repeated turn-on and turn-off of fibroin gene expression is primarily regulated at the transcriptional level (Obara and Suzuki, 1988). A quite unexpected observation from these run-on experiments is that during the fifth intermolt, fibroin gene transcription is resumed initially at the anterior portion of the PSG, which is adjacent to the MSG, and then gradually spreads toward the posterior part. By developmental RNA blot analysis, the temporal expression of the P25 (Couble, Moine, Garel, and Prudhomme, 1983) and fibroin light chain genes (Kimura, Oyama, Ueda, Mizuno, and Shimura, 1985) were shown to be regulated in a manner similar to that of the fibroin gene, suggesting that a common mechanism might be involved in the concerted control of these PSG-specific silk protein genes.

There is now significant evidence that differential splicing is the likely mechanism for generating multiple sericin RNAs of different coding potentials for producing the various sericin proteins found in the silk cocoon (Michaille et al., 1986; Tripoulas and Samols, 1986; Couble et al., 1987; Hamada, Yamashita, and Suzuki, 1987). As described earlier, different regions of the MSG are specialized in forming different sericin layers; thus it may not be surprising to find that the expression of the sericin genes is regulated differentially in these MSG cells (see Table 9.1 for a summary). By developmental RNA blot analysis, the sericin-1 gene was found to be weakly expressed in the third and fourth intermolts, not expressed in the third and the fourth molts, and highly expressed at the late fifth intermolt (Ishikawa and Suzuki, 1985; Tripoulas and Samols, 1986; Michaille, Garel and Prudhomme, 1989). The level of sericin-1 transcripts (Ishikawa and Suzuki, 1985; Michaille, Garel, and Prudhomme, 1989, 1990a), as well as sericin-1 gene transcription (Obara and Suzuki, 1988), increases exponentially during the fifth intermolt. In contrast, the sericin-2 gene was found to be highly expressed in the fourth intermolt, and its transcripts were still detectable in the fourth molt (Michaille et al., 1989). After a rapid accumulation at the beginning of the fifth intermolt, the level of sericin-2 transcripts gradually decreases (Michaille et al., 1989, 1990a). These observations suggest that the expression of the sericin-1 gene, but not sericin-2, might be regulated in a concerted manner similar to that of the fibroin, fibroin light chain, and P25 genes in the PSG.

Sericin-1 and sericin-2 gene transcription is orchestrated with a complex spatial and temporal control of differential splicing during the fifth intermolt (see Table 9.1). Four mature sericin-1 gene transcripts (2.8, 4.0, 9.0, and 10.5 kb) can be detected in fifth intermolt larvae (Michaille et al., 1986; Hamada et al., 1987). Among them, the 2.8 kb RNA accumulates abundantly in the pm region and is barely detected in the pmm region but not found elsewhere in the MSG, suggesting that it might encode a major component in the inner sericin layer. However, evidence relating these transcripts to different sericin proteins is still lacking.

The pmm region appears to have a developmental program of differential splicing abilities; it first accumulates the 4.0 kb RNA, then switches

Table 9.1. *Differential gene expression in the middle silk gland*

Subdivision of MSG	am			amm			pmm			pm		
Hours after 4th ecdysis	36	72	120	36	72	120	36	72	120	36	72	120
Ser-l												
2.8 kb RNA	−	−	−	−	−	−	−	+	++	++	+++	+++
4.0 kb RNA	−	−	−	−	+++	+	+++	+++	+	+/−	−	−
9.0 kb RNA	−	−	−	−	−	+/−	−	+/−	++	−	−	−
10.5 kb RNA	−	−	−	−	++	++	+/−	+++	+	−	−	−
Ser-2												
5.0 kb RNA	+++	+++	+++	+++	++	+	++	−	−	+/−	−	−
3.1 kb RNA	+++	+++	+++	+	+/−	−	+/−	−	−	−	−	−
MSG3	−	ND	−	+/−	ND	+/−	+++	ND	++	++	ND	+++
MSG4	+++	ND	+++	+/−	ND	−	−	ND	−	−	ND	−
MSG5	+++	+++	+++	+	+/−	−	−	−	−	−	−	−

These data are summarized from Couble et al. (1987), Hamada et al. (1987), Ishikawa and Suzuki (1985), and Michaille et al. (1986, 1989, 1990a). ND = not done.

to the 10.5 kb RNA, and finally to the 9.0 kb RNA. In the amm region, only the 4.0 kb and 10.5 kb RNAs, but not the 9.0 kb RNA, could be detected. Sericin-1 transcripts cannot be found in the am region because this gene is not transcribed there, as shown by nuclear run-on experiments (Obara and Suzuki, 1988). In contrast, sericin-2 gene transcripts, which might encode the sericin proteins in the outer sericin layer, are abundantly accumulated in the am region. Interestingly, though a gradient of sericin-2 transcripts could be detected initially in the MSG with a barely detectable level in the pm region increasing to a very abundant level in the am region, the transcripts become detectable only in the am region later in development (Couble et al., 1987).

Other MSG-specific genes have been isolated and shown to share similar RNA accumulation patterns to those of the sericin genes (Michaille et al., 1989). Similar to the sericin-2 gene, the MSG5 gene is actively expressed in the fourth intermolt and its mRNA remains easily detectable during the fourth molt. Transcripts of the MSG5 gene also accumulate at the beginning of the fifth intermolt and then diminish and become detectable only in the am region. The MSG3 gene is weakly expressed during the fourth intermolt and its mRNA becomes undetectable during the fourth molt, similar to that of the sericin-1 gene. Furthermore, MSG3 transcripts could not be detected in the pm region. It has been suggested that the expression of the sericin-1 and MSG3 genes and that of the sericin-2 and the MSG5 genes are regulated by a common mechanism (Michaille et al., 1989). As is described in the section on "Homeobox gene expression," at least four homeobox genes are also actively expressed in the MSG during the fifth intermolt and their mRNA levels are also differentially regulated during the fourth molt/fifth intermolt period. The list of developmentally regulated genes in the MSG is obviously longer, and their systematic analysis warrants a promising avenue toward understanding the complex mechanism(s) involved in the spatial and temporal control of gene expression.

Tissue-specific in vitro transcription in silk gland extracts

Our present knowledge of the transcriptional control elements of the silk protein genes relies mostly on studies using in vitro transcription (for a review, see Suzuki et al., 1990b). Suitable systems for transient or transgenic expression are not yet available (see Iatrou, this volume). Consequently, much effort has been invested in preparing transcriptionally active extracts from different *Bombyx* sources. To date, altogether 14 cell-free transcription systems have been developed from *B. mori* embryos and tissues of various developmental stages (Suzuki et al., 1986, 1990a) since the first establishment of tissue extracts from the PSG and MSG of 2-day-old fifth-instar larvae (Tsuda and Suzuki, 1981). As it is anticipated that the silk gland extracts should contain high concentrations of both basal and regulatory transcription factors that are required for active transcription of the silk protein genes in vivo, these in vitro transcription systems are ideal for the biochemical characterization of the components regulating silk protein gene expression.

In the silk gland extracts, the upstream promoter element (UPE) of the fibroin gene (a 200 base-pair region upstream of the TATA-box) and the sericin-1 gene (a 300 base-pair region upstream of the TATA-box) were found to be required for a maximal level of transcription (Tsuda and Suzuki, 1981, 1983; Suzuki et al., 1986; Hui and Suzuki, 1989; Matsuno et al., 1989; Takiya, Hui, and Suzuki, 1990). Both UPEs can stimulate in vitro transcription from a heterologous adenovirus major late promoter efficiently in PSG and MSG extracts but only weakly in two non–silk gland extracts (Hui and Suzuki, 1989). These studies illustrate that the silk protein genes are transcribed in vitro in a tissue-specific manner (Suzuki et al., 1986).

Interestingly, the fibroin and sericin-1 genes are differentially transcribed in PSG and MSG extracts, mimicking their in vivo expression (Suzuki et al., 1990a). It is likely that a combination of transcriptional regulatory elements in these silk protein genes is responsible for this differential transcription in vitro because the specificities of the two UPEs appear to be quite similar when they are assayed in a heterologous context (Hui and Suzuki, 1989). In this respect, the core promoter (from −37 to +10), which includes the TATA-box and the transcription initiation region, along with an intronic element (from +156 to +454) of the fibroin gene, was recently found to be important for preferential transcription in PSG extracts (Takiya et al., 1990). There is an apparent redundancy of transcriptional regulatory elements in the fibroin gene; both the UPE and the intronic element are functional in stimulating transcription from the core promoter (Takiya et al., 1990; Takiya, unpublished observations). These observations suggest that the core promoter may require both elements for a maximal level of fibroin gene transcription and a higher level of specificity in vivo.

In recent years, similar approaches have also been used to obtain transcriptionally active extracts from *Drosophila* embryos (see Biggin and Tjian, 1989, for a review) and mammalian tissues (Gorski, Carneiro, and

Schibler, 1986) for the characterization of transcriptional regulatory components. These systems can complement the deficiencies of *Drosophila Kc* and human HeLa cell extracts, which have been commonly used for studies of the basic transcription machinery (see Kadonaga, 1990, for a review). Recent studies have further revealed that, in addition to sequence-specific transcription factors and apparently ubiquitous basal transcription factors, some specific components (termed co-activators or adaptors) acting through protein-protein interactions are necessary for transcriptional regulation (for a review, see Lewin, 1990a). Thus it is desirable to have an in vitro transcription system that can be used for the characterization of both basal and regulatory transcription factors as well as other yet unknown regulatory components. In this respect, the silk gland system is advantageous because it is possible to obtain a large quantity of pure tissue of a specific developmental stage that is entirely made up of a single cell type (secretory cells). Besides in vitro transcription assays, silk gland extracts have also been used for characterization of general transcription factors (Tabuchi and Hirose, 1988; Takiya and Suzuki, 1989; Takiya et al., 1990; Takiya, unpublished observations) and sequence-specific transcription factors (Suzuki and Suzuki, 1988; Matsuno et al., 1989, 1990; Hui, Matsuno, and Suzuki, 1990; Ueda and Hirose, 1990; Suzuki et al., 1991a, 1991b; Takiya, unpublished observations), as well as a DNA supercoiling factor (Hirose and Suzuki, 1988; Ohta and Hirose, 1990).

Cis- and *trans*-acting components in silk gene regulation

Protein binding sites in the silk gene promoters

Nucleotide sequence analyses of the 5′ flanking region of the fibroin gene from several strains of *B. mori* (Tsujimoto and Suzuki, 1979a, 1979b; Suzuki and Adachi, 1984; Ueda et al., 1985) and a closely related species, *B. mandarina* (Kusuda, Tazima, Onimaru, Ninaki, and Suzuki, 1986), indicate that the immediate 5′ flanking sequence, which is highly conserved, is probably important for the developmental regulation of fibroin gene transcription. This hypothesis is strongly supported by observations that highly homologous sequence motifs are also found in the 5′ flanking regions of the sericin-1 gene (Okamoto, Ishikawa, and Suzuki, 1982), the P25 gene (Couble et al., 1985) and the fibroin light chain gene (Mizuno, unpublished observations), and that the 5′ flanking sequence or UPE of the fibroin and sericin-1 genes can specify tissue-specific transcription in vitro (see discussion in the previous section). By various DNA-protein binding assays, we have now identified a number of protein factors that interact with these UPE sites in silk gland extracts as putative regulatory components (Suzuki and Suzuki, 1988; Matsuno et al., 1989, 1990; Hui et al., 1990; Suzuki et al., 1991a, 1991b). However, emphasis on these binding proteins in this review should not be interpreted to indicate that other known transcriptional regulatory elements such as the core promoter and the intronic element of the fibroin gene are less important. It is also likely that there are other as yet unknown long-range enhancers involved in the transcriptional control of these silk protein genes.

Five protein-binding regions (A to E) in the UPE of the fibroin gene (Hui et al., 1990; see Figure 9.3) and three protein-binding regions (SA to SC) in the UPE of the sericin-1 gene (Matsuno et al., 1989) have been mapped by DNase I and exonuclease III footprinting assays. The two proximal regions 5′ to the fibroin gene TATA box, the A and B regions, are found to be homologous to the SA region of the sericin-1 gene in binding a silk gland–specific protein, SGF-1. Related sequence motifs with a consensus of A/TGTTT are present in the promoter-proximal regions of other silk protein genes of *B. mori* (Figure 9.4).

Interestingly, the A and B regions are also conserved in the fibroin gene of the Japanese oak silkworm, *Antheraea yamamai,* though the structural parts of these two fibroin genes are very different (Tamura, Inoue, and Suzuki, 1987). It has been shown that the *A. yamamai* fibroin gene can be efficiently transcribed in the PSG extracts of *B. mori* (Tamura et al., 1987), and recent deletion analysis further indicates that the A and B regions of the *A. yamamai* fibroin gene are important for transcription in PSG extracts of both species (Tamura, unpublished observations). Similar to the PSG extracts of *B. mori,* we were able to detect an SGF-1-like protein in the PSG extracts of *A. yamamai* by electrophoretic mobility shift assays using probes of the *Bombyx* A and B regions (Figure 9.5). Furthermore, a protein similar to FBF-A1, a *Bombyx* ubiquitous protein that interacts with the A region, can also be found in the *Antheraea* PSG extract. These evolutionary conservations suggest that SGF-1 is a likely key component involved in the regulation of the fibroin gene and other silk protein genes that possess SGF-1 binding sites. Consistent with this hypothesis, SGF-1 binding sites are important for in vitro transcription of the fibroin and sericin-1 genes (Tsuda and Suzuki, 1983; Matsuno et al., 1989, 1990).

A sequence motif similar to the SGF-1 binding site can also be found in the proximal promoter regions of the *D. melanogaster* Sgs-3 and Sgs-4 genes and the *D. virilis* Lgp-1 gene (see Figure 9.4; for reviews, see Mey-

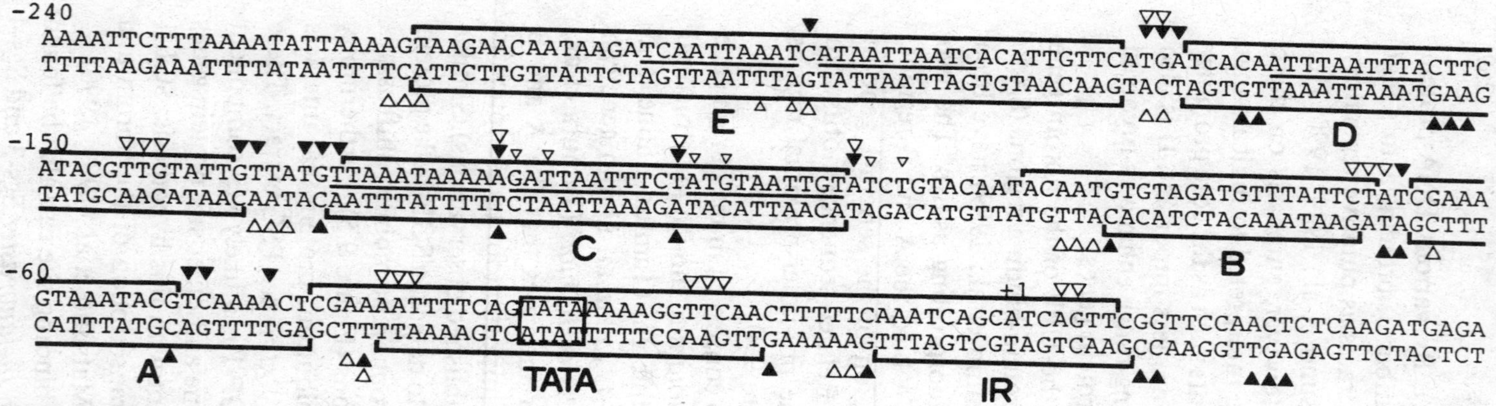

Figure 9.3. Nucleotide sequence of the fibroin gene promoter and protein-binding regions in the PSG extract. The 10 base-pair AT-rich repeats are marked with lines. Brackets indicate the DNase I protected regions and filled triangles represent DNase I hypersensitive sites. Large and small open triangles indicate strong and weak exonuclease III stop sites, respectively.

B. mori	Fb:	A	-44 AGTTTTGAC -52
		B	-77 TGTTTATTC -69
	Fb-L:		-63 TGTTTGATA -55
	P25:		-86 TGTTTCCAC -78
	Ser1:	SA	-97 TGTTTGCAC -89
	Ser2:		-67 AGTTTGACC -59
B. mandarina	Fb:		-44 AGTTTTGAC -52
			-77 TGTTTATTC -69
A. yamamai	Fb:		-69 TGTTTATTT -61
			-120 TGTTTTAAT -112
D. melanogaster	Sgs-3:		-92 TGTTTGCAT -84
	Sgs-4:		-132 TATTTGTAT -124
D. virilis	Lgp-1:		-122 TATTTGCTC -114

Figure 9.4. Putative SGF-1 binding sites in silk protein genes and related motifs in *Drosophila* salivary gland–specific genes.

erowitz, Raghavan, Mathers, and Roark, 1987; Kress and Swida, 1990). These genes encode glue proteins expressed specifically in the *Drosophila* larval salivary gland, and these sequence motifs have been demonstrated to be important for ecdysone-dependent developmental regulation. Recently, it has been found that the 5′ flanking sequence of the *Bombyx* P25 gene can specify the temporally and spatially regulated expression of a beta-galactosidase reporter gene in the anterior part of the *Drosophila* larval salivary gland (Bello and Couble, 1990). These observations suggest that some of the regulatory elements controlling silk protein gene expression in *Bombyx* might be conserved in *Drosophila* for the regulation of the glue protein genes, and argue further for common embryonic origins of these tissues. Evolutionary conservation of regulatory components of chorion genes of silk moths and fruit flies has also been demonstrated using transgenic expression in *Drosophila* (Mitsialis and Kafatos, 1985; see also Kafatos et al., this volume).

In addition to SGF-1 binding sites, a 10 base-pair AT-rich sequence repeat in the C, D, and E regions of the fibroin gene is also conserved in the 5′ flanking regions of various other silk protein genes (see Figure 9.3). These sequence repeats, which contain a core sequence of AATTAA or

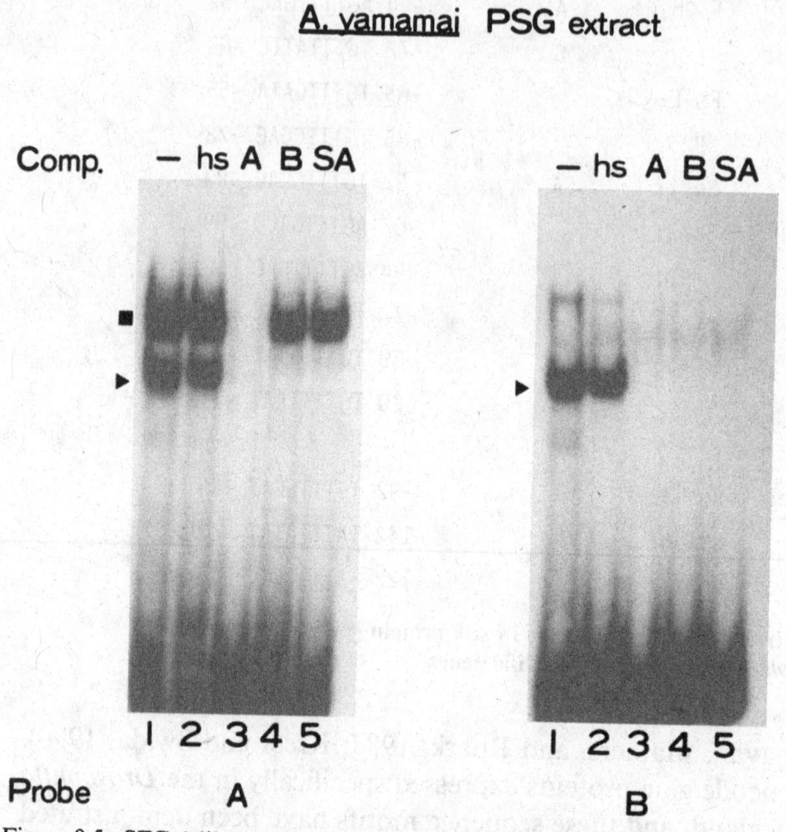

Figure 9.5. SFG-1-like protein in the PSG extract of *Antheraea yamamai.* Oligonucleotide probes of the A and B sites of the *Bombyx mori* fibroin gene were examined in a mobility shift assay using a crude PSG extract of *A. yamamai.* A SGF-1-like protein (filled triangle) was detected by both probes and could be competed specifically by an excess of the A, B, as well as the sericin-1, SA, oligonucleotides but not by a nonspecific hs (heat shock transcription factor) oligonucleotide. The square indicates another specific protein binding to the A site only.

TTAATT, are usually clustered together; in the case of the fibroin gene, six copies are found in the region between −210 and −100. A large number of protein complexes can be detected in PSG extracts with probes of these regions (Hui et al., 1990). Three of them, SGF-2, -3, and -4, have been analyzed to a greater extent by competition and mutational analyses (see Figure 9.6 for a summary). A quite remarkable observation from these studies is that the relative binding affinities of these proteins are very different; for example, the affinity of SGF-4 for the C region is about 50

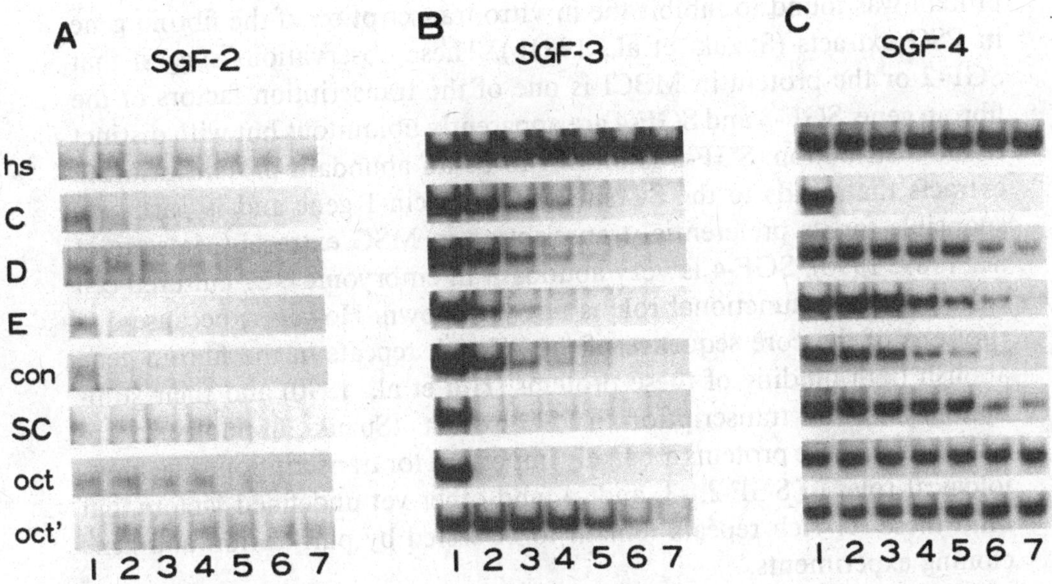

Figure 9.6. Binding specificity of SGF-2, -3, and -4. Relative affinities of various binding sites to SGF-2, -3, and -4 were examined by competition analyses. Lane 1 is the minus control without competitors. Lanes 2–7 contain 0.128, 0.32, 0.64, 1.28, 3.2, and 6.4 nM, respectively, of various competitors. Details of the competitors can be found in Hui et al. (1990).

times higher than for other binding sites. As is discussed in the next section, this property may be important for a fine-tuning of gene activity through the interaction of a related group of transcription factors with a cluster of related binding sites.

Based on their tissue distribution and the results of our recent in vitro transcription experiments (Suzuki, unpublished observations), these proteins are believed to be transcription factors of the silk protein genes. Among them, SGF-2 is likely involved in the PSG-specific transcription of the fibroin gene because it is specific to PSG extracts (Hui et al., 1990). By using a longer probe containing the entire UPE of the fibroin gene, MBCI, a PSG-specific complex consisting of two protein components, was identified (Suzuki and Suzuki, 1988; Suzuki et al., 1991a). MBCI appears to be equivalent to the SGF-2/UPE complex because the formation of MBCI is affected by substitution mutations in the E site, which are known to be important for SGF-2 binding (Suzuki et al., 1991a; see also Hui et al., 1990). One of the components of MBCI has been purified as a 125kDa protein (Suzuki et al., 1991a), and an antiserum raised against the purified protein was found to inhibit the in vitro transcription of the fibroin gene in PSG extracts (Suzuki et al., 1991b). These observations suggest that SGF-2 or the protein in MBCI is one of the transcription factors of the fibroin gene. SGF-3 and SGF-4 are apparently ubiquitous but with distinct tissue distribution. SGF-3 is known to be an abundant protein in MSG extracts that binds to the SC site of the sericin-1 gene and is partly responsible for its preferential transcription in MSG extracts (Matsuno et al., 1989, 1990). SGF-4 is very abundant in embryonic e-21 cultured cell extracts, but its functional role is not yet known. However, because disruptions of the core sequence of the AT-rich repeats in the fibroin gene abolish both binding of these proteins (Hui et al., 1990) and their stimulatory effect on transcription in PSG extracts (Suzuki, unpublished observations), these proteins are likely important for transcription. The physiological roles of SGF-2, -3, and -4, and other yet undefined factors that bind these AT-rich repeats remain to be tested by purification and gene cloning experiments.

Homeodomain binding sites and putative homeodomain proteins

The AT-rich repeats in the silk protein genes are remarkably similar to a consensus sequence, TCAATTAAAT, in the binding sites for a number of *Drosophila* homeodomain-containing proteins (HD proteins) (Scott et al., 1989; Affolter et al., 1990). By DNase I footprinting and electrophoretic mobility shift assays, we have shown that two *Drosophila* HD proteins, the *even-skipped* and *zerknüllt* proteins, bind specifically to the C, D, and E regions of the fibroin gene (Hui and Suzuki, 1990). These two proteins also interact with related sequence motifs in the 5′ flanking regions of the sericin-1 and fibroin light chain genes (Hui and Suzuki, 1990; Hui, Suzuki, et al., 1990; Figure 9.7). Binding analyses using a variety of mutant sites revealed that these two HD proteins and the three silk gland proteins, SGF-2, -3, and -4, possess very similar binding specificities (Hui et al., 1990; Hui and Suzuki, 1990). In addition to the core sequence, which is found to be indispensable for binding, these studies also indicate that the binding of these proteins is differentially affected by base substitutions in these AT-rich repeats.

Several other observations also suggest that SGF-2, -3, and -4 are putative HD proteins. First, an artificial binding site containing three copies of the wild-type consensus sequence (TCAATTAAAT) competes effectively for the formation of these silk gland complexes (Hui et al., 1990). A mutant

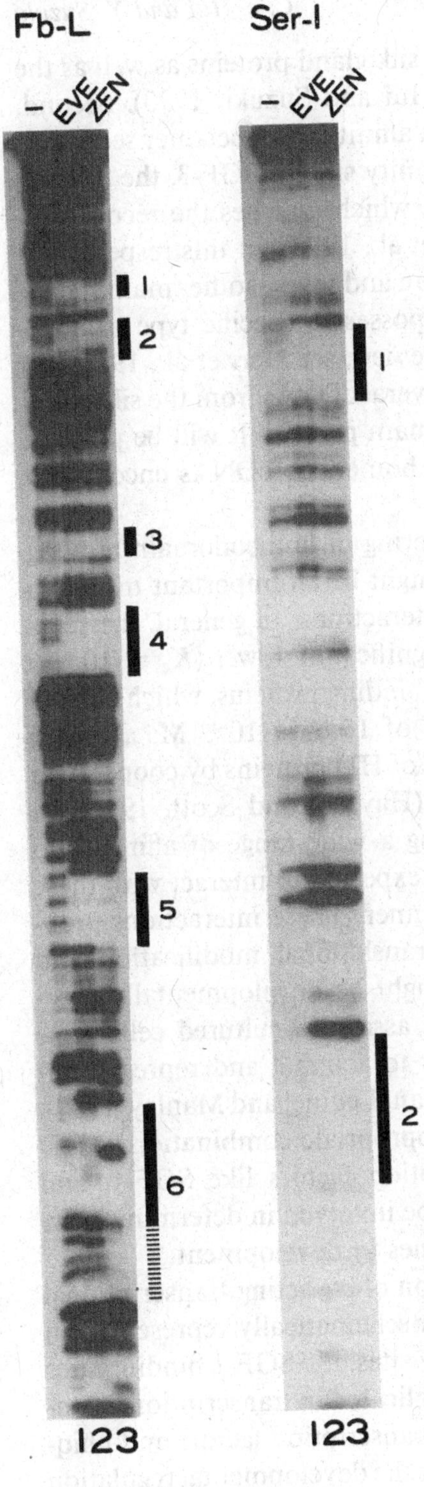

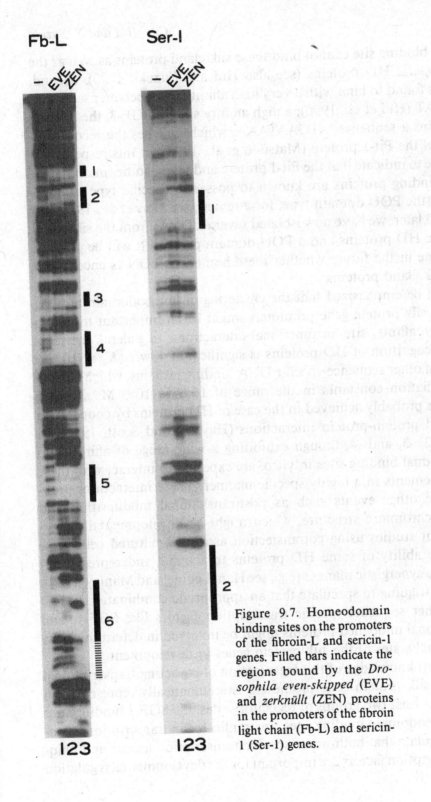

Figure 9.7. Homeodomain binding sites on the promoters of the fibroin-L and sericin-1 genes. Filled bars indicate the regions bound by the *Drosophila even-skipped* (EVE) and *zerknüllt* (ZEN) proteins in the promoters of the fibroin light chain (Fb-L) and sericin-1 (Ser-1) genes.

consensus binding site cannot bind these silk gland proteins as well as the two *Drosophila* HD proteins (see also Hui and Suzuki, 1990). Second, SGF-3 was found to bind with a very high affinity to an octamer sequence, ATGCAAAT (Hui et al., 1990); a high affinity site for SGF-3, the SC site, also contains a sequence, ATGAATAAA, which matches the recognition sequence of the Pit-1 protein (Matsuno et al., 1990). In this respect, it is appropriate to indicate that the Pit-1 protein and several other mammalian octamer binding proteins are known to possess a specific type of homeodomain (the POU domain type; for a review, see Herr et al., 1988). As is described later, we have now isolated several cDNAs from the silk gland that encode HD proteins and a POU domain protein. It will be possible to determine in the future whether these homeobox cDNAs encode any of these silk gland proteins.

It should be emphasized that the clustering of homeodomain binding sites in the silk protein gene promoters might be an important means to create a high affinity site for functional interactions. In general, the specificity of recognition of HD proteins is significantly lower (K_d = 10^{-8} to 10^{-9}M) than other sequence-specific DNA binding proteins, which usually have dissociation constants in the range of 10^{-12} to 10^{-10} M; a higher specificity is probably achieved in the case of HD proteins by cooperative binding and protein-protein interactions (Hayashi and Scott, 1990). In vivo, SGF-2, -3, and -4, though exhibiting a wide range of affinities to these individual binding sites in vitro, are expected to interact with these promoter elements in a highly specific manner. These interactions probably require other events such as posttranslational modification and changes in chromatin structure, which might be developmentally regulated. Recent studies using cotransfection assays in cultured cells have revealed the ability of some HD proteins to activate and repress transcription in a synergistic manner (e.g., see Han, Levine, and Manley, 1989). Thus it is intriguing to speculate that an appropriate combination of HD proteins, other sequence-specific transcription factors like SGF-1, and some additional undefined proteins might be involved in determining the transcriptional states of the silk protein genes in development.

Our present knowledge of the organization of *cis*-acting transcriptional elements in silk protein gene promoters is schematically represented in Figure 9.8. At least two groups of regulatory sites, the SGF-1 binding sites and the homeodomain binding sites, are implicated in transcriptional control. We speculate that both tissue-specific transcription factors and ubiquitous transcription factors are important for the developmental regulation

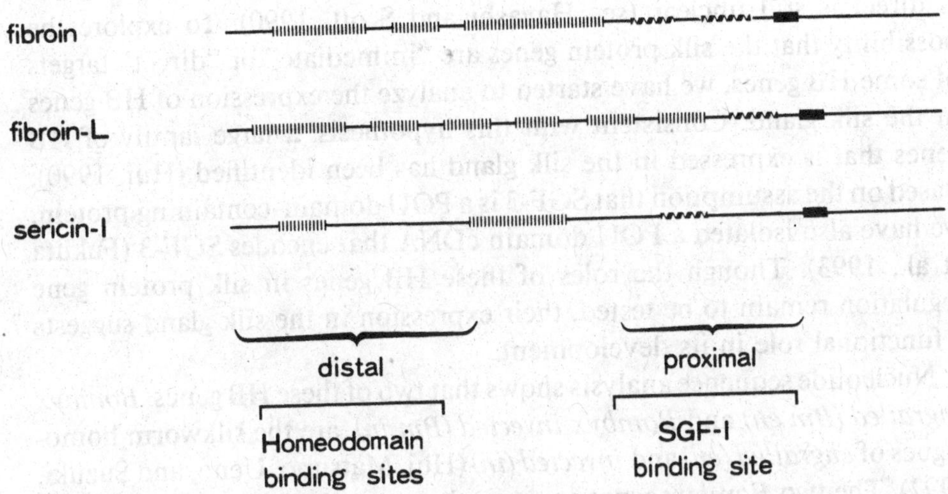

Figure 9.8. Schematic representation of *cis*-acting elements in the silk gene promoters. Homeodomain SGF-1 binding sites are indicated. Filled boxes represent the TATA box of these genes.

of silk protein gene transcription. The modes of transcriptional regulation in the silk gland might resemble those involved in the regulation of liver-specific genes in mammals. For example, the liver-specific albumin promoter also possesses a similar array of proximal and distal *cis*-acting elements; a proximal binding protein, which is liver-specific, plays a key part in the activation of transcription, whereas the more distal binding proteins, which are less tissue-specific, can act cooperatively with the proximal binding protein when its concentration is limiting (for a review, see Johnson, 1990). Transcriptional control in both systems appears to be a combinatorial process requiring more than one "on switch" to allow gene activation. This combinatorial regulation probably provides a more stringent control of transcription by reducing the occurrence of inappropriate gene expression and increases efficiency by exploiting a single protein to regulate different genes in more than one cell type.

Homeobox gene expression in the silk gland

The most interesting observation in our studies is that the distal binding sites in the silk protein genes are possible targets for homeobox (HB) gene action. Though several candidate target genes of HB genes have been identified by genetic approaches in *Drosophila*, whether or not the action is direct is still unclear (see Hayashi and Scott, 1990). To explore the possibility that the silk protein genes are "immediate" or "direct" targets of some HB genes, we have started to analyze the expression of HB genes in the silk gland. Consistent with this hypothesis, a large family of HB genes that is expressed in the silk gland has been identified (Hui, 1990). Based on the assumption that SGF-3 is a POU domain-containing protein, we have also isolated a POU domain cDNA that encodes SGF-3 (Fukuta et al., 1993). Though the roles of these HB genes in silk protein gene regulation remain to be tested, their expression in the silk gland suggests a functional role in its development.

Nucleotide sequence analysis shows that two of these HB genes, *Bombyx engrailed (Bm en)* and *Bombyx invected (Bm in)*, are the silkworm homologues of *engrailed (en)* and *invected (in)* (Hui, Matsuno, Ueno, and Suzuki, 1992). The two *Bombyx* genes are actively transcribed in the MSG during the fourth molt/fifth intermolt period. In *Drosophila, en* is known to be involved in specifying compartments in embryonic segments, as well as in the control of neurogenesis (Kornberg, 1981; Lawrence and Johnston, 1984). Because the silk gland is derived from the ectoderm, we speculate that *Bm en* and *Bm in* might specify compartments in the silk gland. In this respect, it is interesting to find that their transcripts cannot be detected in the PSG, suggesting that the MSG, which expresses both *Bm en* and *Bm in,* might represent part of the posterior developmental compartment. Because, as described in the previous section, the silk gland is compartmentalized into various regions that express different genes, a role of *Bm en* and *Bm in* in silk gland development and/or silk gene regulation can be anticipated.

The silkworm homologue of *Antennapedia (Antp), Bombyx Antennapedia (Bm Antp),* is another HB gene actively expressed in the MSG during the fourth molt/fifth intermolt period (Hui, unpublished observations). The expression of this homeotic gene in the silk gland probably reflects the origin of the silk gland, which is derived from the labial segment where *Antennapedia* is known to be expressed in the *Drosophila* embryo (for a review, see Morata, Macias, Urquia, and Gonzalez-Reyes, 1990). By Northern blot analysis, neither *Bombyx Sex combs reduced* (*Bm Scr*) nor *Bombyx Ultrabithorax,* two homeotic genes that are presumably expressed in more anterior and posterior regions, are found to be expressed in the silk gland at these stages, emphasizing the specific nature of *Bm Antp* expression (Hui, unpublished observations; Ueno, unpublished observations).

By PCR cloning, we have also detected the expression of a large family of HB genes in the PSG of 2-day-old fifth-instar larvae (Hui, 1990; Figure 9.9), which represent sequences like those of *Drosophila Deformed, proboscipedia, labial,* and *caudal.* Several *Antp*-type sequences, as well as a sequence that probably represents a novel type of HB, have also been detected using this approach. At present, the significance of these findings is still unclear, but it should be emphasized that all of these genes are known to be expressed in the head region of *Drosophila.* This would be consistent with the possibility that they might be involved in some aspects of silk gland development in *Bombyx.*

We have also characterized a POU domain-containing protein by PCR and cDNA cloning (Fukuta et al., 1993). By DNase I footprinting and electrophoretic mobility shift assays, this protein was found to be identical to SGF-3. The observation that its transcripts are very abundant in the MSG further suggests that it is encoded by the SGF-3 gene. Interestingly, the C-terminal region of this protein, including the POU domain and the POU homeodomain, is found to be identical to that of the *Drosophila Cf1-a* protein (Johnson and Hirsh, 1990). This *Drosophila* POU domain-containing protein has been suggested to be a regulator in the transcriptional control of dopaminergic neuron-specific genes. It should now be possible to examine the role played by the *Bombyx* POU domain homologue in the transcriptional regulation of the silk protein genes.

Perspectives

Biochemical characterization of transcription factors using in vitro transcription assays remains an indispensable approach to identify more regulatory components and to understand the details of their molecular mechanisms. These assays should be able to test directly the functions of the HB genes in silk protein gene transcription. Additionally, we can analyze how these HB genes work by using cotransfection assays in cultured cells or by microinjection assays in dissected silk glands (Maekawa, unpublished observations). It should be emphasized that some of these HB genes might not act on the silk protein genes. These in vivo experiments are thus essential for testing our working hypotheses by identifying the HB genes that directly regulate the transcription of the silk protein genes.

Though mutations affecting the control of silk gland development are not available, it may be possible now to determine whether any of the

```
Antp     ELEKEFH FNRYLTRRRRIEIAHALCLTERQ IKIWFQN

                                         V V
I        ELEKEFH ----------------------- IKIWFQN   Antp-type
II               -------------------S---               "
III              ---------------V---S---               "
IV               --------------N--------               "
V                --------------N----S---               "
VI               -----A--------N----N---               "
VII              ---------------T---N---               "
VIII             ---------------I-V-S---               "
IX               ---------------T-V-S---           Dfd?
X                Y--------------T-V-S---           Dfd?
XI               -----C-P--V-M-NL-N-----           Dfd?
XII              -----C-P--V-M-NL-N-S---           Hox1.5, pb?
XIII             --H----A---DL-NS---N---           Hox1.5, pb?
XIV              YS--I-I--KA-L-VS-G-S---           lab?
XV               CKK--SLTE-SQ-----K-S-V-           cad?
                                                   new

                 ---Y-----R---A--L-L----
```

Figure 9.9. Deduced amino acid sequence of the "partial" homeobox region identified by reverse PCR from PSG mRNA of 2-day-old fifth instar larvae. Consensus sequences of various types of homeobox genes are taken from Scott et al. (1989).

HB/homeotic genes are involved. In this respect, it is interesting to find that *Drosophila Scr* is involved in the control of embryonic salivary gland development and that mutations in this homeotic gene prevent its formation (Beckendorf, unpublished observations). The placode of the salivary gland appears at the medial border of the labial segment where *Scr* is known to be expressed. It is interesting to note that *en* expression is not detected in the salivary gland (Kornberg, unpublished observations), suggesting that the placode is likely derived from the anterior compartment of the labial segment where *Scr* but not *Antp* is expressed (see Morata et al., 1990). We speculate that *Bm Antp,* instead of *Bm Scr,* might be involved in the control of silk gland development because the placode of the silk gland is probably derived from a more posterior region of the labial segment than the *Drosophila* salivary gland, where both *Bm Antp* and *Bm en* are expressed. The recent demonstration of a partial deletion of *Bm Antp* in the homeotic mutant, *Nc* (Nagata, unpublished observations), should enable us to begin a test of this hypothesis.

The availability of DNA clones encoding these transcription regulators has also made possible a study of their control mechanisms. In this respect, the co-expression of these HB genes in the silk gland provides a very interesting model for studying their possible cross-regulation, as has been shown in *Drosophila* embryogenesis. When compared with the heterogeneous nature of the embryo extracts, the silk gland extracts are better for the biochemical characterization of such interactions because of their high degree of specificity, both in spatial and temporal terms. It will be interesting to test whether any of these genes is autoregulated, and whether some of them form a regulatory cascade among themselves, instead of acting directly on the silk protein genes. Assuming that these HB/homeotic genes are important for silk gland development, we can also explore the regulatory pathway by looking for their downstream genes (see Gould et al., 1990).

The silk gland system appears ideal for the search of keys in development. Starting from two different ends of the regulatory cascade (the regulatory genes or the differentiation-specific genes), these studies should provide better information concerning how homeotic genes regulate cell differentiation and specialization.

Note added in proof

Recent progress made after submission of this chapter will be found in Suzuki, Y. (in press), Genes that are involved in *Bombyx* body plan and silk gene regulation, *International Journal of Developmental Biology*, 38.

Acknowledgments

We would like to thank Drs. J.-C. Prudhomme, P. Couble, S. Mizuno, and S. Hirose for sending us their preprints and reprints, the members of our laboratory for permitting us to cite their unpublished data, and Dr. M. Goldsmith for excellent editorial assistance. C.-c. Hui was supported by a postdoctoral fellowship from the Japan Society for the Promotion of Science. The work in our laboratory was partly aided by a Grant-in-Aid for Research of Priority Areas from the Ministry of Education, Culture and Science of Japan.

10

Control of transcription of *Bombyx mori* RNA polymerase III

KAREN U. SPRAGUE

Background

The silk gland of *Bombyx mori,* the commercial silkworm, provides one of the most striking examples of regulated polymerase III action. Study of this lepidopteran transcription system thus offers the possibility of insight into a particularly interesting aspect of polymerase III function. Moreover, it should illuminate basic features of class III transcription, and given recent indications of mechanistic similarity between polymerases II and III (Lobo and Hernandez, 1989; Murphy, Moorefield, and Pieler, 1989; Lobo, Lister, Sullivan, and Hernandez, 1991; Margottin et al., 1991), it could reveal even more fundamental aspects of eukaryotic transcription.

The fact that polymerase III activity is regulated is not always obvious because many of the polymerase III products are ubiquitous RNAs, such as 5S ribosomal RNA and tRNA. Although it is true that both 5S RNA and tRNA are synthesized by all cells, it is clear that the levels of particular species of these general RNA types vary widely. For instance, the amounts of oocyte type 5S RNA are strikingly different in oocytes and somatic cells of *Xenopus* frogs (Brown and Sugimoto, 1973), and the tRNA population of reticulocytes changes during differentiation to match the requirements for globin synthesis (Hatfield, Matthews, and Rice, 1979; Hatfield, Varricchio, Rice, and Forget, 1982). Specialization of the tRNA population in the *Bombyx mori* silk gland is particularly dramatic (see Hui and Suzuki, this volume, for a description of *B. mori* silk glands). Here, the correlation between amino acid usage and the machinery responsible for protein synthesis is especially clear because of the unusually simple amino acid composition of the major product of the posterior silk gland, fibroin. Because this polypeptide is composed primarily of three amino acids, glycine, alanine, and serine (44 percent, 29 percent, 12 percent of the total residues, respectively) (Lucas, Shaw, and Smith, 1958; Sprague, 1975), specialization of the machinery that synthesizes it might be expected to be particularly obvious. Indeed, the earliest experiments (Matsuzaki, 1966; Garel et al., 1970; Majima, Kawakami, and Shimura, 1975) revealed a remarkable correlation between the abundance of glycine, alanine, and serine in fibroin and the corresponding tRNAs in the posterior silk gland – the tissue that synthesizes fibroin. Although the original quantitations were somewhat indirect because they relied on aminoacylation of total tRNA, the basic findings have been substantiated by direct physical measurement of highly resolved, independently identified, tRNA species (Garel, Hentzen, and Daillie, 1974; Garel, 1976; Chavancy, Chevallier, Fournier, and Garel, 1979).

Specialization of the silk gland translation machinery is not limited to tRNAs. The aminoacyl tRNA synthetase population is likewise highly enriched for certain types. The proportions of glycyl-, alanyl-, and seryl-synthetases increase in the posterior silk gland when that tissue is becoming specialized for fibroin production (Chavancy, Garel, and Daillie, 1975; Majima et al., 1975), and the abundance of these enzymes makes them unusually easy to purify from the silk gland (Dignam and Dignam, 1984; Nishio and Kawakami, 1984; Viswanathan and Dignam, 1988).

The production of reliable antibodies to the purified enzymes, as well as the isolation of cDNA and genomic clones, now permits the study of synthetase gene regulation during silk gland development. This work has already revealed that the period of peak accumulation of the alanyl-, glycyl-, and seryl-synthetases is about day 5 of the fifth instar and that increased levels of the corresponding translatable mRNAs precede the increase in each synthetase by at least three days (Viswanathan, Dignam, and Dignam, 1988). In the case of alanyl-tRNA synthetase, increased steady state levels of mRNA have also been detected directly by hybridization to a well-characterized cDNA clone (Chang and Dignam, 1990). Indeed, the concentration of alanyl-tRNA synthetase mRNA is already so high at the beginning of the instar that it is tempting to imagine that synthetase regulation could be exerted at the level of translation, as well as transcription. Perhaps synthetase production is directly coupled to tRNA accumulation. Current work is aimed at characterizing genomic copies of the synthetase genes to look for both qualitative and quantitative changes in transcripts during silk gland differentiation (Dignam, personal communication). Analysis of the synthetase genes may reveal common signals linking their expression to expression of the fibroin gene and possibly other genes required for efficient fibroin production.

There appear to be two ways in which the enrichment of particular tRNA species is achieved in the silk gland. One is the increased accumulation of ubiquitous isoaccepting species of glycine, alanine, and serine tRNA. Analysis of the *B. mori* glycine tRNA gene family shows that although the members encode the same tRNA, they vary in ability to act as transcription templates in vitro (Taneja, Gopalkrishnan, and Gopinathan, 1992). Some are very efficient; others are almost completely inactive. This observation suggests a possible regulatory strategy. Alterations in the silk gland transcription machinery could allow recruitment of "weak" templates into the active pool, thereby increasing the output of glycine tRNA. Such a mechanism assumes that the "weak" templates are capable of high activity under special conditions. It would be interesting to know whether the inefficient glycine tRNA genes can ever support high level transcription in vitro.

The other way that particular tRNAs are enriched in the silk gland is through synthesis of novel, tissue-specific isoacceptors. The phenomenon of silk gland–specific isoacceptors has been reported for both serine (Hentzen, Chevallier, and Garel, 1981) and alanine tRNAs (Meza et al., 1977; Sprague, Hagenbüchle, and Zuniga, 1977), and has been analyzed in detail for alanine tRNA. One kind of alanine tRNA is expressed in all *Bombyx* tissues and is designated either $tRNA^{Ala}2$ (based on its electrophoretic mobility relative to other tRNAs) or $tRNA^{Ala}C$ (based on its constitutive presence). A second type of alanine tRNA (called $tRNA^{Ala}1$ or $tRNA^{Ala}SG$) is found only in the silk gland indeed, only in the fibroin-producing posterior part of the gland. Figure 10.1 shows that these two tRNAs are extremely similar in primary structure. They have identical anticodons and differ only by one nucleotide in the anticodon stem. Fortunately, for analytical purposes, this substitution confers a distinctive conformation under partially denaturing conditions. Hence the two kinds of alanine tRNA are readily resolved by polyacrylamide gel electrophoresis in the presence of urea and can easily be recognized and quantitated.

The accumulation of the two *Bombyx* alanine tRNAs has been examined in different tissues and over part of the course of silk gland development (Meza et al., 1977; Sprague et al., 1977). The $tRNA^{Ala}SG$ type is undetectable in non–silk gland cells, and if present at all, must be less than 1 percent as abundant as $tRNA^{Ala}C$. Both kinds of alanine tRNA are present in silk gland cells well before the onset of massive fibroin synthesis

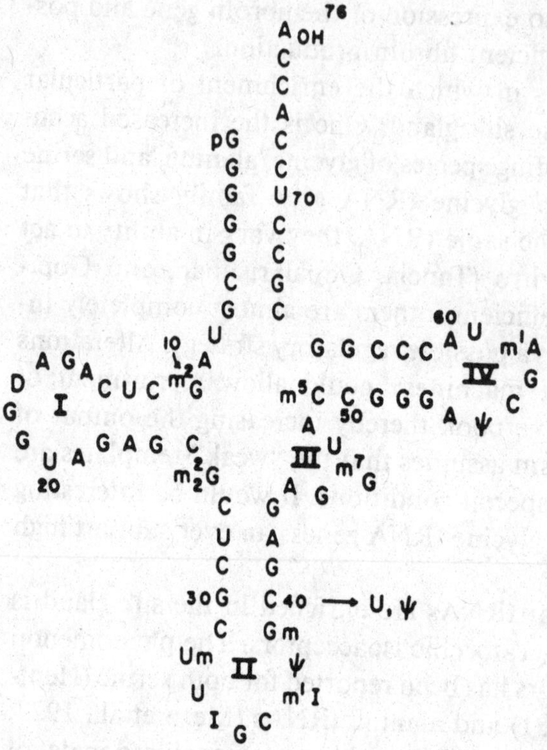

Figure 10.1. Structures of *B. mori* tRNA^{Ala}C and tRNA^{Ala}SG. The two tRNAs differ by a single nucleotide at position 40. tRNA^{Ala}C contains C at this position, whereas tRNA^{Ala}SG contains U (or its derivative, Ψ), as indicated by the arrow. (From Sprague et al., 1977, with permission; © Cell Press.)

at the end of the last larval instar (V). Both increase just before this period of intense biosynthetic activity, but the increase in tRNA^{Ala}SG is particularly striking. At the end of instar IV, the ratio of tRNA^{Ala}SG/tRNA^{Ala}C is only 0.35, but it rises to 2.0 by the middle of instar V.

At present, the biological significance of the silk gland–specific type of tRNA^{Ala} is not completely understood. Clearly, the presence of this isoacceptor augments the total pool of alanine tRNA. Without it, the concentration of alanine tRNA in the silk gland would be threefold lower. Whether this special tRNA also makes a qualitative contribution to protein biosynthesis in the silk gland is not known. There are indications that fibroin synthesis in vitro is stimulated more efficiently by silk gland tRNA than by tRNAs from other sources (Lizardi, Mahdavi, Shields, and Candelas, 1979), but whether this stimulation is due to the presence of special isoacceptors, rather than to high levels of the heavily used ubiquitous tRNAs, is not known.

Recent evidence from another organism actually provides the strongest argument for the biological importance of silk gland–specific alanine tRNA. In particular, spiders have been shown to possess a silk gland–specific alanine tRNA just as silkworms do (Candelas, Arroyo, Carrasco, and Dompenciel, 1990). This finding is significant because, although spiders are silk producers, they differ markedly from silkworms in other respects. As an indication of the phylogenetic distance between them, spiders and Lepidoptera are placed in different subphyla (Chelicerata and Uniramia) that diverged at least 400 million years ago. Thus it seems more than fortuitous that spiders possess both constitutive and silk gland–specific alanine tRNAs that are certainly similar, and possibly identical (sequence information is not yet available), to those in silkworms (Candelas et al., 1990). Presumably, the novel $tRNA^{Ala}SG$ found in these two organisms plays such a key role in silk synthesis that it has either been retained by strong selection or has been invented independently by each of them. In any case, the spider system is extremely interesting. It is also experimentally appealing because the novel $tRNA^{Ala}$ is rapidly synthesized in response to the normal stimulus for silk production and this synthesis can be detected in excised glands (Candelas et al., 1990).

It is also clear that tissue-specific expression of tRNA genes is a general biological phenomenon that occurs in a variety of organisms. In *Xenopus* frogs, for instance, certain tyrosine tRNA genes are expressed only in oocytes, whereas others are expressed both in oocytes and in somatic cells (Stutz, Gouilloud, and Clarkson, 1989). In this case, the mature tRNAs encoded by the two classes of genes are structurally indistinguishable, and analysis of specific gene expression relies on sequences that distinguish their primary transcripts. In the nematode (*Caenorhabditis elegans*), although tissue-specific expression of particular tryptophan tRNA genes has not been demonstrated directly, it provides the most economical explanation for the tissue-specific informational suppression that is observed in that organism (Kondo, Hodgkin, and Waterston, 1988). Finally, a bovine lens–specific phenylalanine tRNA seems likely to be the product of tissue-specific gene expression (Lin et al., 1980).

Differential transcription appears to explain the distribution of $tRNA^{Ala}C$ and $tRNA^{Ala}SG$ in silkworm tissues. Genes corresponding to these two kinds of RNAs have been identified (Hagenbüchle, Larson, Hall, and Sprague, 1979; Young, Takahashi, and Sprague, 1986), and the possibility that differential gene amplification accounts for the phenomenon has been ruled out. The number and distribution of the two classes of $tRNA^{Ala}$ genes in the *B. mori* genome have been determined by molecular hybridization and by cloning frequency. A hybridization protocol that detects $tRNA^{Ala}SG$ genes specifically allowed accurate quantitation of these genes and showed that both silk gland and non–silk gland cells contain 20 copies per haploid genome (Underwood, Knickerbocker, Gardner, Condliffe, and Sprague, 1988). Cloning frequency indicates that $tRNA^{Ala}C$ genes are present in about the same copy number (Hagenbüchle et al., 1979). There is a difference in the distribution of the two kinds of genes, however. Whereas the $tRNA^{Ala}C$ genes are dispersed in the genome, both molecular (Underwood et al., 1988) and classic genetic analyses (Sullivan, Jelinek, and Sprague, unpublished observations) show that the $tRNA^{Ala}SG$ genes are tightly clustered. Whether the physical arrangement of the genes has functional significance is not yet known, and we have focused initially on determining the transcriptional properties of representative individual members of each gene class. Having cloned many genes in each class, we were able to determine the functional and structural properties of several individuals and thus be certain that the genes we are examining in detail are truly representative. In the case of the $tRNA^{Ala}SG$ genes, we took advantage of the clustered arrangement to clone the entire locus and show that all of the genes are nearly identical in structure (Underwood et al., 1988).

Mechanism of constitutive transcription

To discover how differential transcription of particular classes of polymerase III templates is achieved, it is necessary to identify components that direct transcription in *cis* and in *trans*. Because regulated transcription presumably depends on perturbation of some aspect of the basic mechanism, it makes sense to begin by determining the requirements for constitutive transcription.

Cis-acting elements

There has been considerable progress in identifying the *cis*-acting elements that direct *Bombyx* polymerase III transcription. This advance was possible because faithful and highly active polymerase III transcription extracts can be prepared from silkworm tissues. In vitro pol III transcription systems using nuclear extracts from either pupal oocytes or larval silk glands have been developed (Sprague, Larson, and Morton 1980; Fournier, Guerin, Corlet, and Clarkson, 1984; Morton and Sprague, 1984). For most purposes, silk glands are more convenient because larvae yield large amounts of material without investment of the additional effort required to rear pupae. Approximately 1 ml of highly active nuclear extract is obtained from the silk glands of a single larva, and it is a routine matter to rear thousands of larvae. Our laboratory typically has several liters of frozen nuclear extract on hand.

An important result that has emerged from the *Bombyx* analysis is that class III promoters are complex. Early experiments with the *Xenopus* system led to a model in which polymerase III transcription is driven by short sequences that are located downstream from the transcription start site and that are confined to the region that encodes mature RNA. In the case of 5S RNA genes, this internal control region (ICR) is proposed to be a block of contiguous base pairs extending from +55 to +80 (Bogenhagen, Sakonju, and Brown, 1980; Sakonju, Bogenhagen, and Brown, 1980). For tRNA genes, it is considered to consist of two separated sequence blocks known as the A and B boxes (+8 to +19 and +52 to +62, respectively) (Galli, Hofstetter, and Birnstiel, 1981; reviewed by Geiduschek and Tocchini-Valentini, 1988). In contrast to this picture, analysis of several *Bombyx* polymerase III templates showed that sequences upstream of the transcription initiation site also contribute essential promoter function. These results are summarized diagrammatically in Figure 10.2. The *Bombyx* genes in which 5′ flanking sequences are now known to be required for transcription include alanine (Sprague et al., 1980; Larson, Bradford-Wilcox, Young, and Sprague, 1983) and glycine (Fournier et al., 1984) tRNA genes, 5S RNA genes (Morton and Sprague, 1984), and genes encoding a small RNA, called Bm1 or BmX, whose function is not yet known (Adams et al., 1986; Wilson et al., 1988). The 5′ flanking region of the glycine tRNA gene is particularly interesting in that it contains both positive and negative elements (Fournier et al., 1984; Taneja et al., 1992). The positive element is closer to the transcription start site than is the negative element, and it is in approximately the same position as the positive elements in the other *B. mori* class III genes.

The critical nucleotides have not yet been identified by high resolution mutational analysis in any of the upstream promoter elements. Nonetheless, the presence of three short A/T-rich sequences in the positive elements of constitutively expressed genes (see Figure 10.6) is intriguing. Detailed mutagenesis is underway to test the functional significance of these conserved sequences in the $tRNA^{Ala}C$ 5′ flanking region (Palida and Sprague, unpublished observations). We are also interested in the A/T

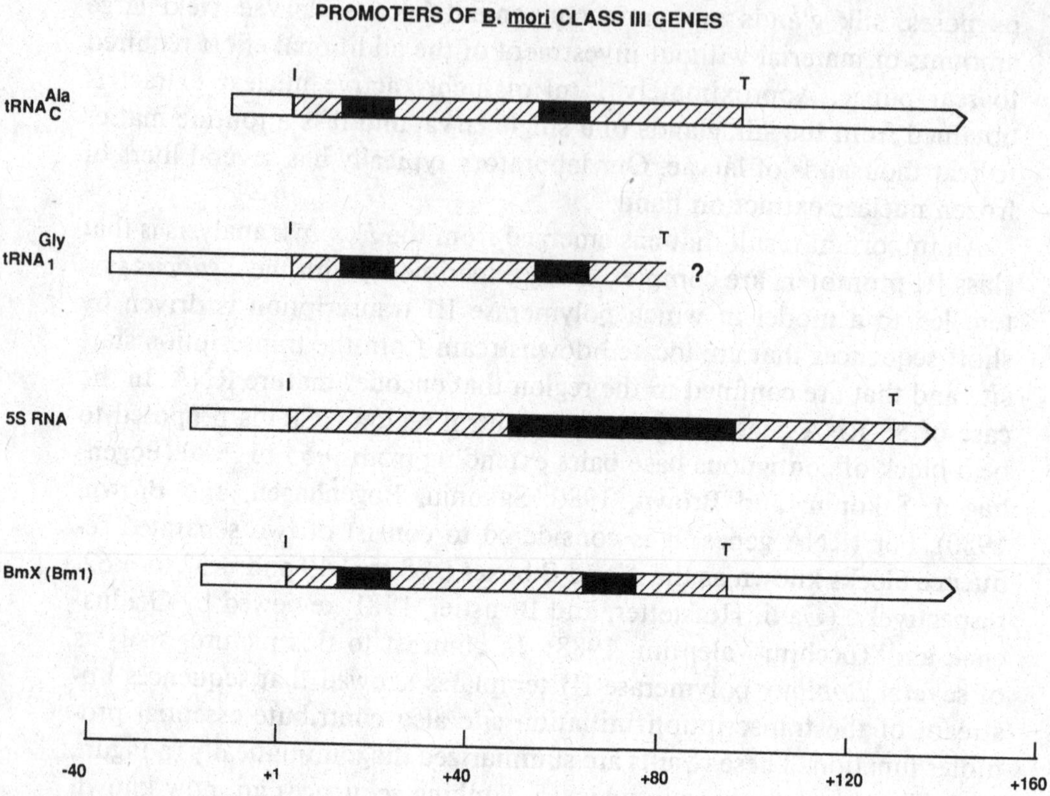

Figure 10.2. Promoter structure in *B. mori* class III genes. The minimum size of each promoter is represented by the extent of an open horizontal arrow relative to the scale in base pairs at the bottom of the diagram. Within each arrow, the coding sequence of the gene is hatched, and sequences resembling the canonical promoter elements described in other systems are black. The sites of transcription initiation (I) and termination (T) are marked. The boundaries of the *B. mori* promoters were established by analysis of a series of deletions entering the genes from each side. In the diagram, the ends of each arrow correspond to the endpoints of the first deletions in these series that affect transcriptional activity. In most cases, the deletions reduce activity. In the case of the $tRNA^{Gly}1$ gene, however, the promoter contains both positive and negative elements, and the 5′ deletion represented here increases transcription. The 3′ boundary of the $tRNA^{Gly}1$ promoter is not yet known.

boxes because of the possibility that they account for the different transcriptional properties of constitutive and silk gland–specific $tRNA^{Ala}$ genes. This idea is discussed more fully in the next section.

Because the results with the *Bombyx* transcription system contrast with those in the more widely studied *Xenopus* system, the *Bombyx* results were at first considered atypical. Work from many laboratories on a broad range of organisms has now established that polymerase III promoters often include upstream sequences. The magnitude of the effect of altering the upstream sequence elements varies among templates and organisms. *Bombyx* and *Drosophila* show extreme upstream sequence dependence for both 5S and tRNA gene transcription (DeFranco, Schmidt, and Soll, 1980; Schaack et al., 1984; Sharp and Garcia, 1988), whereas *Xenopus* and yeast typically exhibit only modest dependence, or none at all (reviewed by Geiduschek and Tocchini-Valentini, 1988). The requirement for upstream sequences is not peculiar to insects, however. Transcription from human and mouse tRNA genes, as well as from nematode and *Neurospora* 5S RNA genes, is severely impaired if these sequences are altered (reviewed by Sprague, Ottonello, Rivier, and Young, 1987). Moreover, the idea that upstream sequences contribute even to yeast class III promoters is supported by the recent demonstration (Kassavetis, Riggs, Negri, Nguyen, and Geiduschek, 1989) that a component of the yeast transcription apparatus, TFIIIB, binds upstream. Finally, recent work with less classic polymerase III templates – particularly those encoding EBER, U6, and 7SK RNAs – shows that in some cases upstream sequences are both necessary and sufficient for transcription (Murphy et al., 1989).

Silkworms have also provided evidence for the promoter function of sequences downstream of the coding region (see Figure 10.2). The *Bombyx* $tRNA^{Ala}C$ and BmX promoters both extend approximately 50 bp beyond the site of transcription termination (Wilson, Larson, Young, and Sprague, 1985; Wilson et al., 1988). The $tRNA^{Arg}$ promoter in *Drosophila* may be similar, as sequences 3′ of the termination site also contribute to its strength (Schaack, Sharp, Dingermann, and Soll, 1983). Recently a striking example of an essential 3′ flanking class III promoter element was reported for the yeast U6 gene (Brow and Guthrie, 1990). So far, these are the only examples of polymerase III promoter elements that lie outside the transcribed region on the downstream side. It is possible that similar elements exist in other genes but have simply escaped detection, as the contribution of downstream elements can be obscured by inappropriate assay conditions (Wilson et al., 1985; Sprague et al., 1987). In any event, the current view is that polymerase III promoters are larger and more complex than was first imagined. They may very well resemble polymerase II promoters, a view that is supported by recent demonstrations that the class II factor, TFIID, is required for transcription of U6 genes by polymerase III (Margottin et al., 1991; Lobo et al., 1991).

Trans-acting components

The *Bombyx* silk gland has proved to be an extraordinarily tractable system for dissecting the polymerase III machinery. Indeed, *Bombyx* polymerase III isolated from the silk gland was one of the first eukaryotic polymerases to be purified (Sklar, Joehning, Gage, and Roeder, 1976). The discovery that specific initiation by eukaryotic nuclear polymerases generally requires accessory proteins prompted further fractionation of the *Bombyx* system in search of these transcription factors (Ottonello, Rivier, Doolittle, Young, and Sprague, 1987). These efforts were facilitated by the ease of obtaining large amounts of material, and the relative absence of potentially troublesome degradative activities in typical extracts of silk glands. Although such extracts may not be entirely free of nucleases and proteases, the levels of these activities do not interfere with in vitro transcription catalyzed either by crude extracts or by partially purified components. Transcription catalyzed by crude extracts is linear for at least 4 hr, and both wild-type and mutant transcripts are stable to incubation in the extracts. Moreover, fractionation has not exposed high levels of nuclease or protease activity that might have been masked in crude extracts (e.g., see Young, Rivier, and Sprague, 1991), and most of the fractions are remarkably tolerant of incubation at room temperature and of repeated freeze-thaw cycles (Sprague and Young, unpublished observations).

Silkworm polymerase III requires multiple transcription factors. Fractionation of the silk gland class III transcription machinery has resolved five distinct components so far, each of which must be present to reconstitute transcription of silkworm tRNA genes in vitro. These components are polymerase III plus four transcription factors designated TFIIIB, TFIIIC, TFIIID, and TFIIIR (Ottonello et al., 1987; Young, Dunstan, et al., 1991). Transcription of silkworm 5S RNA genes requires, in addition to these components, the well-known transcription factor TFIIIA (Sprague and Smith, in preparation). Thus the number of polymerase III transcription components resolved in the silkworm system exceeds that reported for several other systems. A widely held view based on these other systems is that two transcription factors, TFIIIB and TFIIIC, plus polymerase III are sufficient for tRNA transcription, and, with the addition of TFIIIA, are also sufficient for 5S RNA transcription (see, e.g., Lewin, 1990b). Whether this model represents a real difference from the transcription machinery in silkworms is not clear. We are currently testing the ability of heterologous components to substitute for their apparent silkworm counterparts in in vitro transcription reactions. It is already known (Ottonello et al., 1987) that TFIIIA from *Xenopus* frogs can substitute for *Bombyx* TFIIIA in 5S RNA transcription catalyzed by an otherwise *Bombyx* transcription apparatus. Thus the transcription components that interact with TFIIIA cannot be entirely unrelated in silkworms and frogs. Further evidence that the complexity of the silkworm transcription apparatus is not unusual comes from the observation that the yeast TFIIIC fraction contains more than one polypeptide with transcription factor activity (Parsons and Weil, 1990), and that two to four polypeptides in this fraction can be cross-linked to the template (Gabrielson, Marzouki, Ruet, Sentenac, and Fromageot, 1989; Bartholomew, Kassavetis, Braun, and Geiduschek, 1990). Likewise, the human TFIIIC fraction has been resolved into two components that are both required to reconstitute transcriptional activity (Yoshinaga, Boulanger, and Berk, 1987). Whether the components in yeast and human TFIIIC correspond to *Bombyx* TFIIIC, TFIIID, and TFIIIR or to other, as yet unresolved, *Bombyx* components is not known.

TFIIIC and TFIIID form a complex with tRNA genes. The multiplicity of components in the *Bombyx* class III transcription machinery provides a rationale for the complex promoter elements of *Bombyx* tRNA genes. Recently, we have used direct physical methods such as gel retardation and DNase I footprinting to detect the interaction of isolated *Bombyx* transcription factors with *Bombyx* tRNA^{Ala}C genes. Figure 10.3 gives examples of these data, and Figure 10.4 summarizes our conclusions. We find that most of the tRNA^{Ala}C promoter (-1 to at least $+136$) is required for binding TFIIIC and TFIIID and is in direct contact with polypeptides in these fractions (Young, Rivier, and Sprague, 1991). We think it likely that factors C and D bind to the gene as a complex. Earlier experiments in which binding was measured by a functional assay (template exclusion) showed that neither TFIIIC nor TFIIID bound stably to tRNA^{Ala}C genes alone, but that the combined factors did (Ottonello et al., 1987). Recently, gel retardation experiments have confirmed that the combination of TFIIIC and TFIIID forms a specific complex with tRNA genes, but that neither fraction alone binds specifically (Young, Rivier, and Sprague, 1991).

The most striking aspect of the footprint analysis is that it supports the functional importance of the far downstream promoter domain. The 3′ boundary of the DNase I footprint coincides with the 3′ boundary of

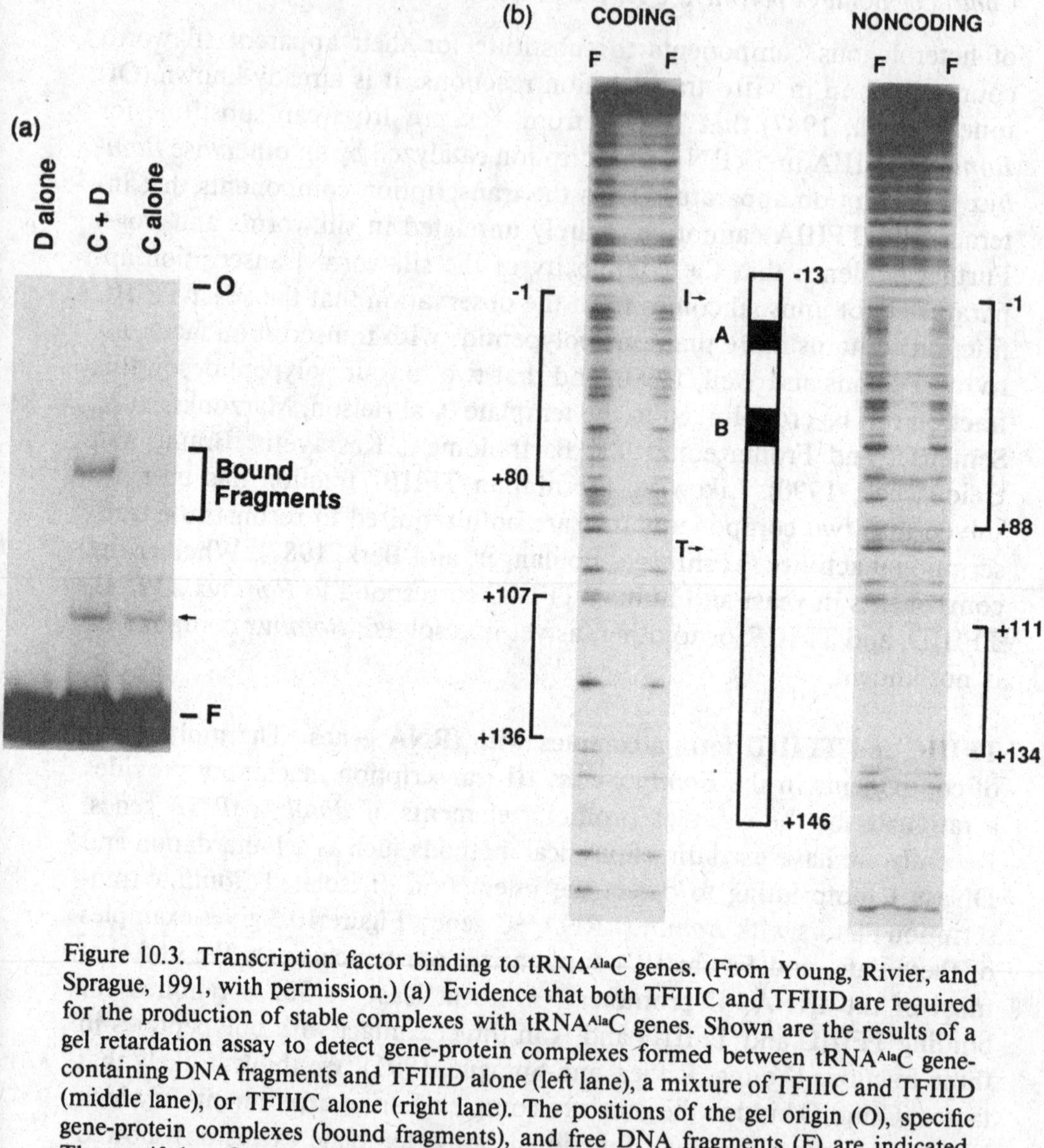

Figure 10.3. Transcription factor binding to tRNA[Ala]C genes. (From Young, Rivier, and Sprague, 1991, with permission.) (a) Evidence that both TFIIIC and TFIIID are required for the production of stable complexes with tRNA[Ala]C genes. Shown are the results of a gel retardation assay to detect gene-protein complexes formed between tRNA[Ala]C gene-containing DNA fragments and TFIIID alone (left lane), a mixture of TFIIIC and TFIIID (middle lane), or TFIIIC alone (right lane). The positions of the gel origin (O), specific gene-protein complexes (bound fragments), and free DNA fragments (F) are indicated. The specificity of the complexes designated "bound fragments" was demonstrated by their susceptibility to competition by gene-containing DNA but not by unrelated DNA (not shown). In contrast, the same competition experiments established that the band marked with an arrow corresponds to a nonspecific complex. (b) DNase I footprint of TFIIIC/D on a tRNA[Ala]C gene. Protection of the coding and noncoding strands by TFIIIC + TFIIID is shown. The extent of sequences required for full transcriptional activity is indicated by a vertical rectangle, within which filled areas represent the A and B boxes. I and T represent the sites of transcription initiation and termination, respectively. In each footprint, the DNase I cleavage products of free DNA (F) and bound DNA are shown in the outer and inner two lanes, respectively.

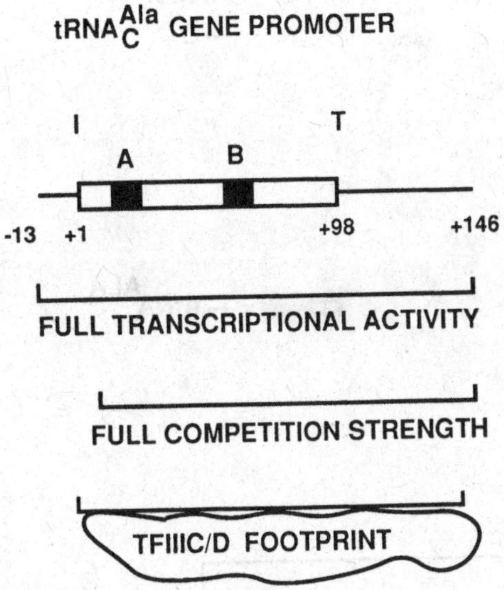

Figure 10.4. Structure and function of the *B. mori* tRNA^{Ala}C promoter. Line 1: Location of the minimum tRNA^{Ala}C promoter (horizontal line) with respect to transcribed sequences (open rectangle +1 to +98) and A and B boxes (filled rectangles). Line 2: Extent of sequences required for wild-type transcriptional activity. Line 3: Extent of sequences required for wild-type competition for TFIIIC/D in unfractionated extracts. Line 4: Extent of sequences protected from DNase I digestion by isolated TFIIIC plus TFIIID.

sequences required for full transcriptional activity. Moreover, mutational alteration of the 3′ flanking promoter element abolishes TFIIIC/D binding (Young, Rivier, and Sprague, 1991). Thus sequences downstream of the transcription termination site have a clear-cut and important role. They provide sequence-specific contacts for one or more essential transcription factors. Future work will be aimed at determining exactly which transcription factors interact with particular parts of the promoter and what interactions among factors influence binding to the promoter. We are particularly interested in the interactions of TFIIID with other components because this factor appears to play a key role in binding. As indicated earlier, TFIIID is required for TFIIIC binding to tRNA genes. It is also required for TFIIIB binding. Thus factor D is apparently not simply a subunit of TFIIIC but is an autonomous factor that can interact with at least two other components.

One of the silkworm transcription factors is composed of RNA. Recently, analysis of the silkworm polymerase III transcription system has revealed something particularly exciting: that the silkworm transcription factors include at least one that is composed of RNA rather than protein (Young, Dunstan, et al., 1991). The existence of this factor, TFIIIR, was demonstrated in two ways. First, the activity of the unfractionated silkworm transcription machinery was shown to be sensitive to nuclease treatment (Figure 10.5a). Second, fractionation of this machinery resolved a com-

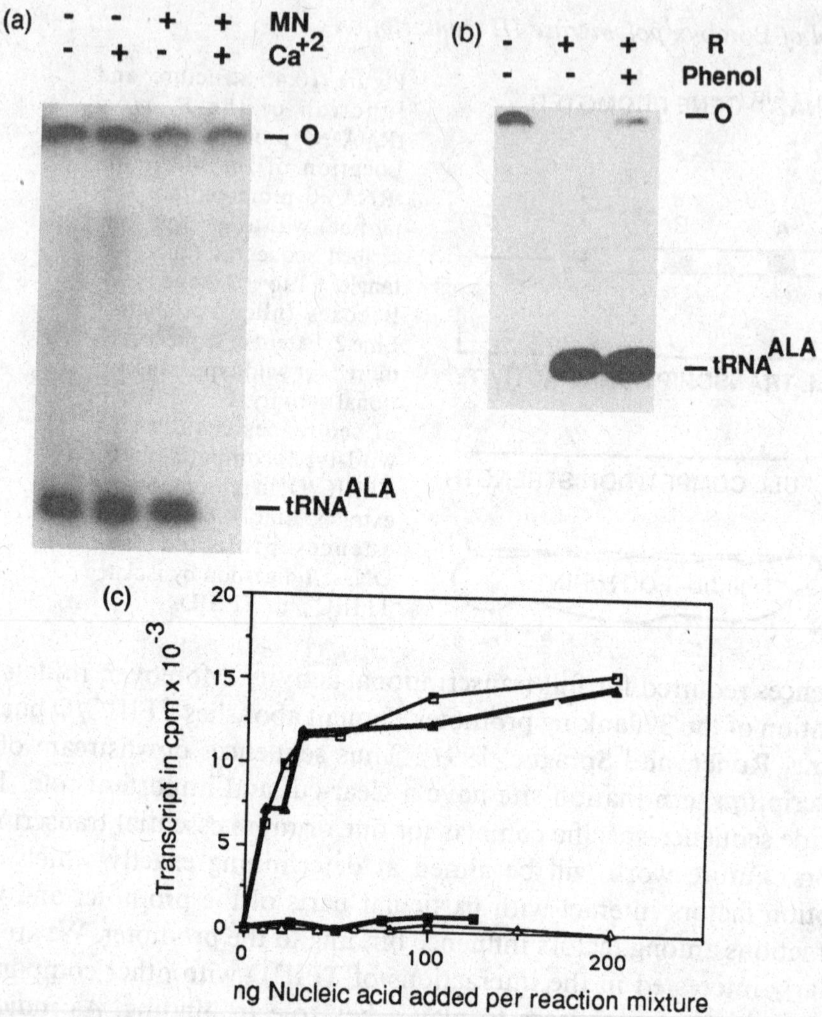

Figure 10.5. Evidence for an RNA transcription factor. (From Young, Dunstan, et al., 1991, with permission from *Science,* © AAAS.) (a) Transcriptional activity of unfractionated extracts is destroyed by treatment with micrococcal nuclease. Silk gland nuclear extract was treated with micrococcal nuclease (+MN) or with an equal amount of bovine serum albumin (−MN) in the presence (+) or absence (−) of $CaCl_2$ (Ca^{+2}). Digestion was stopped by the addition of EGTA to each reaction mixture. To make all subsequent transcription assays equivalent, $CaCl_2$ was added to the reaction mixtures that lacked it initially. The treated samples were tested for transcriptional activity on a $tRNA^{Ala}C$ gene under standard conditions. Transcripts were identified by their electrophoretic mobility on gels. The positions of the gel origin (O) and transcripts ($tRNA^{Ala}$) are shown. (b) TFIIIR is resistant to extraction with detergent and phenol. The transcriptional activity of the native TFIIIR fraction (middle lane) or of the same fraction after extraction with SDS and phenol (right lane) was measured by complementation of fractions of the *B. mori* transcription machinery that lack TFIIIR (left lane). The two forms of TFIIIR were added in amounts that supplied equal concentrations of nucleic acids in the two reaction mixtures. Positions of the gel origin (O) and the transcripts ($tRNA^{Ala}$) after gel electrophoresis of the

ponent that is essential for transcription activity but is sensitive to RNase and is resistant to protease and to extraction with phenol (Figure 10.5b). Although TFIIIR has not yet been purified to homogeneity, it is clear that transcription activity is due to a specific molecule (or molecules) in the TFIIIR fraction. Neither bulk RNA from yeast nor purified ribosomal RNA from silkworms can replace the TFIIIR fraction in a transcription assay (Figure 10.5c).

The existence of an RNA transcription factor raises some questions: (1) What is the role of such an RNA in the transcription reaction? (2) Do analogous factors function in the class III transcription machinery of other organisms? (3) Are RNA transcription factors required by any other classes of polymerase – polymerase I or polymerase II, for example? Our approach to answering the first question is to purify TFIIIR and then to use purified, or possibly synthetic, material to carry out mechanistic analyses. Because the purification effort is still in progress, detailed mechanistic studies are premature. We can draw some conclusions from experiments with partially purified TFIIIR, however. First, TFIIIR appears to be a general pol III transcription factor because it is required for all of the silkworm class III templates we have tested (tRNA^{Ala}C, tRNA^{Ala}SG, 5S RNA, and BmX). Second, preliminary experiments indicate that TFIIIR acts by influencing the capacity of transcription complexes to support multiple rounds of transcription. The possibility that it might act in this fashion was raised by the fact that, in a typical silkworm transcription assay, each template is transcribed hundreds of times. Thus a factor that was not actually required for synthesis of the first transcript but only for productive reinitiation by the polymerase would have a dramatic effect (>100-fold) on the transcription rate observed in a typical multiple round transcription reaction. In contrast, such a factor should have no effect if reinitiation were blocked and a single round of transcription were measured. Indeed, our preliminary results indicate that TFIIIR is not required when transcription is restricted to a single round by the addition of heparin (Chamberlin, Nierman, Wiggs, and Neff, 1979; Kassavetis et al., 1989; Young,

reaction products are shown. (c) Nucleic acid with TFIIIR activity is specific. The transcriptional activity of the native TFIIIR fraction (▲), nucleic acid extracted from the TFIIIR fraction (□), bulk yeast RNA obtained from Sigma (△), or bulk DNA (■) was determined in a standard TFIIIR complementation assay. The amount of radioactivity incorporated into tRNA^{Ala}C transcripts was plotted after subtraction of the negative control (–TFIIIR) value. The native TFIIIR fraction and the TFIIIR fraction that had been extracted with SDS and phenol were added in amounts that supplied equivalent concentrations of nucleic acid.

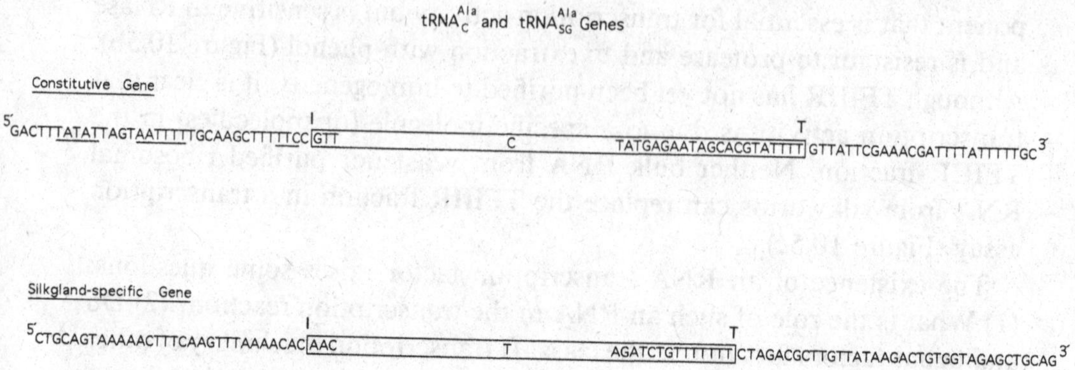

Figure 10.6. Sequences that distinguish constitutive and silk gland–specific tRNA^{Ala} genes. Shown are the flanking sequences and the single internal nucleotide at position 40 (relative to the 5′ end of mature tRNA^{Ala}) that distinguish the two kinds of tRNA^{Ala} genes. Rectangles correspond to the primary transcripts of these genes and the sites of transcription initiation (I) and termination (T) are indicated. Sequences shared by the two genes are represented by open areas within the rectangles. Sequences that are common to the upstream promoter elements of other *B. mori* constitutive class III genes are underlined. (Adapted and corrected from Young et al., 1986, Figure 3.)

unpublished observations). Thus TFIIIR does not influence the number of active transcription complexes but does influence the number of transcripts each complex can produce.

The issue of whether a TFIIIR analogue exists in other organisms or functions with other polymerases is currently under study in our lab. We have preliminary evidence that RNAs in other organisms (e.g., humans and *Neurospora*) are able to supply TFIIIR activity to the fractionated silkworm system (Witte, Young, and Sprague, unpublished observations). Whether these RNAs are structurally similar to silkworm TFIIIR, whether they function as transcription factors with their homologous polymerases, and whether similar RNAs contribute to the pol I or pol II transcription machinery are questions we have not yet answered. We are intrigued, however, by the possibility that RNA transcription factors may not be unique to silkworms.

Mechanism of regulated transcription

Most of the past work on *Bombyx* polymerase III transcription has focused on defining the signals that govern constitutive transcription. A major goal of current work is to discover how these signals are perturbed to achieve regulated expression. In vitro transcription of the two kinds of *Bombyx* alanine tRNA genes (constitutive and silk gland–specific) by homologous machinery should allow mechanistic analysis of this problem. The two kinds of $tRNA^{Ala}$ genes (diagrammed in Figure 10.6) display very different properties in vitro. These properties suggest that the constitutive gene can be expressed under a variety of conditions, whereas the silk gland–specific gene is probably expressed only in a special environment. This conclusion follows from our observation that a wild-type $tRNA^{Ala}C$ gene directs transcription at the same high rate over a range of conditions in vitro. Over this same range, a wild-type $tRNA^{Ala}SG$ gene varies in template efficiency by at least 50-fold (Young et al., 1986). The critical variable appears to be the concentration of nuclear extract used to catalyze transcription. In highly concentrated extracts, both genes direct transcription at equal, high rates, but in standard, less concentrated extracts, or with partially purified components, the transcription rate from a $tRNA^{Ala}SG$ gene can be as low as 2 percent of that from a $tRNA^{Ala}C$ gene. We think that these properties underlie the capacity of the genes to be differentially regulated in vivo. That is, the inefficiency of $tRNA^{Ala}SG$ genes revealed by standard transcription conditions probably accounts for their silence in non–silk gland cells. On the other hand, the capacity to overcome this inefficiency under special conditions could explain their expression in the silk gland.

The goal of our current work is to provide a mechanistic explanation for the distinctive transcription properties of $tRNA^{Ala}SG$ genes. Our analysis has already identified a particular promoter domain that is responsible for the characteristic $tRNA^{Ala}SG$ transcription properties (Young et al., 1986). Chimeric genes comprising all of the possible combinations of upstream, middle, and downstream promoter segments were constructed. As shown in Figure 10.7, the transcription pattern of the chimeric genes is remarkably simple. All of the properties characteristic of the $tRNA^{Ala}SG$ gene are conferred by the 5′ flanking promoter segment. The remaining promoter elements, although important for transcriptional activity per se, do not distinguish the two kinds of genes. The simplicity of this result suggests that a discrete mechanistic feature distinguishes $tRNA^{Ala}SG$ from $tRNA^{Ala}C$ transcription.

The next problem is to identify the part (or parts) of the transcription machinery that discriminates between the two kinds of $tRNA^{Ala}$ genes. Our working hypothesis is that the inefficiency of $tRNA^{Ala}SG$ genes under typical in vitro conditions is the consequence of defective interaction with one or more of the basic transcription components – that is, the com-

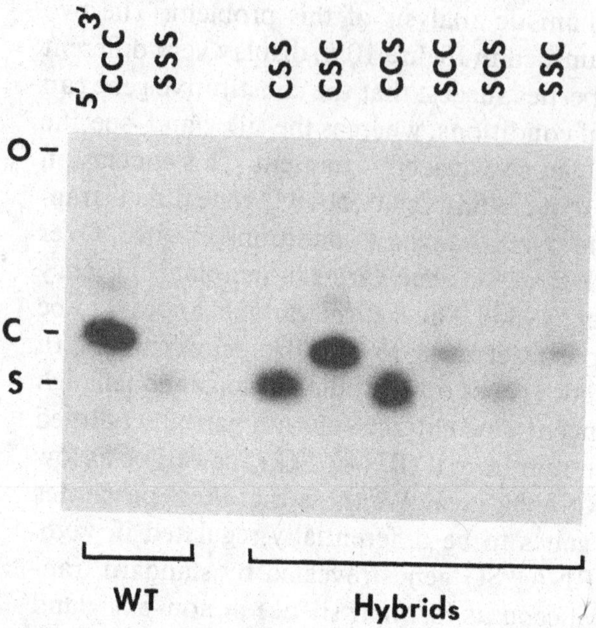

Figure 10.7. Transcriptional activity of wild-type constitutive and silk gland–specific genes and their chimeric derivatives. Symbols at the top of the figure represent the structure of the genes. For example, CCC and SSS correspond to the wild-type constitutive and silk gland–specific tRNA[Ala] genes, respectively, whereas CSS represents a chimeric gene whose 5′ flank is derived from the constitutive gene, but whose middle and 3′ flank are from the silk gland–specific gene. Transcription was catalyzed by a type of extract that discriminates strongly between the parental wild-type genes. The positions of the gel origin (O) and the tRNA[Ala]C (C) and tRNA[Ala]SG (S) transcripts after gel electrophoresis of the reaction products are shown. (From Young et al., 1986.)

ponents that catalyze tRNA[Ala]C transcription. Mixing experiments have ruled out the possibility that the inefficiency of tRNA[Ala]SG transcription is due to a diffusible repressor (Young et al., 1986). We imagine that the defect in tRNA[Ala]SG transcription can be overcome in either of two ways: (1) A novel transcription component (a gene-specific transcription factor, e.g.) could compensate for the defective interaction. In this category, we include modified versions of polymerase III or of the basic transcription factors, if the modification qualitatively alters function. (2) An alternative possibility is that a change in the concentration or specific activity of a basic transcription component, without the introduction of any special components, might be sufficient to overcome the defective interaction.

We have divided the problem of understanding tRNA[Ala]SG transcription into two parts. The first is to determine why tRNA[Ala]SG genes are such inefficient templates under standard in vitro transcription conditions. We are using single round transcription assays (Chamberlin et al., 1979; Kassavetis et al., 1989) to dissect the transcription cycle in order to identify the inefficient step in $tRNA^{Ala}SG$ transcription. We are using competition for particular transcription components, as well as direct binding assays, to identify components that interact more weakly with $tRNA^{Ala}SG$ genes than with $tRNA^{Ala}C$ genes.

The second part of the problem is to determine how the inefficiency of $tRNA^{Ala}SG$ genes is overcome under the special in vitro conditions that allow high rates of transcription from both $tRNA^{Ala}C$ and $tRNA^{Ala}SG$ genes. We are approaching this part of the problem by subjecting the silkworm transcription apparatus to further fractionation in order to resolve the component(s) that specifically turns up $tRNA^{Ala}SG$ transcription. Our next step will be to determine how the action of the critical component is restricted to the posterior silk gland. We hope ultimately to bring the mechanistic analysis of *Bombyx* tRNA genes full circle – to understand how these genes operate in their normal setting in silkworm cells. The puzzle of differential tRNA accumulation in the *Bombyx* silk gland is intriguing, and we think that the possibility of understanding this fascinating piece of lepidopteran biology at the molecular level is within range.

Note added in proof:

TFIIIR. TFIIIR has been identified as an isoleucine tRNA that prevents transcriptional inhibition caused by low level DNA cleavage (Dunstan, Young, and Sprague, 1994a, 1994b). The remarkable functional impact of damage to only ~3% of the template molecules appears to be explained by the capacity of damaged DNA to titrate or inactivate TFIIIC. Although there is no evidence that TFIIIR and the DNase it blocks are parts of the class III transcription machinery in vivo, the specificity of the TFIIIR–DNase interaction argues for some interesting biological role.

Upstream promoters and regulated transcription. Detailed mutagenesis showed that two short A/T-rich blocks are the functionally important elements of the $tRNA^{Ala}C$ upstream promoter (Palida, Hale, and Sprague, 1993). To understand differential $tRNA^{Ala}C$/$tRNA^{Ala}SG$ transcription, we are dissecting the protein–DNA interactions that depend on these blocks. Competition experiments have identified TFIIIB as the fraction that interacts differently with the two genes (Sullivan, Young, White, and Sprague, 1994). Our goal now is to examine individual components of TFIIIB, one of which is likely to be TATA binding protein, to find the key discriminators between $tRNA^{Ala}C$ and $tRNA^{Ala}SG$ genes.

11

Hormonal regulation of gene expression during lepidopteran development

LYNN M. RIDDIFORD

Introduction

After the formation of the early embryo, most of insect development is governed by the hormonal milieu, beginning with the formation of the first instar larva within the egg. Whether maternally derived hormonal signals are also important during early embryogenesis is still unclear. Much is known about the hormones involved and the regulation of their synthesis and secretion, especially in the Lepidoptera (see Gilbert, 1989, Gupta, 1990, and Ohnishi and Ishizaki, 1990, for recent reviews), and about the role of these hormones in directing development on the organismal and cellular levels (see Riddiford, 1985, for a review). Yet how these hormones act at the molecular level is only just beginning to unfold, principally through studies on *Drosophila melanogaster*.

This chapter focuses on the present state of knowledge of the hormonal regulation of gene expression in Lepidoptera. It begins with an overview of the endocrine basis of molting, metamorphosis, and reproduction in Lepidoptera and a description of the mode of action of the insect developmental hormones. This is followed by a short description of the systems in which particular genes have been cloned and studies of their hormonal regulation have been initiated. At the end, focus is on the hormone receptors and their actions, in both *Drosophila* and Lepidoptera, and questions for the future are posed.

Hormonal control of molting and metamorphosis

Because insects have a rigid exoskeleton, they must molt and shed this skeleton in order to increase in size and to change their form at meta-

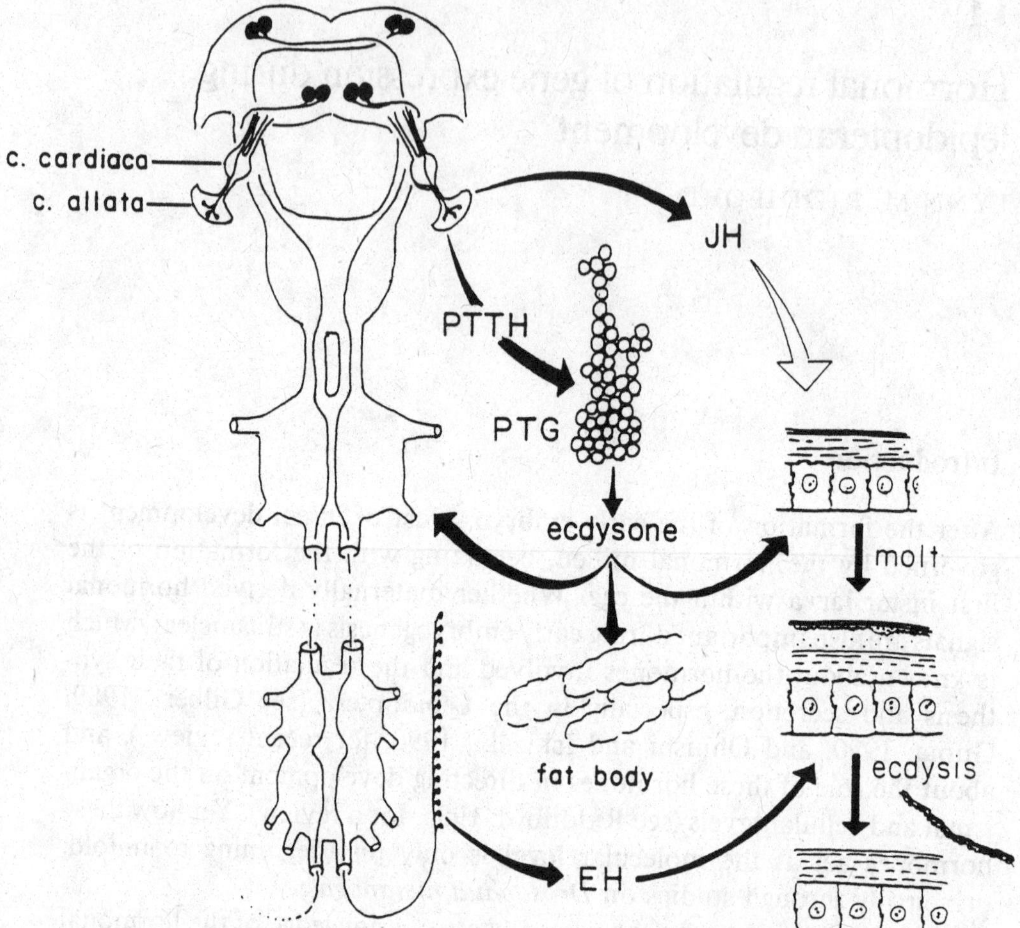

Figure 11.1. Endocrine system of a lepidopteran larva and the hormones controlling development. See text for details. c. allata, corpora allata; c. cardiaca, corpora cardiaca; EH, eclosion hormone; JH, juvenile hormone; PTG, prothoracic gland; PTTH, prothoracicotropic hormone.

morphosis. For Lepidoptera, molting begins in the egg when the new first instar (stage) larval cuticle is formed and the embryonic cuticle is shed. After hatching, these soft-bodied larvae first feed and grow, then molt at a particular size, depending on the species. After going through several such cycles and attaining a certain critical size (Nijhout, 1975), the larva then ceases feeding, begins wandering to find a pupation site, then may or may not spin a cocoon (depending on the species), and subsequently pupates. The pupa then transforms into the adult.

Molting and metamorphosis are controlled by the hormones ecdysone and juvenile hormone (JH) (Riddiford, 1985) (Figure 11.1). When the lepidopteran larva attains a certain size, two pairs of lateral neurosecretory cells in the brain release prothoracicotropic hormone (PTTH) from their neurohemal endings in the corpora allata (O'Brien et al., 1988; Mizoguchi et al., 1990). PTTH then acts on the prothoracic glands to cause the release of ecdysone (and sometimes 3-dehydroecdysone) (Sakurai and Gilbert, 1990), which is converted to 20-hydroxyecdysone (20E) by the fat body, gut, and Malpighian tubules. The 20E then acts on the tissues to initiate the molt, in which the most dramatic set of changes occurs in the epidermis that makes the overlying cuticle. This hormone causes, first, detachment or apolysis of the epidermis from the cuticle, then secretion of both the ecdysial membrane and the molting fluid gel that contains inactive enzymes that will later be activated to digest the old cuticle, and finally production of a new cuticle followed by digestion and reabsorption of the old endocuticle (Locke, 1984). Ecdysis or the shedding of the old cuticle is initiated by another neurosecretory hormone, eclosion hormone (EH). EH is made by a pair of ventromedial cells in the brain, then released within the ventral nervous system and from its neurohemal site in the proctodeal nerve when the ecdysteroid titer has fallen to a low level (Truman, 1990; Hewes and Truman, 1991).

During late embryonic life and through larval life the corpora allata produce JH, with modulations in production occurring during the molt cycle (Granger et al., 1982; Tobe and Stay, 1985; Grossniklaus-Bürgin and Lanzrein, 1990). JH is probably necessary during the intermolt growth phase to maintain larval tissues, such as the epidermis that makes the crochets or hooks (large setae) on the abdominal prolegs (Fain and Riddiford, 1977), and to maintain metabolic activities necessary for growth. At the onset of the molt, JH must be present to ensure that another larval stage will be formed. Later during the molt, JH may also regulate the pigmentation of the next larval stage (Riddiford, 1985).

In the final larval instar in Lepidoptera, the JH titer declines to undetectable levels (Baker, Tsai, Renter, and Schooley, 1987; Plantevin, Bosquet, Calvez, and Nardon, 1987; Rembold and Sehnal, 1987; Grossniklaus-Bürgin and Lanzrein, 1990). This decline allows PTTH release (Rountree and Bollenbacher, 1986; Sakurai, Okuda, and Ohtaki, 1989) and a small rise of ecdysteroid that, in the absence of JH, causes the change from feeding to wandering behavior (Dominick and Truman, 1985), as well as the change in commitment of the epidermis from larval to pupal

(Riddiford, 1976). In species that spin cocoons (Lounibos, 1976; Giebultowicz, Zdarek, and Chroscikowska, 1980), a small rise in ecdysteroid then triggers spinning behavior. Subsequently, PTTH is again released and causes a large rise in ecdysteroid, this time in the presence of JH, which causes the formation of the pupa. In many Lepidoptera, such as *Hyalophora cecropia* (Williams, 1961), *Manduca sexta* (Kiguchi and Riddiford, 1978), and *Mamestra brassicae* (Hiruma, 1980), this prepupal JH is necessary to prevent precocious adult differentiation of the imaginal disks during the pupal molt. Exceptions are *Bombyx mori* and *Galleria mellonella* in which allatectomized larvae, lacking JH, form apparently normal pupae (Bounhiol, 1938; Piepho, 1942).

After pupation, PTTH is again released and triggers the adult molt; this may happen immediately or after a prolonged resting phase called diapause. At this time no JH is present, so the rising ecdysteroid causes the switch to adult commitment followed by adult differentiation (Williams, 1961; reviewed in Riddiford, 1985). JH reappears either in the latter third of adult development in nonfeeding species that initiate egg maturation at this time, such as *H. cecropia* (Williams, 1961), or after adult eclosion in species that feed and mature eggs as adults, such as *M. sexta* (Sroka and Gilbert, 1971; Nijhout and Riddiford, 1974; Sasaki and Riddiford, 1984). In the feeding adult females, JH controls some aspect of vitellogenesis and/or some other aspect of reproductive maturation, depending on the species. By contrast, in the nonfeeding species, JH apparently has no function, as *H. cecropia* adults formed from pupae lacking the corpora allata reproduce normally.

Mode of action of insect developmental hormones

Ecdysteroid action

Steroid hormones readily enter cells. Target cells for a particular hormone have intracellular receptor proteins that specifically bind the hormone with high affinity. The receptor-hormone complex then binds to a specific hormone-response element in the regulatory region of a specific gene and either activates or inactivates it by modifying chromatin structure and directly affecting transcription (Rories and Spelsberg, 1989; Carson-Jurica, Schrader, and O'Malley, 1990). Steroid hormones may also modify mRNA stability (Shapiro, Blume, and Nielsen, 1987). The steroid hormone receptors belong to a protein family characterized by a DNA binding region containing two zinc fingers, each coordinated by four cysteine residues, and a C-terminal ligand binding region (Evans, 1988; Schwabe and Rhodes, 1991). They usually bind as homodimers to a palindromic DNA response element (Beato, 1989), but heterodimeric complexes involving other transcription factors may also be formed (Miner and Yamamoto, 1991; Schüle and Evans, 1991). Whether or not transcription is activated or inactivated by binding of the hormone-receptor complex is dependent on the presence of other transcription factors which may be tissue- or stage-specific or dependent on other environmental signals (Glass, Devary, and Rosenfeld, 1990; Karin, 1991).

Studies on ecdysteroid-induced "puffing" patterns of the giant salivary chromosomes of *Chironomus tentans* (Clever and Karlson, 1960; Clever, 1964) and *D. melanogaster* (Ashburner, Chihara, Meltzer, and Richards, 1974) were the first to show that steroids may act directly on genes and to demonstrate that ecdysteroids initiate a cascade of gene activation. These studies led to a model for ecdysteroid action (Ashburner et al., 1974) in which the ecdysteroid coupled to the ecdysteroid receptor (EcR) acts differentially to regulate several classes of target genes. "Early" genes were thought to be turned on directly by the ecdysteroid-receptor complex, whereas "late" genes were repressed by it. The early genes encoded regulatory molecules that induced secondary responses to hormone by repressing the early genes and activating late genes. This model has been a guiding light over the years for insect endocrinologists trying to understand the cascade of cellular events that ecdysteroid induces to initiate and coordinate a molt.

Several of the early genes are now known to be transcription factors (Burtis, Thummel, Jones, Karim, and Hogness, 1990; Segraves and Hogness, 1990; Thummel, Burtis, and Hogness, 1990; DiBello, Withers, Bayer, Fristrom, and Guild, 1991), and at least one of the early gene products is known to be an important regulator of a late gene complex (Guay and Guild, 1991). These factors have different forms, depending on alternative splicing or alternative promoters, and in at least one case, each form has a different time and dose responsiveness to ecdysteroids (Karim and Thummel, 1991). Thus each tissue, or even individual cells within a tissue, may have its own particular suite of ecdysteroid-induced factors that regulate its particular fate during a molt.

JH action

Juvenile hormone, which is a sesquiterpenoid, appears to have several different modes of action in the cell. In the follicle cells of the bloodsucking bug *Rhodnius prolixus* (Ilenchuk and Davey, 1987; Sevala and Davey, 1989) and in the male accessory gland of *Drosophila* (Yamamoto, Chadarevian, and Pelligrini, 1988), it acts at the membrane level, apparently through activation of protein kinase C. By contrast, in the stimulation of vitellogenin mRNA production in fat body, JH is thought to act as a steroid hormone in direct modulation of gene expression (Wyatt, 1990), and binding proteins with high affinity for JH have been found in fat body nuclei of *Leucophaea maderae* (Engelmann, Mala, and Tobe, 1987; Engelmann, 1990), and *Locusta migratoria* (Braun et al., 1992). In *Drosophila,* resistance of *Met* mutants to the JH analogue methoprene and to JH III has been correlated with reduced binding affinity of a cytosolic JH binder in the fat body (Shemshedini, Lanoue, and Wilson, 1990; Shemshedini and Wilson, 1990). The specific role of these intracellular JH binders in the regulation of vitellogenin synthesis is still unclear and their structure awaits elucidation.

In its morphogenetic action, JH appears to have no specific action of its own, but rather, to alter the molecular responses to ecdysteroids. Thus the insect always molts in response to a high ecdysteroid titer, but the presence of JH ensures that the molt will result in the same form and therefore that the same genes will be expressed after its completion as before its initiation. This role as a modulator of ecdysteroid effects is thought to occur in the nucleus because JH must be present in the cell at the time that ecdysteroids first enter in order to be effective in maintenance of the status quo. JH has no effect on the ecdysteroid-induced puffing response of the larval *Drosophila* salivary glands, but can inhibit that response in late prepupal glands (Richards, 1978). In *Drosophila* Kc cells, JH inhibits the morphogenetic changes and the appearance of acetylcholinesterase in response to 20E, but not its direct induction of EIP 28/29 (Cherbas, Koehler, and Cherbas, 1989). By contrast, in *Drosophila* S3 cells, pretreatment with methoprene inhibits the activation by 20E of a reporter gene containing only a 23 bp piece of the small heat shock protein 27 promoter containing the 15 bp ecdysone response element (Berger, Goudie, Klieger, Berger, and DeCato, 1992). This inhibition is dose-dependent and requires the pretreatment as well as continued exposure to the JH analogue during the ecdysteroid exposure. How this inhibition occurs at the molecular level is still unclear.

In spite of these studies with *Drosophila,* the question of how JH effects its morphogenetic actions remains a mystery. Its well-defined status quo effects in Lepidoptera (Williams, 1961; Riddiford, 1985) make these animals excellent systems in which to assess the action of JH. In order to understand at a molecular level how both ecdysteroid and JH affect the cascade of events during the ecdysteroid-induced molt and metamorphosis, one needs to study genes that are expressed at particular times, such as only during the intermolt or only during the molt, as well as stage-specific genes. Below I focus on genes in the Lepidoptera that are known to be expressed in the epidermis and fat body and on their hormonal regulation, and then return to how the ecdysteroid-induced cascade may control these genes and how JH may modulate that action.

Hormonal regulation of lepidopteran epidermal genes

Figure 11.2 shows the endocrine titers for the tobacco hornworm *Manduca sexta* during the last two larval instars and through metamorphosis, along with a schematic of the events occurring in the epidermis during this time. Because the titer profiles in *Bombyx* (Calvez, Hirn, and De Reggi, 1976; Nagata, Tsuchida, Shimizu, and Yoshitake, 1987; Plantevin et al., 1987), *Calpodes ethlius* (Dean, Bollenbacher, Locke, Smith, and Gilbert, 1980), *Galleria* (Bollenbacher, Zvenko, Kumaran, and Gilbert, 1978; Sehnal, Maroy, and Mala, 1981; Rembold and Sehnal, 1987), and *Trichoplusia ni* (Grossniklaus-Bürgin and Lanzrein, 1990) are quite similar, the *Manduca* titer serves as a guide for the following discussion.

Epidermis

The insect epidermis is a single-celled layer that makes the overlying cuticle, consisting of chitin and protein (see Binnington and Retnakaran, 1991, for detailed reviews). In holometabolous insects, larval, pupal, and adult cuticles are usually quite different, but they contain a mosaic of both conserved and new stage-specific proteins as was first clearly defined in the lepidopteran *H. cecropia* (Cox and Willis, 1985, 1987; see Willis, Wilkins, and Goldsmith, this volume). For example, some proteins are components of rigid cuticles, whereas others are found in flexible cuticles, irrespective of stage. In contrast, other proteins show a clear stage spec-

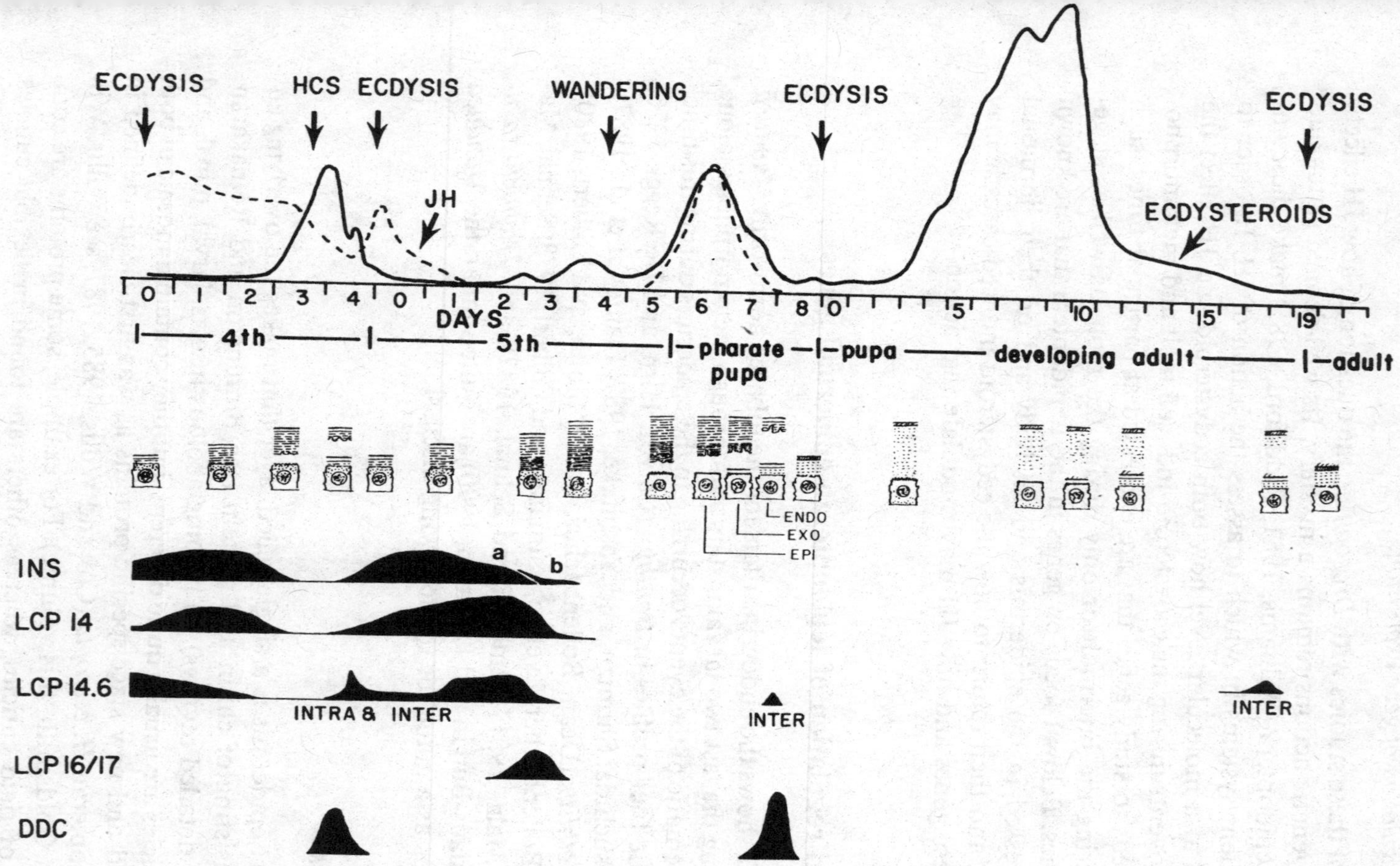

Figure 11.2. Schematic diagram of the ecdysteroid and JH titers in *Manduca sexta* during late larval life and metamorphosis, the developmental events occurring in the epidermis, and the levels of various epidermal RNAs. The ecdysteroid titers are based on Bollenbacher et al. (1981) and Warren and Gilbert (1986); the JH titers are based on Fain and Riddiford (1975) and Baker et al. (1987). See text for details. HCS, head capsule slippage; endo-, exo-, epi- all refer to types of cuticle being deposited; intra- and inter- refer to segmental regions.

ificity (Rebers and Riddiford, 1988; Horodyski and Riddiford, 1989; Apple and Fristrom, 1991).

In the lepidopteran larva, endocuticle is continuously laid down in the body wall to maintain its vital support function as the larva feeds and grows (Condoulis and Locke, 1966; Wolfgang and Riddiford, 1981, 1987). When the ecdysteroid rise initiates the larval molt, endocuticular deposition ceases by the time of head capsule slippage, which signals the end of feeding (see Figure 11.2). Then beginning about 3 hr later, the new epicuticle is deposited followed by the new endocuticle as the ecdysteroid titer falls (Wolfgang and Riddiford, 1986).

At the onset of metamorphosis in response to the "commitment" peak of ecdysteroid in the absence of JH, the epidermis ceases producing lamellate endocuticle and most of the larval endocuticular proteins (Wolfgang and Riddiford, 1981, 1986; Kiely and Riddiford, 1985). It continues to synthesize a few cuticular proteins that are deposited into the cuticle by intersusseption. Then, in response to the molting rise in ecdysteroid, it produces the pupal cuticle, beginning with pupal epicuticle, followed by the exo- and endocuticles as the titer falls (Sedlak and Gilbert, 1976; Kiely and Riddiford, 1985) (see Figure 11.2). Pupal endocuticle is deposited for a short time after pupal ecdysis (Sedlak and Gilbert, 1976). Then the rising titer of ecdysteroid causes the onset of adult development with, first, the division and differentiation of the scale cells followed by the deposition of adult cuticle (Greenstein, 1972a, 1972b; Sedlak and Gilbert, 1976; Nardi and Magee-Adams, 1986) (see Figure 11.2). In addition to the cuticle, the insect epidermis makes the molting fluid enzymes for digestion of the old cuticle and the precursors necessary for hardening of the new cuticle after ecdysis. It also produces various pigments that may serve as camouflage or as communicatory signals.

Larval cuticle genes of *Manduca*

LCP14. The LCP14 gene, which encodes a 14 kDa larval-specific cuticular protein, was isolated from the *Manduca* genomic library using the *Drosophila* larval cuticle protein 1 gene (Rebers and Riddiford, 1988). The LCP14 gene is first expressed in the 60-hr embryo during the formation of the first instar larval cuticle (Nagy and Riddiford, unpublished data) and then throughout larval life except during the molts (Rebers and Riddiford, 1988; see Figure 11.2). At wandering, when the epidermis becomes committed to later formation of the pupal cuticle, LCP14 mRNA disappears and is not seen again. Similar expression patterns are shown by two other cuticular protein genes (LCP 12.8 and 13.3) (Riddiford, Baeckmann, Hice, and Rebers, 1986). In situ hybridization shows that LCP14 is expressed in all epidermal cells except the hair- (seta) forming cells (Riddiford, unpublished data).

Studies with isolated penultimate stage (fourth) epidermis in vitro showed that 4×10^{-6}M 20E (the concentration found at the peak of the ecdysteroid titer during a larval molt [Curtis et al., 1984]) caused a rapid decline of LCP14 mRNA (Hiruma, Hardie, and Riddiford, 1991). This loss appeared to be due both to suppression of transcription and to destabilization of existing mRNA. Importantly, the disappearance was prevented by the presence of the protein synthesis inhibitor cycloheximide. Thus 20E apparently suppresses this gene via the synthesis of a protein(s), presumably a transcription inhibitory factor. Whether the hormone also acts directly on this gene is not known. When 20E was removed from the culture medium, LCP14 mRNA reappeared after a lag period coincident with the production of a new larval cuticle (Hiruma et al., 1991), just as is found in vivo (Rebers and Riddiford, 1988; see Figure 11.2). Presumably, during this lag period the ecdysteroid-induced inhibitory factors disappear so that intermolt syntheses can resume.

During the initial exposure to 20E, JH is necessary for the reappearance of LCP14 mRNA after the removal of 20E (Hiruma et al., 1991). At the time of explantation of the epidermis from the fourth stage intermolt larva, the JH titer is high (Fain and Riddiford, 1975), and immediate exposure to 20E causes the formation of a new larval cuticle. In contrast, culture in hormone-free medium for 48 hr followed by exposure to 20E caused formation of a pupal cuticle. Apparently, JH and its cellular effects decay with time during the preincubation so that the added ecdysteroid can act in the absence of JH to cause first pupal commitment, then pupal differentiation. Importantly, LCP14 mRNA remained high during the preincubation period in the absence of hormones, then disappeared in response to 20E. Under these conditions, however, this mRNA never reappeared when 20E was removed, supporting the idea that LCP14 is a larval-specific gene. Thus JH does not prevent the suppressive action of 20E on this larval-specific intermolt gene, but only prevents its permanent inactivation by the hormone.

LCP 14.6. The gene encoding a 14.6 kDa endocuticular protein is expressed similarly to LCP14 during larval life except for an additional sharp peak of expression just prior to larval ecdysis (Riddiford et al., 1986; see Figure 11.2). However, it is not larval-specific. After pupal commitment, expression is restricted to the intersegmental regions and occurs just before pupal and adult ecdysis. In situ hybridization shows that LCP 14.6 mRNA is present in all larval epidermal cells except for hair cells, but in the pharate pupa it is only in cells underlying the flexible cuticle of the intersegmental membrane (Riddiford, 1991; Riddiford and Poon, unpublished data). In contrast, in the pharate adult, the mRNA is highest in cells around the muscle attachment sites. Thus, through metamorphosis, this gene becomes progressively restricted both spatially and temporally.

Expression of LCP 14.6 mRNA in fifth instar epidermis in vitro was suppressed by low concentrations of 20E equivalent to that of the commitment peak (Riddiford, 1986; Riddiford et al., 1986). Unlike LCP14, it could also be suppressed by the JH analogue methoprene. The significance of this suppression by JH is unclear, although its mRNA levels are relatively low during the penultimate instar when JH is high.

LCP16/17. On the last day of feeding in the final fifth larval instar, in response to a small rise in ecdysteroid in the absence of JH, *Manduca* epidermis begins making a new set of cuticular proteins that are correlated with both a change in the spacing of the cuticular lamellae (Wolfgang and Riddiford, 1986) and an increase in the flexural stiffness of the cuticle (Wolfgang and Riddiford, 1987; see Figure 11.2). Synthesis of these proteins then ceases at the onset of wandering. At least three of these new proteins (16, 16.3, and 17 kDa) are encoded by the multigene family LCP16/17, which consists of five genes (Horodyski and Riddiford, 1989).

The LCP16/17 genes are not expressed during larval life prior to day 2 of the fifth larval instar, or in either the pupa or the adult (Horodyski and Riddiford, 1989; see Figure 11.2). In vitro studies showed that a low concentration of 20E (5×10^{-8}M) such as normally seen on day 2 (Wolfgang and Riddiford, 1986) in the absence of JH, is necessary to induce their expression (Horodyski and Riddiford, 1989), and that the higher levels on day 3 that cause pupal commitment of the epidermis (Riddiford, 1978) suppress this expression (Riddiford and Hiruma, 1990; Riddiford, 1991). Neither of these actions of 20E was seen when protein synthesis inhibitors were present (Riddiford, unpublished data), indicating that, as with LCP14, 20E induces synthesis of proteins that, in turn, regulate LCP16/17 mRNA levels. Importantly, in this case, the concentration of ecdysteroid determined whether transcription was activated or inhibited, suggesting that different factors are induced in response to changing ecdysteroid levels. Such dose-dependent effects of 20E are seen in the induction of the different E74 proteins in *Drosophila*, with E74B appearing at lower levels of 20E than E74A (Karim and Thummel, 1991).

Regulation of the LCP16/17 gene family by JH is typical of that expected for a "metamorphic" gene whose initial activation requires exposure to ecdysteroids in the absence of JH. Because such a gene has never been expressed, the ecdysteroids must stimulate a localized change in chromatin configuration and/or the synthesis of one or several new transcription factors that are necessary for its transcription. Whether other kinds of factors are also involved is not known.

Insecticyanin

Insecticyanin is a blue pigment synthesized and stored by *Manduca* epidermis and also secreted into the hemolymph (Cherbas, 1973; Kiely and Riddiford, 1985; Riddiford et al., 1990). This pigment, together with yellow pigments derived from leaf carotenoids (Kawoova et al., 1985), provides the green coloration that serves as camouflage for larvae feeding on leaves.

During the larval growth phases, the epidermis synthesizes two major forms of insecticyanin, INS-a (pI 5.5) and INS-b (pI 5.7) (Kiely and Riddiford, 1985; Riddiford et al., 1990), which are encoded by different genes (Li and Riddiford, 1992). Both forms are found in storage granules in the apical region of the epidermis, but only INS-b is found in the hemolymph (Riddiford et al., 1990). The fat body has low amounts of insecticyanin mRNA, mainly that of INS-b (Li, 1992). Studies of synthesis of these proteins by the fat body have not been carried out. Just as with LCP14, synthesis ceases during the molt due to ecdysteroid-induced loss of its mRNA and reappears just before ecdysis when the ecdysteroid levels have declined (see Figure 11.2). JH at the time of head capsule slippage during the larval molt delays this reappearance of the insecticyanin mRNA until some time after ecdysis (Li, 1992) and thus prevents the normal accumulation in the fifth instar epidermis (Hori and Riddiford, 1982; Trost and Goodman, 1986; Goodman, Tatham, Nesbit, Bultmann, and Sutton, 1987). How JH regulates this mRNA production is unknown.

On the final day of feeding in the last instar, synthesis of INS-a ceases (Kiely and Riddiford, 1985) due to cessation of its transcription and loss of its mRNA (Li, 1992), whereas INS-b transcription and synthesis continue until wandering (Figure 11.2). In vitro culture experiments with day

1 and day 2 fifth instar epidermis have shown that this differential cessation is due to the fall of JH followed by exposure to the two small peaks of ecdysteroid on days 2 and 3 (Figure 11.2) (Riddiford et al., 1990). The first initiates the disappearance of INS-a mRNA with little effect on INS-b mRNA, whereas the second, larger peak causes the disappearance of INS-b mRNA (Li, 1992). We do not yet know the molecular basis of this differential regulation of the two genes.

Dopa decarboxylase

Ecdysteroids not only initiate the molt but also help to coordinate cellular activities during the entire molting process. Many events that occur at the end of the molt thus depend on the fall of ecdysteroid below a certain threshold level as a signal for initiation, such as the onset of pre-ecdysial pigmentation (Schwartz and Truman, 1983; Curtis et al., 1984) and of muscle degeneration in some moths (Schwartz and Truman, 1983), and the release and action of eclosion hormone (Truman, Rountree, Reiss, and Schwartz, 1983; Morton and Truman, 1988b).

Dopa decarboxylase (DDC) is necessary for cuticular sclerotization (hardening and tanning) after ecdysis (Hopkins and Kramer, 1992), as it catalyzes the conversion of dopa to dopamine, which is a precursor for many of the cross-linking factors. Epidermal DDC activity increases sharply several hours before ecdysis, and this increase reflects new mRNA synthesis once the ecdysteroid level has declined (Clark, Doctor, Fristrom, and Hodgetts, 1986; Hiruma and Riddiford, 1985, 1990). Additionally, in *Manduca*, JH is necessary at the time of head capsule slippage to prevent the melanization of the new larval cuticle just before ecdysis (Truman et al., 1973; Curtis et al., 1984). Because melanization requires dopamine (Hori, Hiruma, and Riddiford, 1984; Hiruma and Riddiford, 1984; Hiruma, Riddiford, Hopkins, and Morgan, 1985), DDC activity is twofold higher in larvae deprived of JH at this critical time (Hiruma and Riddiford, 1985).

The *Manduca* DDC gene encodes a protein (Hiruma, Carter, and Riddiford, in preparation) with 72 percent similarity at the amino acid level to *Drosophila* DDC (Eveleth et al., 1986; Morgan, Johnson, and Hirsh, 1986). DDC mRNA and subsequent protein synthesis occur during the latter part of the molt as the ecdysteroid titer falls, peaking about 10 hr before larval ecdysis and then again just before pupal ecdysis (Hiruma and Riddiford, 1990; Pedersen, Hiruma, and Riddiford, unpublished data;

see Figure 11.2). In melanizing allatectomized larvae, maximal expression was twofold higher and could be returned to normal levels by application of JH I at the time of head capsule slippage (Hiruma and Riddiford, 1990). Thus JH at a critical time during the molt can regulate ecdysteroid-induced gene expression quantitatively as well as qualitatively.

The control of DDC mRNA expression by ecdysteroid has two components, as shown by in vitro studies with day 2 fourth instar *Manduca* epidermis (Hiruma and Riddiford, 1990, 1993; Hiruma, Carter and Riddiford, in preparation). First, the epidermis must be exposed to a molting concentration of 20E, then to hormone-free medium for induction of high levels of DDC mRNA. Under continuous exposure to 20E, no DDC mRNA was made. If a protein synthesis inhibitor such as cycloheximide or anisomycin was added to the 20E-containing medium, DDC mRNA increased to normal maximal levels within 4 hr. This mRNA was newly synthesized because its appearance was prevented by α-amanitin, an RNA polymerase II inhibitor. Similarly, DDC mRNA appeared within 4 hr after injection of cycloheximide at the time of head capsule slippage.

These results indicate that a protein(s) is induced by 20E that acts either directly or indirectly to suppress DDC mRNA production (Riddiford, Palli, and Hiruma, 1990; factor B in Figure 11.3). Moreover, this protein must have a short half-life. The ecdysone receptor–20E–complex may also act as a co-inhibitory factor. Once ecdysteroid is removed, the negative regulatory protein disappears and transcription occurs. Because 20E is required initially for later DDC mRNA expression, the simplest hypothesis is that the hormone also induces a protein(s) that is necessary for the later transcription of DDC mRNA (factor A in Figure 11.3). This protein presumably has a longer half-life in the absence of 20E and indeed may not appear until 20E is removed. The quantitative effect of JH is then hypothesized to be on the production and/or utilization of this factor A (see Figure 11.3).

Gel retardation experiments with crude nuclear extracts have identified only one small region in the first 1 kb of the 5′ region of the DDC gene that specifically binds a factor(s) with the developmental specificity expected for factor B (Hiruma, Carter, and Riddiford, in preparation). By contrast, several regions seem to bind nuclear proteins from stages where one would expect factor A. Further studies are necessary to determine the number of binding proteins present, the DNA sequences involved, and whether these sequences are necessary for the ecdysteroid regulation of DDC expression.

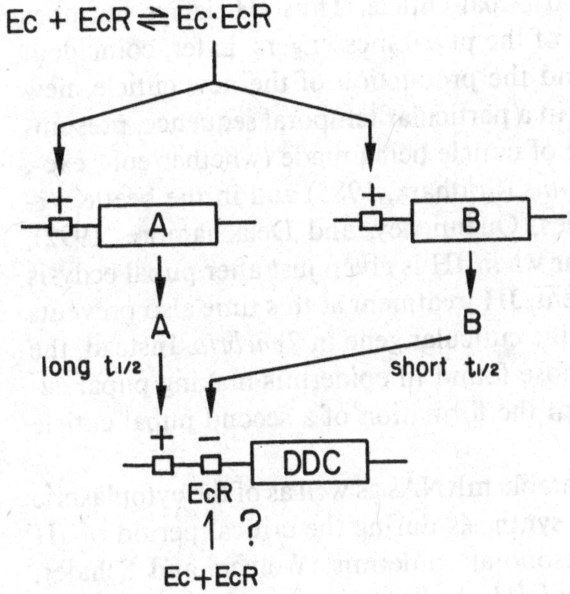

Figure 11.3. Model for the action of ecdysteroid in the regulation of the dopa decarboxylase (DDC) gene. See text for details. Ec, ecdysteroid; EcR, ecdysteroid receptor. (From Hiruma and Riddiford, 1993.)

Pupal and adult cuticle genes

In the Lepidoptera, genes encoding proteins that are found in pupal and/or adult cuticle have been isolated from *H. cecropia* (Willis, Binger, and Lampe, 1991), *Galleria* (Kollberg, Kelber, Obermaier, and Wolbert, 1991), and *Bombyx* (Nakato, Toriyama, Izumi, and Tomino, 1990). The two from *H. cecropia* are not stage-specific but, rather, cuticular type-specific (see also Binger and Willis, 1990). From its pattern of expression only in the first day after pupal ecdysis, the *Galleria* gene apparently encodes a pupal endocuticular protein. In contrast, the gene encoding a pupal cuticle protein of *Bombyx* is also expressed at low levels immediately after both the penultimate and the final larval ecdysis. In the absence of information on the hormonal regulation of these genes, a very brief overview of studies involving changes in translatable epidermal specific mRNAs during the pupal-adult transformation will be given.

Translatable mRNAs from *Antheraea polyphemus* wings (Sridhara, 1985) and from the mesonotum of the waxmoth *Galleria* (see Wolbert and Schafer, 1991, for a review) show changing patterns of expression. Immediately after pupal ecdysis, mRNAs for pupal endocuticular proteins are present, then disappear within a day or two. In *Galleria* this disappearance coincides with the loss of epidermal sensitivity to exogenous JH to cause formation of a second pupal cuticle. Thus JH delays but does not prevent the disappearance of the pupal messengers. Later, coincident with the rise of ecdysteroid and the production of the new cuticle, new mRNAs appear and disappear in a particular temporal sequence, presumably corresponding to the type of cuticle being made (whether epi-, exo-, or endocuticle). In *A. polyphemus* (Sridhara, 1985) and in the beetle, *Tenebrio molitor* (Bouhin, Charles, Quennedey, and Delachambre, 1992), these latter changes fail to occur when JH is given just after pupal ecdysis at the onset of adult development. JH treatment at this time also prevents the expression of an adult-specific cuticular gene in *Tenebrio*. Instead, the mRNAs are more similar to those found in epidermis making pupal cuticle, as would be expected from the formation of a second pupal cuticle under these conditions.

A detailed study of the translatable mRNAs as well as of the cytoplasmic and nuclear proteins and their synthesis during the critical period of JH sensitivity for the *Galleria* mesonotal epidermis (Wolbert and Schafer, 1991) has revealed few effects of JH. At least one cytoplasmic and one nuclear protein, which may be associated with commitment to adult differentiation (based on the timing of their appearance), fail to appear in the presence of JH, even after the characteristic delay caused by the hormone. The identity of these proteins is not yet known. Whether one should expect to see gross differences in mRNAs at this stage of commitment is unclear, as one would not expect the subtle changes in messages for isoforms of ecdysteroid receptors and/or transcription factors or for different chromosomal proteins to be detectable by the methods used.

Hormonal regulation of fat body genes

In the insect, the fat body serves to maintain homeostatic levels of nutrients in the hemolymph, to store excess nutrients for later use, and in the adult female, to make the vitellogenin necessary for egg maturation. Not much storage occurs during the larval feeding stages, as most of the incoming nutrients are utilized for growth. During the final larval instar of Lepidoptera, as in other holometabolous insects, the fat body synthesizes and secretes several large, hexameric proteins (Kanost et al., 1990), most, but not all, of which are sequestered in storage granules in the fat body during the prepupal period. These storage proteins are then utilized during adult development, and in nonfeeding adult Lepidoptera are the source of protein necessary for egg maturation.

Larval proteins

Arylphorin. Arylphorin is perhaps the best known of the storage proteins and is so named because it contains more than 17 percent aromatic residues (tyrosine, phenylalanine) (Telfer et al., 1983). Its sequences and properties show that it is a member of the hemocyanin family of proteins (Fujii, Sakurai, Izumi, and Tomino, 1989; Willot, Wang, and Wells, 1989). In contrast to many of the storage proteins, arylphorin is found in every larval instar in Lepidoptera (see Kanost et al., 1990, for a review). Some appears to be used during the molts for synthesis of the new larval (Webb and Riddiford, 1988a) and pupal (Grün and Peter, 1984) cuticles (see Peter and Scheller, 1991, for a review of the cuticular role of arylphorins); the remainder is used for adult development of various tissues.

Arylphorin is synthesized only during the feeding phase of the larval instars due to disappearance of its mRNA during the molts and at wandering (the onset of metamorphosis) (Riddiford and Hice, 1985; Ray, Memmel, and Kumaran, 1987; Memmel and Kumaran 1988; Memmel, Ray, and Kumaran, 1988; Webb and Riddiford, 1988a, 1988b; Fujii et al., 1989; Leclerc and Miller, 1990; Figure 11.4). Synthesis resumes at some time after each larval ecdysis and is dependent on incoming nutrients (Riddiford and Hice, 1985; Kumaran, Ray, Tertadian, and Memmel, 1987; Memmel et al., 1988; Webb and Riddiford, 1988a, 1988b). At the end of larval feeding, arylphorin mRNA either disappears, as in *Manduca* (Webb and Riddiford, 1988b), or remains at low levels until just after pupation, as in *Galleria* (Kumaran, Memmel, Wang, and Trewitt, 1993). In both cases, molting increases of ecdysteroid in the absence of JH appear to inactivate this gene permanently.

The genes for arylphorin have been isolated from *Bombyx* (SP2) (Fujii et al., 1989), *Manduca* (Willot et al., 1989), and *Galleria* (Memmel, Trewitt, Silhacek, and Kumaran, 1992). The first two share a common sequence in the 5′ upstream region, which is also found in the *Drosophila* larval serum protein 1 and in a *Sarcophaga* arylphorin. None of these upstream regions has yet been tested functionally to determine if it is critical for regulation of this gene by either nutrients or ecdysteroid.

In *Heliothis virescens* the testis sheath makes an arylphorin and another arylphorinlike storage protein beginning in the final larval instar (Miller, Leclerc, Seo, and Malone, 1990). These are secreted into the testicular fluid, and at least the latter has been traced to the fused mitochondria composing the *nebenkern* of the spermatid. Expression of this mRNA is maintained in the testis throughout pupal and adult development, whereas

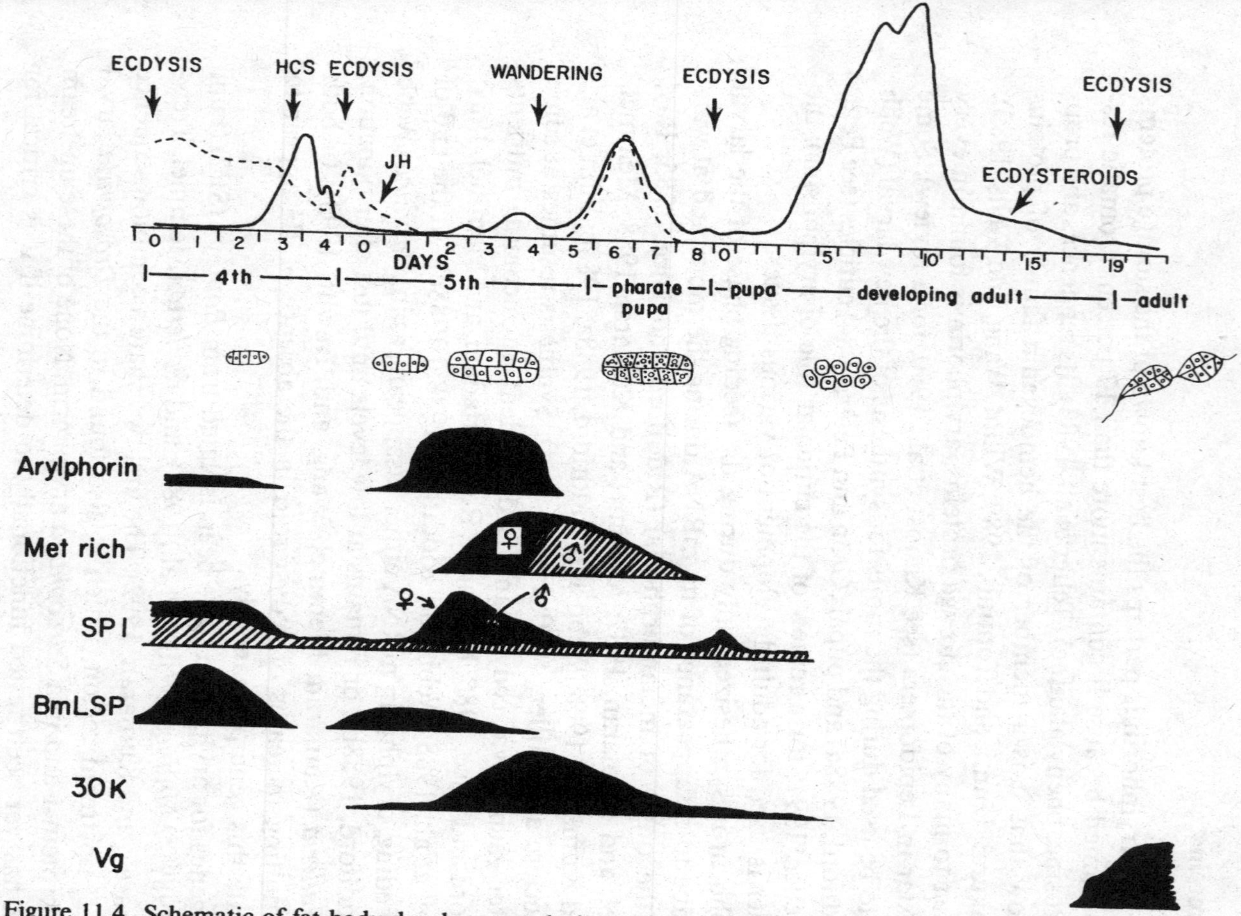

Figure 11.4. Schematic of fat body development during the final two larval instars and metamorphosis and the mRNA levels for various fat body genes. The fat body cartoons are based on *Calpodes* (Locke, 1980). Arylphorin and methionine-rich mRNA levels based on *Manduca* (Webb and Riddiford, 1988b), SP1 on *Bombyx* (Tojo et al., 1981; Izumi et al., 1988); BmLSP and 30K on *Bombyx* larval-specific protein (Fujiwara and Yamashita, 1991) and 30 kDa storage proteins (Sakurai et al., 1988; Bosquet et al., 1989), respectively; Vg (vitellogenin) protein levels based on Imboden and Law (1983). The titers are for *Manduca*, as in Figure 11.2.

that in the fat body ceases during the prepupal period (Leclerc and Miller, 1990). The hormonal control of this interesting system has not been studied. Presumably, the decline of JH in the final larval instar acts as one signal for the onset of testis development, but whether ecdysteroid is also involved is not known. In *Galleria,* arylphorin mRNA has also been detected in the sheath cells of the larval gonad (sex unspecified) (Kumaran et al., 1993).

***Bombyx* larval-specific protein.** A small 30 kDa larval-specific hemolymph protein is also synthesized by fat body in feeding *Bombyx* larvae (Fujiwara and Yamashita, 1990). The protein first appears in the hemolymph in the first instar, and synthesis continues until the final instar, with cessation during each molt like that of arylphorin (see Figure 11.4) (Fujiwara and Yamashita, 1991). The mRNA declines during the final larval instar and disappears by the time of spinning (Fujiwara and Yamashita, 1991), coincident with the appearance of the 30 kDa "storage proteins" (see later discussion), but its hormonal regulation has not been studied. These latter proteins differ in their N-terminal sequences, but they all could be members of the same family. This 30 kDa larval-specific protein is later incorporated into the eggs in an equimolar ratio with vitellogenin and egg-specific protein, which is synthesized by the follicle cells (Sato and Yamashita, 1991a).

JH-regulated proteins

Methionine-rich storage protein. The methionine-rich storage protein is another member of the hemocyanin superfamily of proteins (Sakurai, Fujii, Izumi, and Tomino, 1988; Corpuz, Choi, Muthukrishnan, and Kramer, 1991). It was formerly named "female-specific" storage protein because it was first found in female *Bombyx* (also known as SP1) (Tojo, Kiguchi, and Kimura, 1981) and *Manduca* (Ryan, Keim, Wells, and Law, 1985) larvae. However, both sexes of *Bombyx* synthesize it at low levels in the penultimate instar, whereas in the final instar only females do so, and at high levels (Mine, Izumi, Katsuki, and Tomino, 1983; Izumi, Sakurai, Fujii, Ikeda, and Tomino, 1988; Sakurai et al., 1988). The cessation in the male is under cell-autonomous control, as it occurs only in the male tissues of sexual mosaics (Mine et al., 1983). Yet this sex-specific suppression is dependent on the decline of JH for the final instar, as application of methoprene to penultimate instar larvae prevents suppression in the male (Kajiura and Yamashita, 1989).

In *Manduca* female larvae, the methionine-rich protein is first synthesized on day 2 of the final instar; in males, synthesis begins later at the onset of wandering (Ryan et al., 1985; Webb and Riddiford, 1988a). At least two different mRNAs encoding methionine-rich proteins (SP1A and SP1B) are found in *Manduca* (Corpuz et al., 1991). These have a similar sex differential in timing of appearance, but then each has its own pattern of expression during the prepupal period.

In *Bombyx* (Tojo et al., 1981), *Spodoptera litura* (Tojo, Morita, Agui, and Hiruma, 1985), and *Manduca* (Riddiford and Hice, 1985; Webb and Riddiford, 1988b; Corpuz et al., 1991), the decline of JH is necessary for the appearance of methionine-rich protein mRNAs and the onset of protein synthesis in the final instar. In *Manduca* both transcripts initially are suppressed after treatment with an exogenous JH analogue, then with time as the JH declines, SP1A but not SP1B reappears (Corpuz et al., 1991). Detailed study of this interesting phenomenon has not yet been carried out. Ecdysteroid is also required for the large increase in the methionine-rich mRNAs in isolated abdomens of *Manduca* larvae (Corpuz et al., 1991).

The methionine-rich protein is taken up by the fat body during the prepupal period in response to the molting rise in ecdysteroid (Tojo et al., 1981, 1985; Webb and Riddiford, 1988a; Kajiura and Yamashita, 1989). Uptake of at least some and often all of the storage proteins at this time is common in Diptera as well as in Lepidoptera (Levenbook, 1985). In *Sarcophaga peregrina,* this action of ecdysteroid in inducing storage protein uptake is rapid, independent of protein synthesis, and apparently involves the proteolytic activation of a preformed membrane receptor for the protein (see Natori, 1989, for a review).

JH-suppressible proteins. In *Bombyx* (Plantevin et al., 1987; Bosquet et al., 1989a, 1989b), *Trichoplusia* (Jones, Hiremath, Hellmann, and Rhoads, 1988; Jones, Brown, Manczak, Hiremath, and Kafatos, 1990), and *Galleria* (Kumaran et al., 1987; Memmel and Kumaran, 1988; Memmel et al., 1988), several new serum proteins appear in the midfeeding stage of the final larval instar. In *Bombyx* these are members of a 30 kDa protein family encoded by at least three different genes (Mori et al., 1991a, 1991b) that have similar, but individualized, patterns of expression (Sakurai et al., 1988; Bosquet, Fourche, and Guillet, 1989; Bosquet, Guillet, Calvez, and Chavancy, 1989). The acidic JH-suppressible *Trichoplusia* protein is related to the hemocyanins but has neither high aromatic amino acid nor high methionine content (Jones et al., 1990), as is also true of the *Galleria* 82 kDa protein (LHP82) (Kumaran, personal communication).

These mRNAs normally appear when the JH titer falls, and JH application during or just after the molt to the final instar prevents (*Trichoplusia* and *Galleria*) or retards (*Bombyx*) their appearance. In vivo and in vitro experiments suggest that this suppressive effect of JH acts directly on the fat body (Ray et al., 1987; Memmel et al., 1988; Bosquet, Fourche, and Guillet, 1989). Once these RNAs appear, however, JH is no longer suppressive, and further transcription can only be suppressed by ecdysteroid.

These JH-suppressible fat body proteins, therefore, are very similar in their behavior to the cuticular protein LCP16/17 gene family in that they are one of the first signs of metamorphosis. Whether low ecdysteroid is necessary to activate these genes once the JH titer declines has not been rigorously tested.

Proteins involved in egg maturation

Vitellogenin in the Lepidoptera is synthesized in the adult or developing adult fat body, secreted into the hemolymph, and then taken up by the developing oocyte. In moths that do not feed as adults, such as *H. cecropia*, *Bombyx*, or *Plodia interpunctella*, this synthesis occurs during the last third of adult development and, as such, is dependent on the ecdysteroid that causes this development, as seen in *Bombyx* (Tsuchida, Nagata, and Suzuki, 1987). The initiation of egg maturation occurs during the decline of the ecdysteroid titer and is dependent on that decline, just as is the appearance of DDC in the epidermis at the end of the molt (see Figure 11.2 and earlier discussion). Thus, in *Plodia* a low level of 20E is necessary for maximal vitellogenin synthesis in fat body in vitro, but a high concentration suppresses vitellogenin mRNA (Shirk, Bean, and Brookes, 1990). Although in *H. cecropia* the corpora allata reactivate at about this time, JH apparently has no role in egg maturation, as removal of the corpora allata from the pupa has no effect on this process (Williams, 1961; Truman and Riddiford, 1971).

In *Bombyx*, the follicle cells also make a 72 kDa egg-specific protein at this time that is taken up into the developing oocyte in a 1:1:1 ratio with vitellogenin and the 30 kDa larval-specific protein (Sato and Yamashita, 1991a). The gene for this protein encodes a lipaselike molecule and is first expressed the day before pupal ecdysis (Sato and Yamashita, 1991b). The rise in the ecdysteroid titer for the adult molt causes an increase in mRNA that is maximal at the time of vitellogenesis and is then turned off in the late pharate adult. In *Manduca,* a similar follicle-specific protein is made in the short period between the completion of vitellogenin uptake and chorion formation (Tsuchida, Kawooya, Law, and Wells, 1992). Because this is the stage that is blocked in the absence of JH (see later discussion), this gene may be under hormonal control.

In adult Lepidoptera that feed, such as *Manduca* (Sroka and Gilbert, 1971; Nijhout and Riddiford, 1974, 1979), JH is necessary for vitellogenin production and/or uptake. In *Manduca,* JH does not initiate vitellogenin synthesis or uptake but, rather, increases and maintains high rates of synthesis and uptake so that vitellogenesis can be rapidly completed and hydration and choriogenesis can then begin (Nijhout and Riddiford, 1979; Imboden and Law, 1983). A 31 kDa microvitellogenin that is related to the 30 kDa "storage proteins" of *Bombyx* is synthesized by developing adult *Manduca* fat body and taken up by the maturing oocytes (Kawooya, Osir, and Law, 1987; Wang, Cole, and Law, 1989). The appearance of this RNA can be induced by 20E in late larval female abdomens (Wang et al., 1989). Its expression is apparently unaffected by JH in the adult, but the lack of JH seems to prevent its uptake into the eggs (Kawooya and Law, 1983). These initial studies need to be confirmed and extended to determine how hormones regulate egg maturation in this species.

Molecular aspects of ecdysteroid action in Lepidoptera

Ecdysteroid receptors

Recently, the *Drosophila* EcR gene has been cloned and shown to be a member of the steroid receptor superfamily (Koelle et al., 1991). The EcR protein specifically binds radiolabeled ponasterone (a potent phytoecdysteroid) and binds to both the early and the late chromosomal puff sites as postulated in the Ashburner model. This binding, however, is as a heterodimer with the ultraspiracle protein (USP), another member of the steroid hormone superfamily (Yao, Segraves, Oro, McKeown, and Evans, 1992; Yao et al., 1993). The DNA binding site is an incomplete palindromic consensus sequence of TGAC/ACY in the regulatory regions of several genes that are known to be directly activated by ecdysteroid (Cherbas, Lee, and Cherbas, 1991; Dobens, Rudolph, and Berger, 1991; Luo, Amin, and Voellmy, 1991; Ozyhar, Strangmann-Diekmann, Kiltz, and Pongs, 1991). For both the ecdysone-induced protein 28/29 gene (EIP 28/29) (Cherbas et al., 1991) and the small heat shock protein 27 gene (hsp 27) (Dobens et al., 1991), the unoccupied ecdysteroid receptor suppresses gene transcription, which is then relieved in the presence of the hormone.

Further studies have revealed that there are three different isoforms of the *Drosophila* ecdysteroid receptor that have the same DNA and ligand binding regions but differ in the N-terminal transactivational domain (Talbot, Swyryd, and Hogness, 1993). The expression of the various isoforms appears to be stage- and tissue-specific and, at least in the nervous system, cell-specific and correlated with metamorphic fate (Truman, Talbot, Fahrbach, and Hogness, 1994; see Truman, this volume).

The *Manduca* EcR gene has been isolated using the *Drosophila* EcR cDNA as a probe (Palli and Riddiford, unpublished data). The DNA binding region is 97 percent identical to the *Drosophila* EcR at the amino acid level. In contrast to *Drosophila,* but similar to the vertebrate steroid hormone receptors (Beato, 1989; Carson-Jurica et al., 1990), the *Manduca* gene has an intron between the two zinc fingers in the DNA binding region. It is not yet known whether there are multiple isoforms, as in *Drosophila.*

Preliminary studies show that in *Manduca* epidermis, EcR mRNA is present at low levels during the intermolt feeding periods in the fourth and fifth instars and only increases slightly during the larval molt (Figure 11.5) (Palli and Riddiford, unpublished data). Just after the pupal commitment peak of ecdysteroid, receptor mRNA levels increase 2.5-fold, with a subsequent fall to undetectable levels by the time of pupal ecdysis. A nuclear antigen that cross-reacts with the polyclonal antiserum made against the region between the DNA and ligand binding domains of the *Drosophila* ecdysteroid receptor shows a similar pattern of expression in the epidermis. Nuclei of tracheae, muscle, and subepidermal fat body also have this antigen, although detailed developmental studies have not been performed. By contrast, studies of the *Manduca* ventral nervous system using this same antibody fail to detect this nuclear antigen during larval life except for traces just before ecdysis; high levels then appear at the time of pupal commitment (Riddiford and Truman, 1993; Truman, Talbot, Fahrbach, and Hogness, 1994; Truman, this volume).

The different responses of epidermis and nervous system to ecdysteroid during a larval molt reflect this difference in EcR levels because larval neurons appear unchanged during the molt, whereas the epidermis makes

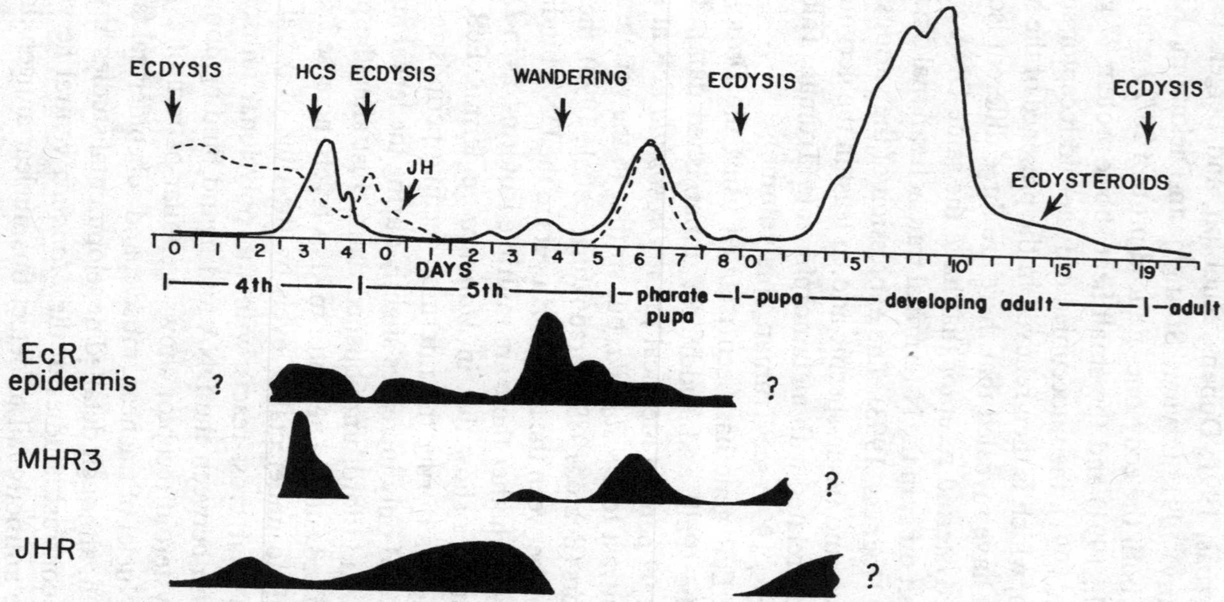

Figure 11.5. Schematic of mRNAs for ecdysteroid receptors (EcR) and MHR3 (a transcription factor) and of the nuclear 29 kDa JH binding protein in *Manduca* epidermis. Hormone titers are the same as in Figure 11.2.

a new cuticle. The finding of a large quantitative increase in EcR at the time of pupal commitment in both tissues suggests that increased levels of receptor are important in programming the metamorphic changes that must occur. As noted earlier, it is not clear whether there is also a qualitative change in the isoform(s) of EcR at this time.

Putative ecdysteroid response elements have been reported for various lepidopteran genes, including the methionine-rich SP1 (Izumi et al., 1988) and the egg-specific protein (Sato and Yamashita, 1991b) genes of *Bombyx* and the microvitellogenin gene of *Manduca* (Wang et al., 1989). These identifications were based solely on sequence similarities to the ecdysteroid-reponsive regions of the genes encoding the small heat shock proteins and not on any direct experimental analysis of function. With the *Drosophila* ecdysteroid response element now experimentally defined by direct binding studies (Cherbas et al., 1991; Luo et al., 1991), it is apparent that these lepidopteran sequences do not, in fact, match the dipteran consensus sequence. Binding studies on lepidopteran genes known to be directly regulated by ecdysteroid are therefore needed.

Ecdysteroid-induced transcription factors

Using a human retinoic acid receptor (hRARα) cDNA (Petkovitch, Brand, Krust, and Chambon, 1987; Palli, Riddiford, and Hiruma, 1991; Palli, Hiruma, and Riddiford, 1992) isolated a member of the steroid hormone superfamily from a *Manduca* genomic library that was expressed during the molts and was ecdysteroid inducible. This gene encodes a protein (Palli et al., 1992) that in the DNA binding region is 67 percent identical with hRARα and 97 percent identical with the *Drosophila* gene for hormone receptor 3 (DHR3) (Koelle, Segraves, and Hogness, 1992). Similarities in the ligand binding region to hRARα and DHR3 are 20 percent and 68 percent respectively, but the ligand is unknown; thus the gene is called MHR3 in view of its similarity to DHR3 (Palli et al., 1992).

MHR3 is expressed in the epidermis in the form of two mRNAs (3.8 and 4.5 kb) during the rise of the ecdysteroid titer for the embryonic, larval, pupal, and adult molts but not during the intermolt period and only at low levels during the commitment peak of ecdysteroids (Palli et al., 1992, and unpublished data) (see Figure 11.5). Its expression can be induced in intermolt epidermis from fourth or fifth stage larvae in vitro by 20E in a dose-responsive manner (Palli et al., 1992). This expression was dependent on the presence of ecdysteroid such that in the absence of hormone, the mRNA disappeared with a half-life of 2 hr. Moreover, induction by 20E was not prevented by the protein synthesis inhibitors cycloheximide or anisomycin, although less MHR3 mRNA accumulated in their presence. Thus expression of MHR3 is directly induced by ecdysteroid, although other ecdysteroid-induced proteins may be necessary for maximal levels. After 12 hr in the continuous presence of 20E, MHR3 mRNA decreased. This disappearance could be prevented by the presence of cycloheximide, indicating that some protein(s) induced by 20E inhibited further MHR3 mRNA synthesis. MHR3 expression in *Manduca* is thus typical of the 20E-induced "early puffs" in *Drosophila* (Ashburner et al., 1974).

In situ analysis of MHR3 mRNA expression in the abdominal epidermis during the molt to the fifth instar has shown that it appears first in the epidermis that forms the crochets (hooked setae) on the abdominal prolegs, then later in the anterior and posterior margins of the segment and still later in the intrasegmental region (Langelan, Hiruma, Palli, and Riddiford, in preparation). The MHR3 protein appears shortly after the mRNA and is localized to the nucleus. This pattern of appearance is the same as that of the spreading of the competence to molt in the absence of further ecdysteroid in abdomens isolated at various times during the rise of ecdysteroid (Truman et al., 1974). Yet the appearance of MHR3 is delayed several hours with respect to the competence to molt.

Homologues of the *Drosophila* E75 gene have also been found in *Manduca* (Segraves and Woldin, 1992) and in *Galleria* (Jindra, Sehnal, and Riddiford, 1994). In both cases they are expressed during the ecdysteroid rise for both the larval and the pupal molts. Preliminary studies with *Manduca* epidermis in vitro have shown that, as in *Drosophila,* E75 is directly induced by 20E (Segraves, Hiruma, and Riddiford, unpublished data).

The roles of MHR3 and E75 in the ecdysteroid-induced molt are not known. The similarity of kinetics of the inductive action of 20E on MHR3 and E75 and of its suppressive effect on LCP14 (Hiruma et al., 1991) makes either a possible candidate for the steroid-induced factor that suppresses LCP14 during the molt. If so, they may be important in the suppression of other genes active during the intermolt in both the epidermis and fat body, such as insecticyanin and arylphorin. Alternatively, and/or additionally, these transcription factors may be important for the activation of genes involved in production of the new cuticle. Important questions that need to be answered concern how these and other ecdysteroid-induced transcription factors are temporally and spatially regulated and how they interact with each other, with the ecdysteroid receptor-DNA complex, and with the structural genes. The presence of several different isoforms of the ecdysteroid receptor and of the ecdysteroid-induced transcription factors suggests the possibility that myriad combinatorial interactions can lead to cellular specificity of the response to the common hemolymph level of ecdysteroid.

Molecular aspects of JH action in lepidoptera

JH receptor

As discussed earlier, JH in its morphogenetic action directs the action of ecdysteroid. To do so, it must be present in the cell at the time that the cell first sees ecdysteroid. Although JH does not interfere with the suppression of ongoing gene expression by ecdysteroid, it prevents permanent suppression. The presence of JH also prevents the ecdysteroid activation of new stage-specific genes. The cellular and molecular bases of this type of action of JH have been best studied in Lepidoptera.

In *Manduca,* JH readily enters the larval epidermal cell, and 32 percent is retained in the nucleus by two binding components with K_d's of 6.6 nM and 88 nM for JH I (Osir and Riddiford, 1988). A 29 kDa nuclear protein that specifically binds photoaffinity analogues of JH I and II is found in both epidermis and fat body (Palli et al., 1990). The cytosol contains some of the 29 kD protein and also a 38 kDa protein that specifically binds these analogues.

The 29 kDa nuclear protein was purified, partially sequenced, and a full length cDNA obtained (Palli et al., 1994). The encoded protein contains two short stretches of 8–12 amino acids toward the C-terminus that have 60–70 percent identity with two regions of the bovine interphotoreceptor retinoid binding protein (Borst et al., 1989). This identity occurs in a fourfold protein repeat region that binds fatty acids; thus it may be a part of the JH binding site. Also, a region of 21 amino acids shows 43 percent identity with the human p68 nuclear protein, an RNA helicase that localizes to the nucleolus just after cell division (Iggo and Lane, 1989). Recombinant 29 kDa protein specifically binds JH I with high affinity (K_d = 10.7nM), and immunocytochemistry confirms its exclusive nuclear location (Palli et al., 1994). Consequently, this protein is likely to be the nuclear receptor for JH.

The nuclear receptor is present in both fourth and fifth instar epidermis, decreases during the larval molt, and disappears at wandering in response to 20E in the absence of JH and then reappears in the pupa (Palli, McClelland, Hiruma, Latli, and Riddiford, 1991; Palli et al., 1994), thus showing the same developmental pattern as the high affinity JH binding component in the nucleus (Riddiford, Osir, Fittinghoff, and Green, 1987). Maintenance of normal levels of the 29 kDa protein in larval epidermis required the presence of JH (Palli et al., 1991), and its reappearance in the pupa required the presence of JH during the pupal molt. Interestingly, a high affinity binding molecule is not found in the pupal wing (Riddiford et al., 1987). These findings correlate with the observation that exogenous JH can cause the formation of a second pupal cuticle on the abdomen, but not a completely pupal wing (Riddiford and Ajami, 1973).

Initial experiments indicated that a 29 kDa nuclear protein bound to both the LCP14 and the LCP16/17 genes with the same developmental pattern as did the 29 kDa JH binding protein (Palli et al., 1990). Yet the deduced sequence of the putative 29 kDa JH binding protein has no known DNA binding motifs (Palli et al., 1994). How this protein interacts with these and other larval structural genes is clearly of great interest.

Influence of JH on ecdysteroid receptors. Preliminary experiments on the *Manduca* ventral nervous system showed that JH may also influence the ecdysteroid receptor level (Renucci and Truman, personal communication). The normal increase in ecdysteroid receptors in motoneurons just after pupal commitment was prevented by isolation of the abdomen before the commitment rise in ecdysteroid. Infusion of 20E restored the increase in ecdysteroid receptors except when the abdomens were treated with the JH analogue methoprene. Studies in progress should reveal whether a similar phenomenon occurs in the epidermis.

Whether there is also a qualitative change in ecdysteroid receptor isoform at this time in *Manduca* is unknown. If so, then this change is likely to be regulated by JH. An analogous situation exists in the hormonal control of amphibian metamorphosis where thyroid hormone induces metamorphosis and prolactin inhibits this action (Dent, 1988; Tata, Kawahara, and Baker, 1991). Normally during larval life, the α isoform of the thyroid hormone receptor is present; then at metamorphic climax the β isoform becomes predominant (Yaoita and Brown, 1990; Kawahara, Baker, and Tata, 1991). Prolactin prevents the appearance of the β isoform (Baker and Tata, 1992). Presumably, the genes that are regulated by the new isoform cannot be activated or repressed so that metamorphosis cannot proceed.

Conclusions

A series of structural genes has been isolated from the Lepidoptera which are hormonally regulated, and progress is being made in the identification of ecdysteroid and JH receptors and several ecdysteroid-induced transcription factors. With various techniques we can determine the details of how these receptors and/or transcription factors interact with the structural genes. Do they do so directly, or is there still another layer of as yet untapped "late genes" that are the true regulatory agents? The significance of these interactions in vivo must then be determined by the use of a germline transformation system, which is yet to be perfected in the Lepidoptera.

Regardless of the details, what is most revolutionary for an understanding of how hormones control molting and metamorphosis are the recent revelations from studies on the isoforms of the ecdysteroid receptor and the ecdysteroid-induced transcription factors in *Drosophila.* Whether or not structural genes in a particular tissue are regulated by ecdysteroid is, first of all, dependent on whether or not that tissue has ecdysteroid receptors at that time. If receptors are present, then the ecdysteroid-induced transcription factors will be induced, the form depending on the concentration of ecdysteroid and possibly on the tissue. Then these factors and/or possibly ones that they in turn generate as part of the ecdysteroid-induced cascade act, in concert with tissue-specific factors and possibly with the ecdysteroid-ecdysteroid receptor complex itself, to regulate the structural genes.

Most of the ecdysteroid-induced transcription factors appear to be activated in every molt, whether larval, pupal, or adult. Nevertheless, they then regulate different stage-dependent genes in polymorphic tissue such as the lepidopteran epidermis or fat body. How is the stage specificity determined? One possibility is that the isoform of the ecdysteroid receptor changes with developmental stage in these polymorphic cells just as it changes in larval neurons of *Drosophila,* depending on their metamorphic fate. Each new form of ecdysteroid receptor may act in concert or as a heterodimer with ecdysteroid-induced transcription factors in activating new genes.

How JH fits into this picture is still a mystery. Does the JH-JH receptor complex interact directly with the active larval structural genes themselves and/or with the inactive pupal and adult genes, or through protein-protein interactions with other transcription factors? Or does it simply influence the type of ecdysteroid receptor and/or specific transcription factor present, and if so, how? Does JH affect RNA metabolism, influencing transcription, translation, or the presence of possible regulatory RNAs, such as has recently been found necessary for production of a silk gland–specific tRNA (Young et al., 1991; Sprague, 1993, this volume)? The answers to these questions are technically feasible and should provide a firm understanding of the hormonal regulation of gene expression in Lepidoptera and in insects in general.

Acknowledgments

I thank Drs. Kiyoshi Hiruma, Subba Reddy Palli, and James Truman for helpful comments on the manuscript and Drs. Hiruma and Truman for preparation of the figures. Cited unpublished work from my laboratory was supported by the National Science Foundation, the National Institutes of Health, and the U.S. Department of Agriculture.

12

Lepidoptera as model systems for studies of hormone action on the central nervous system

JAMES W. TRUMAN

Introduction

Circulating hormones have profound effects on the development and functioning of the central nervous system (CNS) in both vertebrates (Arnold and Gorski, 1984) and invertebrates (Truman, 1988). They act to cause adaptive changes in CNS function, thereby adjusting behavior to meet changing physiological or developmental needs. Classically, hormonal effects have been classified as *organizational* and *activational.* The former are developmental actions that are usually permanent, although the behavioral results of these actions may not be manifest until weeks or months later, long after the hormone has disappeared. The classic example of an organizational effect is the perinatal action of androgens in mammals resulting in the masculinization of the brain so the individual subsequently shows "male" behaviors as an adult. Activational effects, by contrast, represent "immediate" responses to hormones that are in the physiological realm and typically subside once the hormone is withdrawn. For example, estrogen has activational actions in evoking receptive behavior in female rats at the appropriate phase of the estrus cycle.

Study of hormonal effects on behavior has been difficult because of the cellular complexity of the CNS and the nature of some of the cellular responses. Some hormone-evoked responses are biochemical, such as changes in levels of specific neurotransmitters. These can be studied directly at the level of expression of specific genes, typically the enzymes involved in transmitter synthesis or processing. Besides biochemical changes, neurons also respond to hormones by altering their shape, resulting in altered patterns of synaptic connections due to new axon or dendrite growth and synapse formation. These morphological responses are not necessarily measurable in terms of products of single genes and, indeed, it is not clear which are the critical genes to examine. Nevertheless, understanding how hormones influence neuronal shape is a key to unraveling how changing patterns of neuronal connectivity relate to the corresponding behavioral changes. Besides altering the properties of existing neurons, organizational actions may also involve neurogenesis and programmed cell death.

Invertebrate nervous systems in general and insect nervous systems in particular have many large neurons that can be repeatably identified as specific cells with defined morphologies, transmitter content, and patterns of peripheral and central connections. Hence the same cell can be found from individual to individual and its fate followed through the life of the insect. For a given neuron, its fate is invariant and tied to the endocrine cues regulating growth and metamorphosis. The ability to study identified neurons that have known fates and to relate these fates to the well-known endocrinology of Lepidoptera (Riddiford, 1985; Gilbert, 1989) make these animals premier objects for examining the actions of hormones on the CNS.

The major focus of this chapter is on events that occur during metamorphosis when hormones act to alter radically the form and function of the CNS. Because molecular studies of the CNS during this period are in their infancy, the intent of this review is to present a biological framework for future molecular work on the lepidopteran CNS. It deals exclusively with neurons, but it should be noted that glial cells play crucial roles in the morphogenetic changes seen in particular regions of the CNS (cf. Tolbert and Oland, 1989). These cells are undoubtedly important players in hormone-induced changes in the CNS but their roles vis-à-vis hormone action are unclear.

Fates of neurons at metamorphosis

Metamorphosis represents a transition between two stages that are adapted for radically different lifestyles. These changes are dramatic at the level of the whole organism, but they are equally striking when considering individual nerve cells. Embryogenesis results in a nervous system composed of neurons specialized for a larval existence. During the ensuing larval life these neurons increase in size, but their biochemistry, physiology, and morphology change little, if at all. At metamorphosis, though, this larval state becomes "destabilized," larval characteristics are lost, and the neurons acquire a new, adult set of characters. These changes can be profound, affecting dendritic (Casaday and Camhi, 1976; Truman and Reiss, 1976; Kent and Levine, 1988) and axonal (Truman, 1989) arbors, synaptic contacts (Levine and Truman, 1982), transmitter and modulator substances (Tublitz and Sylwester, 1990; Witten and Truman, 1990), and even neuronal survival itself (Truman, 1983).

The most extensively studied aspects of these changes deal with morphology and survival. Figure 12.1 uses examples of abdominal motoneurons from *Manduca* to summarize the types of fates awaiting larval neurons at metamorphosis:

1. The majority of larval motoneurons, such as MN-3, are conserved through metamorphosis and function in the adult. Typically, they lose their larval muscle shortly after the larval-pupal transition and acquire a new target during adult development. In association with this change in target, these neurons undergo a marked remodeling, losing larval-specific dendritic and axonal arbors and subsequently acquiring new, adult-specific branches (Kent and Levine, 1988; Truman and Reiss, 1988).
2. The D-IV motoneurons are representative of larval neurons that persist through metamorphosis (Levine and Truman, 1985) but then die early in adult life. These cells maintain a stable morphology and functional target muscles through metamorphosis but then both muscles and neurons die after emergence of the adult.
3. A few cells, like the proleg motoneuron PPR, have only a larval function. PPR undergoes a dendritic "pruning" coincident with the death of its target muscle during the larval-pupal transition, but unlike MN-3, which subsequently acquires a new adult form, PPR abruptly dies two days after pupal ecdysis (Weeks and Truman, 1985).

These examples show that larval neurons respond to the ecdysteroid surges that bring about metamorphosis in diverse ways: stability versus change or survival versus degeneration. A key issue is how the same hormonal cues can evoke radically different responses in cells that are so similar.

Organizational actions: neuronal remodeling

The larval-pupal transition is when many neurons lose larval-specific features. The pruning back of larval structures has been best studied in the proleg motoneuron PPR (Weeks and Truman, 1985; Weeks, 1987). The

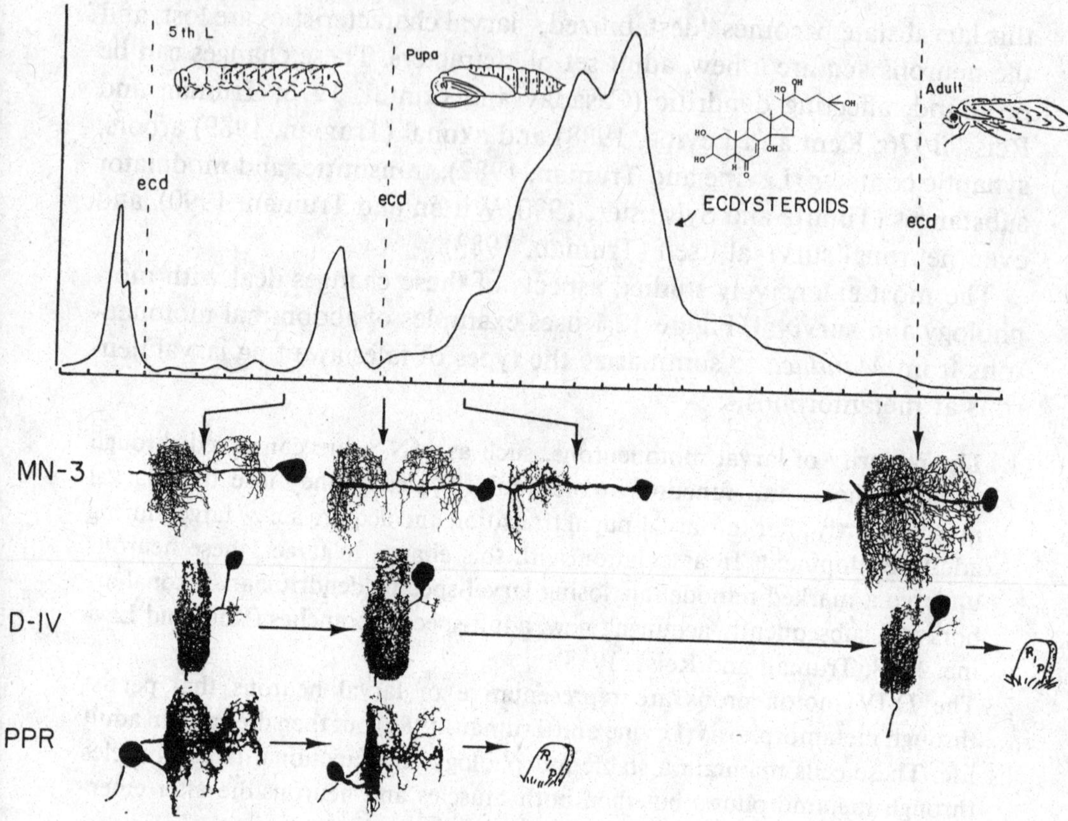

Figure 12.1. Relationship of the ecdysteroid titer to changes in identified neurons during the latter part of larval growth and metamorphosis in *Manduca.* The graph shows the relative changes in ecdysteroids through this period. The drawings show the central dendritic morphology of three motoneurons at indicated times. They represent classes of cells that (1) lose their larval branches but then sprout an adult arbor for use through adult life (MN-3); (2) maintain a stable morphology through metamorphosis but then die after adult emergence (D-IV); and (3) undergo a loss of branches during the larval-pupal transition and then die after pupal ecdysis (PPR). Ecd, ecdysis.

morphology of this cell remains stable through the day after wandering, but then it shows a rapid loss of dendritic branches so that by the time of pupal ecdysis, about three days later, the dendritic arbor has been reduced to 50 percent of its original extent. This loss of dendrites is caused by the prepupal peak of ecdysteroids (Weeks and Truman, 1985; see Riddiford, this volume). When abdomens are isolated prior to this peak, the larval morphology of the neuron is preserved; however, when such preparations are infused with 20-hydroxyecdysone (20E), to mimic the prepupal peak, dendritic loss is then induced. PPR normally degenerates a few days after dendritic regression, but regression does not trigger the death. Ligation experiments during the prepupal peak that limit the amount of ecdysteroid experienced by these neurons show that the two events are independently triggered and that induction of dendritic regression requires shorter exposures to steroid than does the cell death (Weeks, 1987).

In most larval neurons, death does not follow dendritic pruning (Truman and Reiss, 1988; Weeks and Ernst-Utzschneider, 1989). For example, in remodeled motoneurons, such as MN-1 and MN-3, dendrites are lost shortly after pupal ecdysis, but this loss is then followed by the outgrowth of adult-specific processes (Truman and Reiss, 1988). This dendritic sprouting requires the peak of ecdysteroids that drives adult development. Prevention of this ecdysteroid peak, such as during pupal diapause, results in the lack of outgrowth, and cell morphology remains stable. Similarly, the cell does not show adult growth if it is exposed to juvenile hormone mimics (JHMs) early in the period of ecdysteroid action (Truman and Reiss, 1988). Thus ecdysteroids are essential for causing these changes, but the steroid must act in the absence of JH.

Important insights into the regulation of dendritic growth during adult development have been provided by studies of cultured *Manduca* neurons by Levine and colleagues (Prugh, Della Croce, and Levine, 1992). They labeled the cell bodies of thoracic motoneurons by injecting fluorescent tracers into the larval leg musculature and allowing time for transport of the label back to the CNS. The labeled neurons were later recovered from dissociated pupal ganglia and cultured in the presence or absence of 20E. In the absence of steroid, the neurons survive well, growing an axonlike process but having only sparse dendritelike branches. With ecdysteroids, by contrast, the neurons show exuberant dendritic growth that is reminiscent of their profuse growth in vivo.

In contrast to the leg motoneurons, identified neurons that are not remodeled in vivo also fail to show sprouting in response to ecdysteroids in vitro. For example, the intersegmental muscle motoneurons (D-IV cells) show a stable morphology through metamorphosis (see Figure 12.1; Levine and Truman, 1985) and, in culture, their pattern of process outgrowth is the same with or without ecdysteroids (Witten and Levine, 1991). This close parallel between in vivo and in vitro results shows that the induction of dendritic sprouting is a direct response to ecdysteroids but that neurons vary as to whether or not they can show such a response.

The importance of the hormonal milieu in directing growth responses of neurons is also shown by in vivo studies of the sensory neurons. These neurons are advantageous for experimental manipulation because their cell bodies are in the periphery near their sensory structures. The best data are for the neurons that supply the pupal gin-traps – paired cuticular structures found on the anterolateral margins of segments A5 to A7 (Figure 12.2; Bate, 1973). Each trap contains about 20 sensory neurons that project into the CNS and terminate in the ganglion of the next anterior segment. Although the gin-trap is a pupal structure, the sensory neurons are persisting larval cells. In their larval form, they have a sparse terminal axon arbor, but it increases markedly in both length and extent of branching during the transition from larva to pupa (Levine, Pak, and Linn, 1985; Levine, Truman, Linn, and Bate, 1986).

The role of hormones in inducing the pupal-specific sprouting was examined by blocking the metamorphosis of the presumptive gin-trap region by topical application of a juvenile hormone mimic (JHM) (Levine et al., 1986). The resultant pupae had a patch of larval cuticle in place of the gin-trap and the sensory neurons from this patch failed to show axon sprouting despite the fact that they projected into a pupal CNS (see Figure 12.2). The reverse experiment involved using ligatures to produce permanently arrested larval abdomens and then applying ecdysteroids locally to the anterolateral sensory neurons. The neurons from these treated areas grew pupallike axon arbors in an otherwise larval CNS (Levine, 1990). Thus the growth response of these sensory neurons seems dependent on the hormonal conditions experienced by their cell bodies.

Recent studies of neuropeptide expression show that some neurons exhibit biochemical as well as morphological remodeling during metamorphosis. In their larval condition, many *Manduca* motoneurons appear to express and release a peptide related to FMRFamide (Witten and Truman, 1990), in addition to their normal neuromuscular transmitter, glutamate. During the larval-pupal transition, these motoneurons then turn off production of the peptide and apparently release only glutamate in the adult. A similar situation is seen for a set of abdominal neurosecretory cells (Tublitz and Sylwester, 1990). During the larval stages these cells produce one of the cardioacceleratory peptides, but at metamorphosis they then switch over to the production of the peptide bursicon (Taghert and Truman, 1982). The genes for these peptides have not yet been isolated from Lepidoptera but, once in hand, they should provide a useful avenue

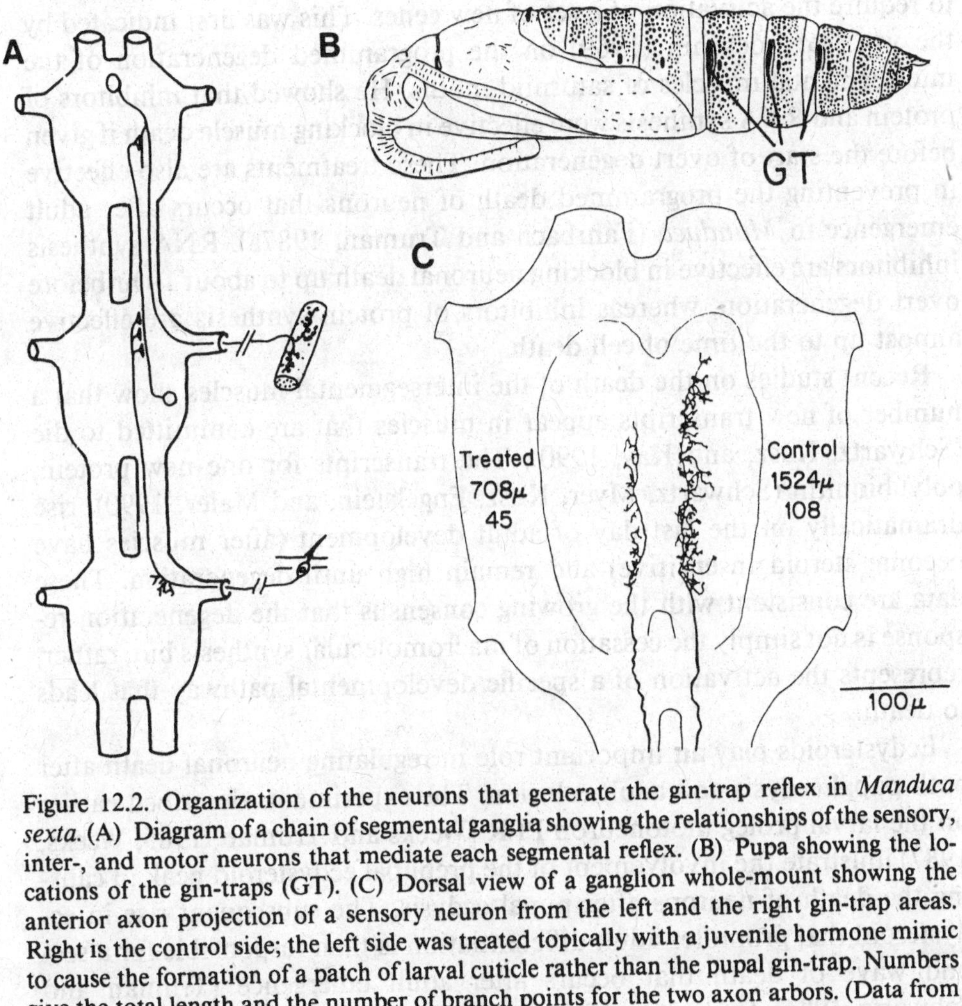

Figure 12.2. Organization of the neurons that generate the gin-trap reflex in *Manduca sexta.* (A) Diagram of a chain of segmental ganglia showing the relationships of the sensory, inter-, and motor neurons that mediate each segmental reflex. (B) Pupa showing the locations of the gin-traps (GT). (C) Dorsal view of a ganglion whole-mount showing the anterior axon projection of a sensory neuron from the left and the right gin-trap areas. Right is the control side; the left side was treated topically with a juvenile hormone mimic to cause the formation of a patch of larval cuticle rather than the pupal gin-trap. Numbers give the total length and the number of branch points for the two axon arbors. (Data from Levine et al., 1986.)

for exploring the basis of stage-specific specializations in individual neurons.

Organizational effects: neuronal death

Metamorphosis is associated with two main periods of neuronal death, the first occurring one to two days following pupal ecdysis and the second during the first two days of adult life. The mechanism of cell death appears to require the activation of a set of new genes. This was first indicated by the work of Lockshin (1969) on the programmed degeneration of the intersegmental muscles of saturniid moths. He showed that inhibitors of protein and RNA synthesis were effective in blocking muscle death if given before the start of overt degeneration. These treatments are also effective in preventing the programmed death of neurons that occurs after adult emergence in *Manduca* (Fahrbach and Truman, 1987a). RNA synthesis inhibitors are effective in blocking neuronal death up to about 10 hr before overt degeneration, whereas inhibitors of protein synthesis are effective almost up to the time of cell death.

Recent studies on the death of the intersegmental muscles show that a number of new transcripts appear in muscles that are committed to die (Schwartz, Kosz, and Kay, 1990). The transcripts for one new protein, polyubiquitin (Schwartz, Myer, Kosz, Engelstein, and Maier, 1990), rise dramatically on the last day of adult development (after muscles have become steroid insensitive) and remain high until degeneration. These data are consistent with the growing consensus that the degeneration response is not simply the cessation of macromolecular synthesis but, rather, represents the activation of a specific developmental pathway that leads to death.

Ecdysteroids play an important role in regulating neuronal death after both pupal ecdysis and adult eclosion. The experiments described earlier on the larval proleg motoneuron PPR (Weeks and Truman, 1985; Weeks, 1987) illustrate the involvement of the prepupal ecdysteroid peak in causing the death of neurons after pupal ecdysis. The subsequent rise in ecdysteroid that promotes adult differentiation is then responsible for a second wave of death that occurs after adult emergence (Truman and Schwartz, 1984). In the case of the latter, ecdysteroids are involved in two aspects of the response: first in setting the fates of the cells and then in the triggering of the death itself.

Ecdysteroids and the triggering of neuronal death

Experimental manipulations indicate that the triggering of neuronal death requires the withdrawal of ecdysteroids (Truman and Schwartz, 1984). For example, the artificial elevation of the steroid titer at the end of metamorphosis delays degeneration for as long as high steroid titers are maintained. Conversely, a premature decline in the ecdysteroid titer results in premature death. Treatment with steroids at various times before the start of overt degeneration identifies a discrete "commitment point" for each neuron when the cell seems to be irreversibly committed to die (Figure 12.3). Prior to this time ecdysteroid treatment delays or prevents the onset of degeneration, but after this time the cell dies even if steroid is subsequently supplied. This commitment point occurs about 10 hr prior to the onset of degeneration, and after this time the death response is also insensitive to inhibitors of transcription (Fahrbach and Truman, 1987a).

The death of the motoneurons is coincident with the death of their target muscles and both are dependent on the withdrawal of ecdysteroids (Schwartz and Truman, 1984). This correlation, however, does not mean that the death of muscles triggers that of the neurons or vice versa. For example, target muscles of the D-IV cells, (the ventral intersegmental muscles or ISMs) have a commitment point that is 16 hr earlier than that for their motoneurons. Ecdysteroid treatment between the two commitment points results in normal muscle death, but neuronal death is blocked, indicating that the death of the ventral ISMs does not trigger the death of their motoneurons (Truman and Schwartz, 1984). The converse conclusion can be drawn from the results of experiments in which JHMs are applied topically to the site of a doomed muscle early in adult development. Because of the local JHM, the muscles do not then die, but their motoneurons, which were not exposed to JHM, do degenerate (Truman, unpublished data). Thus, although the fates of the motoneuron and its target are normally linked, the death of one is neither necessary nor sufficient for the death of the other.

A more difficult assessment is whether presynaptic events are required for neuronal death. For one neuron, MN-12, presynaptic interactions appear to be involved in triggering degeneration. Transection of the ventral connectives immediately anterior to a ganglion containing a pair of MN-12s results in the survival of the MN-12s in that ganglion but not in more posterior ganglia (Fahrbach and Truman, 1987b). Unlike the effects of ecdysteroid injection, which only delays death for as long as the steroid levels are elevated, the rescue that comes about by the transection is permanent. Importantly, the timing of the transection is crucial: Transections prior to the commitment point rescue the cells, but those after the commitment point are without effect. These data suggest that an interganglionic signal is involved in triggering the death of MN-12, and the time of this signal coincides with the commitment point. A possible relationship

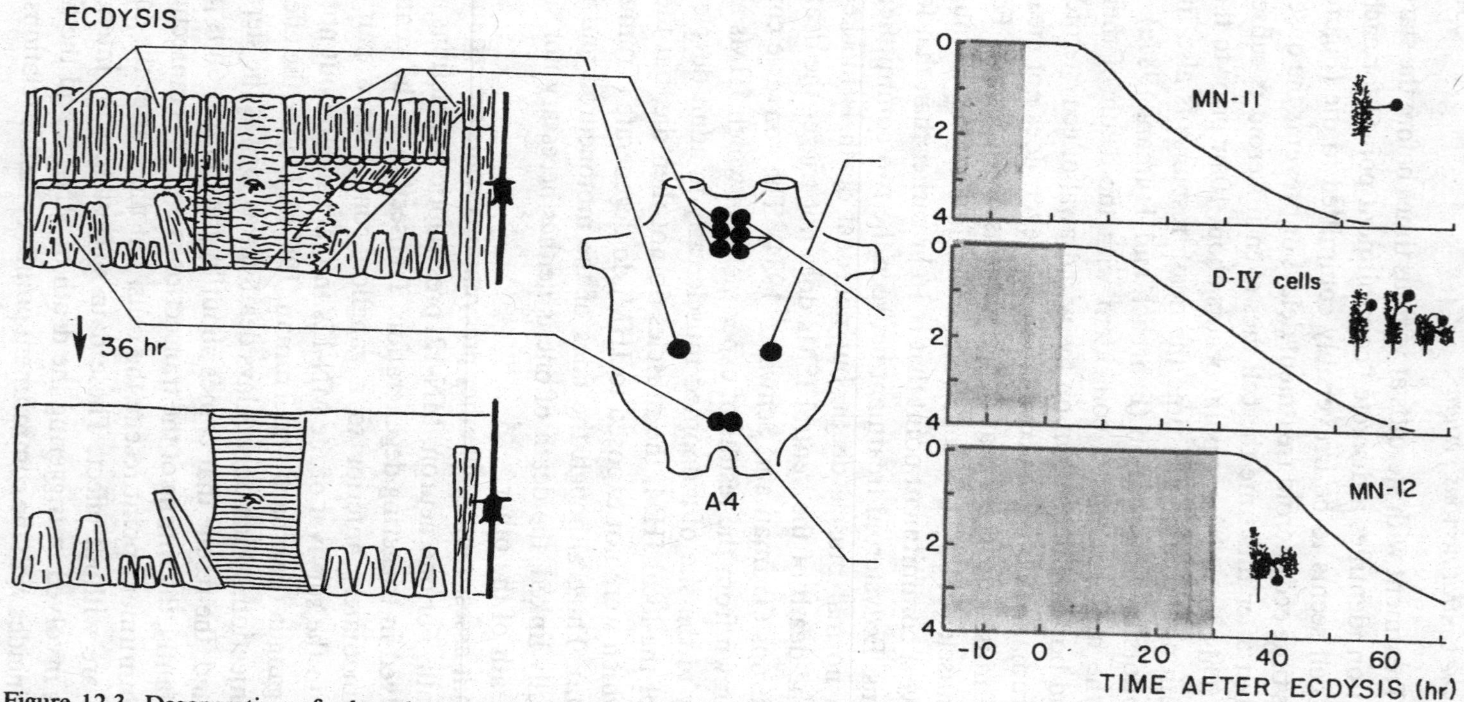

Figure 12.3. Degeneration of selected muscles and their motoneurons after adult ecdysis in *Manduca sexta*. *Left:* A4 hemisegment showing the location of muscles that degenerate after adult ecdysis. *Middle:* Ganglion A4 showing location of motoneurons that innervate selected doomed muscles. *Right:* Time course of degeneration of three sets of motoneurons. The stages of degeneration progress from a normal cell (0), through the rupture of the nucleus (2), to the last remains of the cell (4). The shaded areas show the times when ecdysteroid treatment can delay the death of each neuron. Drawings indicate the central morphology of the cells.

of this transganglionic signal and the withdrawal of ecdysteroids to the subsequent gene expression necessary for the programmed death is presented next.

Ecdysteroids and the establishment of cell fate

Besides being involved in triggering degeneration, ecdysteroids also seem to play a role in setting the fates of the cells that will die. The D-IV motoneurons are present throughout larval life and metamorphosis. In the case of diapausing pupae, these neurons then live for months without signs of degeneration despite the prolonged absence of ecdysteroid. The subsequent ecdysteroid exposure that promotes adult development then results in these cells becoming fated to die. Experiments involving the application of JHM indicate that this effect of ecdysteroid occurs early in the response to the steroid (Truman, unpublished data). Treatment with JHM at the beginning of adult development results in the formation of an animal that is externally a pupal-adult intermediate, and, internally, neither its muscles nor neurons degenerate. These juvenilizing effects of JHM become progressively less as treatment is delayed. By three to four days after pupal ecdysis, the epidermis is insensitive to JHM, and JHM treatment results in a moth with normal external morphology but whose neurons and muscles still fail to die (Truman, unpublished data). Only after day 7 does JHM treatment lose its effectiveness in blocking the eventual fate of the doomed cells. Thus an early action of ecdysteroid, which can be countered by JH, appears to be required for the subsequent death of the cells two to three weeks later.

Death and ecdysteroid receptors

The radiolabeled ecdysteroid ponasterone A was used to map the distribution of ecdysteroid receptors (EcR) in *Manduca* at the time of adult emergence. An important finding from these studies was that many of the neurons fated to die showed high levels of nuclear steroid accumulation, whereas levels of binding in cells that persist in the adult were low (Fahrbach and Truman, 1989). Thus "doomed" neurons appeared to differ quantitatively from other neurons in their EcR levels.

Recent studies on *Drosophila* have better defined the relationship of EcR patterns to neuronal death. As with *Manduca, Drosophila* also has sets of neurons that die after adult emergence (Kimura and Truman, 1990).

EcR was cloned in *Drosophila* (Koelle et al., 1991) and subsequently found to code for three distinct receptor proteins (Talbot, Swyryd, and Hogness, 1993). Two receptor isotypes, EcR-A and EcR-B1, play a prominent role during postembryonic development. They share common DNA and ligand binding domains but vary in the transactivational domains at the N-termini of the proteins. Antibodies specific to the two forms have allowed the levels of each to be followed through metamorphosis in *Drosophila*. During the larval–pupal transition, larval neurons express primarily the B1 isotype. At the time of head eversion (the fly equivalent of pupal ecdysis), a few neurons begin to express very high levels of EcR-A, which they then maintain through the rest of metamorphosis. The remaining larval neurons show only moderate levels of EcR-A (Truman, Talbot, Fahrbach, and Hogness, 1994). The cells that show this high, sustained expression of EcR-A are those fated to die after adult emergence (Robinow, Talbot, Hogness, and Truman, 1993). Presumably, in *Manduca* the high levels of ecdysteroid binding seen in doomed cells (Fahrbach and Truman, 1989) also reflect the EcR-A isotype, but the form-specific probes are not yet available for *Manduca*.

The genes involved in cell death are presumably regulated in a manner similar to "late" genes in the salivary glands of *Drosophila* (Ashburner et al., 1974). According to such a model, ecdysteroids would act at the start of adult development to activate "early" genes that eventually promote or permit the transcription of the death-related genes. Besides positive factors that promote transcription, high levels of EcR-A are also induced and these would serve to silence the death-related genes as long as steroid is present. Whether the suppressive action of JH treatment during this early period results from inhibition of the production of the appropriate transcription factor and/or of high levels of EcR-A is not known. In the case of neurons such as MN-12, the withdrawal of steroid is apparently necessary but not sufficient for the transcription of the death-related genes. For these cells the transsynaptic factors may not be able to work until the suppressive effects of ecdysteroids are removed. Tests of this scheme await the identification of genes that are involved in the death response.

Activational effects: ecdysteroids

Most actions of ecdysteroids in insects are in the organizational realm, in that they cause permanent developmental changes in the CNS. One exception is their role in triggering premetamorphic behaviors. The early work by Piepho (1950) on *Galleria* suggested that ecdysteroids initiated spinning behavior but that JH determined the type of structure that was spun: In the presence of JH the insect spun a simple molting pad, whereas in the absence of JH it constructed an elaborate cocoon. The first experimental demonstration that ecdysteroids could trigger spinning behavior involved its premature induction in *Hyalophora cecropia* by prolonged infusion of low levels of ecdysteroids (Lounibos, 1976). Sphinx moths such as *Manduca* do not spin cocoons, but nevertheless they show defined premetamorphic behaviors (Dominick and Truman, 1984). In preparation for metamorphosis, *Manduca* larvae cease feeding, become positively geotactic, show intense and prolonged locomotor activity ("wandering"), and eventually burrow into the soil, where they construct a subterranean pupation chamber.

Endocrine studies showed that wandering behavior is triggered by ecdysteroids acting in the absence of juvenile hormone. Locomotor activity begins about 10–12 hr after the start of ecdysteroid infusion, and the duration of this behavior is proportional to the length of infusion (Dominick and Truman, 1985). The long delay in the response to the steroid treatment suggests that the behavioral response is mediated through protein and RNA synthesis rather than through rapid, "nongenomic" ("nonclassic") pathways (McEwen, 1991). Studies of the isolated CNS showed that the enhanced CNS excitability associated with wandering was directly induced by ecdysteroids and that targets in the brain were essential for driving phasic activity characteristic of wandering (Dominick and Truman, 1986a, 1986b). Like other hormone-mediated systems (Arnold, Nottabohm, and Pfaff, 1976), though, targets are likely distributed at various levels in the neural hierarchy that mediates wandering behavior. Indeed, neck-ligatured animals respond to infused steroid by showing enhanced levels of spontaneous twitching, suggesting steroid-induced changes in the ventral CNS as well as those in the brain (Dominick and Truman, 1986a). The location of the target cells and the mode of action of ecdysteroids in inducing these effects have yet to be determined.

Activational effects: eclosion hormone

Neuropeptides provide a fertile ground of neuron-specific genes that are only now beginning to be exploited. A growing number of insect peptides have been sequenced, and this has been followed by the isolation of the genes encoding some of them. The list of the latter includes the prothoracicotropic hormone (Kawakami et al., 1990), bombyxin (Iwami et al., 1989), eclosion hormone (Horodyski et al., 1989), adipokinetic hormone (Bradfield and Keeley, 1989; Schulz-Aellen, Roulet, Fischer-Lougheed, and O'Shea, 1989), and FMRFamide (Nambu, Murphy-Erdosh, Andrews, Feistner, and Scheller, 1988; Schneider and Taghert, 1988). This increasing wealth of new genes opens up opportunities to examine issues of temporal and spatial regulation of gene expression, as well as mRNA splicing and peptide processing. Each neuropeptide gene, however, provides an entry to a series of other genes that code for the proteins that mediate the action of the peptide.

In the following paragraphs the eclosion hormone (EH) of moths is used to illustrate some of the complexities that may be encountered in the examination of these mediational cascades.

EH is the best known example of a neuropeptide that has discrete behavioral actions. It is a 62 amino acid neuropeptide that was originally described for its ability to trigger the adult ecdysis behavior of giant silk moths (Truman and Riddiford, 1970). It acts directly on the CNS to trigger ecdysis behaviors, highly specialized behaviors that are used for shedding the old cuticle at the end of a molt (Truman, 1990). In addition to its neural actions, it has ecdysis-related actions on selected peripheral targets such as increasing the plasticity of wing cuticle (Reynolds, 1977), inducing secretion of the cement layer by dermal glands (Hewes and Truman, 1991), and triggering the programmed death of selected abdominal muscles (Schwartz and Truman, 1984).

An intriguing aspect of the action of EH is that the target tissues vary markedly in their sensitivity to the peptide. During each instar, the insect is responsive to EH only for a brief window of 6–8 hr at the very end of each instar (Truman et al., 1983). Once exposed to EH during one of these windows of sensitivity, the tissues then become refractory until the next molt.

The basis for the switch between responsive and refractory states lies in the molecules that mediate EH action (Figure 12.4), which involves the second messenger, cyclic 3′,5′-guanosine monophosphate (cyclic GMP). Cyclic GMP levels increase minutes after EH exposure, before the appearance of any overt neural response (Morton and Truman, 1985). Elevated cyclic GMP levels then activate a cyclic GMP-dependent protein kinase (gPK), which, in turn, phosphorylates two membrane-associated

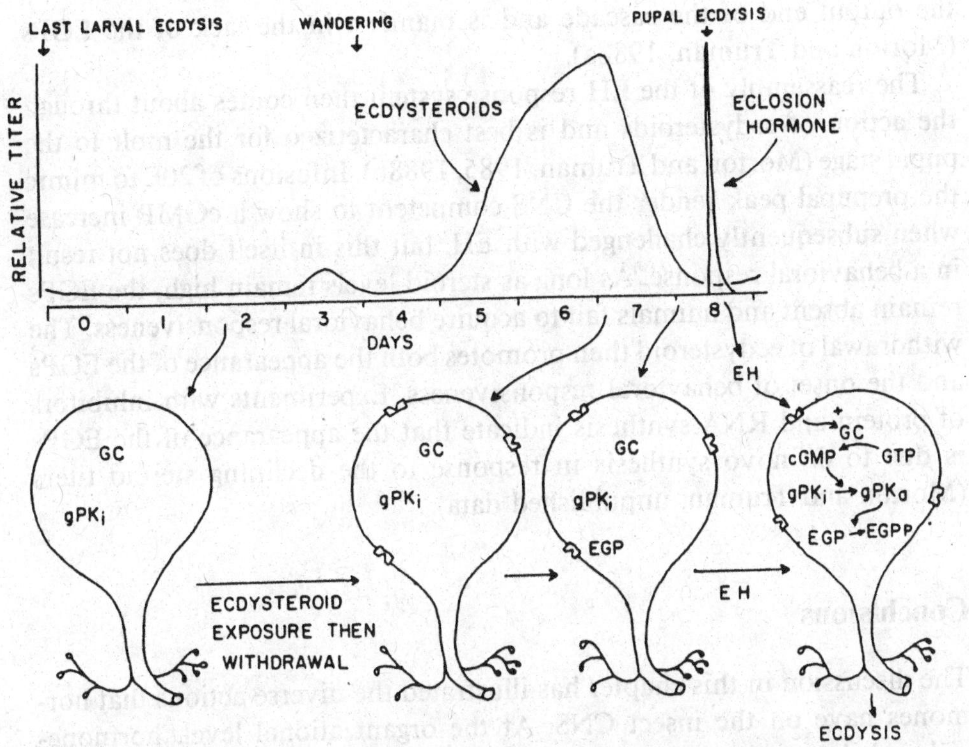

Figure 12.4. Model for ecdysteroid regulation of eclosion hormone (EH) responsiveness during the larval-pupal transition in *Manduca.* Upper diagram indicates the hormone fluctuations during this transition. The lower diagrams show the state of the EH response cascade at various times during this transition. EGP, the cyclic GMP and EH stimulated phosphoproteins in either their dephosphorylated or phosphorylated (P) state; GC, guanylate cyclase; gPK, cyclic GMP dependent protein kinase in either the active (a) or inactive (i) state; the notched box on the membrane represents the components that transduce EH binding to GC activation. See text for details.

proteins (the EGPs) (Morton and Truman, 1988a). The nature of the EGPs and the relationship of their phosphorylation to the expression of ecdysis behavior are not known.

During the intermolt period the CNS lacks two elements of this response pathway. The nature of the first deficit was suggested by the fact that the intermolt nervous system, when challenged with EH, showed no change in the levels of cyclic GMP (Morton and Truman, 1985). Because the guanylate cyclase (GC) levels are normal at this time, we presume that there is a block in the steps between the reception of EH and the activation of GC. The nature of the block is unknown, but it is likely at the level of the EH receptor or the initial transduction steps. The second deficit is at the output end of the cascade and is manifest in the lack of the EGPs (Morton and Truman, 1988a).

The reassembly of the EH response system then comes about through the action of ecdysteroids and is best characterized for the molt to the pupal stage (Morton and Truman, 1985, 1988b). Infusions of 20E to mimic the prepupal peak render the CNS competent to show a cGMP increase when subsequently challenged with EH, but this in itself does not result in a behavioral response. As long as steroid levels remain high, the EGPs remain absent and animals fail to acquire behavioral responsiveness. The withdrawal of ecdysteroid then promotes both the appearance of the EGPs and the onset of behavioral responsiveness. Experiments with inhibitors of protein and RNA synthesis indicate that the appearance of the EGPs is due to de novo synthesis in response to the declining steroid titers (Morton and Truman, unpublished data).

Conclusions

The discussion in this chapter has illustrated the diverse actions that hormones have on the insect CNS. At the organizational level, hormone-induced changes in morphology are complex. In the case of "remodeling," the analysis of specific neurons show that it is carried out in an orderly fashion with, first, the loss of larval features and then the acquisition of adult-specific components. On a molecular level, though, the mechanisms by which a neuron prunes back its axonal and dendritic arbors are unknown. The subsequent sprouting of processes that occurs during adult development appears to be a direct response of the cell to ecdysteroid and undoubtedly involves myriad genes involved with cytoskeleton, synaptogenesis, and transmitter production and reception. How the expression of these genes is coordinated with changing ecdysteroid titers is yet to be established. Besides these morphological changes, other organizational effects may be more tractable to a molecular analysis. Stage-specific shifts in transmitters shown by certain neurons provide target genes for analysis. Also, the degeneration response of specific neurons is especially attractive because it is associated with a unique pattern of EcR expression, the number of genes needed to mediate this response is likely small, and the time of their expression is well circumscribed.

The activational changes are diverse and are typically reversible. From the results of studies of the effects of ecdysteroids in activating wandering behavior and priming the CNS for EH action, these actions likely involve alterations in physiological or biochemical properties of neurons, such as membrane excitability, levels of transmitters, or components of transmitter or neuropeptide activated cascades. The effects of ecdysteroids on the EH system present a concrete example of the latter.

Thus far, the lepidopteran CNS has provided a number of systems in which to study the diverse actions of hormones. These systems provide exquisite cellular resolution for analysis and are poised for the step into the molecular realm.

Acknowledgments

I thank Professor Lynn M. Riddiford for a critical reading of the manuscript. Unpublished results cited in this review were supported by grants from the National Science Foundation and the National Institutes of Health.

13

Molecular genetics of moth olfaction: a model for cellular identity and temporal assembly of the nervous system

RICHARD G. VOGT

Introduction

Critical questions in neurobiology revolve around the issue of specificity. How do neurons receive, process, and coordinate responses to specific information? How do developing neurons organize themselves into specific arrays, and how do growing neuronal processes find their way to and recognize their ultimate targets? At the sensory level, questions of specificity focus on issues of signal recognition. How does an animal perceive a specific signal against a background of apparent noise? These questions are put into sharp focus when we consider them in a specific sensory context, such as olfaction. For example, how do we discriminate the odor of a banana from that of a fragrant flower, a mountain forest, or a summer beach? This chapter presents the moth olfactory system as a model for identifying molecular genetic mechanisms encoding neuronal specificity.

Perception of an odor is a quality of the brain involving the coordination of many biochemical events. The process begins with odor molecules binding to selective receptor proteins in the membranes of first-order olfactory receptor neurons. This binding is then transduced into an electrical signal, and the brain is informed of this event through synaptic connections between first- and second-order neurons. These synapses are organized as distinct glomeruli, grapelike structures, in the olfactory lobe in insects and in the olfactory bulb in vertebrates. The first-order sensory neurons can be viewed as functionally identifiable based on the odorant-sensitive phenotype they express; olfactory receptor neurons express different receptor proteins with selective odorant or ligand binding specificities. A longstanding question has been whether the second-order neurons are functionally identifiable as well, and whether the identities of the primary and secondary neurons are genetically predetermined (that is, the same from animal to animal). This would mean that, during development, a primary neuron with a predetermined odorant-sensitive phenotype targets a specific region and/or second-order neuron in the brain, thus defining a multiple neuron pathway. The emerging answer from two decades of research is that discrete olfactory pathways are functionally identifiable at the cellular level (Hansson, Ljungberg, Hallberg, and Lofstedt, 1992; Shepherd, 1992). We therefore view "olfactory specificity" to be the consequence of the coordination of many processes and not merely the expression of specific types of receptor proteins in the primary neurons. In other words, the identity of olfactory neurons results from regulatory processes that ensure that the right genes are expressed in the right cells; a first-order neuron expressing a specific receptor protein makes synaptic connections with the proper second-order neuron(s).

The moth olfactory system represents a model that allows us to test the hypothesis that gene regulatory mechanisms are the basis of cellular identity of sensory neurons and the resulting ability of the brain to perceive specific odors. Many observations based on both insect and vertebrate systems support the existence of genetically predetermined phenotypic pathways. Insects and vertebrates have both general and specialized olfactory neurons that project to respectively distinct targets in the brain. For example, insect pheromone specific sensory neurons are anatomically distinct from general odorant sensitive neurons and project to a large pheromone specific glomerulus, the macroglomerular complex (MGC), whereas the general odorant neurons project to the 50+ general glomeruli (Ernst and Boeckh, 1983; Hildebrand, 1985; Schneiderman, Matsumoto, and Hildebrand, 1986; Shepherd, 1992). Similarly, certain vertebrates possess a specialized olfactory epithelium, the vomeronasal organ (VO), which is sensitive to sex pheromones. The VO is anatomically separate from the general olfactory epithelium. The VO neurons project to the accessory olfactory bulb, which is associated with but distinct from the main olfactory bulb; neurons of the general olfactory epithelium project to the main olfactory bulb (see reviews by Wysoki and Meridith, 1987; Halpern, 1987; Eisthen, 1992). These specializations, conserved from animal to animal, indicate that broad classes of olfactory neurons of both insects and vertebrates have genetically predetermined projections.

Many studies, utilizing both insects and vertebrates, have attempted to demonstrate topographic neuronal specificity at the cellular or near cellular level. Experiments utilizing vertebrate animals have attempted to determine whether olfactory glomeruli located in defined positions in the bulb possess specific odor phenotypes. Selective accumulation of radioactive 2-deoxyglucose (2-DG) has often been used as an indicator of specific neuronal activity; 2-DG is internalized but not metabolized. A number of studies have shown odorant dependent or electrical stimulation dependent uptake of 2-DG in olfactory bulb glomeruli in discrete patterns that vary for different odorants or electrically stimulated neurons, suggesting that individual glomeruli are functionally defined by the odor phenotype of their primary neurons (Stewart, Kauer, and Shepherd, 1979; Greer, Stewart, Teicher, and Shepherd, 1982; Lancet, Greer, Kauer, and Shepherd, 1982; Astic, Saucier, and Holley, 1987). Similarly, expression of the early transcription factor *c-fos* in olfactory bulb neurons was observed in specific odorant dependent patterns (Guthrie, Anderson, Leon, and Gall, 1991). Other studies have shown odorant dependent morphological changes of vertebrate glomeruli (Royet, Jourdan, Ploye, and Souchier, 1989) and mitral cells (output neurons) (Laing, Panhuber, Pittman, Willcox, and Eagleson, 1985; Panhuber and Laing, 1987) in discrete patterns that depend on the specific odorant regimen applied. Importantly, 2-DG accumulation patterns were shown to be repeatable from animal to animal (Royet, Sicard, Souchier, and Jourdan, 1987), supporting the view that the connections between first- and second-order neurons are genetically predetermined for general odorant processing. However, it has been technically difficult to identify reliably main olfactory bulb glomeruli from one vertebrate animal to the next because of the great number and small size of these glomeruli. The more specific "modified glomerular complex," which processes suckling pheromone odorants in neonatal rats, has provided an olfactory path that can be more readily identified and studied in multiple individuals using 2-DG (Teicher, Stewart, Kauer, and Shepherd, 1980; Greer et al., 1982; Jastreboff et al., 1984; Pedersen, Jastreboff, Stewart, and Shepherd, 1986). These studies again support the view of genetically predetermined relationships between olfactory neurons and their glomerular targets.

The most dramatic demonstration of identified olfactory neurons with specific targets was recently reported in a moth (Hansson et al., 1992). The MGC of *Manduca sexta* was shown to be composed of three distinct domains; second-order neurons were shown to enter these domains in a differential manner, suggesting that the respective domains received inputs from different subsets of first-order neurons (Hansson, Christensen, and Hildebrand, 1991). Using the moth *Agrotis segetum,* Hansson and colleagues (1992) were able to dye-fill first-order neurons from the periphery inward to the brain, obtaining specific labeling by stimulating the neuron with "its" odorant during the filling process. They showed unequivocally and in multiple animals that phenotypically defined neurons always terminate in the same spatial domain within the MGC (Hansson et al., 1992). An odorant "A" neuron always terminates in domain "A," an odorant "B" neuron always terminates in domain "B," and so forth. Clearly, at least for pheromone specific neurons, a neuron with a specific and predetermined phenotype seeks out a specific target in the brain.

In the moth, formation of olfactory synapses occurs relatively early in development, whereas receptor expression is thought to occur much later, near the end of development. Hansson's studies (Hansson et al., 1991, 1992) strongly support the view that the coexpression of these two features, specific receptor and target, represents a complex neuronal phenotype or identity whose expression can be coordinated only through gene regulatory processes that must constitute the basis of neuronal identity and olfactory specificity.

The model: the olfactory antenna of *Manduca sexta*

The olfactory system of an adult male tobacco hawk moth, *M. sexta,* contains approximately 250,000 chemosensory neurons that undergo a synchronous and prolonged development lasting about three weeks (Sanes and Hildebrand, 1976a, 1976b; Hildebrand, 1985, Oland, Tolbert, and Mossman, 1988). These neurons are organized along the antenna in a mosaic of primarily two classes of functionally and morphologically identifiable olfactory sensilla (Figure 13.1), those specific to pheromone and those sensitive to general environmental odorants (Boeckh, Kaissling, and Schneider, 1960, 1965; Keil, 1989; Lee and Strausfeld, 1990). Each *M. sexta* male antenna has about 85,000 neurons distributed among 43,000 pheromone specific sensilla and an additional 166,000 neurons distributed among 55,000 general odorant sensitive sensilla (Sanes and Hildebrand, 1976a). Each female antenna has about 300,000 neurons distributed primarily among a single class of approximately 1–1.5 $\times$ 10^5 general odorant sensitive sensilla (Oland et al., 1988; sensilla number estimated from Lee and Strausfeld, 1990). In the male, the pheromone specific neurons converge on the single tripartite MGC, whereas the general odorant sensitive neurons converge on approximately 50 general glomeruli (Hildebrand,

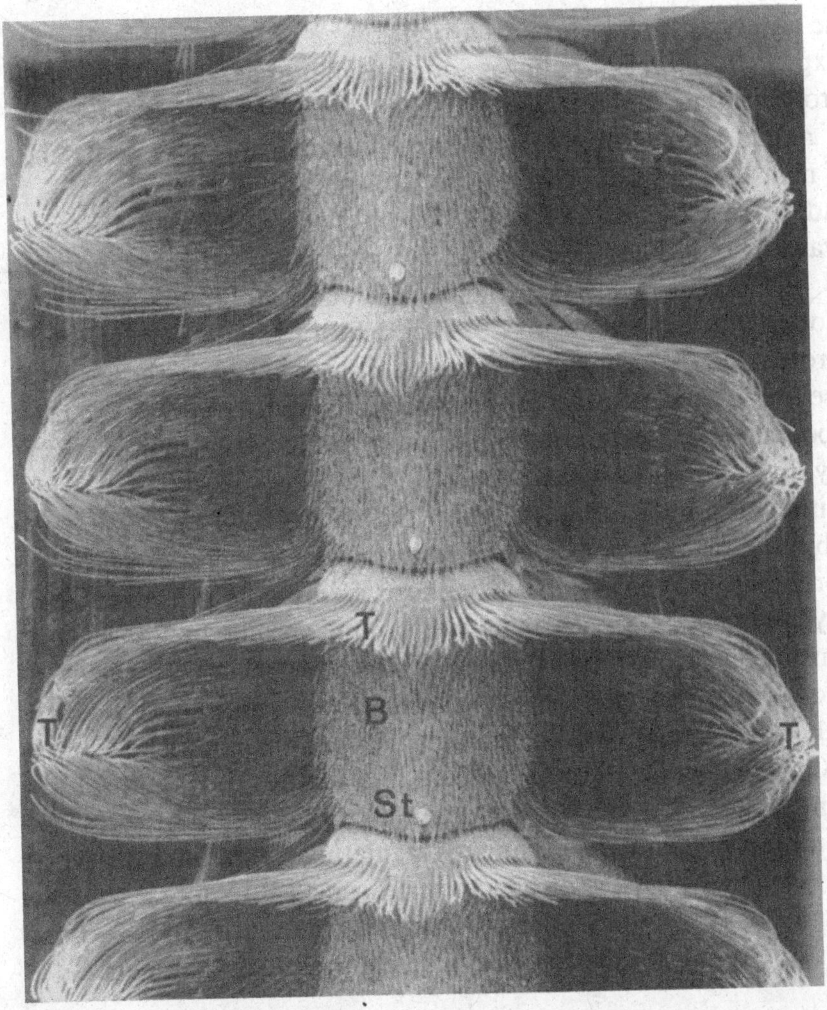

Figure 13.1. Male antenna of *M. sexta,* forward view (scanning electron micrograph), demonstrating the mosaic nature of sensilla arrays. Antenna consists of approximately 80 annuli (segments). Three classes of sensillum are easily recognized in the figure. The long hairs are the pheromone-specific trichoid sensilla (T); each annulus contains about 3,600 neurons distributed among about 1,700 of these sensilla, which are arrayed in bands along the proximal and distal borders of each annulus joining near the dorsal-ventral midlines. Short hairs covering the surface between the long hairs are the general odorant sensitive basiconic sensilla (B); each annulus contains about 1,200 neurons distributed among about 550 of these sensilla. Each annulus contains some additional 100 diverse sensilla, including the single peglike styliform sensilla complex (St). The styliform sensilla respond to temperature and humidity. The longest trichoid sensilla are approximately 400 μm long. Female antennae lack the trichoid sensilla but carry a full complement of basiconic sensilla. (Details are from Lee and Strausfeld, 1990.)

1985). The antennal tissue provides a rich source of neuronally relevant proteins expressed in precise patterns and at developmental stages ranging from neurogenesis to functional maturation. This system represents an accessible model for identifying molecular mechanisms underlying the control of neuronal development during adult metamorphosis.

Our studies have identified three kinds of antennal proteins that are both relevant to olfaction and suitable as tools for the identification of regulatory mechanisms. These proteins are odorant binding proteins (OBP), odorant degrading enzymes (ODE), and a candidate pheromone receptor protein, RP11. They are expressed in association with different olfactory cell types and/or different classes of olfactory neurons. The cellular components of sensilla are composed of a cluster of neurons surrounded by three support cells; all the cells of each sensillum derive from a single epithelial mother cell (see Figure 13.2). The OBPs and ODEs are expressed by support cells and RP11 is expressed by the sensory neurons. Furthermore, three classes of OBP have been identified and cloned that, respectively, associate with either the pheromone specific or the general odorant sensitive neurons. Two different kinds of ODE have been identified, one of which associates with both classes of neuron and the other with only the pheromone specific neurons. Finally, the expression of the OBPs and ODEs is regulated by the steroid ecdysteroids during late adult development. These antennal specific gene products represent tools for identifying molecular genetic mechanisms regulating (1) specificity of gene expression in different classes of olfactory neurons, (2) temporal coordination of neuronal development, and (3) failed regulatory processes underlying dysfunctional or deteriorating (e.g., aging) sensory systems.

Organization of the olfactory epithelium of the moth: anatomy and development

The adult olfactory neurons are organized in clusters of two to several along the antenna (Figure 13.2). Neuronal cell bodies are embedded within the antennal epithelium, with axons projecting inward to the brain and ciliated dendrites projecting outward into the external environment (Steinbrecht, 1980; Steinbrecht and Gnatzy, 1984). The cell bodies of each neuronal cluster are collectively ensheathed in three layers of support cells: thecogen, tormogen, and trichogen. The entire antennal epithelium is covered with cuticle, and the outward projecting dendrites of each cluster are

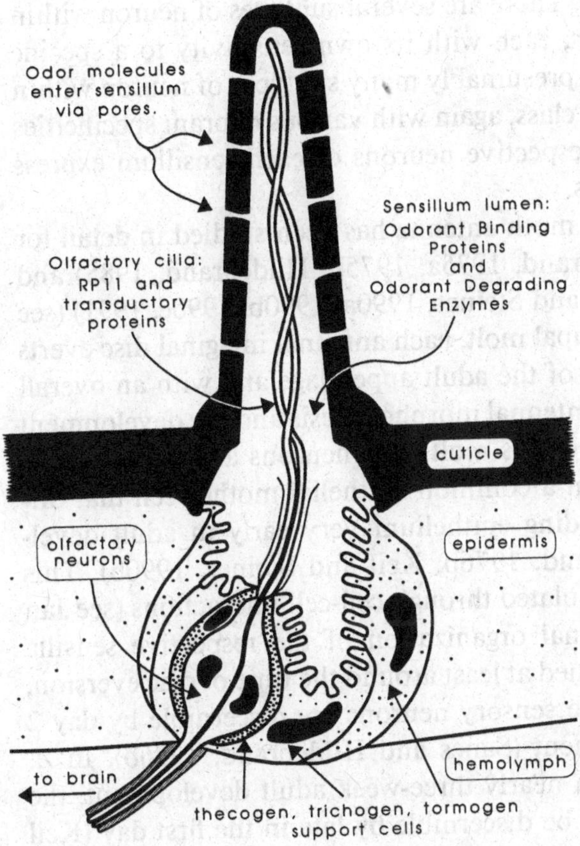

Figure 13.2. Schematic of an olfactory sensillum. Odor molecules have access to the sensillum lumen via some 20,000 pores penetrating the cuticle of each sensillum. The lumen contains water-soluble OBPs and ODEs, as well as the olfactory cilia containing receptor proteins and transduction processes. The lumen is isolated from hemolymph by junctional complexes joining neurons, support cells, and epithelial cells. (Modified from Steinbrecht, 1987.)

ensheathed in a hollow cuticular hair; the hair lumen surrounding the dendrites is filled with proteinaceous fluid containing OBPs and ODEs. The entire structure, including neurons, support cells, and cuticular hair, represents one sensillum; there are typically 10,000 to 100,000 sensilla per antenna, depending on moth species. Olfactory sensilla are represented as two predominant classes, those specific to sex pheromone odorants and those sensitive to general environmental odorants, such as plant volatiles (Boeckh et al., 1965; Kaissling, 1986). In *M. sexta* and many other moths these two sensilla classes are both morphologically distinct and positionally segregated; pheromone specific sensilla are significantly longer than the general odorant sensitive sensilla and are positioned along the proximal and distal borders of a "field" of general olfactory sensilla on each antennal annulus (see Figure 13.1). The two sensilla classes possess neurons expressing olfactory receptor proteins sensitive to distinctly different classes of odorant molecules. There are several subtypes of neuron within the pheromone specific class, each with its own sensitivity to a specific pheromone component, and presumably many subtypes of neuron within the general odorant sensitive class, again with various odorant specificities (Boeckh et al., 1965). The respective neurons of each sensillum express different odorant specificities.

Development of the adult moth antenna has been studied in detail for *M. sexta* (Sanes and Hildebrand, 1975a, 1975b; Hildebrand, 1985) and *Antheraea polyphemus* (Keil and Steiner, 1990a, 1990b, 1990c, 1991) (see Figure 13.6). At the larval–pupal molt, each antennal imaginal disc everts into a sacklike mass the size of the adult appendage and with an overall thickness of two cell layers. Antennal morphogenesis and the development of the olfactory sensilla follow this molt. The neurons and support cells of each sensillum derive from a common epithelial mother cell that differentiates from the surrounding epithelium very early in adult development (Sanes and Hildebrand, 1976b; Keil and Steiner, 1990b). This differentiation is probably regulated through cell-cell interactions (see Jan and Jan, 1990). The positional organization of the respective sensilla classes is presumably determined at least around the time of disc eversion, if not earlier. In *M. sexta,* the sensory neurons are discernible by day 2 of an 18-day adult development (Sanes and Hildebrand, 1976b). In *A. polyphemus,* which also has a nearly three-week adult development, the mother cells were reported to be discernible by late in the first day (Keil and Steiner, 1990a, 1990b, 1991). The sensory axons grow inward to the brain, possibly along guiding nerve tracks, which are reported to be present already at the start of adult development (Sanes and Hildebrand, 1976a; Hildebrand, 1985; Keil and Steiner, 1990a, 1990c), and the arrival of these axons in the brain strongly influences the correct developmental organization of the olfactory lobe (Schneiderman et al., 1986; Tolbert and Sirianni, 1990). Interestingly, the pheromone specific neurons seem to influence synaptic glomerular formation through direct interaction with second-order neurons, whereas the general odorant sensitive neurons accomplish this through initial interaction with glial cells followed by interaction with second-order neurons (Tolbert and Sirianni, 1990). In *M. sexta,* the sensory dendrites were observed to grow outward, along with a process of the trichogen cell that ensheathes the dendrites. The trichogen cell process forms an inner mold of the sensillum hair, secretes the cuticle proteins that form the hair, and then withdraws, leaving behind ciliated dendrites occupying a fluid filled lumen (Sanes and Hildebrand, 1976b).

In *A. polyphemus,* the dendrites were observed to grow outward into the lumen only after withdrawal of the trichogen cell (Keil and Steiner, 1990c, 1991). Initially two tormogen cells are produced (Sanes and Hildebrand, 1976b; Keil and Steiner, 1990b). One of these, the "second tormogen cell," also grows out with the trichogen cell to a short distance and contributes to hair formation before degenerating, around day 9 or 10 in *M. sexta* (Sanes and Hildebrand, 1976b). All of these growth processes are complete by about 60 percent of development. During the final maturation stage, at 80 to 90 percent of development, OBP and ODE expression initiates, apparently within the tormogen and trichogen cells (Steinbrecht, 1992), and the physiological responsiveness to odors appears (Schweitzer, Sanes, and Hildebrand, 1976).

Molecular components of olfaction

Our own biochemical analysis of odor detection in moths initially exploited the large antennae of the silk moth, *A. polyphemus.* As noted, the male antenna is highly enriched with pheromone specific sensilla. At the time that our studies began, this tissue was well characterized physiologically and the sex pheromone had recently been identified (Kochansky et al., 1975), making this an excellent model system for identifying and isolating biochemical components involved in detecting pheromone odorants. The preparation successfully yielded the first odorant binding proteins (OBPs) and odorant degrading enzymes (ODEs) (Vogt and Riddiford, 1981a, 1981b), as well as a candidate pheromone receptor protein (Vogt, Prestwich, and Riddiford, 1988). Subsequent studies have identified OBP homologues in several other moth species, as well as multiple OBP classes (Vogt, Kohne, Dubnau, and Prestwich, 1989; Vogt, Prestwich, and Lerner, 1991; Vogt, Rybczynski, and Lerner, 1991), ODE analogues (Rybczynski, Reagan, and Lerner, 1989, 1990), and a clone of the candidate pheromone receptor RP11 (Vogt and Lerner, unpublished data). These proteins and their tissue localization are summarized in Table 13.1 and are discussed in succeeding sections.

A biochemical model for odor detection

The studies contributed to a biochemical model of insect odor detection that hypothesized that OBPs solubilize and transport odorants to receptor

Table 13.1. *Pattern specific expression of olfactory proteins*

	Sensilla Type		Cell Type		Sex	
	Pheromone	General odorant	Support cell	Neuron	Male	Female
PBP	+	−	+	−	+	−
GOBP	−	+	+	−	+	+
Esterase	+	−	+	−	+	−
Aldehyde oxidase	+	+	+	−	+	+
RP11	+	?	−	+	+	?

Note: See text for description of differential tissue and sex distributions of olfactory proteins.

proteins in the sensory nerve membranes and that ODEs rapidly degrade the odorants in order to maintain neuronal sensitivity (Figure 13.3). One basis for this model concerned the relative aqueous solubility of odorants. Whereas aquatic animals smell water-soluble odorants such as amino acids and nucleotides, terrestrial animals smell volatile lipids, often plant derived, which are relatively insoluble in the extracellular aqueous fluid surrounding olfactory neurons. OBPs may have evolved as a terrestrial adaptation, becoming a major component of the aqueous fluid and essentially turning water-insoluble molecules into water-soluble odorants by binding and transporting the molecules through the aqueous fluid. A second basis for this model concerned the rapid inactivation of insect pheromone odorants as indicated by both physiological and behavioral studies

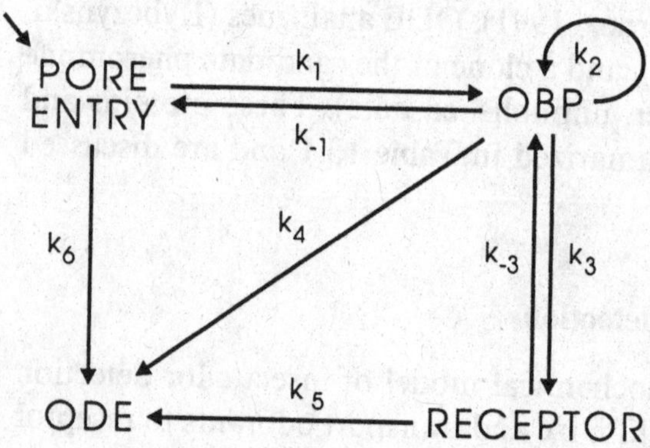

Figure 13.3. Kinetic model of odor fate within a sensillum. Odor molecules enter a sensillum via pores through cuticle. For simplicity, interactions between odor molecule and enzyme (ODE) are considered irreversible and have only forward rate constants. The relative protein concentrations and kinetic parameters ensure that the bulk of odor molecules survives degradation sufficiently long to interact with receptor proteins first (Vogt, 1987).

(Kaissling, 1974; David, Kennedy, and Ludlow, 1983; Baker and Vogt, 1988). Odorant concentrations in the air change rapidly due to local turbulence and shifts in wind direction (Murlis and Jones, 1981; David et al., 1983; Murlis, 1986). In order to follow these changes successfully during flight, insects need a mechanism to inactivate the odorants rapidly within the sensilla to allow maximum temporal sensitivity to changing levels of entering odorants. The ODEs possess kinetic properties consistent with a role of rapidly degrading, and thus inactivating, odorants within the sensilla.

In our working biochemical model for odor detection, odorants enter a sensillum through pores in the hair cuticle and then dissolve in the aqueous fluid that fills the hair lumen and surrounds the sensory dendrites (Figure 13.4). Entry of odorants into this fluid is significantly enhanced through weak interactions with OBPs, which thus act as transporters, essentially keeping the odorants in solution long enough for them to arrive at, bind to, and activate receptor proteins in the dendritic membranes. Following receptor activation, or at any time during the process, odorants may encounter and be degraded by an ODE. An odorant may also be transferred from receptor back to OBP and thus have the opportunity to activate additional receptor proteins. The features that ensure a bulk directional odorant "flow" from entry to receptor to degradation are the kinetic properties and relative concentrations of the individual components. Enzymatic studies suggested that odorant half-life within the sensillum is in the range of 1–10 msec (Vogt, Riddiford, and Prestwich, 1985; Rybczynski et al., 1989), sufficiently fast to accommodate the rapid behavioral changes observed when an odor trail is lost in flight (Baker and Vogt, 1988). Though this model has been the source of some discussion (Kaissling, 1986; Vogt, 1987), recent studies have demonstrated that OBP enhances odorant stimulation of olfactory neurons in vivo, supporting the view that OBP is an odorant transporter (Kaissling, Keil, and Williams, 1991; van den Berg and Ziegelberger, 1991).

Odorant degrading enzymes: an enzyme for any odorant

Although the existence of ODEs had been postulated (Kasang, 1971; Ferkovich, Mayer, and Rutter, 1973; Mayer, 1974), the first such enzyme isolated and characterized was the 55 kDa sensilla esterase (SE) of *A. polyphemus* (Vogt and Riddiford, 1981a, 1981b; Vogt et al., 1985; Klein, 1987). The SE is uniquely expressed in male antennae and is present in the aqueous

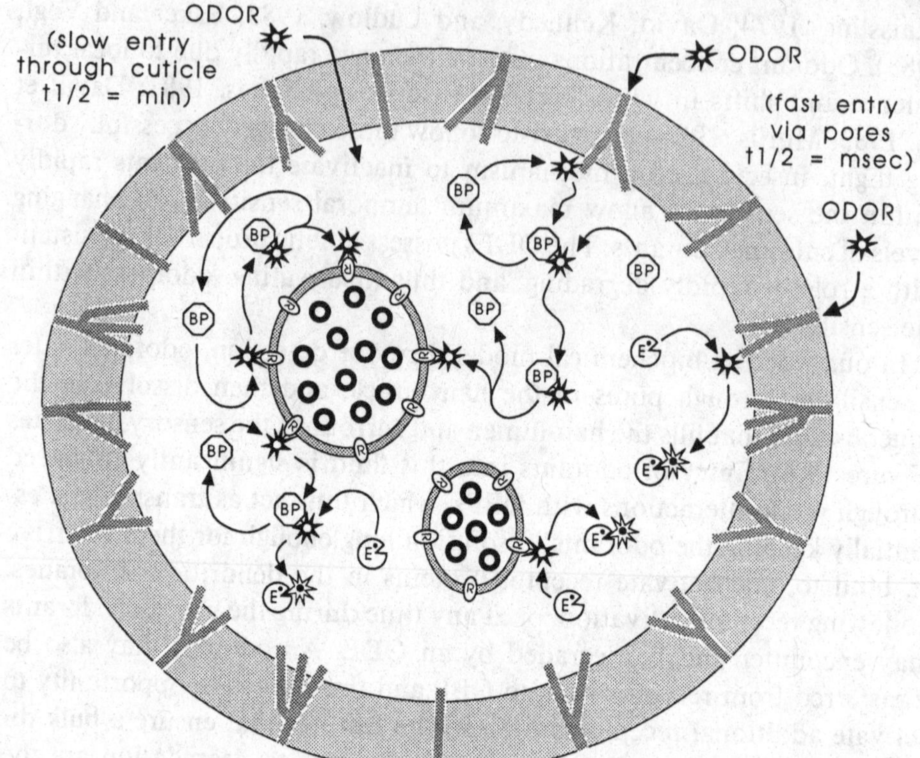

Figure 13.4. A schematic of the working model of the various biochemical pathways possible in odor processing within a sensillum. Drawing represents a cross section through a sensillum, revealing pores through the cuticle and two olfactory cilia containing microtubules. Odorants make rapid entry into the sensillum via pores penetrating the cuticle. Once inside, odorants interact with OBPs (BP) and are thus transported into and through the aqueous lumen to interact further with additional OBPs, with receptor proteins (R) or ODEs (E). The ultimate and biologically rapid fate of an odorant is degradation by an ODE. Active odorant is indicated by black stars with white centers; inactive (degraded) odorant is indicated by white stars with black centers. Odorants may also enter the sensillum by penetrating the cuticle directly (i.e., bypassing a pore), but this passage is presumed to be very slow relative to odorant entry via the pores; odorants entering by this route are presumed to be rapidly degraded especially relative to their rate of passage.

fluid filling the pheromone specific sensilla. *A. polyphemus* utilizes a two-component sex pheromone consisting of the acetate ester and aldehyde (*E,Z*)-6,11-hexadecadienyl acetate and (*E,Z*)-6,11-hexadecadienal (Kochansky et al., 1975). The role of SE as a rapid sex pheromone inactivator, and indeed the entire model, were proposed as a consequence of kinetic studies showing that SE degrades the acetate ester component to an alcohol rapidly, indicating an estimated in situ half-life for pheromone of less than 10 msec (Vogt et al., 1985). However, in the interest of a general theory of odor activation, it was desirable to know the fate of the second pheromone component, the aldehyde.

An aldehyde oxidase (AOX) capable of degrading aldehyde pheromones was subsequently identified in *M. sexta* (Rybczynski et al., 1989). *M. sexta* also utilizes a two-component pheromone, but both components are aldehydes: (*E,Z*)-hexadecadienal (bombykal) (Starratt et al., 1979) and (*E,E,Z*)-10,12,14-hexadecatrienal (Tumlinson et al., 1989). Rybczynski and colleagues (1989) isolated and characterized an antennal specific AOX from *M. sexta* that associated with the olfactory sensilla. Sex pheromone was rapidly degraded by this AOX, at an estimated in situ half-life of about 1 msec (Rybczynski et al., 1989). The AOX is large, about 150 kDa, and sufficiently abundant to be visualized by polyacrylamide gel electrophoresis (PAGE) using Coomassie blue staining (Rybczynski et al., 1989; Rybczynski, Vogt, and Lerner, 1990). A striking difference from the male specific *A. polyphemus* SE is that AOX is present in both male and female antennae, suggesting that it has a dual role of degrading both sex pheromone and general environmental odorants (Rybczynski et al., 1989).

Similar AOXs were subsequently identified as major ODEs in both *A. polyphemus* and *Bombyx mori* (Rybczynski et al., 1990). *B. mori* utilizes a two-component pheromone consisting of the aldehyde bombykal (Kaissling, Kasang, Bestmann, Stransky, and Vostrowsky, 1978) and the alcohol (*E,Z*)-10,12-hexadecadienol (bombykol) (Butenandt et al., 1959). The *A. polyphemus* AOX was directly isolated from the pheromone specific sensilla along with the SE, but as in *M. sexta* the AOX was expressed in both male and female antennae, again supporting a dual role. This was a very important observation in the context of our interest in cellular identity: SE is expressed only in the one class of male specific and sex pheromone specific sensilla, whereas AOX is expressed in both classes of sensilla, pheromone specific and general odorant sensitive. In fact, AOX is the only protein we have in hand that we are certain is olfactory specific but indiscriminate with respect to sensilla class.

Other antennal enzyme activities have been characterized and postulated to function in the degradation of sex pheromones (Prestwich, 1987; Prestwich, Graham, Kuo, and Vogt, 1989; Prestwich et al., 1989). In general, a correlation has been consistently observed between a class of pheromone molecule utilized (e.g., acetate ester, aldehyde, alcohol) and the presence of a corresponding antennal enzymatic activity appropriate for that type of molecule. Thus the existence of potent enzymes for rapid degradation of pheromone odorants appears to be the rule rather than the exception. ODEs presumably have a central role in critical insect behaviors, including mating, egg dispersal, and feeding. It seems feasible to alter insect behavior by chemically interfering with the ODEs, significantly affecting the perception of certain complex odors. For example, moth pheromones are known to consist of precise ratios of two or more compounds. Altering the rate of degradation of one of these compounds by inhibiting its ODE with a volatile inhibitor might shift this chemical ratio within the sensillum, rendering the pheromone nonattractive.

Odorant binding proteins: specific associations with neuron classes

Our initial biochemical interest in moth olfaction was, admittedly, to isolate a pheromone receptor protein. Our approach was to search for male antennal specific proteins in *A. polyphemus* that bound radioactive pheromone (Vogt and Riddiford, 1981a, 1981b). These studies yielded the small, water-soluble, and very abundant pheromone binding protein (PBP). This male specific protein is located in the aqueous fluid surrounding the sensory dendrites of the pheromone specific sensilla (Vogt and Riddiford, 1981b) at concentrations of about 10 mM (Klein, 1987). PBPs from both *A. polyphemus* and the gypsy moth *Lymantria dispar* have been shown to bind sex pheromone (Vogt and Riddiford, 1981b; Vogt et al., 1989; Prestwich, 1991). Although it has not proven possible to obtain accurate binding constants, given the difficulty of maintaining a stable free concentration of hydrophobic pheromone molecules in aqueous solution, indirect measurements have suggested relatively weak binding between pheromone and PBP (Vogt and Riddiford, 1986), possibly in the range of 10^{-7} K_D (de Kramer and Hemberger, 1987). Recent immunocytochemical studies have shown that, in addition to its presence in the extracellular fluid, PBP is present in the tormogen and trichogen cells of sex pheromone specific sensilla but not in general odorant sensitive sensilla (Steinbrecht, 1992), suggesting that the former cells are the sites of PBP expression.

At least 15 insect OBPs have been characterized from seven moth species. Among them, three distinct but homologous OBP classes have been identified (Vogt and Lerner, 1989; Breer, Krieger, and Raming, 1990; Vogt, Prestwich, and Lerner, 1991; Vogt, Rybczynski, and Lerner, 1991) that differentially associate with the two major classes of olfactory sensilla. The

PBPs represent one class of OBP and are generally uniquely expressed in male antennae (Vogt and Riddiford, 1981b; Vogt, Prestwich, and Lerner, 1991). The general odorant binding proteins GOBP1 and GOBP2 represent the other two OBP classes and associate with the general odorant sensitive olfactory neurons common to both male and female antennae (Vogt, Prestwich, and Lerner, 1991). PBPs have been cloned and sequenced from *M. sexta* (Gyorgyi, Roby-Shemkovitz, and Lerner, 1988), *A. polyphemus* (Raming, Krieger, and Breer, 1989), and *A. pernyi* (Raming, Krieger, and Breer, 1990; Krieger, Raming, and Breer, 1991). A member of each GOBP class has been cloned and sequenced from *M. sexta* (Vogt, Rybczynski, and Lerner, 1991), and a GOBP2 has been cloned and sequenced from *A. pernyi* (Breer, Krieger, and Raming, 1990; identified as such in Vogt, Rybczynski, and Lerner, 1991). In general, the OBPs are 16 kDa proteins deriving from mRNAs in the range of 0.8–1.6 kb. The proteins possess conserved amino acid motifs that substantiate their homology, a series of conserved cysteines that may have structural significance, and typical and cleaved signal peptides utilized for protein secretion (see Vogt, Rybczynski, and Lerner, 1991).

In the initial study identifying the three OBP classes, N-terminal amino acid sequences were compared among 13 OBPs from six species representing four phyletic families (Vogt, Prestwich, and Lerner, 1991). A striking difference was observed among these species with respect to the degree of within-class sequence variation: PBP sequences were highly variable, whereas the respective GOBPs were highly conserved. The variation in PBP sequences was consistent, with each species and each PBP interacting with a different ligand or sex pheromone; selective evolutionary pressures have either forced or allowed the PBPs to drift toward species unique sequences. Interestingly, the most similar PBP sequences were those of *M. sexta* (Sphingidae) and *B. mori* (Bombycidae); these are distantly related species but share bombykal as a pheromone component. The conservation in the GOBP sequences suggested that the relevant ligands, or general odorants, of each species were the same. In other words, the same kinds of general odorant molecules are of equal importance to all of the species examined, regardless of differences in olfactory habitats; selective evolutionary pressures have forced the GOBPs to remain conserved with behavioral features common to each of these species. This was unexpected because some of the species showed different degrees of specificity with respect to their habitats: *B. mori* eat and oviposit on mulberry, whereas *M. sexta* are equally restricted to tobacco and *A. polyphemus* inhabit the mixed environment of deciduous forests. One additional interpretation of these differences is that the PBPs are tested by few and rather specific selective pressures, whereas the GOBPs are tested by relatively general selective pressures.

The variations just discussed pertain to each of the three OBP classes. Sequence comparisons within the classes show that the classes are quite dissimilar while retaining certain conserved sequence motifs: Any PBP versus any GOBP possess about 30 percent amino acid sequence identity and any GOBP1 versus any GOBP2 possess about 50 percent amino acid sequence identity. On an empirical basis alone, such differences indicate very different ligand binding properties for each class of OBP. Thus the existence of two classes of GOBP implies that natural selection has somehow divided general odorants into two fundamentally different chemical categories.

The OBPs represent molecular markers for the two classes of olfactory sensilla. We do not yet know whether GOBP1 and GOBP2 are coexpressed in the same sensilla, or whether these OBPs mark further subtypes of sensilla. We also do not know how many more classes of OBP might be present. We do know, however, that specific OBPs are expressed in association with specific neuron classes by virtue of the intimate relationship between sister groups of support cells and neurons. Furthermore, as is discussed in a later section, expression of the OBPs is regulated during development by ecdysteroids. The OBPs thus represent an outstanding and accessible model for identifying molecular mechanisms regulating spatial and temporal organization of the olfactory system during development.

RP11, an abundant protein of the olfactory dendrite membrane

The basic process in odor recognition is odor-receptor interaction: The odorant molecule binds to a receptor protein located in the membrane of the olfactory dendrite cilia. To attempt visualization of a pheromone receptor protein, we utilized a tritium labeled photoaffinity analogue of the acetate ester component of the *A. polyphemus* sex pheromone that was synthesized by Glenn Prestwich (Vogt, Prestwich, and Riddiford, 1988). This probe revealed a 69 kDa protein that appeared to be uniquely represented in the membranes of olfactory dendrite cilia of male *A. polyphemus* antennae. "Real" pheromone competed off this probe, and this competition, together with the apparent tissue specificity, suggested that the 69 kDa protein was a pheromone receptor.

In as yet unpublished studies, we (Vogt and Lerner) isolated and partially sequenced a 69 kDa protein from dendrite membranes of isolated pheromone specific sensilla of 800 male *A. polyphemus* antennae. This was by far the most abundant protein in the preparation and certainly the only one present in sufficient amount to obtain N-terminal amino acid sequence data. We subsequently used this sequence to design oligonucleotide primers for the amplification of N-terminal encoding DNA from antennal cDNA by polymerase chain reaction (PCR). The resulting PCR product was sequenced to confirm its relation to the protein, then used to screen both a Northern blot and a random primed cDNA library made from antennal mRNA. The Northern blot revealed a 6 kb antennal specific message and the library screen yielded about 20 positive clones from 250,000 plaques. One of these, referred to as RP11, contained a 2.7 kb insert, including the entire and expressible coding region.

The RP11 open reading frame is 1,575 bases, encoding a protein 525 amino acids long with a predicted molecular weight of 59.9 kDa. This mass is consistent with that protein expressed by in vitro translation of RNA transcribed from the RP11 insert. The difference between this molecular weight and the original 69 kDa may be due to posttranslational glycosylation in vivo, a common modification for membrane proteins. Data base searches for sequence homology, including comparisons with known hormone receptor or neurotransmitter receptor sequences, were completely negative, leaving the function of RP11 an unanswered question.

Olfactory transduction mechanisms, a divergence

Our interest in cloning RP11 was driven by our interest in identifying a pheromone receptor protein. An understanding of RP11 function must therefore include what is known of olfactory receptor/transduction mechanisms. Thus it is helpful to digress into a discussion of the current view of these mechanisms (Figure 13.5).

The identification of receptor-transductory proteins has been a central focus of chemosensory studies. A general view has emerged from studies of olfactory mechanisms in a variety of vertebrates as well as insects and lobsters. The mechanisms in these seemingly disparate groups are surprisingly similar. Two major transduction pathways operate in olfaction,

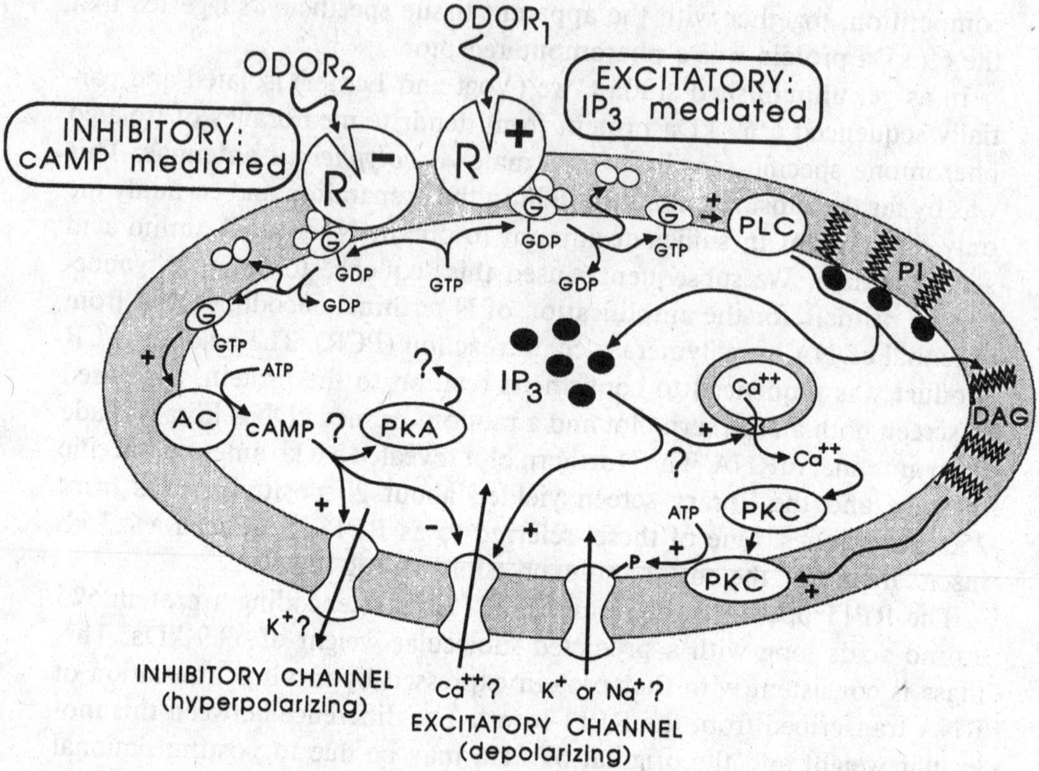

Figure 13.5. Biochemical scheme of odor signal transduction in arthropods. Parallel excitatory (IP_3) and inhibitory (cAMP) pathways coexist in general odorant sensitive neurons, whereas only an excitatory (IP_3) pathway has been identified in pheromone-specific neurons. Whereas in many cell types IP_3 is conventionally thought to liberate Ca^{++} from internal vesicular stores with liberated Ca^{++} partially activating protein kinase C (PKC), in olfactory cilia IP_3 is thought to act directly on inward oriented (depolarizing) cation channels situated in the outer membrane of the olfactory cilia. Thus Ca^{++} may be responsible for excitatory (depolarizing) currents in these neurons. Cyclic AMP mediation of the inhibitory (hyperpolarizing) pathway has been identified in lobster where cAMP is thought to act directly on outward oriented (hyperpolarizing) cation channels (K^+?). R, receptor; G, G-protein; GDP, GTP, cAMP, ATP, nucleotides; PLC, phospholipase C; PI, phosphatidylinositol; IP_3, inositol trisphosphate; DAG diacylglycerol; AC, adenylyl cyclase; PKA, protein kinase A; K^+, Ca^{++}, Na^+, potassium, calcium, and sodium cations.

one utilizing cAMP as a second messenger and the other utilizing inositol trisphosphate (IP_3) as a second messenger. Both pathways are thought to be driven by receptor proteins belonging to the class known as G-protein coupled receptors represented by homologues of rhodopsin and the muscarinic acetylcholine receptor (Buck and Axel, 1991). These findings developed from an identification of unusually high levels of adenylate cyclase activity in vertebrate olfactory tissue (Kurihara and Koyama, 1972), which led to experiments where odorants were shown to stimulate either cAMP production or IP_3 production in isolated olfactory cilia of vertebrates (Huque and Bruch, 1986; Sklar, Anholt, and Snyder, 1986; Boekhoff, Tareilus, Strotmann, and Breer, 1990; Breer, Boekhoff, and Tareilus, 1990; Breer and Boekhoff, 1991). In the cockroach, accumulation of IP_3 but not cAMP was observed following sex pheromone stimulation of antennal tissue (Boekhoff, Raming, and Breer, 1990; Breer, Boekhoff, and Tareilus, 1990). In the lobster, accumulation of both IP_3 and cAMP in response to odor stimulation has been observed, occurring in the same neuron but in response to different odorants. Production of IP_3 led to neuronal excitation (Fadool and Ache, 1992), whereas production of cAMP led to inhibition (Michel, Fadool, and Ache, 1992). Thus, in the lobster, olfactory transduction involves parallel transductory pathways, one excitatory (IP_3) and one inhibitory (cAMP).

It is likely that these same parallel pathways also exist in insects. The cockroach studies mentioned earlier involved only sex pheromone activation, but, as noted earlier, insects also detect general odorants and do this via a separate class of sensilla/olfactory neurons. Neuronal responses to general odorants are very similar in both lobsters and insects; in both the lobster and the silk moth, general odorant neurons are either excited or inhibited, depending on the odorant, and an odorant that excites one neuron may inhibit another (Boeckh et al., 1965; Fadool and Ache, 1992; Michel et al., 1992). Studies to date therefore suggest that sex pheromone transduction may involve the IP_3 pathway alone, whereas general odorant transduction may involve the opposing and parallel IP_3 and cAMP pathways.

The molecular targets of cAMP and IP_3 are thought to be ion channels. An ion channel directly activated by cAMP was demonstrated in toad olfactory cilia (Nakamura and Gold, 1987), and a channel with similar properties was subsequently cloned from rat olfactory epithelium (Ludwig, Margalit, Eismann, Lancet, and Kaupp, 1990). Recent views suggest that IP_3 similarly has a direct action on plasma membrane Ca^{++} channels in both vertebrates and invertebrates (Reed, 1992).

The observation that odorants stimulate production of cAMP and IP_3 is suggestive of the nature of the odorant receptor protein. Cyclic AMP and IP_3 are produced by receptor mediated activation of adenylate cyclase and phospholipase C, respectively, and the coupling between receptor and either enzyme is via G-proteins. G-proteins, tetrameric proteins with an alpha subunit that binds GDP (hence the term G-protein), are common intermediaries between receptors and their effects (see Lamb and Pugh, 1992), and an olfactory specific G-protein was identified and cloned from rat olfactory epithelia (G_{olf}) (Jones and Reed, 1989). A large number of homologous receptor proteins have been identified that couple with G-proteins. Buck and Axel (1991) used conserved sequence information from these G-protein coupled receptors to identify a large family of homologous olfactory specific cDNA clones encoding proteins with three properties expected of olfactory receptor proteins: (1) They are members of the G-protein receptor family; (2) they are uniquely expressed in olfactory tissue; and (3) they are represented by a diverse and large class of more than 100 different genes.

The prevailing working model for odor transduction (see Figure 13.5) involves odors binding to and activating receptor proteins situated in the olfactory cilia membrane. Receptor activation leads to second messenger production (IP_3 or cAMP) through G-protein coupling. The second messenger acts directly on an ion channel situated in the olfactory cilia membrane. This model currently applies to humans as well as insects, though the existence of parallel transduction pathways coexisting in the same neuron has so far only been implicated for arthropods.

Possible functions of moth RP11

RP11 shares no discernible homology with the proteins Buck and Axel (1991) cloned nor with any other member of this G-protein coupled receptor family. However, a data base search has suggested that RP11 does share homology, if weakly, with a human protein, CD36, which is expressed in endothelial cell membranes and functions during malaria response by binding *Plasmodium*-infected red blood cells (Oquendo, Hundt, Lawler, and Seed, 1989). In a broad sense, CD36 appears to function as a cell–cell recognition molecule of sorts. Our work to date suggests two possible functions for RP11. First, it may represent a pheromone receptor protein of unanticipated form, as suggested by our initial ligand binding studies (Vogt et al., 1988). Alternatively, RP11 may represent a previously undescribed cell–cell recognition protein, suggested by the function of its apparent homologue human CD36, possibly involved in some developmental process. Regardless of the uncertainty about RP11 function, the protein is likely to be important by virtue of its uniqueness and its abundance in neuronal membranes. Furthermore, in the context of our interest in cellular identity, RP11 represents a neuronal gene product, whereas the OBPs and AOX represent support cell gene products. This difference allows us to consider RP11 as a useful tool in concert with the OBPs and AOX to explore differences in programming between the closely related olfactory neurons and their support cells. To this end we are directing efforts at obtaining an *M. sexta* homologue of the *A. polyphemus* RP11.

Ecdysteroid regulation of OBP expression: a model for the temporal regulation of olfactory development

Development of the *M. sexta* antennal system is both prolonged and coordinated. Final maturation is marked by expression of the OBPs and AOX, which coincides with the onset of odor dependent electrophysiological activity of the olfactory neurons (Schweitzer et al., 1976; Gyorgyi et al., 1988; Rybczynski et al., 1989; Vogt, Rybczynski, and Lerner, 1990; Vogt, Rybczynski, Cruz, and Lerner, 1993). In particular, the OBPs represent accessible models for identifying regulatory mechanisms controlling temporal maturation in the olfactory system, as well as temporal regulation of development in the metamorphosing nervous system in general. Considerable information is available on *M. sexta* concerning neural development and its hormonal regulation at the cellular level during adult metamorphosis (Truman, 1988; Levine and Weeks, 1990; see also Truman, this volume) and development of the olfactory system during this same period (Hildebrand, 1985; Schneiderman et al., 1986; Oland and Tolbert, 1987; Oland et al., 1988; Oland, Orr, and Tolbert, 1990; Tolbert and Sirianni, 1990) (Figure 13.6; see also Table 13.2). We have begun using the OBPs as models for investigating the role of ecdysteroids in regulating olfactory development.

The *M. sexta* ecdysteroid levels increase during the early phases of adult development and then decrease gradually to adult emergence (Bollenbacher et al., 1981; see Riddiford, this volume). Schwartz and Truman (1983; Truman and Schwartz, 1984) demonstrated that the declining ecdysteroids late in adult development trigger the death of abdominal intersegmental muscles (ISMs), as well as the death of about 50 percent of the interneurons and motoneurons of the abdominal ganglia shortly following adult eclosion (see Truman, this volume). We reasoned that, because the *M. sexta* PBP first appeared about one day before adult eclosion, PBP expression might also be under the control of declining ecdysteroids.

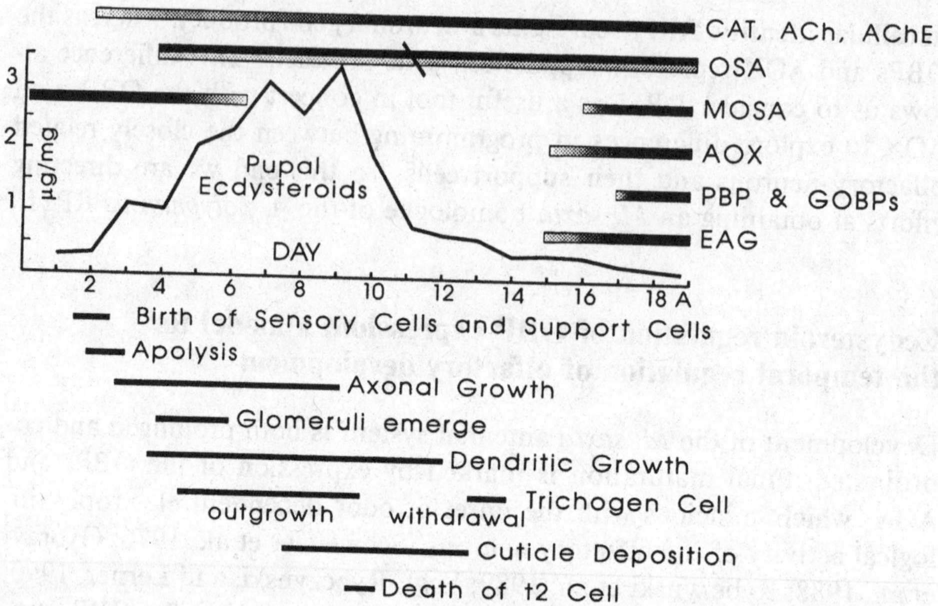

Figure 13.6. Summary of antennal development and the pupal ecdysteroid profile. Ecdysteroid profile is from Bollenbacher et al., 1981. Morphological events of the developing *M. sexta* antenna are reviewed by Hildebrand, 1985. Data for CAT, ACh, and AChE (choline acetyl transferase, acetylcholine, and acetylcholine esterase) are from Sanes and Hildebrand, 1976c; OSA and MOSA (olfactory specific antigen and male olfactory specific antigen) are from Hishinuma et al., 1988; AOX, PBP, and GOBP are from Vogt et al., 1993; EAG (electroantennogram) is from Schweitzer et al., 1976.

Using the same developmental staging criteria employed by Schwartz and Truman (1983), we established that all three OBPs first appear around stage 10, 36–40 hr before adult eclosion. We then cultured antennal tissue from staged animals for 20 hr in the presence or absence of 1.5 μg/ml 20-hydroxyecdysone (20E) followed by 2 hr continued culture in the presence of ^{35}S-methionine, using culture conditions established by Riddiford, Curtis, and Kiguchi (1979). Cultured tissue was thus effectively isolated from its endogenous hormone source, and hormone was replaced under controlled conditions. OBP expression was assayed by visualization on X-ray film following nondenaturing PAGE of homogenates of cultured tissue. OBP expression was strongly and prematurely released in the absence of hormone at least two days earlier than its normal time of expression; premature expression of OBP did not occur in tissue cultured in the presence of 20E. The hormone had no discernible effect on tissue isolated from stage 11 or greater; OBP is normally observed in animals older than stage

Table 13.2. *Summary of neural developmental events influenced by ecdysteroids during adult metamorphosis in* Manduca sexta

A. Events induced by prepupal hormones
- Death of proleg muscles and motoneurons (Weeks and Truman 1985; Weeks, 1987)
- Selective death and initial growth of imaginal nest cells (Booker and Truman, 1987)
- Respecification of sensory neurons in gin-traps (Levine et al., 1986; Levine, 1989)

B. Events induced by rising pupal ecdysteroids
- Growth of motoneuron dendrites (Truman and Reiss, 1988)
- Maturation of respecified motoneurons of proleg and VNL (Weeks and Ernst-Utzschneider, 1989)
- Growth of surviving imaginal nest cells (Booker and Truman, 1987)
- Neurotransmitter expression in imaginal nest cells (Witten & Truman, in Truman, 1988)

C. Events induced by falling pupal ecdysteroids
- Selective muscle death (Schwartz and Truman, 1983)
- Selective neuron death (Truman and Schwartz, 1984)
- Death of VNL (Weeks and Ernst-Utzschneider, 1989)
- Expression of OBPs (Vogt et al., 1993)

Note: Hormone titers and their influence on neuronal development are reviewed by Weeks and Truman (1985), Truman (1988), and Levine and Weeks (1990); see also Riddiford (this volume) and Truman (this volume).

10, and OBP mRNA would have already been expressed in these tissues at the time of removal for culture. These experiments indicate that OBP expression is under the control of the declining ecdysteroid level and suggest that ecdysteroids may be responsible for coordinating many of the events of olfactory development.

As might be expected, there are a great many similarities between events of the developing olfactory system and the developing central nervous system of *M. sexta*. Early in adult development, during the time when olfactory neurons are growing into the brain and establishing synapses and beginning to produce neurotransmitter, rising levels of ecdysteroids regulate growth of imaginal nest cells (Booker and Truman, 1987) and abdominal motoneurons (Truman and Reiss, 1988; Weeks and Ernst-Utzschneider, 1989), as well as neurotransmitter production in imaginal nest cells (Witten and Truman in Truman, 1988). Later, at about 60 percent of development, death of the "second (t_2) olfactory tormogen cell" (Sanes and Hildebrand, 1976b) and change in expression of an olfactory specific antigen (OSA) (Hishinuma, Hockfield, McKay, and Hildebrand,

1988) coincide with completion of dendritic growth for abdominal motoneurons (Truman and Reiss, 1988) and the onset of ecdysteroid decline. Toward the end of adult development, ecdysteroid regulated expression of the OBPs (Vogt et al., 1993) and death in the neuromuscular system (Truman and Schwartz, 1984; Bennett and Truman, 1985) coincide with expression of olfactory AOX (Rybczynski et al., 1989), a male olfactory specific antigen (MOSA) (Hishinuma et al., 1988), and odor dependent electrophysiological activity (Schweitzer et al., 1976). It seems plausible that ecdysteroids represent a major coordinating mechanism in olfactory development.

The mode of action of the ecdysteroids in antennal development is presumably complex in order to account for cellular responses to both rise and fall of steroid. Like other steroids, the active ecdysteroid interacts with a nuclear receptor that binds to specific regulatory DNA sequences (reviewed by Thummel, 1990; see also Riddiford, this volume). In the most studied case, ecdysteroid-regulated larval gene expression in *Drosophila,* the hormone-receptor complex induces the expression of additional regulatory proteins, which alone or together with the ecdysteroid receptor regulate the later expression of structural genes (Ashburner et al., 1974; Thummel, 1990). Recent cloning studies have identified both the *Drosophila* ecdysteroid receptor and a family of homologous transcription factors with conserved steroid binding and DNA binding sites but with as yet no known steroid ligands (Koelle et al., 1991; Segraves, 1991). In *M. sexta,* Fahrbach and Truman (1989) have demonstrated the dynamic appearance of the ecdysteroid receptor in specific neurons and at different developmental stages, using cellular accumulation of radiolabeled ponasterone A, an ecdysteroid homologue, as a receptor assay. More recent studies by Truman (personal communication) indicate that there is a very complex pattern of expression of multiple forms of the ecdysteroid receptors in the metamorphosing nervous system. Thus the cellular responses to ecdysteroids in the developing antenna are probably regulated not just by the changing ecdysteroid levels but by a dynamic interplay of transcription factors as well. Given this complexity, a neuronal model that can offer analysis on the molecular level throughout development is desirable. The OBPs offer an opportunity to identify ecdysteroid-regulated transcription factors controlling final maturation of the olfactory system and possibly other events in the nervous system that are responsive to the late developmental ecdysteroid decline.

The moth antenna: a model system for identifying mechanisms encoding spatial and temporal specificity in the nervous system

Like the rest of the nervous system, the developing olfactory tissue of *M. sexta* is organized as a spatial and temporal mosaic. The antenna is represented by specific cellular identities (see Figures 13.1 and 13.2), and the assembly of this tissue involves a progressive expression of an array of gene products required for morphogenesis, neuronal growth, and functional maturation. The strength of this system as a model for examining developmental regulation lies in the fact that the antenna is composed of (1) large numbers of nearly identical neurons/sensilla that are (2) organized predominantly in two major classes: pheromone specific and general odorant sensitive. The system lends itself to a biochemical approach, that is, the direct isolation and characterization of proteins and and mRNAs, and thus complements systems whose strengths lie primarily in either genetics or histology. Furthermore, the system is attractive because processes can be studied at the molecular level but in the context of a well-characterized behavior, olfactory coordination of flight.

We have described a variety of gene products that are expressed in different cellular patterns in the antenna (see Table 13.1). The OBPs, AOX, and RP11 represent molecular markers of the spatial and temporal mosaic in the antenna. A variety of developmental pattern-based decisions must be made to coordinate the correct expression of these antennal proteins (Figure 13.7). These decisions determine whether expression will be antennal specific, in an epithelial or sensillar cell, in a neuron or support cell, in a pheromone specific or general odorant sensitive sensilla, and in either or both sexes, and when expression will occur during the developmental sequence. The mechanism underlying the mosaic expression of these proteins plausibly involves individual DNA response elements for each criterion, such that if the proper array of transcription factors is present, expression will occur. However, the mosaic itself is apparently established long before these proteins are expressed. For example, *Drosophila* mutations have been described that manifest at the time of imaginal disc eversion and that regulate the formation of specific sensillar cell types (see Jan and Jan, 1990). Presumably, positional information at or before disc eversion determines sensillar class. Thus it is likely that a hierarchical decision is made at disc eversion that determines whether a

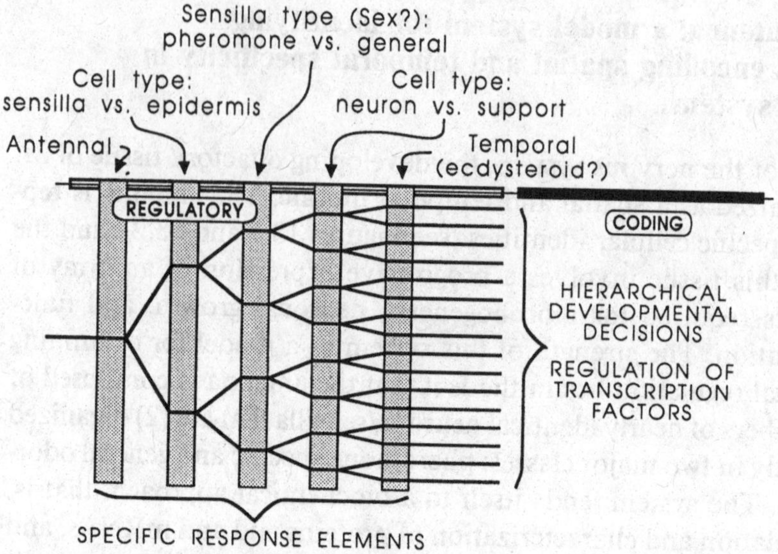

Figure 13.7. A model for spatially/temporally controlled gene expression in the olfactory system. Sequentially made hierarchical decisions determine the pattern of gene expression by regulating the cellular complement of transcription factors, which in turn regulate such features as sensilla class and cell type. Appropriate transcription factors bind to specific DNA sequences or response elements in the regulatory domains of olfactory genes.

sensillum will be pheromone specific or general odorant sensitive, and this decision results in the proper expression of these late developmental proteins more than two weeks later in the case of *M. sexta*. We can exploit the antennal system by using the genes encoding the late proteins to identify transcription factors that link the hierarchical decision with the actual expression of a pattern.

The moth antenna has provided a rich source of behaviorally relevant gene products that can be evaluated in the context of a well-studied behavior. This connection between molecular biology and an animal's behavioral interaction with the environment offers a rare opportunity to evaluate mechanisms underlying the evolution of behavior, how various selective pressures have acted to shape behavior by stabilizing certain molecular forms. These gene products also allow us to examine mechanisms that regulate the pattern-specific expression of olfactory proteins, providing insight into the regulation of olfactory specificity or the ability of an organism to perceive and recognize specific odors. Developmental decisions are made which establish that patterns will emerge. Utilizing structural end points such as the OBPs to identify transcription regulators offers a strategy to follow a transcriptional transduction pathway backward to these developmental decisions. For example, the availability of the nucleic acid sequences encoding the OBPs (Vogt, Rybczynski, and Lerner, 1991) makes the OBPs especially attractive and accessible molecular tools for the identification of genomic sequences responsive to declining ecdysteroid regulation, of novel transcription factors regulating the maturation of the olfactory system and late adult development, and of regulatory mechanisms determining sensilla class and neuronal specificity.

Abbreviations

AOX	aldehyde oxidase
CD36	cloned membrane protein from human endothelial cells
2-DG	2-deoxyglucose
EAG	electroantennogram
GOBP	general odorant binding protein
20E	20-hydroxyecdysone
IP_3	inositol trisphosphate
MGC	macroglomerular complex in insect olfactory lobe
MOSA	male olfactory specific antigen in *M. sexta*
OBP	odorant binding protein
ODE	odorant degrading enzyme
OSA	olfactory specific antigen in *M. sexta*
PBP	pheromone binding protein
RP11	cloned membrane protein from *A. polyphemus* olfactory cilia
SE	sensilla esterase
t_2	second olfactory tormogen cell
VNL	lateral (L) branch of the ventral nerve (VN) in *M. sexta*
VO	vomeronasal organ in certain vertebrates

Acknowledgments

I would like to thank the editors, and especially M. R. G., for their efforts and patience. Support to R. G. V. was from NIH DC00588.

14

Molecular biology of the immune response

AMY B. MULNIX and PETER E. DUNN

Introduction

The cell populations that compose all multicellular organisms share the need to sense their environment and survey features of their cellular neighbors. These surveillance functions serve to ensure appropriate cell sorting and cellular association during embryonic development and metamorphosis, trigger repair processes that eliminate damaged or dysfunctional cells, and detect the presence of foreign (nonself) cells and elicit appropriate responses to eliminate the intruders. Collectively, the surveillance and effector mechanisms that serve the latter repair and defense functions are referred to as the organism's immune system.

We are most familiar with the antibody-based immune systems of vertebrates. These systems are capable of efficiently detecting "foreignness," either in the form of nonself cells, aberrant self-cells or nonself macromolecules. They also provide for specific memory of previous encounters with nonself (immunization) so that a faster and more aggressive defensive response may be mobilized in the event of a secondary encounter. It is widely believed that among all living organisms, only the small group of animals that synthesize immunoglobulins (vertebrates) are capable of mounting a classic immune response in the presence of nonself challenges.

Recently, evidence has begun to accumulate that demonstrates that a variety of nonvertebrate organisms are capable of detecting foreign cells and foreign macromolecules, of eliciting effective and specific defensive responses to these challenges, and of acquiring "primed" states in which they are capable of surviving challenges that otherwise would be lethal to a "naive" individual. Among these organisms, which we now recognize as possessing functional immune responses, are both plants and invertebrate animals, including insects (Dunn, 1986, 1990, 1991; Boman and Hultmark, 1987; Duwel-Eby and Karp, 1990; Karp and Duwel-Eby, 1991).

The pioneering studies that form the basis of our understanding of insect immunity utilized larvae of the waxmoth *Galleria mellonella* (Briggs, 1958; Stephens, 1959, 1962; Hink and Briggs, 1968). These studies demonstrated that insects challenged with attentuated or killed pathogenic bacteria acquire a protected state that allows them to survive a second, otherwise lethal, dose of living pathogenic bacteria. Subsequent investigations of the antibacterial responses of diapausing pupae of the silk moth, *Hyalophora cecropia* (Boman and Hultmark, 1987), and both larvae and pupae of the tobacco hornworm, *Manduca sexta* (Dunn, 1986, 1990, 1991; Spies, Karlinsey, and Spence, 1986a), have produced a broad base of fundamental information on insect antibacterial responses. In the last several years, this information has stimulated investigations of the antibacterial responses of nonlepidopteran insects, including dipteran (primarily the flesh fly, *Sarcophaga peregrina* [Natori, 1977], the blowfly, *Phormia terranovae* [Keppi, Zachary, Robertson, Hoffmann, and Hoffmann, 1986], and the fruit fly, *Drosophila melanogaster* [Robertson and Postlethwait, 1986; Flyg, Dalhammar, Rasmuson, and Boman, 1987; Kylsten, Samakovlis, and Hultmark, 1990]); coleopteran (the tenebrionid beetles, *Eleodes* [Spies, Karlinsey, and Spence, 1986b] and *Zophobas atratus* [Bulet et al., 1991]); hemipteran (*Rhodnius prolixus* [De Azambuja, Freitas, and Garcia, 1986]); and hymenopteran (the honeybee, *Apis mellifera* [Casteels, Ampe, Jacobs, Vaeck, and Tempest, 1989]) species. Specific graft rejection and adaptive humoral immunity in cockroaches, primarily the American cockroach, *Periplaneta americana* (Duwel-Eby and Karp, 1990; Karp and Duwel-Eby, 1991), have also been investigated. Additionally, the original lepidopteran investigations have produced vigorous model systems for studies elucidating mechanisms of eukaryotic cell surveillance and mechanisms regulating specific gene expression.

It is the goal of this chapter to provide an interpretive, but not exhaustive, review of molecular aspects of the immune response of lepidopterans. We draw heavily on data derived from studies of *M. sexta.* Reference is also made to critical developments in dipteran and cockroach systems to provide a richer, more balanced understanding of the status of molecular insect immunology.

Overview of immune responses in insects

The primary focus of studies of insect immune responses has been the description and characterization of responses elicited by the injection or implantation of foreign materials (abiotic particles, cells, and solutions of molecules derived from cells) into the insect's hemocoel. Among the components of the immune response, wound healing and hemostasis (hemolymph clotting) have received little attention, the contributions of circulating hemocytes (blood cells) have been the focus of largely descriptive studies, and the synthesis of defensive proteins has received intense analysis at the molecular level. Currently, emphasis is on providing greater insight into specific recognition and regulatory mechanisms that mobilize immune defenses and into molecular mechanisms that underlie hemocytic responses. The brief overviews that follow summarize the status of knowledge relating to wound healing/hemostasis, hemocyte-mediated sequestration of foreign particles, and adaptive humoral immunity. The discussion then concentrates on the intensely studied bacteria-elicited synthesis of hemolymph antibacterial proteins.

Wound healing and hemostasis

Wound healing and hemostasis are important components of insect immune responses because the entry of foreign organisms into the hemocoel is generally associated with a wound (bacteria) or with some active penetration process (fungi, parasitoids). Investigations of wound healing and hemostatic processes via light microscopy have described both cellular coagulation and plasma coagulation processes in representatives of various insect groups (Gregoire, 1974; Brehelin, 1979; Barwig and Bohn, 1980; Minnick, Rupp, and Spence, 1986). Studies of hemostasis in Lepidoptera have suggested that hemocytes encountering wound sites extend long filopodia that associate to form extended fibrous networks that entrap other hemocytes present in the hemolymph flowing through the wound (Gregoire, 1974). This process leads ultimately to the formation of a hemocyte-derived plug, which is subsequently stabilized through melanization. Melanin deposition results from oxidation of hemolymph phenols to form reactive cross-linking agents. This oxidation is catalyzed by a plasma phenoloxidase, which is present as a proenzyme prior to clot formation.

Recent studies of hemolymph coagulation at wound sites on the surface of *M. sexta* larvae have demonstrated that the fibrous network (coagulum), which forms in hemolymph flowing from a wound, is not composed of aggregated hemocyte filopodia but, rather, of assembled filaments formed from a normally soluble plasma protein (Geng, 1990). This protein has been named hemofibrin (Geng and Dunn, 1988), and its characterization is in progress. Further studies of *M. sexta* hemocytes collected using a procedure that does not stimulate coagulation have shown that hemocytes play an essential role in initiating the assembly of hemofibrin into its insoluble filamentous form. They have also shown that a soluble macromolecular factor extractable from cuticle stimulates hemocytes to initiate coagulum formation in vitro (Geng, 1990).

Hemocyte-mediated sequestration of particles

Insect hemolymph contains a number of morphologically distinguishable populations of circulating hemocytes (Gupta, 1979). The number and nomenclature of hemocyte morphotypes vary among species. Lepidopterans typically possess five primary morphotypes, referred to, in order of decreasing abundance, as granular cells or granulocytes, plasmatocytes, spherulocytes, prohemocytes, and oenocytoids.

Particles, abiotic or biotic, that enter the body cavity are rapidly sequestered from circulation via a variety of hemocyte-mediated mechanisms (Ratcliffe and Rowley, 1979). Particles smaller than individual hemocytes are phagocytosed by plasmatocytes or granulocytes. When bacteria enter the hemocoel of a lepidopteran in numbers larger than can be sequestered by phagocytosis alone, they are sequestered in large hemocytic aggregates called nodules. Foreign matter larger than individual hemocytes is encapsulated in multilayered cellular capsules composed primarily of plasmatocytes. It has been suggested that both nodule formation and encapsulation processes may be initiated by release of materials from the granules of granulocytes (Ratcliffe and Gagen, 1977). In Lepidoptera, both nodule formation and encapsulation are often accompanied by phenoloxidase-catalyzed melanization of the resulting aggregate (Ratcliffe, Leonard, and Rowley, 1984). Mechanisms whereby sequestered biotic agents are killed have not been characterized. It has been observed in studies of *M. sexta* hemocytes that both pathogenic and nonpathogenic bacteria elicit rapid and effective sequestration processes; however, pathogenic forms may not be killed after entrapment but, rather, multiply within the protected environment of the hemocyte aggregates (Horohov and Dunn, 1983).

Responses mediated by circulating hemocytes represent the first line of defense in the case of bacterial infection, as these mobile cells are the first to encounter the invaders. Hemocyte responses are rapid and effective, sequestering 95 percent or more of the invaders at or near the site of infection within 15–30 min (Horohov and Dunn, 1983). They are also the only immediate defenses of naive insects (see later discussion). It is important to recognize that the circulating hemocyte population may be severely depleted through hemocyte aggregation associated with nodule formation and encapsulation responses when the initial bacterial infection involves large numbers of invaders. Thus insects that have successfully responded to an initial bacterial infection are potentially vulnerable until the first-line hemocytic defenses are restored (Dunn, 1990, 1991).

Adaptive humoral immune responses

Empirical demonstrations of specific allograft rejection, of adaptive humoral immunity to soluble foreign proteins, and of specific immunologic memory after challenge by either grafts or foreign proteins among the Insecta have been confined to the American cockroach (reviewed in Karp and Duwel-Eby, 1991). Although these phenomena parallel those characteristic of the vertebrate immune response, no information is available on the molecular components mediating these cockroach defense responses. It is unclear whether the lack of empirical evidence for these immunologic phenomena in Lepidoptera and other insect groups is due to the absence of an adaptive immune response characterized by specific memory or, rather, that performing appropriate experiments to demonstrate these phenomena is difficult with individuals from rapidly metamorphosing, short-lived species (Dunn, 1990). Clearly, all insects share the need to recognize and to sort cell populations appropriately during organ and tissue formation during embryogenesis and metamorphosis. Thus, at least some specific recognition mechanisms are most certainly conserved across the Insecta.

Bacteria-elicited synthesis of hemolymph antibacterial proteins

Whereas the hemolymph of naive lepidopteran insects is not bactericidal (it is, in fact, an excellent medium for bacterial growth), the hemolymph of bacteria-challenged individuals becomes intensely bactericidal (Stephens, 1962; Horohov and Dunn, 1982; Boman and Hultmark, 1987).

The development of bactericidal activity in the hemolymph after challenge by bacteria requires the synthesis of RNA and protein, and coincides with the appearance or increase in concentration of specific hemolymph proteins and peptides, several of which have antibacterial activity (Faye, Pye, Rasmuson, Boman, and Boman, 1975; Boman and Hultmark, 1987; Dunn, 1991). Also coincident with the appearance of these proteins is the acquisition of a protected or primed (immunized) state in which the individual insect is able to survive challenge doses of bacteria that would otherwise be lethal (Briggs, 1958; Stephens, 1959).

The appearance of potent antibacterial activity occurs 6–8 hr after the onset of infection of a previously naive insect. At this time in the defensive response, the majority of invading bacteria have been sequestered and killed by hemocyte-mediated mechanisms, and the population of circulating hemocytes has been depleted through hemocyte aggregate formation. These observations have suggested that the major biological roles of the hemolymph bactericidal factors are to eliminate bacteria escaping entrapment and to protect the vulnerable insect while hemocytic defenses are being restored (Dunn, 1990, 1991).

The synthesis of bacteria-elicited hemolymph proteins was first described in diapausing pupae of *H. cecropia* (reviewed in Boman and Hultmark, 1987), but has since been demonstrated in *M. sexta* (Spies et al., 1986a; Dunn, 1991), *G. mellonella* (Hoffmann, Hultmark, and Boman, 1981), *Antheraea pernyi* (Qu, Steiner, Engström, Bennich, and Boman, 1982), *Samia cynthia* (Boman and Hultmark, 1987), and *Bombyx mori* (Teshima, Ueki, Nakai, Shiba, and Kikuchi, 1986; Morishima, Suginaka, Ueno, and Hirano, 1990). In addition, bacteria-elicited hemolymph proteins have subsequently been described from several dipterans (*S. peregrina* [Natori, 1977], *P. terranovae* [Keppi et al., 1986], and *D. melanogaster* [Robertson and Postlethwaite, 1986; Flyg et al., 1987; Kylsten et al., 1990]); two coleopterans (*Eleodes* [Spies et al., 1986b] and *Z. atratus* [Bulet et al., 1991]), and the honeybee (*A. mellifera,* [Casteels et al., 1989]). The principal bacteria-elicited hemolymph proteins of Lepidoptera and other insects are listed in Table 14.1.

Antibacterial proteins: protein and gene structure and function

The next sections describe the structures of bacteria-elicited lepidopteran antibacterial proteins, the structures of genes encoding the proteins, the functions of these proteins in the antibacterial response, and the antibacterial proteins isolated from other insect groups.

Lysozyme

Protein structure and genes. The protein and nucleic acid sequences for lysozyme from *H. cecropia* and *M. sexta* have been determined. The secreted lysozymes from these species (Engström, Xanthopoulos, Boman, and Bennich, 1985; Rosenthal and Dahlman, 1991; Dunn, unpublished data) are very similar to each other and share features common to chicken-type lysozymes (see Engström et al., 1985). Furthermore, the overall structure of the lysozyme genes (Mulnix, 1991; Sun, Åsling, and Faye, 1991) has been conserved; the positions of the two introns are identical in the two insect genes and correspond to the positions of the first two introns in vertebrate lysozymes.

The nucleic acids encoding the insect lysozymes differ from each other in several respects (Engström et al., 1985; Mulnix, 1991; Sun, Åsling and Faye, 1991). The mRNA from *M. sexta* is larger ($\sim$950 bases) than that observed in *H. cecropia* ($\sim$820 bases) and contains a relatively long 3′-nontranslated region that apparently is absent in *H. cecropia.* In addition, the first intron of the *H. cecropia* gene contains a repetitive element of unknown function that is not present in the *M. sexta* gene. The significance (if any) of these differences has not been ascertained.

Biological activity and role in the antibacterial response. Lysozyme has been suggested to contribute to the immune response of lepidopteran insects in a number of ways. The enzyme hydrolyzes the peptidoglycan polymer of bacterial cell walls and thus may be involved in bacterial killing, clearance of bacterial debris from the hemolymph, and/or generation of elicitors of the humoral response. The limited bactericidal spectrum of lysozyme led Boman and Hultmark (1987) to suggest that, in *H. cecropia,* lysozyme acts synergistically with the more potent cecropins and attacins to kill bacteria and that the major role of lysozyme is to clear bacterial debris by degrading the peptidoglycan sacculus remaining after bacteriolysis.

Lysozyme may serve an additional and central role in the defense response of *M. sexta* (discussed in detail in Dunn, 1991). The low levels of lysozyme maintained in the hemolymph of naive larvae are thought to act on the cell walls of invading bacteria (or cell wall fragments released

Table 14.1. *Hemolymph antibacterial proteins from insects*

Insect order	Protein	Molecular mass (kDa)	Biological activity	Representative species
Lepidoptera	Lysozyme	14	muramidase	*Galleria mellonella* *Bombyx mori* *Spodoptera littoralis* *Hyalophora cecropia* *Manduca sexta*
	Cecropin	4	bactericidal	*Hyalophora cecropia* *Antheraea pernyi* *Manduca sexta* *Bombyx mori*
	Attacin	22	bacteriostatic (?)	*Hyalophora cecropia* *Manduca sexta*
	Hemolin	48	hemocyte recognition (?)	*Hyalophora cecropia* *Manduca sexta*
	M13	36	hemocyte coagulation (?)	*Manduca sexta*

Diptera	Lysozyme	14	muramidase	*Drosophila melanogaster*
	Cecropin (sarcotoxin I)	4	bactericidal	*Drosophila melanogaster* *Sarcophaga peregrina* *Phormia terranovae*
	Sarcotoxin II (attacinlike)	26	unknown	*Sarcophaga peregrina*
	Sarcotoxin III	7	bactericidal	*Sarcophaga peregrina*
	Defensin	4	bactericidal	*Sarcophaga peregrina* *Phormia terranovae* *Drosophila melanogaster*
	Diptericin	9	bactericidal	*Phormia terranovae*
Coleoptera	Coleoptericin	8	bactericidal	*Zophobas atratus*
	Defensin	4	bactericidal	*Zophobas atratus*
Hymenoptera	Apidaecin	2	bacteriostatic	*Apis mellifera*

during phagocytosis and nodule formation). Peptidoglycan fragments released by this enzymatic action act as elicitors and trigger the fat body to synthesize and secrete additional antibacterial proteins. Thus lysozyme may play a key role in initiating and sustaining the humoral antibacterial response in *M. sexta* larvae.

Whether lysozyme plays a similar role in the defense response of *H. cecropia* pupae has not been determined. The immune response triggered by wounding of this species argues against such a role, as it indicates that a mechanism independent of bacteria exists for triggering the response. However, bacterial components, such as lipopolysaccharide (LPS), may be required to amplify the response. Lysozyme is present in the naive hemocytes of *H. cecropia* pupae and could contribute to the generation of peptidoglycan fragments in the initial phases of the defense response. It would be interesting, therefore, to test peptidoglycan fragments as elicitors in this species.

Cecropins and cecropinlike peptides

Peptide structure and genes. The first insect antibacterial proteins to be isolated and structurally characterized were cecropins A, B, and D (originally designated immune proteins P9) from hemolymph of the diapausing pupae of *H. cecropia* (reviewed in Boman and Hultmark, 1987). The cecropins are a family of homologous basic peptides, 35 to 37 amino acids in length, with most charged and polar residues clustered in the amino terminal one-third of the sequence. The sequence of amino acids in this N-terminal sequence led to the prediction that this region would form an amphipathic helix. Analysis of the circular dichroism spectra of the cecropins has provided data consistent with this hypothesis (Steiner, 1982). The remaining two-thirds of the cecropin structure consists of an extended hydrophobic sequence culminating in an amidated C-terminal glycine. Bactericidal peptides homologous to the cecropins have been characterized from hemolymph of *A. pernyi* (Qu et al., 1982), *M. sexta* (Dickinson, Russell, and Dunn, 1988), and *B. mori* (Teshima et al., 1986; Morishima et al., 1990) and from hemolymph of the flesh fly, *S. peregrina* (Okada and Natori, 1985a). The cecropinlike peptides from *B. mori* (lepidopterans) are unique due to the presence of a residue of hydroxylysine at position 21 in the amino acid sequence (Teshima et al., 1986; Morishima et al. 1990).

cDNAs encoding cecropins and homologous cecropinlike peptides have been characterized from *H. cecropia* (van Hofsten et al., 1985; Lidholm, Gudmundsson, Xanthopoulos, and Boman, 1987) and *M. sexta* (Dickinson et al., 1988) and from the flesh fly, *S. peregrina* (Matsumoto et al., 1986). Analysis of the structures of cDNAs encoding the lepidopteran peptides revealed extended N-terminal preprosequences consisting of putative hydrophobic signal sequences terminating in an Ala-Ala sequence (positions 4 and 5 in proprocecropins A and B and positions 2 and 3 in preprocecropin D). Processing of the procecropins to generate the N-terminus observed in the hemolymph form of the peptide would occur via cleavage after a proline residue in each of the sequences, as occurs for the precursor of the honeybee venom peptide melittin. The structural features directing posttranslational processing of the N-terminus of the cecropins are conserved in the sequence of a cecropin D-like bactericidin characterized from *M. sexta* larvae (Dickinson et al., 1988).

Genomic sequences encoding cecropins B, A, and D cloned and sequenced from *H. cecropia* revealed the presence of a single intron inserted between the sequences encoding amino acids 8 and 9 of the mature cecropin (Xanthopoulos et al., 1988; Gudmundsson, Lidholm, Åsling, Gan, and Boman, 1991). A similar genomic structure with a single intron has been observed for sequences encoding cecropinlike peptides from *M. sexta* (Dickinson and Dunn, unpublished data). Genes encoding cecropinlike peptides have also been characterized from the dipterans *S. peregrina* (Kania and Natori, 1989) and *D. melanogaster* (Kylsten et al., 1990). A clustering of cecropin-encoding genomic sequences has been observed in *M. sexta* (Dickinson and Dunn, unpublished data) and *D. melanogaster* (Kylsten et al., 1990).

Biological activity and role in the antibacterial response. The cecropins and cecropinlike peptides exhibit broad but discrete bactericidal activity directed primarily against gram negative bacteria, but including less potent activity against some gram positive species (Boman and Hultmark, 1987). Notably, bactericidal activity directed against several economically important plant and animal pathogenic bacteria has been described (Jaynes, Xanthopoulos, Destefano-Beltran, and Dodds, 1987). However, the cecropins are not toxic to mammalian cells or to yeast (Steiner, Hultmark, Engström, Bennich, and Boman, 1981; Boman and Hultmark, 1987). Studies of structural requirements for bactericidal activity via direct synthesis of peptide homologues and truncated analogues have revealed a remarkable plasticity that allows both significant modification without loss of activity and the potential to change activity spectrum through protein engineering (Merrifield, Vizioli, and Boman, 1982; Andreu, Merrifield, Steiner, and Boman, 1983, 1985; Jaynes et al., 1988).

Studies of the mechanism of action of cecropins and their homologues have suggested a detergentlike action involving insertion of the peptide into the membrane of the target cell. A consequence of this insertion is a rapid loss of membrane potential and leakage of cellular constituents (Okada and Natori, 1984, 1985b; Nakajima, Qu, and Natori, 1987; Christensen, Fink, Merrifield, and Mauzerall, 1988).

The potent, broad spectrum bactericidal activity of the cecropins and their homologues strongly suggests that they are the primary factors responsible for the bactericidal activity observed in hemolymph after bacterial challenge. Likewise, sustained synthesis of the cecropins for several days after a primary infection has been eliminated is consistent with the hypothesis that they are responsible for the induced protected state observed after an initial infection of a naive insect by bacteria.

Attacins

Protein structure and genes. Attacins, originally called immune protein P5 (Pye and Boman, 1977), were the second bacteria-elicited hemolymph proteins to be characterized extensively from diapausing pupae of *H. cecropia.* They were originally isolated as a family of related 22–24 kDa proteins (Hultmark et al., 1983; Engström, Engström, Tao, Carlsson, and Bennich, 1984), subsequently shown to be generated by posttranslational proteolytic processing of the protein products of two genes that have been characterized (Engström, Engström et al., 1984; Kockum et al., 1984; Sun, Lindström, Lee, and Faye, 1991). These genes encode one basic and one acidic-neutral protein.

Attacinlike proteins and cDNAs encoding them have been isolated from *M. sexta* (Dai, 1988) and from the flesh fly, *S. peregrina* (Ando, Okada, and Natori, 1987; Ando and Natori, 1988).

Biological activity and role in the antibacterial response. Unlike the cecropins, the antibacterial spectrum of the attacins is limited to a small number of gram negative bacteria (Hultmark et al., 1983). Further, the attacins are not bactericidal but, rather, possess bacteristatic activity against actively dividing bacteria (Hultmark et al., 1983) via an interaction with a component of the outer membrane of susceptible gram negative bacteria (Engström, Carlsson et al., 1984).

Hemolin

Protein structure. Hemolin, also known as P4, is a major component of the set of induced antibacterial proteins in both *H. cecropia* pupae (Faye et al., 1975; Faye, 1990; Sun et al., 1990) and *M. sexta* larvae (Ladendorff and Kanost, 1990). Several forms of the protein have been identified in each of these species (Hurlbert, Karlinsey, and Spence, 1985; Andersson and Steiner, 1987); the protein from *M. sexta* is glycosylated with a high mannose carbohydrate structure (Ladendorff and Kanost, 1990). Hemolin is unique among the antibacterial proteins in that it is present at low levels in the naive insect of both of these species. This constitutive expression is elevated in the postlarval stages of *M. sexta* (Ryan, Cole, Kawooya, Wells, and Law, 1988).

Recent reports provide exciting evidence that hemolin belongs to the immunoglobulin (Ig) superfamily (Sun et al., 1990; Ladendorff and Kanost, 1991). Although several other insect proteins have been found to belong to this superfamily (see Sun et al., 1990), hemolin is the first defense protein with similarity to Igs. The protein contains four domains, 90 to 110 amino acids in length, most similar in amino acid sequence to the C2-type Ig domains. Circular dichroism (Andersson and Steiner, 1987) and Chou-Fasman (Sun et al., 1990) analyses confirm that, like Igs, hemolin is folded into a β-sheet structure.

Role in the antibacterial response. Although the function of hemolin remains speculative, several lines of evidence suggest that the protein is involved in recognition of nonself.

1. Sun et al. (1990) found that hemolin binds to the surface of bacteria in conjunction with other proteins. One of these other proteins appears to be structurally related to hemolin, but does not cross-react with antihemolin antibodies.
2. The homologue of hemolin in *Ephestia kuehniella* appears to be involved in the hemocytic response to parasitoid eggs (Berg, Schuchmann-Federsen, and Schmidt, 1988; also see Dunn, 1991).

3. The presence of the protein in the naive insect and its developmental regulation is consistent with this role (see the following discussion).
4. This function is consistent with the cellular recognition function of other members of the Ig superfamily (Williams and Barclay, 1988).

Inasmuch as the defense response of lepidopteran insects has not been shown to have the specificity or memory that is characteristic of the vertebrate Ig-based humoral response, Sun et al. (1990) have suggested that hemolin may play a role in nonclonal forms of recognition termed pattern recognition (Janeway, 1989). It has been suggested that these more primitive defense mechanisms recognize classes of molecules rather than specific antigens. Several of the vertebrate acute phase proteins, including C-reactive protein, lipopolysaccharide binding protein, and mannose binding proteins, have been suggested to be involved in vertebrate versions of nonclonal recognition (Ezekowitz, Sastry, Bailly, and Warner, 1990).

Antibacterial proteins from nonlepidopteran species

The original isolation of cecropins and attacins from lepidopteran species was followed by investigations of hemolymph antibacterial factors produced by representatives of other insect taxa. These investigations resulted in the isolation of cecropin and attacin homologues, as mentioned earlier, but also revealed several additional classes of insect antibacterial proteins. Studies of the flesh fly, *S. peregrina,* revealed the sapecins, a family of 4.1 kDa peptides homologous to the vertebrate defensins (Matsuyama and Natori, 1988a, 1988b). Defensinlike proteins have also been described from the blowfly, *P. terranovae* (Lambert et al., 1989), and from a tenebrionid beetle (Bulet et al., 1991). Studies of *P. terranovae* also led to the description of the diptericins, a group of 9 kDa antibacterial peptides related structurally to the attacins (Keppi et al., 1986; Dimarcq et al., 1988).

Induction and cellular aspects

Soon after the description of bacteria-elicited hemolymph antibacterial proteins, attention was focused on identification of the signals that elicit the synthesis of these proteins in lepidopteran model systems. Much early work on this question utilized larvae of the waxmoth, *G. mellonella,* as a model system and suggested that antibacterial protein synthesis is induced nonspecifically by wounding or by injection of any foreign material (reviewed by Kanost, 1983). With the availability of the amino acid and nucleic acid sequences of bacteria-elicited antibacterial proteins, this interest in regulatory mechanisms has intensified, and the focus has shifted to specific elicitors regulating antibacterial gene expression in various individual tissues.

More recent studies on this aspect of lepidopteran immunity have focused on the role of bacterial cell wall peptidoglycan as an elicitor of antibacterial protein synthesis by *M. sexta,* on bacteria- and wound-induced antibacterial protein synthesis in *H. cecropia,* and on the tissue specificity of these responses. Subsequent studies have investigated antibacterial gene expression during development of lepidopterans and bacteria-induced, wound-induced, and developmental synthesis of the proteins in various dipteran species. These studies have revealed developmental regulation of antibacterial gene expression in both lepidopteran and dipteran models.

Elicitors of antibacterial protein synthesis

Studies of the regulation of antibacterial protein synthesis in *M. sexta* larvae demonstrated that the synthesis of lysozyme (and cecropinlike and attacinlike proteins) was a specific response to the presence of bacteria or bacterial cell wall–derived materials within the hemocoel (Kanost, Dai, and Dunn, 1988). Viable and nonviable bacteria and gram positive and gram negative bacteria elicited statistically indistinguishable elevated levels of lysozyme activity in hemolymph. Injection of filter-sterilized saline and of various nonbacterial particles and cells did not elicit accumulation of the proteins. This was of interest because most cellular and particulate materials tested elicited strong hemocytic responses; this clearly distinguishes mechanisms regulating hemocyte responses from those inducing the subsequent accumulation of hemolymph proteins.

However, several interesting correlates between hemocyte responses and protein synthesis were noted: (1) a threshold dose of greater than 10^3 bacteria was required to elicit antibacterial protein accumulation; (2) a dose-dependent rate and level of protein accumulation were observed when doses of bacteria above threshold were injected; and (3) a maximal accumulation at doses in excess of 10^8 bacteria per insect was exhibited (Dunn, Kanost and Drake 1987). This threshold and dynamic range correlate closely with doses of bacteria required to elicit nodule formation and its concomitant hemocyte depletion in lepidopterans (Geng and Dunn, 1989). Thus the data were consistent with the hypothesis that antibacterial proteins accumulate in direct proportion to the level of hemocyte depletion and the resulting vulnerability of the insect to secondary infection.

Peptidoglycan. Of the various components of the bacterial cell wall that were assayed independently as elicitors of antibacterial protein accumulation in *M. sexta* larvae, non-cross-linked peptidoglycan and peptidoglycan fragments (obtained by depolymerizing *Micrococcus luteus* peptidoglycan by lysozyme digestion) elicited the strongest response (Kanost et al., 1988). It is noteworthy that these studies found a significantly lower response to injection of various lipopolysaccharides or of the lipid A component of lipopolysaccharide. Subsequently, low molecular weight, soluble fragments of peptidoglycan were isolated and shown to mimic all aspects of the regulation of antibacterial protein synthesis previously observed with intact bacteria and bacterial cell walls, including the dose-dependent response phenomenon.

Low molecular weight fragments of peptidoglycan generated by lysozyme digestion have also been shown to be strong elicitors of antibacterial protein synthesis in larvae of the silkworm, *B. mori* (Morishima, Yamada, and Ueno, 1992). In *B. mori,* fragments from *Escherichia coli* and *Bacillus megaterium* peptidoglycan were stronger elicitors than fragments from *M. luteus* peptidoglycan.

Studies in which peptidoglycan on the surface of *B. subtilis,* a gram positive bacterium, was depleted by protoplast preparation demonstrated a significant reduction in the capacity of the protoplasts to elicit antibacterial protein accumulation (Dunn and Dai, 1990). Further, mixtures of intact bacteria and *M. sexta* lysozyme, adjusted to the lysozyme concentration found in naive larvae and incubated in vitro, generate soluble fragments containing reducing sugar termini. Supernatants containing fragments from these digests were sufficient to elicit antibacterial protein accumulation in previously naive larvae (Dunn and Dai, 1990). Finally, hemolymph taken from larvae at short intervals after injection of intact bacteria contain low molecular weight materials that are also capable of eliciting enhanced antibacterial protein accumulation when injected into naive larvae (Dai, 1986; Dunn and Dai, 1990). Together, these observations support the hypothesis that peptidoglycan is both a necessary and sufficient stimulus to elicit antibacterial protein synthesis in *M. sexta.*

Fat body (Dunn et al., 1985) and other tissues responsible for antibacterial protein synthesis (see the next section) also produce antibacterial proteins in response to peptidoglycan fragments, as demonstrated by in vitro experiments. Most recently, it has been shown that peptidoglycan fragments elicit enhanced expression from antibacterial protein encoding genes in MRRL-CH-1, an established *M. sexta* cell line (Liu, 1992).

Other elicitors of antibacterial protein synthesis. It should be noted that the specificity of induction observed with *M. sexta* larvae has not been observed in other systems. Studies of the induction of antibacterial protein accumulation in other lepidopteran models have presented a much less clear picture. Investigations of antibacterial protein synthesis by *H. cecropia, G. mellonella,* and *Spodoptera eridania* have shown a consistently strong response to bacteria (for a review, see Dunn, 1991). In addition, simple wounding and injection of sterile saline and soluble proteins induce varying levels of antibacterial proteins. However, induction following wounding or injection of saline is typically of lower magnitude and of shorter duration than responses induced by bacteria. In addition to bacteria, LPS and phorbol esters have been shown to elevate the level of lysozyme and acidic attacin transcripts in *H. cecropia* (Sun, Åsling, and Faye, 1991; Sun et al., 1991). Studies of antibacterial protein induction in dipteran models consistently report elevated levels of antibacterial proteins following wounding or saline injection (Komano, Mizuno, and Natori, 1980; Keppi et al., 1986; Robertson and Postlethwait, 1986). Indeed, wounding is a sufficient stimulus for a consistently strong induction of antibacterial proteins in *S. peregrina* (Komano et al., 1980). Hence, the type and magnitude of response to various inducing agents differ in a species-specific manner.

Tissue specificity

Studies in which various tissues were cultured in vitro identified the fat body as the major tissue synthesizing and secreting the antibacterial proteins in *H. cecropia* (Faye and Wyatt, 1980; Abu-Hakima and Faye, 1981; Trenczek and Faye, 1988) and *M. sexta* (Dunn et al., 1985). Hemocytes from challenged insects have also been shown to express RNAs encoding several antibacterial proteins (Dickinson et al., 1988; Mulnix, 1991) and to synthesize the antibacterial proteins (Trenczek and Bennich, 1987). Examination of the tissue specificity of antibacterial gene expression in

M. sexta demonstrated that expression of the RNAs encoding the antibacterial proteins is a global response. In addition to fat body and hemocytes, Malpighian tubules, salivary glands, muscle, heart/pericardial cell complex, epidermis, and midgut have been shown to contain varying amounts of transcripts encoding lysozyme (Mulnix, 1991) and cecropinlike peptides (Dickinson et al., 1988) in the peptidoglycan-treated insect.

Although the expression of antibacterial protein genes in the fat body and hemocytes makes physiological sense in light of these tissues' roles in hemolymph protein synthesis and the cellular defense response, respectively, the widespread expression in tissues of *M. sexta* is more puzzling. The fact that lysozyme protein is synthesized by the pericardial cells (Russell and Dunn, 1990) and midgut (Russell and Dunn, 1991) of induced larvae suggests that the RNAs in these tissues are translated. Perhaps the generalized expression of the antibacterial proteins serves to limit the extent of infection at all possible points (alimentary canal, exoskeleton, hemolymph).

Developmental regulation of expression

In addition to being induced by bacterial challenge, several of the *M. sexta* antibacterial proteins are apparently responsive to developmental cues; both lysozyme and hemolin are differentially expressed during the life cycle of a naive insect. Russell and Dunn (1990) demonstrated that lysozyme activity in the pericardial cell/heart complex increases during the later part of the fifth stadium. In addition, during metamorphosis of naive larvae, large quantities of lysozyme are found in an apical vacuole in the differentiating pupal midgut epithelial cells and subsequently released into the lumen of the pupal midgut. The release of an antibacterial protein into the midgut lumen at the onset of the restructuring of the larval midgut has been hypothesized to prevent the gut microflora from entering the hemocoel during metamorphosis. This lysozyme synthesis in the metamorphosing midgut appears to be regulated by some developmental, perhaps endocrinologic, signal (Russell and Dunn, 1991).

Hemolin in *M. sexta* was originally identified as a protein expressed at high levels in naive pupae and adults (Ryan et al., 1988; Ladendorff and Kanost, 1990). The postlarval form (PLP) of the protein differs from the larval form in that PLP behaves as a dimer, lacks the high mannose glycosylation, and has a blocked amino terminus; the larval form behaves as a monomer, is glycosylated, and does not have a blocked amino terminus. The significance of these differences has not been determined, but it is attractive to speculate that they reflect variations in the function of the protein in the different developmental stages. This is especially attractive because, as discussed earlier, hemolin has been shown to belong to the immunoglobulin superfamily and has been suggested to play a role in recognition of self. As a bacteria-elicited protein in larvae, perhaps hemolin may be distinguishing between self and bacteria; as a protein expressed during postlarval development, perhaps it has a role in tissue reconstruction during metamorphosis.

Cecropinlike (Nanbu, Nakajima, Ando, and Natori, 1988; Samakovlis, Kimbrell, Kylsten, Engström, and Hultmark, 1990) and defensinlike (Matsuyama and Natori, 1988b) genes are also regulated differentially during various developmental stages in some dipteran species.

Regulation of antibacterial gene expression

The lepidopteran systems have been extremely useful in identifying the mechanisms responsible for regulating the induction of the antibacterial proteins. Early studies with inhibitors indicated that both RNA and protein synthesis were required. The availability of cDNAs and antibodies targeting the antibacterial proteins have allowed direct demonstration of increases in abundance of the transcripts and proteins. In addition, structural features of the genes, mRNAs, and proteins have provided clues to possible regulatory mechanisms. Subsequent investigations have suggested that regulation of antibacterial protein expression occurs at a variety of levels, including transcription, posttranscription, and co/posttranslation.

Regulation of transcript abundance

Kinetics of changes in transcript abundance. Investigations of the antibacterial response indicate that regulation occurs at the nucleic acid level. The abundance of transcripts encoding cecropins (Dickinson et al., 1988; Gudmundsson et al., 1991), attacins (Lee, Edlund, Ny, Faye, and Boman, 1983; Dai, 1988; Sun et al., 1991), and lysozyme (Mulnix, 1991; Sun, Åsling, and Faye, 1991) increases in response to simulated bacterial infection. Elevation of the transcripts encoding the antibacterial proteins in dipteran species has also been demonstrated (Matsumoto et al., 1986;

Takahashi, Komono, and Natori, 1986; Ando and Natori, 1988; Reichhart et al., 1989; Dimarcq, Zachary, Hoffmann, Hoffmann, and Reichhart, 1990; Kylsten et al., 1990; Samakovlis et al., 1990; Wicker et al., 1990).

Although the kinetics of transcript accumulation differs slightly within and between species, in all cases investigated so far, the response is rapid and sustained. Transcripts generally appear within several hours after bacterial challenge and remain elevated for several days. This pattern is parallel to that observed with proteins and suggests that the primary mechanism of regulation may be at pretranslational levels.

Regulatory sequences. The accumulation of antibacterial protein encoding transcripts could be due to a specific decrease in the rate of mRNA degradation, an increase in the rate of synthesis, or both. Although direct measurements of synthetic and degradation rates have not been performed, analysis of the nucleic acid sequences has provided clues to possible regulatory mechanisms. An AU-rich sequence element in the 3′ nontranslated region that has similarity to elements shown intrinsically to destabilize mRNAs encoding proteins involved in the vertebrate inflammatory response (lymphokines and cytokines) (Malter, 1989) is found in the cecropin B (Xanthopoulos et al., 1988), acidic attacin (Sun et al., 1991), and lysozyme (Sun, Åsling, and Faye, 1991) genes from *H. cecropia* and in the lysozyme gene from *M. sexta* (Mulnix, 1991). The role of these elements in regulating message stability has not yet been directly demonstrated. However, phorbol esters, which moderate message stability in vertebrates (Shaw and Kamen, 1986), are inducers of the accumulation of transcripts encoding antibacterial proteins in *H. cecropia* (Sun, Åsling, and Faye, 1991; Sun et al., 1991).

With the exceptions of *M. sexta* lysozyme and hemolin, none of the transcripts encoding the antibacterial proteins is detected in naive insects, but they accumulate rapidly following bacterial challenge. This induction suggests a second mechanism by which transcripts could accumulate: a specific increase in the rate of transcription. Identification of the putative promoter elements regulating the unique expression of the genes is just beginning. It is reasonable to suggest that the genes share some regulatory elements because they are coordinately regulated. One such element, found in a number of the *H. cecropia* antibacterial genes and suggested to play a regulatory role (Faye, 1990; Sun, Åsling, and Faye, 1991), is a sequence motif similar to the NF-κB binding site.

The NF-κB binding site was originally identified in the vertebrate immunoglobulin κ-light chain enhancer and has since been identified in a variety of other genes involved in the vertebrate defense response (Lenardo and Baltimore, 1989). Although this sequence element has been identified in the lysozyme (Sun, Åsling, and Faye, 1991), cecropin B (Xanthopoulos et al., 1988), and attacin (Sun et al., 1991) genes from *H. cecropia* and in the lysozyme gene from *M. sexta,* it is absent from the *H. cecropia* cecropin A and D genes (Gudmundsson et al., 1991). Interestingly, the activity of the factor has been shown to be induced by LPS and phorbol esters (Sen and Baltimore, 1986), both of which are elicitors of the antibacterial response in *H. cecropia.* Working with the promoter region of the lysozyme gene from *H. cecropia,* Sun, Åsling, and Faye (1991) have demonstrated that a nuclear factor from bacteria-induced pupae binds to the NF-κB-like sequence. This nuclear regulatory protein complex, Cecropia immunoresponsive factor, has been purified by DNA affinity chromatography and characterized (Sun and Faye, 1992a, 1992b). These results indicate that an NF-κB-like element may play a major role in the regulation of the antibacterial genes in *H. cecropia.*

Co/posttranslational regulation

Intracellular sorting. The lepidopteran antibacterial proteins (lysozyme, cecropins, attacins, hemolin) were originally identified in the hemolymph of immunized insects, and, as predicted, all contain appropriate aminoterminal signal sequences required for targeting to the rough endoplasmic reticulum. However, a number of pieces of evidence suggest that several of the antibacterial proteins follow alternative or additional intracellular pathways.

Based on the small size of the cecropins, Lidholm et al. (1987) suggested that the preprocecropins are translocated by a mechanism similar to that described for prepromelittin (Muller and Zimmermann, 1987), which is independent of the signal recognition particle and docking protein. Rather, import into the endoplasmic reticulum appears to depend on the recognition of a specific conformation of the entire prepropeptide by ATP-dependent translocation machinery. Posttranslational (as well as cotranslational) translocation has subsequently been demonstrated (Schlenstedt, Gudmundsson, Boman, and Zimmerman, 1990). The high degree of conservation of the preproleader sequences for members of the cecropin family (Dunn, 1991) may reflect the conformational requirements for the alternative translocation apparatus.

In addition to being secreted, at least one of the antibacterial proteins appears to be stored intracellularly. Lysozyme has been found in granules in the pericardial cells of *M. sexta* (Russell and Dunn, 1990). Although the origin of the granular lysozyme is unclear (synthesis or pinocytosis) (Russell and Dunn, 1990), the presence of the protein in granules suggests that multiple pathways for intracellular trafficking of lysozyme exist in the pericardial cells. The mechanisms controlling these alternative pathways are yet to be elucidated.

Covalent modification. Several posttranslational modifications of the antibacterial proteins have been described, including glycosylation (Ladendorff and Kanost, 1990), hydroxylation (Teshima et al., 1986), and proteolysis. In addition to cleavage of the signal peptide, maturation of several of the antibacterial proteins involves processing of the proprotein to the mature hemolymph form. Two types of proteolytic enzymes have been implicated: dipeptidylaminopeptidase (DPAP) and trypsinlike endopeptidases. Posttranslational processing of the amino termini of the major lepidopteran antibacterial proteins is summarized in Figure 14.1.

Isolation of cDNA clones encoding the cecropins (van Hofsten et al., 1985; Lidholm et al., 1987) indicated that a number of amino acid residues with the sequence X-Pro or X-Ala were present between the putative signal peptide cleavage site and the amino terminus of the peptide found in the hemolymph. This observation suggested that the peptides were synthesized as preproproteins and that processing involved both signal peptidase and a DPAP similar to the one activating melittin (see Kreil, 1990, for a review). The cleavage of Ala-Pro and Glu-Pro from preprocecropin B by a DPAP partially purified from the hemolymph of *H. cecropia* has been demonstrated (Boman et al., 1989). The preprocecropinlike peptide from *M. sexta* contains the amino acid sequence Ser-Ala-Phe-Ala-Met-Ala-Ser-Ala-Ala-Pro (Dickinson et al., 1988) and has been suggested to be processed in a multistep manner by DPAP (Kreil, 1990). Interestingly, members of the attacin/diptericin family from dipteran species (Wicker et al., 1990) may also be processed by DPAP.

The second type of posttranslational proteolysis observed for the defense proteins involves processing by a trypsinlike endopeptidase that

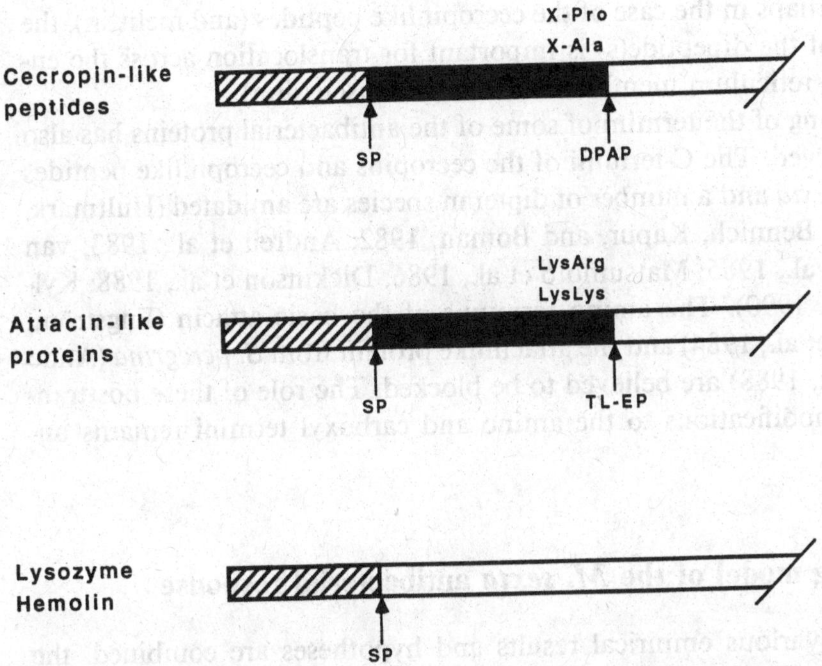

Figure 14.1. Posttranslational processing of the amino termini of the major lepidopteran antibacterial proteins. Cecropinlike and attacinlike proteins are synthesized as preproproteins; lysozyme and other antibacterial proteins are synthesized as preproteins. The signal peptide is cleaved from preproteins by the enzyme signal peptidase (SP) in the rough endoplasmic reticulum. The secreted proproteins are processed to their mature forms by diaminopeptidylpeptidase (DPAP) or trypsinlike endopeptidase (TL-EP).

cleaves the proprotein after a Lys-Arg or Arg-Arg dipeptide. This type of processing is suggested for both the basic and acidic attacins from *H. cecropia* (Kockum et al., 1984; Sun et al., 1991), the defensin from *P. terranovae* (Dimarcq et al., 1990), and sapecin (defensinlike) from *S. peregrina* (Matsuyama and Natori, 1988b).

The significance of the synthesis of the antibacterial proteins as preproproteins is not clear, especially in light of the different strategies employed by different species to process homologous peptides. Boman et al. (1989) have demonstrated an increase in the antibacterial activity with processing of the procecropins. However, unlike melittin, the cecropins (and attacins) are not active against eukaryotic cells and should not need to be synthesized as precursors to protect the synthetic cells from their action. Perhaps in the case of the cecropinlike peptides (and melittin), the presence of the dipeptide(s) is important for translocation across the endoplasmic reticulum membrane (see earlier discussion).

Processing of the termini of some of the antibacterial proteins has also been observed. The C-termini of the cecropins and cecropinlike peptides from *M. sexta* and a number of dipteran species are amidated (Hultmark, Engström, Bennich, Kapur, and Boman, 1982; Andreu et al., 1983; van Hofsten et al., 1985; Matsumoto et al., 1986; Dickinson et al., 1988; Kylsten et al., 1990). The amino terminus of the basic attacin (Engström, Engström et al., 1984) and the attacinlike protein from *S. peregrina* (Ando and Natori, 1988) are believed to be blocked. The role of these posttranslational modifications to the amino and carboxyl termini remains unknown.

A working model of the *M. sexta* antibacterial response

When the various empirical results and hypotheses are combined, the following working model of the antibacterial responses of naive *M. sexta* emerges. Following wounding of the larval integument, bacteria gain access to the insect's hemocoel. Injury to the epidermis triggers activation of hemocytes, which leads to assembly of hemolymph hemofibrin into its filamentous form. The resulting hemofibrin meshwork traps hemocytes in the hemolymph flowing out of the wound, thus forming a hemocytic aggregate that plugs the opening. Within minutes, the cellular plug condenses and is stabilized through melanization catalyzed by a plasma polyphenoloxidase also activated as a result of the wound.

Circulating hemocytes in the vicinity of the wound sequester and kill invading bacteria by a combination of phagocytosis and nodule formation, with a concomitant depletion of the circulating pool of hemocytes. Nodules that form are also stabilized through polyphenoloxidase-catalyzed melanization reactions.

The cell walls of phagocytized bacteria and of bacteria remaining in circulation are partially hydrolyzed by the lysozyme present in naive larvae. This releases soluble fragments of peptidoglycan into circulation. Tissues exposed to the hemolymph bind the soluble peptidoglycan fragments and this induces the synthesis of the full suite of antibacterial proteins at a rate and to a level proportional to the inducing concentration of peptidoglycan. After 6–8 hr, the level of antibacterial proteins in circulation increases dramatically. The resulting bactericidal and bacteristatic activity of the cecropin- and attacinlike proteins, together with enhanced hemocytic defense due to an increased level of hemolin, eliminates the remaining invaders. During the next several days, the preinfection levels of hemocytes are restored, antibacterial protein synthesis ceases, and the circulating titer of antibacterial proteins is reduced by turnover of the proteins.

The early regulatory phase of this antibacterial response is summarized in Figure 14.2.

New horizons for research and innovative applications of existing knowledge

The last 10 years have witnessed a dramatic explosion of research and new information on the biochemistry and cell and molecular biology of insect immune responses. There is every reason to anticipate that this rapid growth in our understanding of these key processes in insect biology will continue. This period of active research has been dominated by attempts to isolate and to characterize structurally the array of antibacterial proteins and peptides elicited in insects in response to bacterial infection, and to determine the structures of nucleic acids encoding these factors. The resulting data, together with early studies of elicitors and responsive tissues, have set the stage for a period of intense investigation of receptors, signal transduction pathways, and molecular regulatory mechanisms responsible for the now well-known induction of antibacterial proteins. Continuing descriptive studies of defensive responses by hemocytes, together with advances in our understanding of cell adhesion and increasingly intense efforts to develop antibodies that recognize components of specific hemocyte morphotypes, have prepared the field for dramatic advances in the cell and molecular biology of hemocytic immune responses.

Just as studies of lepidopteran species were seminal in initiating the current wave of new understanding of insect immunology, these systems are poised for further exploitation. Of particular interest, however, is the accumulating evidence for specific adaptive immunity in cockroaches. It is critical to apply the tools developed in the study of lepidopterans to explore this fascinating observation and to investigate anew the capacity of lepidopteran species for specific recognition and specific immunologic memory.

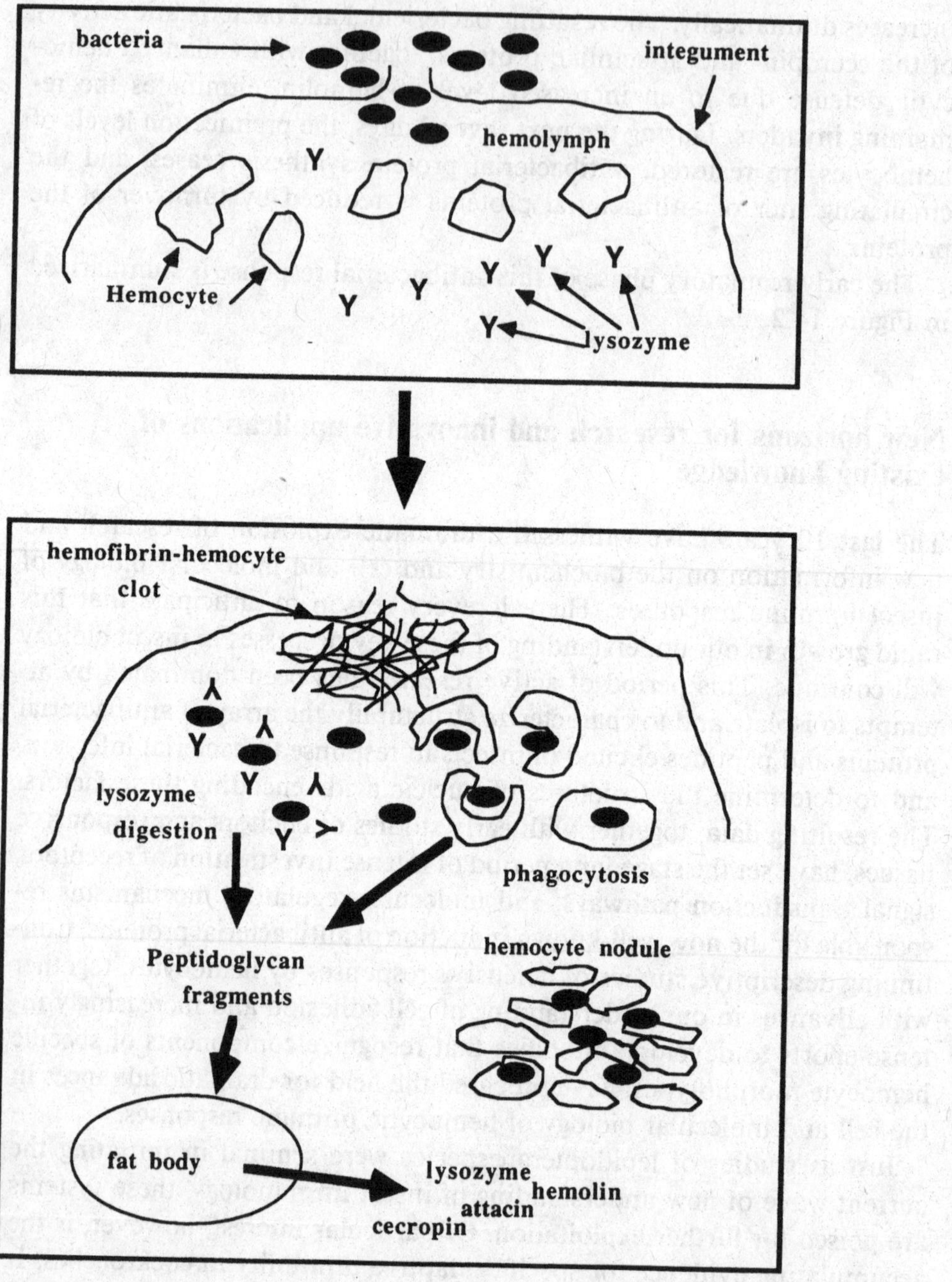

Figure 14.2. Summary of the antibacterial responses of *Manduca sexta*.

In addition to the promising vista of continuing progress, the future may also witness applications of our current knowledge. Of particular interest has been the knowledge of new classes of peptide antimicrobial agents encoded by simple insect genes. Insect antibacterial peptides are already known to include several important plant and animal pathogens within their spectrum of biological activity. Further, initial attempts at protein engineering have demonstrated the capacity to increase the activity spectrum of the cecropinlike peptides. Together, these observations suggest increased potential agricultural uses of the cecropin genes expressed by transgenic plants to provide genetic resistance to plant pathogens, as well as additional potential applications as antimicrobial agents in human and veterinary pharmacology.

Acknowledgments

Research from the authors' laboratory was supported by grants from the National Science Foundation (DCB-8416991), the USDA Competitive Grants Program (59-2182-0-1-429-0, 85-CRCR-1-1773, 88-37263-3502), the National Institutes of Health (1R01-GM41753), and the Indiana Corporation for Science and Technology. The authors thank our colleagues Jeffry Stuart and Jonathan Neal for their critical reading of the manuscript. This is Journal Paper No. 13,763 from the Indiana Agricultural Experiment Station.

15

Engineered baculoviruses: molecular tools for lepidopteran developmental biology and physiology and potential agents for insect pest control

KOSTAS IATROU

Introduction

As previous chapters in this book have demonstrated, considerable progress has been achieved during the last decade in the study of a variety of developmentally regulated processes that play key roles in lepidopteran insect cell differentiation and organismic development. Recent advances in recombinant DNA methodology have permitted the cloning of many new structural and several regulatory genes encoding DNA binding proteins that may control the transcriptional properties of other genes. In addition, the adaptation of functional assays, such as cell transfection and in vitro transcription, to lepidopteran systems has resulted in the acquisition of significant new information regarding the functional properties of cloned genes. Despite this progress, however, our understanding of the regulatory processes that control differential gene function during cell differentiation or the precise role that specific gene products assume in complex physiological functions that determine normal growth and development remains sketchy. This has been due to two main reasons: first, lack of expression systems that could allow introduction of cloned genes into insect cells in vivo and subsequent assessments of the consequences of normal or abnormal gene expression during the lifespan of the organism; and second, difficulty in obtaining pure, physiologically relevant proteins in quantities that can permit the undertaking of extensive biochemical studies.

The importance of in vivo expression systems can be best exemplified by the progress that has been achieved in the fruit fly, *Drosophila melanogaster,* following the development of P element–based embryo transformation methods (Rubin and Spradling, 1982). The information explosion that has occurred as a result has provided such an increased understanding of the molecular makeup of regulatory pathways and the role of specific gene products as determinants of complex physiological functions that *Drosophila* is now used in many cases as a prototype for relevant processes occurring in a variety of other animal species, including mammals.

In contrast to *Drosophila,* only minimal progress has been made with respect to in vivo expression of cloned genes in other insect systems, including lepidopteran ones. Injection of foreign DNA into developing preblastoderm stage lepidopteran embryos, whose early embryogenesis parallels that of the fruit fly embryo, has been achieved recently using the domesticated silkworm *Bombyx mori* as a model system, and transient transgene expression in injected embryos has been reported (Tamura et al., 1990). The major disadvantage of this method, particularly with respect to the investigation of genes expressed during the terminal differentiation of adult tissues, is the unequal distribution of exogenous genes during cell division and organogenesis and their eventual disappearance during development because of lack of autonomously functioning origins of replication.

The development of vectors that could direct chromosomal integration of transgenes remains elusive for the moment, as no transposable elements analogous to the *Drosophila* P element system have been found in Lepidoptera. Furthermore, attempts to employ P element–based vectors in lepidopteran embryo transformation experiments have proven unsuccessful, because this transposition system does not function in the cells of other insect species. The lack of lepidopteran embryo transformation systems has forced developmental biologists working with nondrosophilid insect gene systems into investigations of gene function in transformed *D. melanogaster* (Mitsialis and Kafatos, 1985; Mitsialis et al., 1987, 1989; Bello and Couble, 1990). As described in previous chapters, these studies can provide significant insights into regulatory processes that have been conserved during evolution. They cannot, however, elucidate the macromolecular interactions that control the expression of most specific lepidopteran genes in vivo.

As is detailed in this chapter, baculoviruses,* a class of DNA viruses

*The term *baculoviruses* is used throughout this chapter as an alternative to *nuclear polyhedrosis viruses (NPVs).* It encompasses viruses classified under subgroup A of the Baculoviridae family. Viruses such as granulosis viruses and nonoccluded baculoviruses, which are classified into subgroups B and C of the same family, are not considered in this chapter.

that infect mainly lepidopteran insect species, represent tools that could help overcome the fundamental difficulties faced by molecular entomologists in both transgene expression in vivo and acquisition of insect proteins in quantities that are amenable to structural and functional studies. These pathogens can be used as convenient vectors for introducing chromosomal genes of interest into all cell types of lepidopteran insect hosts at any time during organismic development, and baculovirus-transduced genes can be expressed in the cells of the infected species for a long enough time to allow an evaluation of their functional properties (Iatrou and Meidinger, 1990). Although, at their current stage of development, baculovirus-based expression systems have some of the limitations of the DNA embryo injection technique, such as maintenance of transgenes in an episomal rather than a chromosomally integrated state and capability for transient rather than stable transgene expression, the simplicity of manipulations in their use makes them very attractive. Furthermore, information derived from the study of various baculovirus genes can be exploited to design engineered versions that ultimately may be used as vectors, allowing stable genetic transformation of infected lepidopteran hosts.

Baculoviruses can also be used as expression vectors for large-scale production of nonviral proteins of any origin, including those of insects, in infected lepidopteran cells in vitro and in vivo. Although few proteins of insect origin have been produced to date through the use of baculovirus expression vectors, where cloned gene sequences are available a great potential exists for their increased production in quantities that could allow undertaking of detailed biochemical and pharmacological studies and elucidation of their physiological roles.

In this chapter, I discuss in some detail how baculoviruses can serve as powerful tools to facilitate the study of problems of lepidopteran insect developmental biology and physiology. The progress that has been achieved and future possibilities arising from the utilization of these viruses as expression vectors directing efficient synthesis of proteins of insect origin are reviewed first, together with various aspects of their employment as vehicles for introducing insect genes into the cells of infected hosts for elucidation of transcriptional and posttranscriptional regulatory mechanisms.

A corollary of the application of baculovirus vectors for the study of insect genes and proteins is the utilization of recombinant viruses to achieve the disruption of predetermined insect functions and unbalancing of normal insect physiology. Because baculoviruses can be used as effective, environmentally friendly replacements of chemical pesticides in lepidopteran insect pest management programs, I also review the recent developments in the field of baculovirus-based insect pest control and discuss avenues of investigation that could lead to the generation of recombinant viruses that would effect a speedy incapacitation of infected insect pests in the field.

Recombinant baculoviruses as tools for the molecular entomologist

Baculoviruses are members of a large family of DNA viruses (more than 400 reported species) that infect exclusively invertebrate species. They possess a circular, superhelical, double-stranded genome, ranging in size from 80 to 200 kb, which is packaged in rod-shaped nucleocapsids. The latter are surrounded by a membrane envelope of cellular origin derived from the plasma membrane or the nuclear membrane of infected cells, depending on the stage of infection. The family of baculoviridae has been subdivided into three subgroups, A, B, and C. Subgroup A comprises members known as nuclear polyhedrosis viruses (NPVs), which are enclosed or "occluded" in large numbers into paracrystalline protein matrices, termed occlusion bodies or polyhedra, during the last phase of the infection process. Occlusion takes place within the nucleus and is triggered by the synthesis and intranuclear accumulation of copious amounts of a single viral polypeptide, polyhedrin, during late infection. Polyhedrin constitutes by far the main component of polyhedra, but a few other proteins of viral origin are also required for their formation.

The structural features and infectious properties of NPVs, as well as progress in their molecular characterization and their development as vectors for expression of foreign proteins in insect cells in vitro (tissue culture) or in live insects in vivo, have been extensively reviewed recently in a number of detailed monographs (Friesen and Miller, 1986; Miller, 1988, 1989, 1991; Maeda, 1989a; Summers, 1989, 1990; Webb and Summers, 1991), and the reader is referred to these reviews for background information. Two salient features of nuclear polyhedrosis viruses do, however, deserve further comment because they are relevant to the topics considered in this chapter: host range and potential as foreign gene expression vectors.

NPVs are extremely specific with respect to the hosts that they can infect. With the notable exception of the NPV of the alfalfa looper *Autographa californica* (AcNPV), which has been shown to be capable of infecting productively several lepidopteran families (Vail and Jay, 1973; Witt and Janus, 1976; Kaya, 1977; Capinera and Kanost, 1979; Danyluk and Maruniak, 1987; McIntosh and Ignoffo, 1989; Scheepens and Wysoki, 1989), most NPVs can only replicate in vivo or in vitro into the cells of their normal host and into a few additional insect species usually belonging to the same family (Crawford, 1981; McIntosh and Ignoffo, 1981; Biever and Andrews, 1984; McIntosh, Ignoffo, and Andrews, 1985; Stairs, 1989). Most important, baculoviruses are incapable of infecting productively other arthropod, invertebrate, or vertebrate organisms (Tjia, zu Altenschildesche, and Doerfler, 1983; Volkman and Goldsmith, 1983; Couch and Martin, 1984; Gröner at al., 1984; Brusca, Summers, Couch, and Courtney, 1986; Carbonell and Miller, 1987), and for this reason, they are viewed as environmentally acceptable alternatives to chemical-based methods of lepidopteran insect pest control. Progress and future prospects for this type of baculovirus application are discussed in the section on "Biological control."

Because of their organization into rod-shaped nucleocapsids, baculoviruses can sustain insertions of foreign DNA of considerable lengths into their genomes. This has allowed the prediction that these viruses may be engineered into expression vehicles, carriers of foreign genes to be expressed in infected insect cells. This prediction has been proven to be correct and, as a result, two types of baculovirus vectors have already been developed for foreign gene expression, expressing and transducing vectors, and a third class, transforming vectors, is currently under development. All three types of vectors can be employed as molecular tools by insect molecular and developmental biologists and insect physiologists to enhance the effectiveness of their investigations. The potential applications of recombinant baculoviruses in basic and applied insect science are summarized in Table 15.1.

Expression vectors are capable of directing high level transcriptional expression of foreign gene sequences linked to strong promoter elements derived from "nonessential" viral genes, such as polyhedrin or p10, which are expressed abundantly late in the infection cycle (Miller, 1988, 1989; Vlak et al., 1988; Weyer and Possee, 1988, 1989; Maeda, 1989a; Summers, 1989, 1990; Weyer, Knight, and Possee, 1990). The generation of expres-

Table 15.1. *Baculovirus vectors for molecular entomologists*

Type of vector	Application	Baculovirus systems reported
Expression vectors	Large-scale synthesis of proteins of interest in cultured insect cells or in vivo; synthesis directed by strong viral promoters	many
Transducing vectors	Transient expression of insect genes in vivo; expression driven by authentic or mutagenized insect gene promoters	BmNPV
Transforming vectors	Stable expression of insect genes in vivo through organismic transformation; expression driven by authentic or mutagenized insect gene promoters	not yet reported (under development)

sion vectors is based on the molecular properties of the wild-type viruses. Because all of the genetic information required for the generation of a complete baculovirus is present in the viral genome and the onset of immediate early viral functions depends on *trans*-activation by cellular factors, fully functional baculoviruses can be obtained by transfection of cultured insect cells with intact viral DNA. Functional virus can be harvested from the medium in which these cells are maintained four to five days posttransfection because, at the initial stages of infection, the virus can exit the cell without destroying it in the process. Recombinant baculoviruses expressing foreign gene products under viral promoter control can, accordingly, be obtained by co-transfecting insect cells with a solution containing DNA of the wild-type virus and DNA of a recombinant bacterial plasmid, termed a transfer or transplacement vector. This vector contains a portion of the viral genome, typically the polyhedrin gene and adjacent 5′ and 3′ flanking sequences, with the foreign gene sequences whose expression is sought cloned under the control of the polyhedrin promoter. Progeny virus obtained through the co-transfection consists of a mixture of wild-type and foreign gene-containing recombinant baculoviruses, the latter being generated through allelic exchanges between common regions of the wild-type viral genome and the transfer vector (homologous recombination resulting from double crossovers in the flanking regions of the polyhedrin gene). Recombinant viruses in which the polyhedrin/foreign gene fusion has substituted for the normal polyhedrin gene can be subsequently selected from the progeny virus mixture by a relatively simple plaque purification scheme on the basis of their inability to form occlusion bodies in the infected cells at the end of the infectious cycle.

Stability of foreign mRNA and codon utilization-dependent translatability are factors affecting the yield of recombinant proteins produced in insect cells infected with baculovirus expression vectors: Although levels as high as 500 mg per liter of infected insect cell culture can be achieved, recombinant protein yields as low as 1 to 10 μg per liter of cell culture are not unusual. Most important, however, insect cells have been shown to be capable of covalently modifying the recombinant proteins in a variety of ways, including cleavage of signal peptide sequences required for extracellular secretion, phosphorylation, palmitoylation, myristylation, C-terminus amidation, sulfation, N-linked glycosylation, and O-linked glycosylation (Miller, 1988; Maeda, 1989a; Summers, 1990); and this, in many cases, results in the expression of foreign proteins that are biologically active.

Recently, a second type of baculovirus vector has been reported, termed a transducing vector (Iatrou and Meidinger, 1990). In contrast to expression vectors, which utilize viral promoters linked to a protein coding sequence, transducing vectors merely utilize the virus as means for introducing a complete chromosomal gene of nonviral origin into insect cells, thereby enabling investigation of the expression pattern of the transduced gene. These results suggest that it may become feasible to develop baculovirus-based transforming vectors that could be employed for achieving stable genetic transformation of lepidopteran insects.

The employment of these three types of baculovirus vectors in studies of insect gene systems is discussed in the following sections.

Expression vectors

The primary purpose of the expression vectors is to achieve production of large quantities of biologically active insect proteins for basic research. The transcriptional efficiency of the viral promoters used to direct transcription from insect gene sequences linked to them and the presence in the transcribed mRNAs of codons that are efficiently utilized by the insect cell translational machinery can be predicted to result in high yields of recombinant insect mRNAs, high translational rates, and correspondingly high yields of the relevant proteins. The ability of infected cells to carry out all insect-specific post-translational protein modifications should also contribute to increased yields of recombinant insect proteins due to modification-dependent stabilization of newly made polypeptides. Most important, however, recombinant proteins of insect origin, particularly those whose activity is dependent on specific modifications, are also predicted to be biologically active due to insect-specific posttranslational processing and modification that are carried out by the infected cells.

The number of insect proteins that have been produced to date through the utilization of NPV expression vectors (summarized in Table 15.2) is limited; nevertheless, the results obtained from the expression of these model genes have confirmed the predictions regarding the advantages offered by baculovirus systems over other available expression systems. For example, expression of a cDNA sequence encoding the enzyme juvenile hormone esterase (JHE) of the corn-ear worm, *Heliothis virescens* (Hammock, Bonning, Possee, Hanzlik, and Maeda, 1990), resulted in the accumulation of approximately 75 mg of secreted JHE per liter of medium of infected *Spodoptera frugiperda* cultured cells (the JHE cDNA encodes an N-terminal signal peptide sequence that directs extracellular secretion of the recombinant enzyme). Furthermore, the recombinant enzyme was found to be biologically active both in vivo and in vitro (see discussion in "Biological control").

The eclosion hormone cDNA of the tobacco hornworm, *Manduca sexta* (Horodyski et al., 1989), was similarly expressed in cultures of *S. frugiperda* and yields of biologically active hormone on the order of 20 mg per liter were obtained from the culture medium of the infected cells.

Baculovirus-mediated synthesis of biologically active diuretic hormone of *M. sexta* has also been reported following expression of a synthetic gene in infected silkworm cells. To achieve extracellular secretion and biological activity, a sequence encoding a 17 amino acid signal peptide sequence derived from the cuticle protein CP2 of *D. melanogaster* was ligated 5′ to the 41 codons of the mature *Manduca* hormone, and an additional C-terminal glycine codon was added at the 3′-terminus of the synthetic gene to allow for C-terminal amidation, which was predicted from the presence of an amidated peptide in the diuretic hormone of the cecropia moth *Hyalophora cecropia.* No detailed biochemical characterization or information about yields of the recombinant hormone has been presented in this case, but indirect measurements on silkworm larvae infected with the recombinant virus have suggested a 30 percent reduction in hemolymph volume relative to noninfected control larvae, larvae infected with wild-

Table 15.2. *Insect proteins produced by baculovirus expression vectors*

Protein	Origin	Viral promoter	Reference
Diuretic hormone	*Manduca sexta*	BmNPV polyhedrin	Maeda, 1989b
Chorion proteins	*Bombyx mori*	BmNPV polyhedrin	Iatrou et al., 1989
Juvenile hormone esterase	*Heliothis virescens*	AcNPV/BmNPV polyhedrin	Hammock et al., 1990
Eclosion hormone	*M. sexta*	AcNPV polyhedrin	Eldridge et al., 1991

type BmNPV, or larvae infected with another, hormone-unrelated, recombinant virus.

Finally, two silk moth chorion proteins must also be included in the list of insect proteins that have been expressed using NPV expression vectors. These were expressed in cultured silk moth cells from chromosomal genes rather than cDNA copies. Two polypeptides with properties indistinguishable from those of two authentic chorion proteins were found in infected cells, suggesting that cleavage of signal peptide sequences as well as additional protein modification took place. The main finding was that protein expression can be achieved directly from intron-containing chromosomal genes, thus eliminating the need to use more complex methods of protein identification, such as cDNA cloning and hybridization-mediated selection of specific mRNAs for translation in cell lysates or in *Xenopus* oocytes.

The employment of NPV expression vectors for the production of biologically functional proteins of lepidopteran insect origin should increase as researchers become aware of the advantages offered by these vectors. Recombinant proteins can, for example, be purified in bulk for structural studies or used as antigens for the generation of antibodies, which, in turn, can be used as probes for immunocytochemical localization. Finally, it is expected that, as additional insect proteins are expressed from cloned gene sequences, the use of NPV vectors will expand to studies of protein structure–function relationships based on expression of in vitro mutagenized gene sequences and evaluation of the functional properties of mutant polypeptides.

Transducing vectors

Baculoviruses may also be used as vectors allowing the transfer of insect chromosomal genes into various cell types for transcriptional and posttranscriptional control studies. These vectors have been termed transducing vectors, to indicate that exogenous genetic information is introduced into the cells of the host and that the role of the baculovirus is simply to act as a vehicle for transporting foreign genes into the host cell nucleus. By analogy to the prokaryotic terminology for abortive transduction, the term *transducing* also indicates that the genetic information introduced into the host cell nucleus is maintained in an episomal rather than integrated form throughout the infection cycle.

Transducing vectors represent powerful tools of experimentation for insect molecular and developmental biologists because they allow models of differential insect gene expression to be tested in vivo. They may also be utilized to express foreign proteins under the control of regulatory sequences of host origin for the purpose of generating effective bioinsecticides (see "Biological control").

A model system for transducing vectors has recently been developed using BmNPV and chromosomal genes of *B. mori* (Iatrou and Meidinger, 1990). A 4 kb fragment of *Bombyx* genomic DNA encompassing two chorion genes that are divergently and coordinately transcribed under the control of a shared promoter element (Iatrou and Tsitilou, 1983; Kafatos et al., this volume) was inserted into a transplacement vector containing a polyhedrin gene that had been rendered transcriptionally inactive by deletion mutagenesis (Iatrou and Meidinger, 1989), and recombinant BmNPVs containing the silk moth chorion genes and their flanking sequences were selected because of their inability to produce polyhedra in infected silk moth tissue culture cells. Chorion/BmNPV recombinant virus was injected into the hemocoel of Gr^B mutant silkworm pupae that lacked the injected chorion genes because of a chromosomal deletion (Iatrou et al., 1980; Durnin-Goodman and Iatrou, 1989; Eickbush and Izzo, this volume), resulting in infection of all pupal tissues, including ovaries.

The RNA of infected follicular cells, in which chorion genes are normally expressed, was examined for the presence of transcripts originating from the transduced chorion genes, and it was found that these genes were abundantly transcribed. Transcription initiation occurred at the authentic mRNA start sites and, most important, splicing of intervening sequences of the chorion gene primary transcripts was found to be quantitative and indistinguishable in terms of splice junction selection from that occurring in vivo in wild-type control ovaries. Further quantitative studies revealed that the ratio of accumulation of the transcripts derived from the two transduced chorion genes matched the ratio existing in the follicular cells of wild-type control pupae. Whether translation of the mRNA generated from the transduced genes and synthesis of the corresponding polypeptides had taken place in the infected follicular cells was not investigated in detail, but both were assumed to have occurred; in separate experiments it had been shown that silk moth tissue culture cells infected with two recombinant BmNPVs expressing the same genes under the control of the polyhedrin promoter expressed the corresponding chorion polypeptides (Iatrou et al., 1989).

Whether the transcriptional activation of the transduced chorion genes in infected follicular cells occurred at the same stage of oogenesis as in normal follicular cells was not examined. However, it is known that the expression of all late chorion genes (including those that were BmNPV-transduced) depends on the availability of specific transcription factors that are themselves spatially and temporally regulated (Skeiky and Iatrou, 1991; Drevet, Skeiky, and Iatrou, 1994; see also Kafatos et al., this volume). Therefore, in all probability, faithful transcription of the transduced genes should only be possible at the correct developmental (choriogenic) stage, when the entire complement of factors required for their transcriptional activation is present in follicular cell nuclei.

Similar analyses of the RNA contents of fat body cells of the same infected pupae revealed the presence of aberrantly initiated transcripts originating from the transduced chorion genes (Iatrou and Meidinger, 1990). Correctly initiated RNA was undetectable in fat body cells. Thus, in contrast to the endogenous chromosomal chorion genes, which are normally completely repressed in fat body cells, the transduced genes were leaky, and incorrectly transcribed. This was probably a consequence of their episomal state, which precluded a supranucleosomal configuration in which inactive chromosomal loci are known to exist (Felsenfeld, 1992). Inaccurate transcriptional initiation was also found to occur when chorion genes were transfected as recombinant plasmids in cultured silk moth cells whose endogenous chorion genes are transcriptionally silent (Skeiky and Iatrou, unpublished results).

The studies described here have demonstrated that BmNPV can be employed as a transducing vector for introducing silk moth chromosomal genes of interest into the cells of the host through infection, and that the transduced genes are expressed with correct spatial specificity and, probably, temporal specificity as well. The key feature of the BmNPV transducing vector, as well as that of equivalent ones that can be developed from NPVs that infect other lepidopteran hosts, is that the genes of interest can be introduced into all cell types of the host at any developmental stage. Viral replication inside the infected cells ensures the presence of high copy numbers and easy detection of transcription and translation products. Thus, when contrasted with transient expression systems based on embryo injections with nonreplicating and nonintegrating plasmid vectors, the NPV transducing vector system offers many advantages for the study of gene regulation during terminal cell differentiation. This system also possesses the major technical advantage of its simplicity of infecting the experimental organism by injection into the hemocoel, as opposed to the more demanding DNA injection into early embryos.

When coupled with in vitro mutagenesis, the NPV transducing vector system should allow the identification of *cis*-acting promoter determinants controlling tissue, temporal, and quantitative expression of any insect gene of interest. Analysis of determinants for tissue and temporal specificity can be achieved directly because it is based on qualitative assessments. Quantitative analysis of the effects of specific mutations on the expression of transduced genes may be undertaken more effectively by introducing a reference insect gene into the viral vector together with the wild-type or mutant gene whose properties are to be investigated. The cytoplasmic actin gene of *B. mori* (Mounier and Prudhomme, 1986, 1991), which is constitutively expressed in all cell types of the organism, is an excellent candidate for use as normalization reference for quantitative assessments of expression of mutated genes relative to wild-types.

Transforming vectors

Although, as discussed earlier, baculoviruses can be used as vehicles for transducing manipulated insect chromosomal genes into infected host cells, without further refinement they cannot fulfill two essential requirements: Stable transgene integration into the host genome and transmission into the progeny. Thus research efforts should focus on two aspects: The identification of post-replication viral functions that are important for completion of the infection cycle and associated lysis of infected cells and the identification of elements of insect origin that can be used for enhancing recombination between the DNA of the vector and the chromosomes of the infected cells.

The rationale for the identification of gene products that regulate post- rather than prereplication functions relates to the need to generate sufficient copy numbers of recombinant baculovirus genomes in infected cells.

This should, in turn, facilitate the analysis of transgene expression and increase the probability for integration into the host chromosomes. Blocking of critical viral functions that occur after viral DNA replication should also allow survival of infected cells.

Several late viral genes have been already identified that could prove to be relevant to the control of late viral gene functions. Most prominent among them appears to be a recently cloned gene, *zlz* (Thiem and Miller, 1989). This gene encodes a 30 kDa protein that is expressed both early and late in infection from two distinct promoter elements. Although no specific function has been assigned to this protein, it encompasses two zinc finger motifs and a leucine zipper domain reminiscent of DNA binding and protein dimerization elements, respectively (Johnson and McKnight, 1989), and it may prove to be a transcriptional regulator of early as well as late genes. It is, therefore, conceivable that mutagenesis of the late promoter of the *zlz* gene may abolish important postreplication functions. The function of another late viral gene, *cor* (Wilson, Mainprize, Friesen, and Miller, 1987; Maeda, Kamita, and Kataoka, 1991), which encodes a protaminelike basic polypeptide intimately associated with the viral genome, should also be investigated. It is possible that this polypeptide binds to specific viral sequences and does not simply represent a ubiquitous masking component of the viral genome. Finally, an additional gene encoding a late polypeptide of unknown function, p34.8 (Wu and Miller, 1989), should also be examined as a possible target for inactivation of late functions, because previous studies have suggested that insertional mutagenesis of its coding region causes an apparent instability of the corresponding mutant NPV (Wu and Miller, 1989).

Studies have already been initiated to determine whether any of these genes are necessary for completion of the infection cycle and suitable for designing a transformation system (Lu and Iatrou, unpublished results). A selection scheme based on *trans*-complementation of obligate late viral functions in cell lines stably transformed with some of these genes is employed in an effort to generate viruses deficient in the respective functions. The properties of such defective viruses could then be assessed in nontransformed host cells.

The development of engineered NPVs that can proceed through DNA replication but are unable to complete the infection cycle will be the biggest challenge confronting baculovirologists interested in the development of transforming vectors. Once viruses deficient in late functions become available, it should be relatively easy to incorporate additional elements in their genomes that could facilitate recombination and integration into the host genomes. In the case of BmNPV and *Bombyx,* for example, we anticipate that insertion of a copy of a *Bm1* element (midrepetitive sequence present in more than 2×10^4 copies per haploid genome; Adams et al., 1986; see Eickbush, this volume) into the viral genome could facilitate homologous recombination and chromosomal integration. Although no direct biochemical evidence demonstrates that homologous recombination takes place in insect cell nuclei, the high frequency at which allelic exchanges occur between various transplacement baculovirus vectors and wild-type viral DNA following co-transfection of cultured insect cells (the standard methodology employed for generating recombinant baculoviruses), the presence of multigene families such as chorion (Eickbush and Izzo, this volume) and bombyxin (Iwami, 1990) in lepidopteran species, the ability of such families to expand and contract through unequal crossover, and finally, the demonstrated genetic instability of the *Bm1* elements in silk moth cells (Fotaki and Iatrou, 1988) – all collectively suggest that homologous recombination occurs at high frequency in lepidopteran cell nuclei. Injection of incapacitated viruses containing a copy of a repetitive element as target for homologous recombination into early lepidopteran embryos offers a promising approach for germ cell transformation. Their major advantage over plasmid-based vectors would be their ability to replicate within the injected embryos, providing a potential for a corresponding increase in the probability of homologous recombination in cells of the germ line.

Recombinant baculoviruses for biological control of insect pests

Because of their limited host range and effectiveness in killing their hosts, baculoviruses have long been considered environmentally safe alternatives to chemical agents for control of insect (mainly lepidopteran) pest infestations. Several wild-type baculovirus species have already been licensed for use as insecticides in North and South America, Asia, and Europe. However, their usefulness as control agents has been limited by the fact that, in contrast to chemical pesticides, which usually act on the target insect (and other nontarget) species immediately upon application, baculovirus infection–mediated reduction in the size of the population of the targeted insects (usually feeding larvae) occurs in the field only one to two weeks after application. The demonstration that proteins of insect and noninsect origin can be expressed in cultured lepidopteran cells and in live hosts upon infection with recombinant NPVs has led to investigations on the feasibility of generating recombinant baculoviruses possessing enhanced insecticidal properties. The basic premise for the generation of recombinant baculoviruses for use as improved insecticidal agents has been that abnormal expression of physiologically relevant insect or noninsect proteins may accelerate the rate at which infected insect pests are incapacitated. Already, several reports have appeared in the literature describing the properties of recombinant baculoviruses designed for use as insecticides. These are summarized in Table 15.3.

Two recombinant viruses that allow expression of insect proteins of physiological relevance under the control of the polyhedrin promoter have been tested for their potency in causing accelerated host death. One case was that of a recombinant BmNPV expressing the diuretic hormone of *M. sexta* (Maeda, 1989b). Its effectiveness as a silkworm larval pathogen has been reported to be higher than that of the wild-type. As mentioned earlier, larvae infected with the diuretic hormone recombinant virus by abdominal injection were found to have 30 percent less hemolymph than uninfected controls or larvae similarly infected with either wild-type or other recombinant BmNPVs. Furthermore, larvae injected with the hormone-expressing recombinant BmNPV died one day earlier than those injected with wild-type or other recombinant BmNPVs. It appears, therefore, that in this case, accelerated death of the host insect had been achieved through the unregulated expression of an insect protein that controls an important physiological function.

The other case was that of a JHE-expressing AcNPV (Hammock et al., 1990). A significant reduction in the rate of growth and overall larval size was observed when first instar larvae of the cabbage looper *Trichoplusia ni* were infected per os with this recombinant virus, and the development of infected larvae was arrested at the third instar stage. However, when the same virus was fed to older larvae, no differential effects were observed relative to those caused by feeding with wild-type virus, probably because the levels of JHE accumulated in the hemolymph of these larvae were insufficient to inactivate the high levels of circulating JH.

The search for proteins whose expression may result in increased insecticidal activity of NPVs has been expanded to include proteins of noninsect origin that are toxic to insects. In an early effort to obtain recombinant baculoviruses that express proteins toxic to insect cells, a

Table 15.3. *Recombinant baculoviruses tested for enhanced insecticidal activity*

Host incapacitating products	Source organisms	Gene sequences	Viral promoters	Test hosts	References
Host Proteins					
Diuretic hormone	*Manduca sexta*	synthetic	BmNPV polyhedrin	*Bombyx mori*	Maeda, 1989b
Juvenile hormone esterase	*M. sexta*	cDNA	AcNPV polyhedrin	*Trichoplusia ni*	Hammock et al., 1990
Insectotoxins					
Toxin Be-It (insectotoxin-1)	*Buthus eupeus*	synthetic	AcNPV polyhedrin	*T. ni, Galleria mellonella, Sarcophaga*[a]	Carbonell et al., 1988
δ-endotoxin	*Bacillus thuringiensis*	gene	AcNPV polyhedrin AcNPV p10	*T. ni* *T. ni*	Merryweather et al., 1990
Toxin TxP-I	*Pyemotis tritici*	cDNA	AcNPV polyhedrin AcNPV P_{synXIV}	*T. ni, G. mellonella*[a] *T. ni*	Tomalski and Miller 1991
Toxin AaIT	*Androctonus australis*	synthetic	BmNPV polyhedrin	*B. mori, Sarcophaga falculata,*[a]	Maeda et al., 1991
		synthetic	AcNPV p10	*M. sexta, Heliothis virescens, Spodoptera exigua, S. falculata*[a]	McCutchen et al., 1991
		synthetic	AcNPV p10	*T. ni, Musca domestica*[a]	Stewart et al., 1991

[a]Insects injected with lysates or media of infected tissue culture cells or with hemolymph from infected hosts.

recombinant AcNPV expressing insectotoxin-1 of the scorpion *Buthus eupeus* under the control of the polyhedrin promoter was constructed (Carbonell, Hodge, Tomalski, and Miller, 1988). Although recombinant toxin mRNA sequences and the corresponding polypeptides were detected in infected *S. frugiperda* cells, no neurotoxic activity was observed when *T. ni, Galleria mellonella,* or *Sarcophaga bullata* larvae were injected abdominally with recombinant virus, lysates of infected tissue culture cells, or medium from infected cultures. The failure to detect biological activity has been attributed to the nature of the scorpion gene construct: A synthetic gene fusion containing the sequences of the mature toxin linked to the signal peptide sequences of human β-interferon, rather than the authentic scorpion gene, which encodes a protoxin protein molecule, was cloned adjacent to the polyhedrin promoter, and this may have affected the three-dimensional folding of the mature toxin.

The δ-endotoxin gene of *Bacillus thuringiensis* subsp. *kurstaki* (reviewed by Höfte and Whiteley, 1989) was subsequently expressed in *T. ni* larvae under the control of two late AcNPV promoters, polyhedrin and p10 (Merryweather et al., 1990). The encoded protein is normally produced as a 130 kDa precursor polypeptide that is proteolytically processed by the gut cells of susceptible insect orders (usually Lepidoptera and Diptera; see Höfte and Whiteley, 1989, for more details) to produce the active form of the toxin, which is approximately 62 kDa. The active toxin acts on the target cells through its interaction with specific cell surface receptors and mediates the generation of pores in the cell membranes. The presence of these pores results in a net uptake of ions and water by the cells, unbalancing of the osmotic pressure, and lysis (Knowles and Ellar, 1987; Schwartz, Garneau, Masson, and Brousseau, 1991). Despite the fact that mature toxin was produced by *S. frugiperda* cells infected with the recombinant viruses in vitro, experiments with larvae proved disappointing. *T. ni* larvae exposed to large quantities of nonoccluded recombinant virus (toxin gene under polyhedrin promoter control) through their diet refused to consume the virus-containing diet, presumably because the virus was surface contaminated by the protoxin. This precluded an assessment of the effectiveness of the recombinant virus in terms of speed of action. On the other hand, larvae fed with occluded AcNPV/toxin recombinant virus (toxin gene under p10 promoter control) exhibited nearly identical LD_{50} values as those fed wild-type virus. In this case, however, no 130 kDa precursor or 62 kDa mature toxin molecules were evident in the hemolymph of infected larvae. It is worth noting that the gut cells of infected larvae were not examined for the presence of the toxin or its mRNA. Thus the possibility has not been excluded that the recombinant toxin gene failed to be transcribed in the gut cells, the only cell type that may secrete the protoxin into the midgut lumen where processing into its active form can occur (Höfte and Whiteley, 1989).

Successful employment of recombinant baculoviruses expressing insect-specific toxins was reported recently by four different groups. Complementary DNA clones or synthetic genes, the latter linked to sequences encoding insect signal peptides to allow for posttranslational processing and extracellular secretion, encoding paralytic insectotoxins present in the venom of female mites (Tomalski and Miller, 1991) or the venom of a North African scorpion (Maeda et al., 1991; McCutchen et al., 1991; Stewart et al., 1991) were linked to various late viral gene promoters and introduced into the genome of AcNPV or BmNPV. When live insects were infected with such recombinant viruses, either by abdominal injection or orally, they showed clear toxin-induced symptoms, such as early distress, cessation of feeding, and paralysis.

The most relevant experiments were those that examined the effects of the toxin recombinant viruses on larvae that had been infected per os. These experiments revealed that the recombinant viruses had an advantage over wild-types with respect to speed of initial incapacitation and eventual death of the host. In the case of neonate *T. ni* larvae infected with the mite toxin-expressing virus (Tomalski and Miller, 1991), more than 50 percent of them were paralyzed or died by day 3 postfeeding (compared to 10 percent deaths of control larvae fed with wild-type virus), and that number rose to more than 80 percent by day 4 (for the wild-type virus this figure was not achieved until day 5). Thus, at least in the case of neonate larvae, the toxin-expressing recombinant virus was shown to have a 1.5 to 2-day advantage over its wild-type counterpart with respect to rate of host incapacitation.

Similar results were reported when recombinant viruses expressing the scorpion toxin were tested. When third instar larvae of *H. virescens* were infected per os (McCutchen et al., 1991), the first toxic signs, manifested by feeding cessation, became evident 3.5 days after feeding; the larvae became paralyzed by day 4 and died by day 5. Considering that control larvae similarly infected by wild-type virus died on day 6, host incapacitation (feeding cessation) through infection with the toxin-expressing recombinant virus was achieved two days faster. Testing of a recombinant virus expressing the same toxin on neonate *T. ni* larvae (Stewart et al.,

1991) revealed a one-day advantage over the wild-type virus with respect to host survival time (ST_{50} of 3.6 days versus 4.7 days), and third instar larvae infected with the toxin-expressing virus consumed 40 percent less foliage than did larvae infected with the wild-type virus. Thus the feasibility of improving the effectiveness of baculoviruses as insect pest control agents by incorporating genes into their genomes that can contribute to an accelerated demise of their hosts has been demonstrated.

Future prospects

Following the demonstration that recombinant baculoviruses can be designed that can cause incapacitation of their hosts faster than their wild-type counterparts, the question arises as to whether it is possible to exploit additional properties of baculoviruses and their target insects to design recombinant viruses capable of incapacitating their hosts at an even faster rate without compromising environmental safety. In this context, two major aspects of recombinant viruses should be examined, the type of genes to be expressed in the infected hosts and, perhaps more important, the promoter elements that should be used to drive their expression.

Choice of promoter sequences

An important question that relates to the efficacy of recombinant NPVs as biological control agents is whether the use of late viral promoters, such as polyhedrin, p10, or even the synthetic one that was used recently in the design of one of the toxin-expressing viruses (Tomalski and Miller, 1991; Wang, Ooi, and Miller, 1991), to express toxic or other host-incapacitating proteins is appropriate for achieving fast incapacitation of the infected hosts. At the time of maximal polyhedrin or p10 promoter utilization, infected cells are only a few hours away from their final demise (lysis) and, as noted in the preceding discussion, expression of incapacitating molecules late in the infection cycle may offer only modest advantages, on the order of one to two days, in terms of accelerating cessation of feeding, retardation of growth, and death relative to infection with a wild-type virus. Furthermore, as was elegantly demonstrated by Volkman and her collaborators, in larvae infected per os with low doses of polyhedra, there is a 36 hr lag between the time of appearance of the virus at the primary site of infection, the midgut columnar and regenerative cells, and the generalized spread of the virus to other tissues of the infected host (Keddie, Aponte, and Volkman, 1989). Most important, these studies have also shown that the connective layer of the midgut, which contains, among other things, muscle cells controlling midgut contraction, becomes infected only a few hours later than the midgut cells themselves. In view of the fact that ferocious feeding is the major destructive activity of most lepidopteran pests, these observations suggest that the design of recombinant baculoviruses that can effect an immediate incapacitation of infected insect pests should be based on the utilization of specific "incapacitating" genes and promoter elements that can be expressed in the cells of the midgut and its surrounding connective sheath and upset their normal function.

In this context, it is also worth considering that although inclusions resembling polyhedra have been observed in infected columnar midgut cells, these are infrequent and, when present, they are distinct from bona fide polyhedra by virtue of the fact that they do not contain viral nucleocapsids (Harrap and Robertson, 1968; Tanada and Hess, 1976; Tinsley and Harrap, 1977; Granados and Lawler, 1981). Furthermore, the question of whether the polyhedrin and/or p10 genes are expressed in midgut cells and, if so, to what extent has not been addressed biochemically. Thus it is formally possible that one or both of these two promoter elements are not fully or even partially active in these cells. These considerations prompt the suggestion that the utilization of other promoter elements that can be expressed immediately upon infection in all insect tissues, including midgut cells and their surrounding connective layer, may prove more effective in terms of developing recombinant baculoviruses that cause an accelerated incapacitation of infected hosts.

Two types of promoter elements should be considered as candidates for directing an immediate and ubiquitous expression of host-incapacitating genes incorporated into the genomes of recombinant baculoviruses: early viral promoters and ubiquitous cellular ones.

The study of baculoviruses, primarily AcNPV, has resulted in the identification and structural characterization of approximately 25 of the 150 or so genes which the viral genome is thought to contain (for a recent compilation, see Miller, 1991). The transcriptional properties of most of the cloned AcNPV genes and the function of the polypeptides encoded by some of them have been deduced. At least six of the characterized genes, listed in Table 15.4, have been classified as immediate early, in that they are expressed prior to viral DNA replication, which occurs about 7–10 hr postinfection, and are active in the absence of viral *trans*-activators,

Table 15.4. *Immediate early baculovirus genes*

Gene	Function	References
IE-1	transcriptional activator	Guarino and Summers, 1986, 1987; Chisholm and Henner, 1988
IE-N	transcriptional activator (enhancer)	Carson et al., 1988, 1991
PE-38	transcriptional factor (putative)	Krappa and Knebel-Mörsdorf, 1991
dnapol	DNA polymerase	Tomalski, Wu, and Miller, 1988
pcna	DNA polymerase ancillary factor (putative)	Crawford and Miller, 1988; O'Reilly et al., 1989
"35 kD"	unknown	Friesen and Miller, 1987; Nissen and Friesen, 1989

Note: Classification is on the basis of their ability to be expressed in infected cells, although not necessarily at maximal levels, in the absence of viral *trans*-activators.

although not necessarily at maximal levels. All of them can be transcribed in the absence of ongoing protein synthesis, indicating that their transcriptional activation is mediated by preexisting cellular factors or, at most, by viral factors present in mature, infective nucleocapsids. Although their promoter elements could, in principle, be used for driving an immediate and ubiquitous expression of foreign genes, the probable indispensability of the proteins that they encode precludes their utilization for the construction of simple transplacement vectors analogous to those developed from the nonessential polyhedrin and p10 genes. The only exception appears to be the *pcna* gene product, which is apparently not required for viral maintenance; however, substitution mutant NPVs containing a β-galactosidase gene instead of *pcna* coding sequences were shown to propagate slower than the wild-type virus in infected cells (Crawford and Miller, 1988).

The possibility of constructing transplacement vectors containing early gene promoters in the context of other nonessential genes, such as polyhedrin and p10, or in the context of other regions of the genome that can tolerate insertions of foreign DNA as the means for generating recombinant viruses containing partial early gene duplications and capable of expressing foreign genes at the early stages of infection is, technically, a viable alternative. The principle of promoter duplication has been demonstrated: Recombinant AcNPVs containing partial duplications of the genes for p10 (Weyer et al., 1990), vp39 (encoding the major viral capsid protein) (Thiem and Miller, 1990) and the basic protein *cor* (Hill-Perkins and Possee, 1990) have been generated, and expression from these duplicated promoter elements has been achieved. However, the stability of duplication-containing recombinant viruses over repeated serial passaging in vitro or in vivo has not been assessed, leaving open the possibility for a major drawback of this strategy. Considering that, as a rule, wild-type baculovirus genomes derived from infected hosts are devoid of pronounced sequence duplications, one might expect that duplication-containing recombinants may prove unstable upon serial passaging and lead to an enhanced generation of defective interfering viral particles (for other considerations on the generation of such particles, see Wickham et al., 1991).

The use of promoter elements derived from nonessential viral genes expressed early after infection for constructing simple transplacement vectors that do not create sequence duplications appears to be a promising alternative, provided their functions can truly be dispensed with. An AcNPV gene, *egt,* encoding the enzyme ecdysteroid UDP-glucosyl transferase, was recently identified and characterized (O'Reilly and Miller, 1989, 1990). Although the function of this enzyme, transfer of glucose from UDP-glucose to ecdysteroid molting hormones and concomitant blocking of larval molting and pupation, appears to be important, there is no a priori selective advantage for the maintenance of such a gene in the virus. In fact, the gene is thought to have been acquired from the host genome. Its replacement with an *egt* promoter-driven transgene would be predicted to result in a recombinant virus that is indistinguishable from wild-type in terms of viability, infectivity, and virulence. The replacement of the *egt* coding sequences with unrelated ones has already been reported, along with the expression of foreign gene sequences (O'Reilly and Miller, 1990). Again, it is critical to evaluate the overall behavior of the resultant recombinant AcNPV relative to that of the wild-type virus over prolonged passage in tissue culture cells and in live hosts.

The utilization of promoter elements of cellular origin to direct the expression of foreign genes in infected hosts is a promising alternative to viral promoter elements. The demonstration that cellular promoter elements integrated into NPV genomes can function properly in infected cells and that RNA transcribed under their control is subject to normal tissue-specific processing (Iatrou and Meidinger, 1990) supports the notion that such promoter elements may be employed for the design of fast-acting or "rapid destruction" baculoviruses for biological control. Incorporation into baculovirus genomes of expression cassettes consisting of abundantly, constitutively, and ubiquitously expressed cellular promoter elements under whose control host-incapacitating gene sequences of choice are placed could ensure transgene expression immediately upon infection and, most important, expression at the primary site of infection, the cells of the midgut. Such cellular expression cassettes would have the extra advantage of representing unique sequences within the recombinant baculovirus genomes, thus minimizing recombination-induced instability and generation of defective interfering particles.

The concept of constitutive cellular cassettes as parts of recombinant baculovirus insecticides is currently being tested, using as a model system a *Bombyx* cytoplasmic actin promoter/chloramphenicol acetyl transferase (CAT) gene expression cassette incorporated into the genome of BmNPV (Johnson and Iatrou, unpublished results). Although we are examining the timing of onset of CAT gene expression in infected larvae fed with this recombinant BmNPV relative to control larvae infected with another recombinant virus containing CAT gene sequences under polyhedrin promoter control, our main objective is to see whether a difference can be observed between the two recombinant viruses with respect to CAT gene expression in cell types of the midgut. If these studies show that this model expression cassette does indeed direct earlier transgene expression and/or preferential expression in midgut cells relative to the polyhedrin promoter, the concept of cellular promoter utilization in recombinant baculovirus insecticides could be applied to all known insect/baculovirus combinations and expanded through the utilization of additional regulatory elements with properties similar to those of the cytoplasmic actin promoter, including those of insect ribosomal, tRNA, snRNA, and other housekeeping genes that are expressed constitutively in all cell types.

Choice of host-incapacitating gene products

As evidenced in the previous discussion, the key to the successful development of fast-acting NPVs appears to be related to the type of host-incapacitating protein products that are expressed in infected insects, as well as their site, timing, and levels of expression. Because the destructive properties of lepidopteran pests relate to their feeding ability, the most effective recombinant NPV bioinsecticides should be primarily those that result in the immediate incapacitation of functions of the gut. The insect midgut could either be targeted for destruction through expression of toxic proteins or incapacitated through unregulated expression of molecules that normally exist in the host and control important physiological functions.

Toxin molecules thought to act exclusively on insect cells, such as the crystal protein of *B. thuringiensis* and those present in the venoms of various arthropods, are promising candidates for achieving the desirable action. Both classes of toxins have the extra advantage that, although the insects in which they are expressed are incapacitated relatively quickly, these insects survive for an additional period that allows replication of toxin-expressing virus in nonaffected cell types. This, in turn, contributes to an increase in the number of infectious particles within the host and to enhanced horizontal spread of the infection in an insect population following the death of infected individuals. The paralytic insectotoxins under investigation, as well as new ones that may be found in other arthropod venoms, are considered to be environmentally safe, because existing studies on their functional specificity suggest that they have no effects on mammalian muscle and nervous system (Zlotkin, Rochat, Kopeyan, Miranda, and Lissitzky, 1971; Walther, Zlotkin, and Rathmayer, 1976; Teitelbaum, Lazarovici, and Zlotkin, 1979; Darbon, Zlotkin, Kopeyan, van Rietschoten, and Rochat, 1982; Zlotkin, 1983; Gordon, Jouer, Couraud, and Zlotkin, 1984; de Dianous, Hourau, and Rochat, 1987; Tomalski et al., 1989).

Despite the findings of these studies, the use of insectotoxin-expressing baculoviruses should be carefully scrutinized with respect to their impact on ecosystems. The dispersion methods employed for application of baculovirus insecticides in pest-infested areas, the fate of infected hosts after death (decomposition or ingestion by predators), and the route of delivery of these toxins, which act through uptake by target cells from the circulation rather than affecting exclusively the cells in which they are synthesized, warrant more extensive studies on the effects of the toxins on various animal and plant species and the overall impact of toxin-expressing recombinant viruses on contained microenvironment ecosystems.

A potentially safer alternative to toxin-expressing baculoviruses is viruses directing expression of neuropeptides of host origin. At least 50 insect neuropeptides have been characterized to date that have been shown or postulated to have myotropic, visceral, and/or skeletal muscle stimulatory or inhibitory activity (for recent reviews on insect and other invertebrate neuropeptides, see De Loof and Schoofs, 1990, and Holman, Nachman, and Wright, 1990a, 1990b). Insect myotropic peptides that could be considered for use in the design of baculovirus insecticides are listed in Table 15.5. Their classification in different families is based on sequence similarities at the amino acid level. Unregulated expression of such peptides

Table 15.5. *Insect neuropeptides with demonstrated or suspected myotropic function*

Family	Members	References
Adipokinetic hormone	CCI, CCII	O'Shea et al., 1984; Scarborough et al., 1984
Myotropins	proctolin	Starrat and Brown, 1975
	leucokinins	Holman et al., 1990
	achetakinins	Holman et al., 1990
	drosulfakinins	Nichols et al., 1988
	locustakinin	Schoofs et al., 1992
	locustamyotropins	Schoofs et al., 1990b, 1990e
	leucopyrokinin	Hollman et al., 1986a
	locustapyrokinin	Schoofs et al., 1991a
	leucosulfakinins	Nachman et al., 1986a; Nachman, Holman, Cook et al., 1986
	locustasulfakinins	Schoofs et al., 1990a
FMRFamide	FMRFamide-related peptides	Evans and Myers, 1986; Nambu et al., 1988; Schneider and Taghert, 1990; Meyrand and Marder, 1991
	leucomyosuppressin	Holman et al., 1986b
	schistoFLRFamide	Robb et al., 1989; Lange et al., 1991
Enkephalinlike	met-enkephalin-arg-phe	Duve et al., 1991
Tachykinins	locustatachykinins	Schoofs et al., 1990c, 1990d
Unclassified	locustamyoinhibiting peptide	Schoofs et al., 1991b
	cardioacceleratory peptides	Tublitz et al., 1991

in virus-infected animals could result in severe, persistent perturbations of normal physiological functions, including but not restricted to gut functions, and host incapacitation.

The major obstacle for the generation of myotropic peptide–expressing baculoviruses is the unavailability of relevant cloned gene sequences. These neuropeptides are quite short, from 4 to 14 amino acids, and are products of post-translational processing of longer precursor polypeptides for which, in most cases, no primary structure information exists. Furthermore, many of these neuropeptides contain modified residues, such as amidated carboxyl termini, N-terminal pyroglutamic acid, and internal sulfated tyrosine moieties, and are biologically active only in their modified forms. Most modifications, including cleavages to mature sizes, occur on the larger precursor polypeptides, and it is unlikely that expression of synthetic genes encoding only the final peptide products will prove productive in terms of addition of modifying groups. The possibility that cells expressing synthetic neuropeptide genes may be unable to secrete the peptides, even if the gene constructs are linked to sequences encoding signal peptides, is also a potential problem, because small polypeptides are exported from cells via secretory pathways distinct from those utilized for the secretion of larger proteins. For these reasons, it is important that for the candidate peptides their precursors be identified and the genes cloned.

Cloning of neuropeptide-encoding sequences by traditional hybridization screening of cDNA and genomic libraries or even by reverse transcription-coupled, polymerase chain reaction-mediated amplification of mRNA sequences (Saiki et al., 1988; Rappolee, Wang, Mark, and Werb, 1989) may prove difficult due to the limited length of the mature peptides, and antibody screening of tissue-specific cDNA expression libraries may represent the most effective approach for obtaining neuropeptide precursors. To date only two *Drosophila* neuropeptide precursor cDNAs have been cloned, one encoding several FMRFamide-related peptides (Nambu et al., 1988; Schneider and Taghert, 1990) and one encoding three drosulfakinin peptides (Nichols, Schneuwly, and Dixon, 1988). These could be tested for biological activity in the context of recombinant baculoviruses. Hopefully, additional cloned neuropeptide precursors will become available soon and utilized for similar studies.

Other classes of insect proteins and peptides that control additional physiological functions could also be explored for use as components of baculovirus insecticides. These classes, summarized in Table 15.6, include a number of key peptide hormones and their receptors, neuropeptides that control hormone biosynthesis, and enzymes catalyzing modification-mediated hormone inactivation. In contrast to the neuropeptides discussed earlier, many of the polypeptides listed in Table 15.6 have been cloned, and their gene sequences are available for engineering relevant recombinant viruses. Again, the prediction can be made that persistent disruptions of normal hormonal balance resulting from ectopic or unregulated expression of these types of gene products immediately upon infection of various insect species by the corresponding recombinant viruses should prove catastrophic for the hosts.

As discussed earlier, two polypeptides of this type, diuretic hormone and juvenile hormone esterase, have already been tested for biological effects on insects infected with relevant recombinant baculoviruses, with significant results obtained. These proteins can exert their effects in various insect species – that is, their function is not limited by cross-species bar-

Table 15.6. *Insect growth regulators: candidates for baculovirus-mediated expression*

Polypeptides	Function	References
Peptide Hormones		
Bombyxins	insulinlike growth factors? inducers of ecdysteroid biosynthesis?	Adachi et al., 1989; Kawakami et al., 1989; Iwami et al., 1990
Diuretic hormone	control of water excretion	Kataoka, Troetschler et al., 1989; Maeda, 1989b; Kay et al., 1991
Eclosion hormone	trigger of eclosion behavior	Horodyski et al., 1989; Kono et al., 1990
Prothoracicotropic hormone	inducer of ecdysteroid biosynthesis	Kawakami et al., 1990
Hormone/Neuropeptide Receptors		
Ecdysone receptor	transcriptional regulator	Koelle et al., 1991
Juvenile hormone receptor	transcriptional regulator	Palli et al., 1990, 1991
Tachykinin receptor	tachykinin binding	Li et al., 1991; Monnier et al., 1992
Hormonal Neuromodulators		
Allatostatins	suppressor of juvenile hormone biosynthesis	Woodhead et al., 1989; Pratt et al., 1990, 1991; Kramer et al., 1991
Allatotropin	inducer of juvenile hormone biosynthesis	Kataoka, Toschi et al., 1989
Hormone Modifying Enzymes		
Ecdysteroid UDP-glucosyl transferase	inactivation of ecdysteroid hormones	O'Reilly and Miller, 1989, 1990
Juvenile hormone esterase	inactivation of juvenile hormone	Hanzlik et al., 1989; Philpott and Hammock, 1990
Juvenile hormone epoxide hydrolase	inactivation of juvenile hormone	Casas et al., 1991

riers. Although baculovirus-mediated expression was found to produce only modest results with respect to accelerated host incapacitation, the outcome of these experiments demonstrated that the principle of utilization of such genes as parts of baculovirus insecticides is fundamentally correct. Further progress in this area depends on the ingenuity of researchers in devising improved versions of recombinant viruses expressing more stable incapacitating proteins or combinations of proteins that can target specific insect functions more effectively than individual ones. Coexpression of juvenile hormone esterase and allatostatin genes in baculovirus-infected insects, for example, could result in reducing circulating juvenile hormone titers to levels that would be insufficient for normal development even during late larval life.

Because of the high reproductive rates of insects, it is likely that mutations will eventually arise that confer resistance to particular toxic or incapacitating molecules expressed by recombinant baculoviruses. For this reason, it would be desirable to develop a wide variety of viruses, each expressing only one or two incapacitating genes. Each virus could be used for a limited time before switching to another, thus minimizing the chances for selection of resistance mutations. Obviously, the development of recombinant baculoviruses as biopesticides that fulfill the requirements of effectiveness and environmental safety will require close collaboration among insect physiologists, developmental and molecular biologists, and baculovirologists.

Conclusion

Recombinant baculoviruses hold considerable promise for at least four distinct but interrelated areas of basic and applied research: First, as expression vectors, they can be used to direct the synthesis of large quantities of proteins of various origins; second, as transducing vectors, they can be employed for achieving somatic transformation and analysis of transient transgene expression in vivo; third, they could be developed into vectors that mediate permanent organismic transformation with any chosen genes; and fourth, they have the potential for being effective, environmentally safe bioinsecticides. The feasibility and effectiveness of the first two applications have been demonstrated. It will be interesting to see whether the results to be derived from future work will confirm the predicted usefulness of recombinant baculoviruses in the last two areas.

Acknowledgments

I would like to express my appreciation to Dr. R. R. Johnson for his critical comments during the preparation of this chapter and to Drs. P. Cherbas, L. Cherbas, S. Maeda, L. K. Miller, L. M. Riddiford, M. D. Summers, and J. W. Truman for communicating results of their work prior to publication. The generous contributions of materials from the laboratories of Drs. E. B. Carstens, P. Couble, M. R. Goldsmith, B. D. Hammock, M. Iwami, F. C. Kafatos, J.-C. Prudhomme, P. H. Taghert, and D. Uderfriend are gratefully acknowledged. I am also grateful to all trainees in my lab who contributed to the work on BmNPV that has been cited in this chapter. Special thanks go to my past and present research assistants, D. Schmiel, A. Vipond, and R. Meidinger, for their contributions and to the following summer students who invested long hours of their time to see their experiments come to fruition: M. H. M. Tran, E. M. Molnar, A. V. Vieira, A. S. Ahluwalia, M. J. Mitsch, and M. Slater. Work on BmNPV at the University of Calgary has been supported by operating grants from the Alberta Cancer Board, the Medical Research Council of Canada and Insect Biotech Canada (one of the Federal Networks of Centres of Excellence), and through the award of several summer and graduate studentships and postdoctoral fellowships to members of my lab by the Alberta Heritage Foundation for Medical Research.

16

Epilogue: Lepidopterans as model systems – questions and prospects

ADAM S. WILKINS and MARIAN R. GOLDSMITH

Lepidopterans have a threefold significance for biological science as a whole. First, they show certain features and phenomena that are either unique to this group or virtually so. Such aspects invite investigation simply through their novelty and intrinsic interest. Second, particular lepidopteran groups have features that make them excellent model systems for investigating certain widely shared biological properties or phenomena. Third, the Lepidoptera provide an excellent set of "out groups" for various comparative studies, particularly in evolutionary, molecular, and developmental matters but also for physiological phenomena. In this final chapter we revisit the material presented in the preceding chapters, in terms of the special advantages of lepidopteran research and indicate some possible future lines of investigation that are likely to be of interest.

Unique or special features

As noted, the Lepidoptera exhibit certain features in a more diverse or exaggerated form than is seen in any other animal group. One of these, and a classic area of lepidopteran research, is butterfly wing patterns; their beauty and diversity are apparent to all. Recent years have seen these patterns receiving renewed attention, in terms of general theories of pattern formation. Unfortunately, there is still relatively little molecular biology in this area, and we have not attempted to cover it in this book. For the interested reader, however, Nijhout's recent *Development and Evolution of Butterfly Wing Patterns* (1991) provides an excellent discussion and review of this subject.

Another lepidopteran special feature, seen in *Bombyx mori* and in the saturniids, is silk production. Although silk is also produced by the Chelicerata (spiders), the economic importance of silk production by *B. mori* in particular has made this species the premier model system for studying both the development of the silk gland (SG) (a modified salivary gland) and the molecular biology of silk production. One set of properties that have allowed the extreme specialization of the SG for the large-scale production of silk is its accentuated production of those tRNAs – including certain SG-specific ones – that are used extensively for the main silk protein, fibroin, namely, glycine, alanine, and serine. This specialization of the translational apparatus is underlain by transcriptional specializations of the gland that provide its unique translational capacities (see Sprague, this volume).

One of the many intriguing discoveries in this work is that one of the transcription factors required by RNA polymerase III for transcribing the needed silk-gland-specific tRNA genes is itself a tRNA molecule (Young, Dunstan, et al., 1991). Whether this molecule is needed in this or in some other capacity in vivo is one of the puzzles to be solved. Future paths in this research will involve the elucidation of the complete mechanism that gives preferential transcription to certain tRNA genes in the SG and the nature of the requirements of these SG-unique tRNAs in the translation of fibroin.

Other transcriptional and developmental aspects of the SG have been reviewed in this book by Hui and Suzuki, who concentrate on the factors that directly control the transcription of fibroin and sericin, the protein that coats fibroin. One of the most striking findings in this work is that certain homeodomain proteins are involved in the regulation of these proteins, directly or indirectly. These include the *Bombyx* homologues of the *Drosophila engrailed (en)* and *Antennapedia (Antp)* genes. The latter may be involved in the development of the SG itself, and the *en* gene product may be a direct regulator of transcription of some of the structural genes of the SG. Because one of the major quests in *Drosophila* development is a search for the "downstream" genes of the homeotic gene transcriptional factors (Gould et al., 1990), this work on the *Bombyx* SG seems likely to complement and augment the *Drosophila* work. On the other hand, the comparative developmental aspects in the use of the homeotic genes in development of the SG will add another dimension to our understanding of the system (see "Comparative studies").

Beyond butterfly wing patterns and the biology of the SG, a third highly developed feature of the lepidopterans is their production of and response to pheromones, the sex attractant odorants that are emitted by female moths and detected by males of the same species at remarkably slight concentrations. This field is described in this volume by Vogt. Pheromones were, indeed, first discovered in Lepidoptera, and the growing integration of behavioral, molecular, neurological, and evolutionary information about moth pheromone utilization is creating one of the most satisfying stories in neurobiology. Undoubtedly, the information will also have wider application in understanding the general biology of olfaction, in particular, in vertebrate systems.

Lepidopterans as model systems for general insect phenomena

Beyond the highly specialized features that recommend lepidopteran systems for particular study, certain species exhibit characteristics that make them especially favorable model systems for more generally observed phenomena, seen both in other insect classes and in vertebrates. Chief among these are various hormonal changes associated with metamorphosis (reviewed in the chapters by Willis, Wilkins, and Goldsmith and by Riddiford) and the construction of the insect eggshell, the chorion (reviewed in the chapters by Regier et al. and Kafatos et al.). In addition, the analysis of changes in hemolymph proteins and cuticular proteins, facilitated by contemporary molecular approaches, promises to be highly informative, particularly in comparison with parallel studies in *Drosophila.* This work is described in the chapters by Willis, Wilkins, and Goldsmith and by Riddiford. The two general characteristics that have made the chosen model systems especially suitable are the large size of these animals and their comparatively slow developmental pace. Both features facilitate experimental intervention and analysis.

For endocrinologic work, *Manduca sexta* has been the insect system of choice for many investigators. These studies have revealed the relationships between juvenile hormone and ecdysteroid concentrations in regulating molting and are now entering the molecular stage in which the roles of specific hormone receptors in triggering particular gene activations can now be investigated (as described by Riddiford). In addition, the roles of hormonal changes in remodeling the nervous system during development have been characterized in detail at the cellular and developmental levels. This work, reviewed by Truman, is now at the threshold of molecular analysis. In the near future the ways in which particular neuronal deaths (and births) are triggered or influenced by changes in the hormonal milieu may be increasingly well understood in terms of precise molecular mechanisms.

Although not treated in depth in this book, the large size of the caterpillar in some of the more extensively studied lepidopteran systems has also facilitated the isolation of peptide hormones and investigation of their mechanisms of action, along with many of the other physiological and biochemical changes that occur during development. One may anticipate interesting developments in the peptide hormone work in the near future.

Comparative studies: from flies to moths to mammals

An important and rapidly growing set of studies in lepidopteran systems derives much value from their comparative aspects, both with respect to other insects, in particular *Drosophila,* and to other animal systems, including vertebrates. This work often provides intriguing evolutionary perspectives that are of direct interest and, in turn, provide assistance in evaluating claims of general significance. The importance of comparative studies for such work is carefully set out in this volume by Regier et al.

One such area, whose impact on developmental studies is only beginning to be felt, concerns the homeotic genes of the Lepidoptera. Both the similarities and the differences in the organization and expression of these genes have been reviewed by Ueno, Nagata, and Suzuki, who describe the evidence for lepidopteran homologues of the *Drosophila Ubx, abd-A, Abd-B,* and *Antp* genes. One of the little understood, but undoubtedly significant, findings is that mutational alterations of expression of some of these genes create reverse directional transformations in the dorsal and ventral parts of the animal, a phenomenon not seen in the fruit fly. Ultimately, these differences may be better understood in terms of the underlying differences in embryonic development between the Lepidoptera and the Diptera. As discussed in detail by Nagy in this volume, the Lepidoptera show an unusual mode of development that is somewhat similar, though not identical, to the intermediate germ band form and thereby distinctly different from the classic long germ band mode of *Drosophila.*

In addition, several of the homologues of the *Drosophila* segmentation genes have also been cloned and their expression patterns are being studied. As might be expected, the pair-rule genes of *Drosophila,* whose early expression is essential for the long germ band mode, appear, on the basis of present evidence, to behave differently in short and intermediate germ band insects, as noted by Nagy. It may be that the original functions of several of the segmentation genes were in neural development (Patel, Ball, and Goodman, 1989) and then were subsequently employed in segmentation during the evolution of the long germ band insects.

Another area of comparative studies concerns chorion construction within the different lepidopteran families. As emphasized by Regier et al. in this volume, such comparative studies are most informative when one has accurate phylogenies to work from. The construction of such phylogenies – often regarded in the past by many biologists as a problem unto itself – thus becomes a matter of direct interest to developmental biologists. Although this area of study is still comparatively new, the data on expression patterns already permit some direct inferences about the relationships between changes in gene expression patterns and chorion structural differences between the different groups. From this evolutionary perspective, the detailed analysis of transcriptional controls of the chorion genes in the Lepidoptera (Kafatos et al., this volume) acquires an added significance beyond its importance as a problem in control of gene expression.

Comparative work on genome structure and evolution is also providing much information of interest. The mobile elements found in lepidopteran genomes display, as a group, certain shared properties with both the Diptera and mammals, as described in the chapter by Eickbush, and the detailed analysis of chorion gene cluster evolution has unearthed some significant evidence of gene conversion involving sequences with recombinational hotspots, reviewed here by Eickbush and Izzo. Such instances of "molecular drive" (Dover, 1982) raise important questions about the roles of selection and genome recombinational mechanisms in creating long-term genetic changes within gene families. In particular, it will be of interest to determine whether the peptide sequence encoded by the hotspot plays an important functional role in the function of chorion proteins, or whether it is selectively neutral in itself. Even if the particular encoded sequence should prove unimportant in function, the existence of such a hotspot is presumably of selective value in helping to maintain uniformity of family members and driving concerted evolution.

Finally, and not least, analysis of lepidopteran response to microbial invasion has not only revealed several novel families of proteins required for and involved in immune defense systems in insects but has led to the discovery that several of these proteins are also present and expressed in mammals, as discussed by Mulnix and Dunn in this volume. Not for the first time (see Willis, Wilkins, and Goldsmith, this volume) has a discovery in lepidopteran systems revealed a phenomenon of much wider provenance in biology. Furthermore, the lepidopteran immune systems exhibit not only the sharing of certain molecular species with vertebrates, but the findings, taken as a whole, hint at the possibility of greater specificity in the recognition process in insect immune systems than was dreamt of as recently as two decades ago (Salt, 1970).

Technical challenges and the future

Much of the progress in the study of lepidopteran systems from the standpoint of molecular biology hinges on future technical developments. A crucially important area is the further development of the conventional genetics of these organisms, in particular *Bombyx*, and the correlative development of a molecular map, as discussed in the chapter by Goldsmith. Aided and abetted by the continuing use of *Drosophila* gene probes to fish out lepidopteran homologues, expansion of *Bombyx* genetics can be expected to be rapid in the coming decade. Should gene replacement techniques comparable to those now available for *Drosophila* (Gloor, Nassif, Johnson-Schlitz, Preston, and Engels, 1991) be developed, the future of genetic manipulation in lepidopterans is bright.

In addition, determining the structural elements and mechanisms underlying the distinctive or unique features of the genetic system of the Lepidoptera – their holocentric chromosomes, lack of crossing over in females, somatic inactivation of the W chromosome (see discussion by Goldsmith) – will contribute to a better understanding of lepidopteran systematics and evolution. Investigation of these phenomena on the molecular level is just beginning and should be able to take advantage of the silkworm as a readily available laboratory model. As seen in the characterization of the telomere, this work is likely to benefit from technical breakthroughs in such systems as yeast and mammals, where progress has been more rapid.

The other important lines of technical development are the further expansion of recombinant baculovirus technology as expression and transducing vectors, and the possible development of a germ line transformation system. The progress in expression vector technology, for both basic and applied research, the latter involving their potential as insecticides, and the development of transducing vectors for somatic cell transformation, have been reviewed by Iatrou in this volume. The development of a germ line transformation system is still far from certain but is being actively pursued in several laboratories.

Immunologic probes, although not discussed in this book, will also undoubtedly play a part in the detailed tracking of individual gene products of interest. The rate-limiting steps here, however, seem solely to be ones needing transfer of existing methodologies to lepidopteran systems rather than special technical hurdles to be overcome.

Altogether, when one surveys the field of basic research in Lepidoptera and compares its present state to that of, say, a decade ago, progress has clearly been great. More importantly, prospects for exciting work ahead seem excellent. We hope that the day is not too far distant when investigators proposing a new *Drosophila* project will be greeted with the response: "Why not one of the Lepidoptera?"

References

Abu-Hakima, R., & Faye, I. (1981). An ultrastructural and autoradiographic study of the immune response in *Hyalophora cecropia* pupae. *Cell and Tissue Research,* 217, 311–320.

Adachi, A., & Chikushi, H. (1977). Bio-histochemical analysis of gene action in flimsy cocoon mutant of *Bombyx mori. Scientific Bulletin of the Faculty of Agriculture, Kyushu University,* 31, 159–173. In Japanese with English summary.

Adachi, T., Takiya, S., Suzuki, Y., Iwami, M., Kawakami, A., Takahashi, S. Y., Ishizaki, H., Nagasawa, H., & Suzuki, A. (1989). cDNA structure and expression of bombyxin, an insulin-like brain secretory peptide of the silkmoth *Bombyx mori. Journal of Biological Chemistry,* 264, 7681–7685.

Adachi-Yamashita, N., Sakaguchi, B., & Chikushi, H. (1980). Fibroin secretion in the posterior silk gland cells of a flimsy cocoon mutant of *Bombyx mori. Cell Structure and Function,* 5, 105–108.

Adams, D. S., Eickbush, T. H., Herrera, R. J., & Lizardi, P. M. (1986). A highly reiterated family of transcribed oligo (A)-terminated, interspersed DNA elements in the genome of *Bombyx mori. Journal of Molecular Biology,* 187, 465–478.

Addison, W. R., Astell, C. R., Delaney, A. D., Gilliam, I. C., Hayashi, S., Miller, R. C., Rajput, B., Smith, M., Taylor, D. M., & Tener, G. M. (1982). The structures of genes hybridizing with $tRNA_4^{Val}$ from *Drosophila melanogaster. Journal of Biological Chemistry,* 257, 670–673.

Affolter, M., Schier, A., & Gehring, W. J. (1990). Homeodomain proteins and the regulation of gene expression. *Current Opinion in Cell Biology,* 2, 485–495.

Agerberth, B., Lee, J.-Y., Bergman, T., Carlquist, M., Boman, H. G., Mutt, V., & Jörnvall, H. (1991). Amino acid sequence of PR-39. Isolation from pig intestine of a new member of the family of proline-arginine-rich antibacterial peptides. *European Journal of Biochemistry,* 202, 849–854.

Akai, H. (1984). The ultrastructure and functions of the silk gland cells of *Bombyx mori.* In *Insect Ultrastructure,* vol. 2, ed. R. C. King & H. Akai, pp. 323–364. New York: Plenum.

Akam, M., (1987). The molecular basis for metameric pattern in the *Drosophila* embryo. *Development,* 101, 1–22.

Akam, M. (1989). Hox and HOM: homologous gene clusters in insects and vertebrates. *Cell,* 57, 347–349.

Akam, M., & Dawes, R. (1992). More than one way to slice an egg. *Current Biology,* 2, 395–398.

Aksoy, S., Williams, S., Chang, S., & Richards, F. F. (1990). SLACS retrotransposon from *Trypanosoma brucei gambiense* is similar to mammalian LINEs. *Nucleic Acids Research,* 18, 785–792.

Alexopoulou, M. G., Spoerel, N., Sakaguchi, B., Zetlan, S., Nelson, S. P., & Goldsmith, M. R. (1988). Analysis of two sex-limited translocation strains of *Bombyx mori* that cover chorion structural genes. *Sericologia,* 28, Supplement, 45–46.

Andersen, S. O. (1979). Biochemistry of insect cuticle. *Annual Review of Entomology,* 24, 29–61.

Andersen, S. O. (1985). Sclerotization and tanning of the cuticle. In *Comprehensive Insect Physiology, Biochemistry and Pharmacology,* vol. 3, ed. G. A. Kerkut & L. I. Gilbert, pp. 59–74. Oxford: Pergamon.

Anderson, D. T. (1972). The development of holometabolous insects. In *Developmental Systems: Insects,* ed. S. J. Counce & C. H. Waddington, pp. 165–242. London: Academic Press.

Anderson, D. T. (1973). *Embryology and Phylogeny in Annelids and Arthropods.* Oxford: Pergamon.

Anderson, D. T., & Wood, E. C. (1968). The morphological basis of embryonic movements in the light brown apple moth, *Epiphyas postvittana* (Walk.) (Lepidoptera: Tortricidae). *Australian Journal of Zoology,* 16, 763–793.

Anderson, K. V. (1987). Dorsal-ventral embryonic pattern genes of *Drosophila. Trends in Genetics,* 3, 91–97.

Anderson, K. V. (1989). *Drosophila:* the maternal contribution. In *Genes and Embryos, Frontiers in Molecular Biology Series,* ed. D. M. Glover & B. D. Hames, pp. 1–37. Oxford and New York: IRL Press.

Anderson, K. V., Bokla, L., & Nüsslein-Volhard, C. (1985). Establishment of dorsal-ventral polarity in the *Drosophila* embryo: the induction of polarity by the *Toll* gene product. *Cell,* 42, 791–798.

Anderson, K. V., & Nüsslein-Volhard, C. (1984). Information for the dorsal-ventral pattern of the *Drosophila* embryo is stored as maternal mRNA. *Nature,* 311, 223–227.

Andersson, K., & Steiner, H. (1987). Structure and properties of protein P4, the major bacteria-inducible protein in pupae of *Hyalophora cecropia. Insect Biochemistry,* 17, 133–140.

Ando, H., & Kobayashi, Y. (1978). The formation of germ rudiment in the primitive moth, *Neomicropteryx nipponensis* ISSIKI (Micropterygidae, Zeugloptera, Lepidoptera) and its phylogenetic significance. *Proceedings of the Japanese Society of Systematic Zoology,* 15, 47–50.

Ando, H., & Tanaka, M. (1976). The formation of germ rudiment and embryonic membranes in the primitive moth, *Endoclita excrescens* Butler (Hepialidae, Monotrysia, Lepidoptera) and its phylogenetic significance. *Proceedings of the Japanese Society of Systematic Zoology,* 12, 52–55.

Ando, H., & Tanaka, M. (1980). Early embryonic development of the primitive moths, *Endoclyta signifer* Walker and *E. excrescens* Butler (Lepidoptera: Hepialidae). *International Journal of Insect Morphology and Embryology,* 9, 67–77.

Ando, K., & Natori, S. (1988). Molecular cloning, sequencing, and characterization of cDNA for sarcotoxin IIA, an inducible antibacterial protein of *Sarcophaga peregrina* (flesh fly). *Biochemistry* 27, 1715–1721.

Ando, K., Okada, M., & Natori, S. (1987). Purification of sarcotoxin II, antibacterial proteins of *Sarcophaga peregrina* (flesh fly) larvae. *Biochemistry,* 26, 226–230.
Andreu, D., Merrifield, R. B., Steiner, H., & Boman, H. G. (1983). Solid-phase synthesis of cecropin A and related peptides. *Proceedings of the National Academy of Sciences, U.S.A.,* 80, 6475–6479.
Andreu, D., Merrifield, R. B., Steiner, H., & Boman, H. G. (1985). N-terminal analogues of cecropin A: synthesis, antibacterial activity, and conformational properties. *Biochemistry,* 24, 1683–1688.
Apple, R. T., & Fristrom, J. W. (1991). 20-Hydroxyecdysone is required for, and negatively regulates, transcription of *Drosophila* pupal cuticle protein genes. *Developmental Biology,* 146, 569–582.
Arbogast, R. T., Chauvin, G., Strong, R. G., & Byrd, R. V. (1983). The egg of *Endrosis sarcitrella* (Lepidoptera: Oecophoridae): fine structure of the chorion. *Journal of Stored Products Research,* 9, 63–68.
Arnheim, N. (1983). Concerted evolution of multigene families. In *Evolution of Genes and Proteins,* ed. M. Nei & R. K. Koehn, pp. 38–61. Sunderland: Sinauer.
Arnold, A. P., & Gorski, R. (1984). Gonadal steroid induction of structural sex differences in the brain. *Annual Review of Neuroscience,* 7, 413–442.
Arnold, A. P., Nottabohm, F., & Pfaff, D. W. (1976). Hormone concentrating cells in vocal control and other areas of the brain of the zebra finch (*Poephila guttata*). *Journal of Comparative Neurology,* 165, 487–512.
Aruga, H., & Shigemi, K. (1952). Tyrosinase activity and melanin formation of some larval marking in the silkworm. *Japanese Journal of Breeding,* 1, 137–140. In Japanese with English summary.
Asaoka, K., & Mano, Y. (1992). Breeding of polyphagous silkworms by early selection for feeding ability on LP-1 artificial diet. *Journal of Sericultural Science of Japan,* 61, 1–5. In Japanese with English summary.
Ashburner, M. (1992). Mapping insect genomes. In *Insect Molecular Science,* ed. J. M. Crampton & P. Eggleston, pp. 51–75. London: Academic Press.
Ashburner, M., Chihara, C., Meltzer, P., & Richards, G. (1974). Temporal control of puffing activity in polytene chromosomes. *Cold Spring Harbor Symposia on Quantitative Biology,* 38, 655–662.
Astic, L., Saucier, D., & Holley, A. (1987). Topographic relationships between olfactory receptor cells and glomerular foci in the rat olfactory bulb. *Brain Research,* 424, 144–152.
Bailey, D. W. (1971). Recombinant-inbred strains. *Transplantation,* 11, 325–327.
Bailey, D. W. (1981). Recombinant inbred strains and bilineal congenic strains. In *The Mouse in Biomedical Research,* vol. 1, ed. H. L. Foster, J. D. Small, & J. G. Fox, pp. 223–239. New York: Academic Press.
Baker, B. S., & Tata, J. R. (1992). Prolactin prevents the autoinduction of thyroid hormone receptor mRNAs during amphibian metamorphosis. *Developmental Biology,* 149, 463–467.
Baker, F. C., Tsai, L. W., Renter, C. C., & Schooley, D. A. (1987). In vivo fluctuation of JH, JH acid, and ecdysteroid titer, and JH esterase activity during development of fifth stadium *Manduca sexta. Insect Biochemistry,* 17, 989–996.

Baker, T. C., & Vogt, R. G. (1988). Measured behavioral latency in response to sex-pheromone loss in the large silk moth *Antheraea polyphemus. Journal of Experimental Biology,* 137, 29–38.
Balinsky, B. I. (1986). Early differentiation in the egg of the butterfly, *Acraea horta* under normal conditions and after ultraviolet irradiation. *Acta Embryologiae et Morphologiae Experimentalis,* 6, 103–141.
Baltimore, D. (1981). Gene conversion: some implications for the immunoglobulin genes. *Cell,* 24, 592–594.
Barbier, R., & Chauvin, G. (1974a). Ultrastructure et rôle des aéropyles et des envelopes de l'oeuf de *Galleria mellonella. Journal of Insect Physiology,* 20, 809–820.
Barbier, R., & Chauvin, G. (1974b). The aquatic egg of *Nymphula nymphaeata* (Lepidoptera: Pyralidae). On the fine structure of the egg shell. *Cell and Tissue Research,* 149, 473–479.
Barth, F. G. (1973). Microfiber reinforcement of an arthropod cuticle. *Zeitschrift für Zellforschung und mikroskopische Anatomie,* 144, 409–433.
Bartholomew, B., Kassavetis, G. A., Braun, B. R., & Geiduschek, E. P. (1990). The subunit structure of *Saccharomyces cerevisiae* transcription factor IIIC probed with a novel photocrosslinking reagent. *The EMBO Journal,* 9, 2197–2205.
Barwig, B., & Bohn, H. (1980). Evidence for presence of two clotting proteins in insects. *Naturwissenschaften,* 67, 447–448.
Bassand, D. (1965). Contribution à l'étude de la diapause embryonnaire et de l'embryogenèse de *Zeiraphera griseana* Hubner (=Z. *diniana* Guenée [Lepidoptera: Tortricidae]). *Revue Suisse de Zoology,* 72, 431–542.
Bassi, A. (1958). Del mal del segno. In *Phytopathological Classics,* vol. 10, transl. P. J. Yarrow, ed. G. C. Ainsworth & P. J. Yarrow. Baltimore: American Phytopathological Society.
Bastian, H., & Gruss, P. (1990). A murine *even-skipped* homologue, *Evx 1,* is expressed during early embryogenesis and neurogenesis in a biphasic manner. *European Journal of Molecular Biology,* 9, 1839–1852.
Bate, C. M. (1973). The mechanism of the pupal gin-trap. I. Segmental gradients and the connections of the triggering sensilla. *Journal of Experimental Biology,* 59, 95–107.
Bateson, W. (1894). *Materials for the Study of Variation Treated with Especial Regard to Discontinuity in the Origin of Species.* London and New York: Macmillan.
Bauer, B. J., & Waring, G. L. (1987). 7C female sterile mutants fail to accumulate early eggshell proteins necessary for later chorion morphogenesis in *Drosophila. Developmental Biology,* 121, 349–358.
Bauer, H. (1967). Die kinetische Organisation der Lepidopteren-Chromosomen. *Chromosoma* (Berl.), 22, 101–125.
Beadle, G. W. (1977). Genes and chemical reactions in *Neurospora.* In *Nobel Lectures in Molecular Biology 1933–75,* pp. 51–63. New York: Elsevier.
Beadle, G. W., & Ephrussi, B. (1936). The differentiation of eye pigments in *Drosophila* as studied by transplantation. *Genetics,* 21, 225–247.
Beames, B., & Summers, M. D. (1988). Comparisons of host cell DNA insertions and altered transcription at the site of insertions in few polyhedra baculovirus mutants. *Virology,* 162, 206–220.
Beato, M. (1989). Gene regulation by steroid hormones. *Cell,* 56, 335–344.

Beck, S. D. (1960). Growth and development of the greater wax moth, *Galleria mellonella* (Lepidoptera: Galleriidae). *Transactions of the Wisconsin Academy of Sciences, Arts and Letters,* 49, 137–148.

Beeman, R. W. (1987). A homeotic gene cluster in the red flour beetle. *Nature,* 327, 247–249.

Beeman, R. W., Stuart, J. J., Haas, M. S., & Denell, R. E. (1989). Genetic analysis of the homeotic gene complex (HOM-C) in the beetle *Tribolium castaneum. Developmental Biology,* 133, 196–209.

Bello, B., & Couble, P. (1990). Specific expression of a silk-encoding gene of *Bombyx* in the anterior salivary gland of *Drosophila. Nature,* 346, 480–482.

Belyaeva, E. S., Vlassova, I. E., Biyasheva, Z. M., Kakpakov, V. T., Richards, G., & Zhimulev, I. F. (1981). Cytogenetic analysis of the 2133-4–2B11 region of the X chromosome of *Drosophila melanogaster.* II. Changes in 20-OH ecdysone puffing caused by genetic defects of puff 2B5. *Chromosoma* (Berl.), 84, 207–219.

Bender, W., Akam, M., Karch, F., Beachy, P. A., Peifer, M., Spierer, P., Lewis, E. B., & Hogness, D. S. (1983). Molecular genetics of the bithorax complex in *Drosophila. Science,* 221, 23–29.

Bender, W., Spierer, P., & Hogness, D. S. (1983). Chromosomal walking and jumping to isolate DNA from the *Ace* and *rosy* loci and the bithorax complex in *Drosophila melanogaster. Journal of Molecular Biology,* 168, 17–33.

Bennett, K. L., & Truman J. W. (1985). Steroid-dependent survival of identifiable neurons in cultured ganglia of the moth *Manduca sexta. Science,* 229, 58–60.

Benz, G. A. (1986). Introduction: historical perspectives. In *The Biology of Baculoviruses,* vol. I, ed. R. A. Granados & B. A. Federici, pp. 1–35. Boca Raton, FL: CRC Press.

Berendes, H. B., & Ashburner, M. (1978). The salivary glands. In *The Genetics and Biology of Drosophila,* vol. 2b, ed. M. Ashburner & T. R. F. Wright, pp. 453–498. London: Academic Press.

Berg, G. J., & Gassner, G. (1978). Fine structure of the blastoderm embryo of the pink bollworm, *Pectinophora gossypeilla* (Saunders) (Lepidoptera: Gelechiidae). *International Journal of Insect Morphology and Embryology,* 7, 81–105.

Berg, R., Schuchmann-Feddersen, I., & Schmidt, O. (1988). Bacterial infection induces a moth (*Ephestia kuhniella*) protein which has antigenic similarity to virus-like particle proteins of a parasitoid wasp *(Venturia canescens). Journal of Insect Physiology,* 34, 473–480.

Berger, E. M., Goudie, K., Klieger, L., Berger, M. N., & DeCato, R. (1992). The juvenile hormone analogue, methoprene, inhibits ecdysterone induction of small heat shock protein gene expression. *Developmental Biology,* 151, 410–418.

Bhagirath, T., Kundu, S. C., & Ibotombi, N. (1988). Pachytene synaptonemal complex analysis in an Indian silkworm, *Philosamia ricini. Sericologia,* 28, 99–104.

Biever, K. D., & Andrews, P. L. (1984). Susceptibility of lepidopterous larvae to *Plutella xylostella* nuclear polyhedrosis virus. *Journal of Invertebrate Pathology,* 44, 117–119.

Bigger, T. R. L. (1975). Karyotypes of some Lepidoptera chromosomes and changes in their holokinetic organisation as revealed by new cytological techniques. *Cytologia,* 40, 713–726.

Bigger, T. R. L. (1976). Karyotypes of three species of Lepidoptera including an investigation of B-chromosomes in *Pieris. Cytologia,* 41, 261–282.

Biggin, M. D., & Tjian, R. (1989). Transcription factors and the control of *Drosophila* development. *Trends in Genetics,* 5, 377–383.

Binger, L. C., & Willis, J. H. (1990). In vitro translation of epidermal RNA from different anatomical regions and metamorphic stages of *Hyalophora cecropia. Insect Biochemistry,* 20, 573–583.

Binnington, K., & Retnakaran, A., eds. (1991). *Physiology of the Insect Epidermis.* East Melbourne: CSIRO Publications.

Blackman, R. K., & Gelbart, W. M. (1989). The transposable element hobo of *Drosophila melanogaster.* In *Mobile DNA,* ed. D. E. Berg & M. M. Howe, pp. 523–530. Washington, DC: American Society of Microbiology.

Blau, H. M., & Kafatos, F. C. (1978). Secretory kinetics in the follicular cells of silkmoths during eggshell formation. *Journal of Cell Biology,* 78, 131–151.

Bock, S. C., Campo, K., & Goldsmith, M. R. (1986). Specific protein synthesis in cellular differentiation VI. Temporal expression of chorion gene families in *Bombyx mori* strain C108. *Developmental Biology,* 117, 215–225.

Bodenstein, D. (1971). Introduction. In *Milestones in Developmental Physiology of Insects,* ed. D. Bodenstein, pp. 1–5. New York: Meredith.

Boeckh, J., Kaissling, K.-E., & Schneider, D. (1960). Sensillen und Bau der Antennengeissel von *Telea polyphemus* (Verleiche mit weiteren Saturniden: *Antheraea, Platasamia,* und *Philosamia). Zoological Journal of Anatomy,* 78, 559–584.

Boeckh, J., Kaissling, K.-E., & Schneider, D. (1965). Insect olfactory receptors. *Cold Spring Harbor Symposia on Quantitative Biology,* 30, 263–280.

Boeke, J. D. (1989). Transposable elements in *Saccharomyces cerevisiae.* In *Mobile DNA,* ed. D. E. Berg & M. M. Howe, pp. 335–374. Washington, DC: American Society of Microbiology.

Boeke, J. D., & Corces, V. G. (1989). Transcription and reverse transcription of retrotransposons. *Annual Review of Microbiology,* 43, 403–434.

Boekhoff, I., Raming, K., & Breer, H. (1990). Pheromone-induced stimulation of inositol-triphosphate formation in insect antennae is mediated by G-proteins. *Journal of Comparative Physiology,* B, 160, 99–103.

Boekhoff, I., Tareilus, E., Strotmann, J., & Breer, H. (1990). Rapid activation of alternative second messenger pathways in olfactory cilia from rats by different odorants. *The EMBO Journal,* 9, 2453–2458.

Bogenhagen, D. F., Sakonju, S., & Brown, D. D. (1980). A control region in the center of the 5S RNA gene directs specific initiation of transcription. II. The 3′ border of the region. *Cell,* 19, 27–35.

Bollenbacher, W. E., Smith, S. L., Goodman, W., & Gilbert, L. I. (1981). Ecdysteroid titer during larval-pupal-adult development of the tobacco hornworm, *Manduca sexta. General and Comparative Endocrinology,* 44, 302–306.

Bollenbacher, W. E., Zvenko, H., Kumaran, A. K., & Gilbert, L. I. (1978). Changes in ecdysone content during postembryonic development of the wax moth, *Galleria mellonella:* the role of the ovary. *General and Comparative Endocrinology,* 34, 169–179.

Boman, H. G., Boman, I. A., Andreu, D., Li, Z-q., Merrifield, R. B., Schlenstedt, G., & Zimmerman, R. (1989). Chemical synthesis and enzymic processing of precursor forms of cecropins A & B. *Journal of Biological Chemistry,* 264, 5852–5860.

Boman, H. G., Faye, I., Gudmundsson, G. H., Lee, J.-Y., & Lidholm, D. A. (1991). Cell-free immunity in Cecropia. A model system for antibacterial proteins. *European Journal of Biochemistry,* 201, 23–31.

Boman, H. G., & Hultmark, D. (1987). Cell-free immunity in insects. *Annual Review of Microbiology,* 41, 103–126.

Booker, R., & Truman, J. W. (1987). Postembryonic neurogenesis in the CNS of the tobacco hornworm, *Manduca sexta.* II. Hormonal control of imaginal nest cell degeneration and differentiation during metamorphosis. *Journal of Neuroscience,* 7, 4107–4114.

Borst, D. E., Redmond, T. M., Elser, J. E., Gonda, M. A., Wiggert, B., Chader, G. J., & Nickerson, J. M. (1989). Interphotoreceptor retinoid-binding protein: gene characterization, protein repeat structure, and its evolution. *Journal of Biological Chemistry,* 264, 1115–1123.

Bosquet, G., Fourche, J., & Guillet, C. (1989). Respective contributions of juvenile hormone and 20-hydroxyecdysone to the regulation of major haemolymph protein synthesis in *Bombyx mori* larvae. *Journal of Insect Physiology,* 35, 1005–1015.

Bosquet, G., Guillet, C., Calvez, B., & Chavancy, G. (1989). The regulation of major haemolymph protein synthesis: changes in mRNA content during the development of *Bombyx mori* larvae. *Insect Biochemistry,* 19, 29–39.

Botstein, D., White, R. L., Skolnick, M., & Davis, R. W. (1980). Construction of a genetic linkage map in man using restriction fragment length polymorphisms. *American Journal of Human Genetics,* 32, 314–331.

Boudreaux, H. B. (1981). About the panorpoid complex. *Annals of the Entomological Society of America,* 74, 155–157.

Bouhin, H., Charles, J-P., Quennedey, B., & Delachambre, J. (1992). Developmental profiles of epidermal mRNAs during the pupal-adult molt of *Tenebrio molitor* and isolation of a cDNA clone encoding an adult cuticular protein: effects of a juvenile hormone analogue. *Developmental Biology,* 149, 112–122.

Bounhiol, J. J. (1938). Recherches expérimentales sur le déterminisme de la métamorphose chez les Lépidoptères. *Biological Bulletin,* 24, Supplement, 1–199.

Bowers, W. S. (1985). Antihormones. In *Comprehensive Insect Physiology, Biochemistry and Pharmacology,* vol. 8, ed. G. A. Kerkut & L. I. Gilbert, pp. 551–564. Oxford: Pergamon.

Bowers, W. S., Thompson, M. J., & Uebel, E. C. (1965). Juvenile and gonadotropic hormone activity of 10, 11-epoxyfarnesenic acid methyl ester. *Life Sciences,* 4, 2323–2331.

Bradfield, J. Y., & Keeley, L. L. (1989). Adipokinetic hormone gene sequence from *Manduca sexta. Journal of Biological Chemistry,* 264, 12791–12793.

Bradfield, J. Y., Locke, J., & Wyatt, G. R. (1985). An ubiquitous interspersed DNA sequence family in an insect. *DNA,* 4, 357–363.

Braun, R. P., Edwards, G. C., Wyatt, G. R., Yagi, K. J., & Tobe, S. S. (1992). Juvenile hormone binding proteins and receptors in locust fat body. *Abstracts, XIX International Congress of Entomology,* 123.

Bray, S. J., & Kafatos, F. C. (1991). Developmental function of Elf-1: an essential transcription factor during embryogenesis in *Drosophila. Genes and Development,* 5, 1672–1683.

Breer, H., & Boekhoff, I. (1991). Odorants of the same odor class activate different second messenger pathways. *Chemical Senses,* 16, 19–29.

Breer, H., Boekhoff, I., & Tareilus, E. (1990). Rapid kinetics of second messenger formation in olfactory transduction. *Nature,* 345, 65–68.

Breer, H., Krieger, J., & Raming, K. (1990). A novel class of binding proteins in the antennae of the silk moth *Antheraea pernyi. Insect Biochemistry,* 20, 735–740.

Brehelin, M. (1979). Hemolymph coagulation in *Locusta migratoria:* evidence for a functional equivalent of fibrinogen. *Comparative Biochemistry and Physiology,* B, 62, 329–334.

Bridges, C. B., & Morgan, T. H. (1923). The third-chromosome group of mutant characters of *Drosophila melanogaster. Publications of the Carnegie Institute, Washington,* 327, 202.

Briggs, J. D. (1958). Humoral immunity in lepidopterous larvae. *Journal of Experimental Zoology,* 138, 155–188.

Brinkley, B. R., Valdivia, M. M., Tousson, A., & Balczon, R. D. (1989). The kinetochore: structure and molecular organization. In *Mitosis, Molecules and Mechanisms,* ed. J. S. Hyams & B. R. Brinkley, pp. 77–118. San Diego: Academic Press.

Britten, R. J., Baron, W. F., Stout, D. B., & Davidson, E. H. (1988). Sources and evolution of human *Alu* repeated sequences. *Proceedings of the National Academy of Sciences, U.S.A.,* 85, 4770–4774.

Britten, R. J., & Kohne, D. E. (1968). Repeated sequences in DNA. *Science,* 161, 529–540.

Broadie, K. S., Bate, M., & Tublitz, N. (1990). Quantitative staging of embryonic development of the tobacco hawkmoth, *Manduca sexta. Roux's Archives of Developmental Biology,* 199, 327–334.

Brow, D. A., & Guthrie, C. (1990). Transcription of a yeast U6 snRNA gene requires a polymerase III promoter element in a novel position. *Genes and Development,* 4, 1345–1356.

Brown, A. J. L., & Ish-Horowicz, D. (1981). Evolution of the 87A and 87C heat-shock loci in *Drosophila. Nature,* 290, 677–682.

Brown, D. D., & Sugimoto, K. (1973). The structure and evolution of ribosomal and 5S DNAs in *Xenopus laevis* and *Xenopus mulleri. Cold Spring Harbor Symposia on Quantitative Biology,* 38, 501–505.

Brunelle, A., & Schleif, R. (1987). Missing contact probing of DNA-protein interactions. *Proceedings of the National Academy of Sciences, U.S.A.,* 84, 6673–6676.

Brusca, J., Summers, M., Couch, J., & Courtney, L. (1986). *Autographa californica* nuclear polyhedrosis virus efficiently enters but does not replicate in poikilothermic vertebrate cells. *Intervirology,* 26, 207–222.

Brutlag, D. L. (1980). Molecular arrangement and evolution of heterochromatic DNA. *Annual Review of Genetics,* 14, 121–144.

Bryant, P. J. (1993). The polar coordinate model goes molecular. *Science,* 259, 471–472.

Buck, L., & Axel, R. (1991). A novel multigene family may encode odorant receptors: A molecular basis for odor recognition. *Cell,* 65, 175–187.

Buckner, J. S., Henderson, T. A., Ehresmann, D. D., & Graf, G. (1990). Structure and composition of the urate storage granules from the fat body of *Manduca sexta*. *Insect Biochemistry,* 20, 203–214.

Bulet, P., Cociancich, S., Dimarcq, J. L., Lambert, J., Reichhart, J. M., Hoffmann, D., Hetru, C., & Hoffmann, J. A. (1991). Insect immunity: isolation from a coleopteran insect of a novel inducible antibacterial peptide and of new members of the insect defensin family. *Journal of Biological Chemistry,* 266, 24520–24525.

Burke, W. D., Calalang, C. C., & Eickbush, T. H. (1987). The site-specific ribosomal insertion element Type II of *Bombyx mori* (R2Bm) contains the coding sequence for a reverse transcriptase-like enzyme. *Molecular and Cellular Biology,* 7, 2221–2230.

Burke, W. D., & Eickbush, T. H. (1986). The silkmoth late chorion locus. I. Variation within two paired multigene families. *Journal of Molecular Biology,* 190, 343–356.

Burr, B., & Burr, F. A. (1991). Recombinant inbreds for molecular mapping in maize: theoretical and practical considerations. *Trends in Genetics,* 7, 55–60.

Burr, B., Burr, F. A., Thompson, K. H., Albertson, M. C., & Stuber, C. W. (1988). Gene mapping with recombinant inbreds in maize. *Genetics,* 118, 519–526.

Burtis, K. C., Thummel, C. S., Jones, C. W., Karim, F. D., & Hogness, D. S. (1990). The *Drosophila* 74EF early puff contains *E74,* a complex ecdysone-inducible gene that encodes two *ets*-related proteins. *Cell,* 61, 85–99.

Butenandt, A., Beckman, R., Stamm, D., & Hecker, E. (1959). Über den sexuallockstoff des Seidenspinners *Bombyx mori,* Reidarstellung und Konstitution. *Zeitschrift für Naturforschung,* 14b, 283–284.

Butenandt, A., & Karlson, P. (1954). Über die Isolierung eines Metamorphose-Hormons der Insekten in kristallisierter Form. *Zeitschrift für Naturforschung,* 9b, 389–391.

Calvez, B., Hirn, M., & De Reggi, M. (1976). Ecdysone changes in the hemolymph of two silkworms (*Bombyx mori* and *Philosamia cynthia*) during larval and pupal development. *FEBS Letters,* 71, 57–61.

Candelas, G. C., Arroyo, G., Carrasco, C., & Dompenciel, R. (1990). Spider silk-glands contain a tissue-specific alanine tRNA that accumulates in vitro in response to the stimulus for silk protein synthesis. *Developmental Biology,* 140, 215–220.

Capinera, J. L., & Kanost, M. R. (1979). Susceptibility of the zebra caterpillar to *Autographa californica* nuclear polyhedrosis virus. *Journal of Economic Entomology,* 72, 570–572.

Cappello, J., Handelsman, K., & Lodish, H. F. (1985). Sequence of *Dictyostelium* DIRS-1: an apparent retrotransposon with inverted terminal repeats and an internal circle junction sequence. *Cell,* 43, 105–115.

Carbonell, L. F., Hodge, M. R., Tomalski, M. D., & Miller, L. K. (1988). Synthesis of a gene coding for an insect-specific scorpion neurotoxin and attempts to express it using baculovirus vectors. *Gene,* 73, 409–418.

Carbonell, L. F., & Miller, L. K. (1987). Baculovirus interaction with nontarget organisms: a virus-born reporter gene is not expressed in two mammalian cell lines. *Applied Environmental Entomology,* 53, 1412–1417.

Carpenter, A. T. C. (1975). Electron microscopy of meiosis in *Drosophila melanogaster* females. II. The recombination nodule–a recombination associated structure at pachytene? *Proceedings of the National Academy of Sciences, U.S.A.*, 72, 3186–3189.

Carpenter, A. T. C. (1979). Recombination nodules and synaptonemal complex in recombination-defective females of *Drosophila melanogaster. Chromosoma* (Berl.), 75, 259–292.

Carpenter, A. T. C. (1987). Gene conversion, recombination nodules, and the initiation of meiotic synapsis. *BioEssays,* 6, 232–236.

Carr, J. N., & Taghert, P. H. (1989). Pair-rule expression of a cell surface molecule during gastrulation of the moth embryo. *Development,* 107, 143–152.

Carroll, S. B., & Vavra, S. H. (1989). The zygotic control of *Drosophila* pair-rule gene expression. II. Spatial repression by gap and pair-rule gene products. *Development,* 107, 673–683.

Carson, D. D., Guarino, L. A., & Summers, M. D. (1988). Functional mapping of an AcNPV immediate early gene which augments expression of the IE-1 trans-activated 39K gene. *Virology,* 162, 444–451.

Carson, D. D., Summers, M. D., & Guarino, L. A. (1991). Molecular analysis of a baculovirus regulatory gene. *Virology,* 182, 279–286.

Carson-Jurica, M. A., Schrader, W. T., & O'Malley, B. W. (1990). Steroid receptor family: Structure and functions. *Endocrine Reviews,* 11, 201–220.

Carstens, E. B. (1987). Identification and nucleotide sequence of the regions of *Autographa californica* nuclear polyhedrosis virus genome carrying insertion elements derived from *Spodoptera frugiperda. Virology,* 161, 8–17.

Cary, L. C., Goebel, M., Corsaro, B. G., Wang, H., Rosen, E., & Fraser, M. J. (1989). Transposon mutagenesis of baculoviruses: analysis of *Trichoplusia ni* transposon *IFP2* insertions within the FP-locus of nuclear polyhedrosis viruses. *Virology,* 171, 156–169.

Casaday, G. M., & Camhi, J. M. (1976). Metamorphosis of flight motoneurons in the moth *Manduca sexta. Journal of Comparative Physiology,* 112, 143–158.

Casas, J., Harshman, L. G., & Hammock, B. D. (1991). Epoxide hydrolase activity on juvenile hormone in *Manduca sexta. Insect Biochemistry,* 21, 17–26.

Caspari, E. (1933). The action of a pleiotropic gene in the flour moth *Ephestia kuhniella* Zeller. In *Milestones in Developmental Physiology of Insects,* ed. D. Bodenstein, pp. 65–95. New York: Meredith.

Casteels, P., Ampe, C., Jacobs, F., Vaeck, M., & Tempst, P. (1989). Apidaecins: antibacterial peptides from honeybees. *The EMBO Journal,* 8, 2387–2391.

Chadwick, J. S. (1970). Relation of lysozyme concentration to acquired immunity against *Pseudomonas aeruginosa* in *Galleria mellonella. Journal of Invertebrate Pathology,* 15, 455–456.

Chamberlin, M. J., Nierman, W. C., Wiggs, J., & Neff, N. (1979). A quantitative assay for bacterial RNA polymerases. *Journal of Biological Chemistry,* 254, 10061–10069.

Chang, P. K., & Dignam, J. D. (1990). Primary structure of alanyl-tRNA synthetase and the regulation of its mRNA levels in *Bombyx mori. Journal of Biological Chemistry,* 265, 20898–20906.

Chauvin, G. (1977). Contribution à l'étude des insectes keratophages (Lepidoptera: Tineidae). Leurs principales adaptations à la vie en milieu sec. Thesis, Université de Rennes. (Cited in Fehrenbach et al., 1987.)

Chauvin, G., & Barbier, R. (1972a). Perméabilité et ultrastructures des oeufs des deux lépidoptères Tineidae: *Monopis rusticella* et *Trichophaga tapetzella*. *Journal of Insect Physiology*, 18, 1447–1462.
Chauvin, G., & Barbier, R. (1972b). Développement des oeufs en fonction de l'humidité et structure de leurs enveloppes chez quatre lépidoptères Tineidae: *Monopis rusticella* Clerck, *Trichophaga tapetzella* L., *Tineola bisselliella* Hum. et *Tinea pellionella* L. *97è Congrès National des Sociétés Savantes, Nantes, Sciences*, 3, 627–643. (Cited in Fehrenbach et al., 1987.)
Chauvin, G., & Barbier, R. (1974). Ultrastructure des oeufs parthenogénétiques de *Luffia ferchaultella* Steph. et de *Fumea casta* Pallas (Lepidoptera, Psychidae). *Bulletin biologique de la France et de la Belgique*, 108, 245–252.
Chauvin, G., & Barbier, R. (1979). Morphogenèse de l'enveloppe vitelline, ultrastructure du chorion et de la cuticle sérosale chez *Korscheltellus lupulinus* L. (Lepidoptera: Hepialidae). *International Journal of Insect Morphology and Embryology*, 8, 375–386.
Chauvin, G., Rahn, R., & Barbier, R. (1974). Comparison des oeufs des Lépidoptères *Phalera bucephala* L. (Ceruridae), *Acrolepiaassectella* Z. et *Plutella maculipennis* Curt. (Plutellidae): morphologie et ultrastructures particulières du chorion au contact du support végétal. *International Journal of Insect Morphology and Embryology*, 3, 247–256.
Chauvin, J. T., & Chauvin, G. (1980). Formation des reliefs externes de l'oeuf *Micropteryx calthella* L. (Lepidoptera: Micropterigidae). *Canadian Journal of Zoology*, 58, 761–766.
Chavancy, G., Chevallier, A., Fournier, A., & Garel, J. P. (1979). Adaptation of iso-tRNA concentration to mRNA codon frequency in the eukaryotic cell. *Biochimie*, 61, 71–78.
Chavancy, G., Garel, J. P., & Daillie, J. (1975). Functional adaptation of aminoacyl-tRNA synthetases to fibroin biosynthesis in the silkgland of *Bombyx mori*. *FEBS Letters*, 49, 380–388.
Cherbas, L., Koehler, M. M., & Cherbas, P. (1989). Effects of juvenile hormone on the ecdysone response of *Drosophila* Kc cells. *Development Genetics*, 10, 177–188.
Cherbas, L., Lee, K., & Cherbas, P. (1991). Identification of ecdysone response elements by analysis of the *Drosophila* Eip28/29 gene. *Genes and Development*, 5, 120–131.
Cherbas, P. (1973). Biochemical studies of insecticyanin. Ph. D. thesis, Harvard University.
Chikushi, H. (1972). *Genes and Genetical Stocks of the Silkworm*. Tokyo: Keigaku.
Chinnaswamy, K. P., & Devaiah, M. C. (1987). Susceptibility of different larval instars of the silkworm *Bombyx mori* Linnaeus to *Aspergillus tamarii* Kita. *Sericologia*, 27, 399–404.
Chino, H., Murakami, S., & Harashima, K. (1969). Diglyceride-carrying lipoproteins in insect hemolymph: isolation, purification and properties. *Biochimica et Biophysica Acta*, 176, 1–26.
Chippendale, G. M., & Yin, C.-M. (1973). Endocrine activity retained in diapause insect larvae. *Nature*, 246, 511–513.
Chisholm, G. E., & Henner, D. J. (1988). Multiple early transcripts and splicing of the *Autographa californica* nuclear polyhedrosis virus IE-1 gene. *Journal of Virology*, 62, 3193–3200.

Christensen, B., Fink, J., Merrifield, R. B., & Mauzerall, D. (1988). Channel-forming properties of cecropins and related model compounds incorporated into planar lipid membranes. *Proceedings of the National Academy of Sciences, U.S.A.*, 85, 5072–5076.

Christensen, P. J. H. (1943). Serosa-und Amnionbildung der Lepidopteren. *Entomologiske Meddelelser*, 23, 204–223.

Christensen, P. J. H. (1953). The embryonic development of *Cochlidion limacodes* Hufn. (Fam., Cochlididae, Lepidoptera). A study of living dated eggs. *Kongelige Danske Videnskabarnes Selsk Biologiske Skrifterr*, 6, 1–46.

Clare, J., & Farabaugh, P. (1985). Nucleotide sequence of a yeast Ty element: evidence for a novel mechanism of gene expression. *Proceedings of the National Academy of Sciences, U.S.A.*, 82, 2829–2833.

Clark, W. C., Doctor, J., Fristrom, J. W., & Hodgetts, R. B. (1986). Differential responses of the dopa decarboxylase gene to 20-OH-ecdysone in *Drosophila melanogaster*. *Developmental Biology*, 114, 141–150.

Clem, R. J., Fechheimer, M., & Miller, L. K. (1991). Prevention of apoptosis by a baculovirus gene during infection of insect cells. *Science*, 254, 1388–1390.

Clever, U. (1964). Actinomycin and puromycin: effects on sequential gene activation by ecdysone. *Science*, 146, 794–795.

Clever, U., & Karlson, P. (1960). Induktion von Puff-Vernderungen in den Speicheldrsen-chromosomen von *Chironomus tentans* durch Ecdyson. *Experimental Cell Research*, 20, 623–626.

Cock, A. G. (1964). Dosage compensation and sex-chromatin in non-mammals. *Genetical Research* (Camb.), 5, 354–365.

Coen, E., Strachan, T., & Dover, G. (1982). Dynamics of concerted evolution of ribosomal DNA and histone gene families in the *melanogaster* species subgroup of *Drosophila*. *Journal of Molecular Biology*, 158, 17–35.

Coen, E. S., Robbins, T. P., Almeida, J., Hudson, A., & Carpenter, R. (1989). Consequences and mechanisms of transposition in *Antirrhinum majus*. In *Mobile DNA*, ed. D. E. Berg & M. M. Howe, pp. 413–436. Washington, DC: American Society of Microbiology.

Colleaux, L., d'Auriol, L., Betermier, M., Cottarel, G., Jacquier, A., Galibert, F., & Dujon, B. (1986). Universal code equivalent of a yeast mitochondrial intron reading frame is expressed into *E. coli* as a specific double strand endonuclease. *Cell*, 44, 521–533.

Comings, D. E. (1978). Mechanisms of chromosome banding and implications for chromosome structure. *Annual Review of Genetics*, 12, 25–46.

Condoulis, W. V., & Locke, M. (1966). The deposition of endocuticle in an insect, *Calpodes ethlius* stoll (Lepidoptera, Hesperiidae). *Journal of Insect Physiology*, 12, 311–323.

Corpuz, L. M., Choi, H., Muthukrishnan, S., & Kramer, K. J. (1991). Sequences of two cDNAs and expression of the genes encoding methionine-rich storage proteins of *Manduca sexta*. *Insect Biochemistry*, 21, 265–276.

Couble, P., Chevillard, M., Moine, A., Ravel-Chapuis, P., & Prudhomme, J.-C. (1985). Structural organization of the P25 gene of *Bombyx mori* and comparative analysis of its 5′ flanking DNA with that of the fibroin gene. *Nucleic Acids Research*, 13, 1801–1814.

Couble, P., Michaille, J.-J., Garel, A., Couble, M.-L., & Prudhomme, J.-C. (1987). Developmental switches of sericin mRNA splicing in individual cells of *Bombyx mori* silkgland. *Developmental Biology*, 124, 431–440.

Couble, P., Moine, A., Garel, A., & Prudhomme, J.-C. (1983). Developmental variations of a nonfibroin mRNA of *Bombyx mori* silkgland, encoding for a low-molecular-weight silk protein. *Developmental Biology,* 97, 398–407.

Couch, J. A., & Martin, S. M. (1984). A simple system for the preliminary evaluation of infectivity and pathogenesis of insect virus in a nontarget estuarine shrimp. *Journal of Invertebrate Pathology,* 43, 351–357.

Coutagne, G. (1902). Recherches expérimentales sur l'héridité chez les vers à soie. *Bulletin Scientifique de la France et de la Belgique,* 37, 1–194.

Cox, D. L., & Willis, J. H. (1985). The cuticular proteins of *Hyalophora cecropia* from different anatomical regions and metamorphic stages. *Insect Biochemistry,* 15, 349–362.

Cox, D. L., & Willis, J. H. (1987). Analysis of the cuticular proteins of *Hyalophora cecropia* with two dimensional electrophoresis. *Insect Biochemistry,* 17, 457–468.

Crain, W. R., Boshar, M. F., Cooper, A. D., Durica, D. S., Nagy, A., & Steffen, D. (1987). The sequence of a sea urchin muscle actin gene suggests a gene conversion with a cytoskeletal actin gene. *Journal of Molecular Evolution,* 25, 37–45.

Crain, W. R., Davidson, E. H., & Britten, R. J. (1976). Contrasting patterns of DNA sequence arrangement in *Apis mellifera* (honeybee) and *Musca domestica* (housefly). *Chromosoma* (Berl.), 59, 1–12.

Crampton, H. E. (1900). An experimental study upon Lepidoptera. *Wilhelm Roux' Archiv für Entwicklungsmechanik der Organismen,* 9, 293–318.

Crawford, A. M. (1981). Attempts to obtain *Orocytes* baculovirus replication in three insect cell cultures. *Virology,* 112, 625–633.

Crawford, A. M., & Miller, L. K. (1988). Characterization of an early gene accelerating expression of late genes of the baculovirus *Autographa californica* nuclear polyhedrosis virus. *Journal of Virology,* 62, 2773–2781.

Cruickshank, W. J. (1972). Ultrastructural modifications in the follicle cells and egg membranes during development of flour moth oocytes. *Journal of Insect Physiology,* 8, 485–498.

Curtis, A. T., Hori, M., Green, J. M., Wolfgang, W. J., Hiruma, K., & Riddiford, L. M. (1984). Ecdysteroid regulation of the onset of cuticular melanization in allatectomized and black mutant *Manduca sexta* larvae. *Journal of Insect Physiology,* 30, 597–606.

D'Amato, F. (1989). Polyploidy in cell differentiation. *Caryologia,* 42, 183–211.

Dai, W. (1986). Regulation of the synthesis and secretion of lysozyme by fat body from the tobacco hornworm, *Manduca sexta,* in vitro. M.S. thesis, Purdue University.

Dai, W. (1988). Structure and regulation of the synthesis and secretion of an attacin-like protein from larvae of *Manduca sexta.* Ph.D. thesis, Purdue University.

Daniels, G. R., & Deininger, P. L. (1985). Repeat sequence families derived from mammalian tRNA genes. *Nature,* 317, 819–822.

Danilevskii, A. S. (1965). *Photoperiodism and Seasonal Development of Insects.* Edinburgh: Oliver and Boyd.

Danyluk, G. M., & Maruniak, J. E. (1987). In vivo and in vitro range of *Autographa californica* nuclear polyhedrosis virus and *Spodoptera frugiperda* nuclear polyhedrosis virus. *Journal of Invertebrate Pathology,* 50, 207–212.

Darbon, H., Zlotkin, E., Kopeyan, C., van Rietschoten, J., & Rochat, H. (1982). Covalent structure of the insect toxin of the North African scorpion *Androctonus australis.* Hector. *International Journal of Peptide Protein Research,* 20, 320–330.

Datta, R. K., & Pershad, G. D. (1988). Combining ability among multivoltine × bivoltine silkworm (*Bombyx mori* L.) hybrids. *Sericologia,* 28, 21–29.

Datta, R. K., Sengupta, K., & Das, S. K. (1978). Induction of dominant lethals with ethyl methane-sulfonate in male germ cells of mulberry silkworm, *Bombyx mori* L. *Mutation Research,* 56, 299–304.

David, D. T., Kennedy, J. S., & Ludlow, A. R. (1983). Finding of a sex pheromone source by gypsy moths released in the field. *Nature,* 303, 804–806.

Davidson, E. H., Hough, B. R., Amenson, C. S., & Britten, R. J. (1973). General interspersion of repetitive with non-repetitive sequence elements in the DNA of *Xenopus. Journal of Molecular Biology,* 61, 615–627.

Davis, D. R. (1986). A new family of monotrysian moths from Austral South America (Lepidoptera: Palaephatidae), with a phylogenetic review of the Monotrysia. *Smithsonian Contributions to Zoology,* 434.

Dawid, I. B., & Rebbert, M. L. (1981). Nucleotide sequence at the boundaries between gene and insertion regions in the rDNA of *Drosophila melanogaster. Nucleic Acids Research,* 9, 5011–5020.

Dawid, I. B., Wellauer, P. K., & Long, E. O. (1978). Ribosomal DNA and related sequences in *Drosophila melanogaster.* Isolation and characterization of cloned fragments. *Journal of Molecular Biology,* 126, 749–768.

Dean, R., Bollenbacher, W. E., Locke, M., Smith, S. L., & Gilbert, L. I. (1980). Hemolymph ecdysteroid levels and cellular events in the intermoult/moult sequence of *Calpodes ethlius. Journal of Insect Physiology,* 26, 267–280.

Dean, R. L., Locke, M., & Collins, J. V. (1985). Structure of the fat body. In *Comprehensive Insect Physiology, Biochemistry and Pharmacology,* vol. 3, ed. G. A. Kerkut & L. I. Gilbert, pp. 155–210. Oxford: Pergamon.

DeAzambuja, P., Freitas, C. C., & Garcia, E. S. (1986). Evidence and partial characterization of an inducible antibacterial factor in the haemolymph of *Rhodnius prolixus. Journal of Insect Physiology,* 32, 807–812.

de Dianous, S., Hourau, F., & Rochat, H. (1987). Re-examination of the specificity of the scorpion *Androctonous australis* Hector insect toxin towards arthropods. *Toxicon,* 25, 411–417.

DeFranco, D., Schmidt, O., & Soll, D. (1980). Two control regions for eukaryotic tRNA gene transcription. *Proceedings of the National Academy of Sciences, U.S.A.,* 77, 3365–3368.

Deininger, P. (1989). SINEs: short interspersed repeated DNA elements in higher eukaryotes. In *Mobile DNA,* ed. D. E. Berg & M. E. Howe, pp. 619–636. Washington, DC: American Society of Microbiology.

Deininger, P. L., & Daniels, G. R. (1986). The recent evolution of mammalian repetitive DNA elements. *Trends in Genetics,* 2, 76–80.

de Kramer, J. J., & Hemberger, J. (1987). The neurobiology of pheromone reception. In *Pheromone Biochemistry,* ed. G. D. Prestwich & G. L. Blomquist, pp. 433–472. Orlando, FL: Academic Press.

De Loof, A., & Schoofs, L. (1990). Homologies between the amino acid sequences of some vertebrate peptide hormones and peptides isolated from invertebrate sources. *Comparative Biochemistry and Physiology,* B, 95, 459–468.

Denell, R. E., Hummels, K. R., Wakimoto, B. T., & Kaufman, T. C. (1981). Developmental studies of lethality associated with the *Antennapedia* gene complex in *Drosophila melanogaster. Developmental Biology,* 81, 43–50.

Denlinger, D. L. (1985). Hormonal control of diapause. In *Comprehensive Insect Physiology, Biochemistry and Pharmacology,* vol. 8, ed. G. A. Kerkut & L. I. Gilbert, pp. 353–412. Oxford: Pergamon.

Dent, J. N. (1988). Hormonal interaction in amphibian metamorphosis. *American Zoologist,* 28, 297–308.

DeRosier, D. J., Tilney, L. G., & Egelman, E. (1980). Actin in the inner ear: the remarkable structure of the steriocilium. *Nature,* 287, 291–296.

DiBello, P. R., Withers, D. A., Bayer, C. A., Fristrom, J. W., & Guild, G. M. (1991). The *Drosophila* Broad-Complex encodes a family of related, zinc finger-containing proteins. *Genetics,* 129, 385–397.

Dickinson, L., Russell, V., & Dunn, P. E. (1988). A family of bacteria-regulated, cecropin D-like peptides from *Manduca sexta. Journal of Biological Chemistry,* 263, 19424–19429.

Dietrich, W., Katz, H., Lincoln, S. E., Shin, H.-S., Friedman, J., Dracopoli, N. C., & Lander, E. S. (1992). A genetic map of the mouse suitable for typing intraspecific crosses. *Genetics,* 131, 423–447.

Digan, M. E., Spradling, A. C., Waring, G. L., & Mahowald, A. P. (1979). The genetic analysis of chorion morphogenesis in *Drosophila melanogaster.* In *Eucaryotic Gene Regulation,* ed. R. Axel, T. Maniatis, & C. F. Fox, pp. 171–181. New York: Academic Press.

Dignam, S. S., & Dignam, J. D. (1984). Glycyl- and alanyl-tRNA synthetases from *Bombyx mori:* purification and properties. *Journal of Biological Chemistry,* 259, 4043–4048.

Dimarcq, J. L., Keppi, E., Dunbar, B., Lambert, J., Reichhart, J.-M., Hoffmann, D., Rankine, S., Fothergill, J. E., & Hoffmann, J. A. (1988). Insect immunity: purification and characterization of a family of novel inducible antibacterial proteins from immunized larvae of the dipteran *Phormia terranovae* and complete amino-acid sequence of the predominant member, diptericin A. *European Journal of Biochemistry,* 171, 17–22.

Dimarcq, J.-L., Zachary, D., Hoffmann, J. A., Hoffmann, D., & Reichhart, J.-M. (1990). Insect immunity: expression of the two major inducible antibacterial peptides, defensin and diptericin, in *Phormia terranovae. The EMBO Journal,* 9, 2507–2515.

Dobens, L., Rudolph, K., & Berger, E. M. (1991). Ecdysterone regulatory elements function as both transcriptional activators and repressors. *Molecular and Cellular Biology,* 11, 1846–1853.

Doe, C. Q., Chu-LaGraff, Q., Wright, D. M., & Scott, M. P. (1991). The *prospero* gene specifies cell fates in the *Drosophila* central nervous system. *Cell,* 65, 451–464.

Doira, H. (1978). Genetic stocks of the silkworm. In *The Silkworm: An Important Laboratory Tool,* ed. Y. Tazima, pp. 53–81. Tokyo: Kodansha.

Doira, H. (1983). Linkage maps of *Bombyx mori*–status quo in 1983. *Sericologia,* 23, 245–256.

Doira, H. (1992). *Genetical Stocks and Mutations of* Bombyx mori: *Important Genetic Resources.* Fukuoka, Japan: Institute of Genetic Resources, Kyushu University.

Dominick, O. S., & Truman, J. W. (1984). The physiology of wandering behaviour in *Manduca sexta.* I. Temporal organization and the influence of the internal and external environments. *Journal of Experimental Biology,* 110, 35–51.
Dominick, O. S., & Truman, J. W. (1985). The physiology of wandering behaviour in *Manduca sexta.* II. The endocrine control of wandering behaviour. *Journal of Experimental Biology,* 117, 45–68.
Dominick, O. S., & Truman, J. W. (1986a). The physiology of wandering behaviour in *Manduca sexta.* III. Organization of wandering behaviour in the larval nervous system. *Journal of Experimental Biology,* 121, 115–132.
Dominick, O. S., & Truman, J. W. (1986b). The physiology of wandering behaviour in *Manduca sexta.* IV. Hormonal induction of wandering behaviour from the isolated nervous system. *Journal of Experimental Biology,* 121, 133–152.
Doncaster, L., & Raynor, G. H. (1906). Breeding experiments with Lepidoptera. *Proceedings of the Zoological Society of London,* 1, 125–133.
Doolittle, R. F., Feng, D.-F., Johnson, M. S., & McClure, M. A. (1989). Origins and evolutionary relationships of retroviruses. *Quarterly Review of Biology,* 64, 1–30.
Doolittle, W. F., & Sapienza, C. (1980). Selfish genes, the phenotype paradigm and genome evolution. *Nature,* 284, 601–604.
Dorn, A., Bishoff, S. T., & Gilbert, L. I. (1987). An incremental analysis of the embryonic development of the tobacco hornworm, *Manduca sexta. International Journal of Invertebrate Reproduction and Development,* 11, 137–158.
Dover, G., & Coen, E. (1981). Spring cleaning ribosomal DNA: a model for multigene evolution. *Nature,* 290, 731–732.
Dover, G. A. (1982). Molecular drive: a cohesive mode of species evolution. *Nature,* 299, 111–117.
Dow, R. C., Carlson, S. D., & Goodman, W. G. (1988). A scanning electron microscope study of the developing embryo of *Manduca sexta* (L.) (Lepidoptera: Sphingidae). *International Journal of Insect Morphology and Embryology,* 17, 231–242.
Drevet, J. R., Skeiky, Y. A. W., & Iatrou, K. (1994). GATA-type zinc finger motif-containing sequences and chorion gene transcription factors of the silkworm *Bombyx mori. Journal of Biological Chemistry,* 269, 10660–10667.
Driever, W., & Nüsslein-Volhard, C. (1988a). A gradient of *bicoid* protein in *Drosophila* embryos. *Cell,* 54, 83–93.
Driever, W., & Nüsslein-Volhard, C. (1988b). The *bicoid* protein determines position in the *Drosophila* embryo in a concentration-dependent manner. *Cell,* 54, 95–104.
Dugdale, J. S. (1974). Female genital configuration in the classification of Lepidoptera. *New Zealand Journal of Zoology,* 1, 127–146.
Duncan, I. (1987). The bithorax complex. *Annual Review of Genetics,* 21, 285–329.
Dunn, P. E. (1986). Biochemical aspects of insect immunology. *Annual Review of Entomology,* 31, 321–339.
Dunn, P. E. (1990). Humoral immunity in insects. *BioScience,* 40, 738–744.
Dunn, P. E. (1991). Insect antibacterial proteins. In *Phylogenesis of Immune Functions,* ed. G. W. Warr & N. Cohen, pp. 19–44. Boca Raton, FL: CRC Press.

Dunn, P. E., & Dai, W. (1990). Bacterial peptidoglycan: a signal for initiation of the bacteria-regulated synthesis and secretion of lysozyme in *Manduca sexta.* In *Defense Molecules,* ed. J. J. Marchalonis & C. L. Reinisch, pp. 33–46. New York: Wiley-Liss.

Dunn, P. E., Dai, W., Kanost, M. R., & Geng, C. (1985). Soluble peptidoglycan fragments stimulate antibacterial protein synthesis by fat body from larvae of *Manduca sexta. Developmental and Comparative Immunology,* 9, 559–568.

Dunn, P. E., Kanost, M. R., & Drake, D. R. (1987). Increase in serum lysozyme following injection of bacteria into larvae of *Manduca sexta.* In *Molecular Entomology,* ed. J. H. Law, pp. 381–390. New York: Alan R. Liss.

Dunstan, Heather M., Young, L. S., & Sprague, K. U. (1994a). TFIIIR is an isoleucine tRNA. *Molecular and Cellular Biology,* 14, 3588–3595.

Dunstan, Heather M., Young, L. S., & Sprague, K. U. (1994b). $tRNA^{Ile}$IAU (TFIIIR) plays an indirect role in silkworm class III transcription in vitro and inhibits low-frequency DNA damage. *Molecular and Cellular Biology,* 14, 3596–3603.

Durnin-Goodman, E. M., & Iatrou, K. (1989). The Gr^B deletion of the chorion locus of the silkmoth *Bombyx mori:* localization of the left breakpoint and isolation of the deletion junction. *Journal of Molecular Biology,* 205, 633–645.

Dush, M. K., & Martin, G. M. (1992). Analysis of mouse *evx* genes: *evx-1* displays graded expression in the primitive streak. *Developmental Biology,* 151, 273–287.

Duve, H., Sewell, J. C., Scott, A. G., & Thorpe, A. (1991). Chromatographic characterisation and biological activities of neuropeptides immunoreactive to antisera against Met5-enkephalin-Arg6-Phe7 (YGGFMRF) extracted from the blowfly *Calliphora vomitoria* (Diptera). *Regulatory Peptides,* 35, 145–159.

Duwel-Eby, L. E., & Karp, R. D. (1990). The inducible humoral immune response to soluble proteins in the American cockroach. In *Defense Molecules,* ed. J. J. Marchalonis & C. L. Reinisch, pp. 63–78. New York: Wiley-Liss.

Eastham, L. E. S. (1927). A contribution to the embryology of *Pieris rapae. Quarterly Journal of Microscopical Science,* 71, 353–393.

Ebinuma, H., & Yoshitake, N. (1981). The genetic system controlling recombination in the silkworm. *Genetics,* 99, 231–245.

Efstratiadis, A., Crain, W. R., Britten, R. J., Davidson, E. H., & Kafatos, F. C. (1976). DNA sequence organization in the lepidopteran *Antheraea pernyi. Proceedings of the National Academy of Sciences, U.S.A.,* 73, 2289–2293.

Eickbush, T. H., & Burke, W. D. (1985). Silkmoth chorion gene families contain patchwork patterns of sequence homology. *Proceedings of the National Academy of Sciences, U.S.A.,* 82, 2814–2818.

Eickbush, T. H., & Burke, W. D. (1986). The silkmoth late chorion locus. II. Gradients of gene conversion in two paired multigene families. *Journal of Molecular Biology,* 190, 357–366.

Eickbush, T. H., Jones, C. W., & Kafatos, F. C. (1981). Organization and evolution of the developmentally regulated silkmoth chorion gene families. In *Developmental Biology Using Purified Genes,* vol. 23, ed. D. Brown & C. F. Fox, pp. 135–153. New York: Academic Press.

Eickbush, T. H., & Kafatos, F. C. (1982). A walk in the chorion locus of *Bombyx mori*. *Cell,* 29, 633–643.

Eickbush, T. H., & Robins, B. (1985). *Bombyx mori* 28S genes contain insertion elements similar to the type I and II elements of *Drosophila melanogaster*. *The EMBO Journal,* 4, 2281–2285.

Eickbush, T. H., Rodakis, G. C., Lecanidou, R., & Kafatos, F. C. (1985). A complex set of early chorion DNA sequences from *Bombyx mori*. *Developmental Biology,* 112, 368–376.

Eissenberg, J. C., Cartwright, I. L., Thomas, G. H., & Elgin, S. C. R. (1985). Selected topics in chromatin structure. *Annual Review of Genetics,* 19, 485–536.

Eisthen, H. L. (1992). Phylogeny of the vomeronasal system and of receptor cell types in the olfactory and vomeronasal epithelia of vertebrates. *Microscopy Research and Technique,* 23, 1–21.

Eldridge, R., Horodyski, F. M., Morton, D. B., O'Reilly, D. R., Truman, J. W., Riddiford, L. M., & Miller, L. K. (1991). Expression of an eclosion hormone gene in insect cells using a baculovirus vector. *Insect Biochemistry,* 21, 341–351.

Endoh, H., & Okada, N. (1986). Total DNA transcription in vitro: a procedure to detect highly repetitive and transcribable sequences with tRNA-like structures. *Proceedings of the National Academy of Sciences U.S.A.,* 83, 251–255.

Engelmann, F. (1990). Hormonal control of arthropod reproduction. In *Progress in Comparative Endocrinology,* ed. A. Epple, C. G. Scanes, & M. H. Stetson, pp. 357–364. New York: Wiley-Liss.

Engelmann, F., Mala, J., & Tobe, S. S. (1987). Cytosolic and nuclear receptors for juvenile hormone in fat bodies of *Leucophaea maderae*. *Insect Biochemistry,* 17, 1045–1052.

Engels, W. (1989). P element in *Drosophila melanogaster*. In *Mobile DNA,* ed. D. E. Berg & M. M. Howe, pp. 437–484. Washington, DC: American Society of Microbiology.

Engels, W. R., Johnson-Schlitz, D. M., Eggleston, W. B., & Sved, J. (1990). High frequency P element loss in *Drosophila* is homolog dependent. *Cell,* 62, 515–525.

Engström, Å., Engström, P., Tao, Z.-j., Carlsson, A., & Bennich, H. (1984). Insect immunity. The primary structure of the antibacterial protein attacin F and its relation to two native attacins from *Hyalophora cecropia*. *The EMBO Journal,* 3, 2065–2070.

Engström, Å., Xanthopoulos, K. G., Boman, H. G., & Bennich, H. (1985). Amino acid and cDNA sequences of lysozyme from *Hyalophora cecropia*. *The EMBO Journal,* 4, 2119–2122.

Engström, P., Carlsson, A., Engström, Å., Tao, Z.-j., & Bennich, H. (1984). The antibacterial effect of attacins from the silk moth *Hyalophora cecropia* is directed against the outer membrane of *Escherichia coli*. *The EMBO Journal,* 3, 3347–3351.

Ephrussi, A., Dickinson, L. K., & Lehmann, R. (1991). *oskar* organizes the germ plasm and directs localization of the posterior determinant *nanos*. *Cell,* 66, 37–50.

Ernst, J. F., Stewart, J. W., & Sherman, F. (1981). The *cyc-11* mutation in yeast reverts by recombination with a nonallelic gene: composite genes determining the iso-cytochromes c. *Proceedings of the National Academy of Sciences U.S.A.*, 78, 6334–6338.

Ernst, K.-D., & Boeckh, J. (1983). A neuroanatomical study on the organization of the central antennal pathways in insects. III. Neuroanatomical characterization of physiologically defined response types of deutocerebral neurons in *Periplaneta americana. Cell and Tissue Research*, 229, 1–22.

Evans, P. D., & Myers, C. M. (1986). The modulatory actions of FMRFamide and related peptides on locust skeletal muscle. *Journal of Experimental Biology*, 126, 403–422.

Evans, R. M. (1988). The steroid and thyroid hormone receptor superfamily. *Science*, 240, 889–895.

Evans, T., & Felsenfeld, G. (1989). The erythroid-specific transcription factor Eryf1: a new finger protein. *Cell*, 58, 877–885.

Eveleth, D. D., Gietz, R. D., Spencer, C. A., Nargant, F. E., Hodgetts, R. B., & Marsh, J. L. (1986). Sequence and structure of the DOPA decarboxylase gene of *Drosophila:* evidence for novel RNA splicing variants. *The EMBO Journal*, 5, 2663–2672.

Ezekowitz, R. A. B., Sastry, K., Bailly, P., & Warner, A. (1990). Molecular characterization of the human macrophage mannose receptor: demonstration of multiple carbohydrate recognition-like domains and phagocytosis of yeasts in COS-1 cells. *Journal of Experimental Medicine*, 172, 1785–1794.

Fadool, D. A., & Ache, B. W. (1992). Plasma membrane inositol 1,4,5-triphosphate-activated channels mediate signal transduction in lobster olfactory receptor neurons. *Neuron*, 9, 907–918.

Fahrbach, S. E., & Truman, J. W. (1987a). Mechanisms of programmed cell death in the nervous system of a moth. In *Selective Neuronal Death*, vol. 126, ed. G. Bock & M. O'Connor, *CIBA Foundation Symposium*, pp. 65–81. Chichester: Wiley.

Fahrbach, S. E., & Truman, J. W. (1987b). Possible interactions of a steroid hormone and neuronal inputs in controlling the death of an identified neuron in the moth *Manduca sexta. Journal of Neurobiology*, 18, 497–508.

Fahrbach, S. E., & Truman, J. W. (1989). Autoradiographic identification of ecdysteroid-binding cells in the nervous system of the moth *Manduca sexta. Journal of Neurobiology*, 20, 681–702.

Fain, M. J., & Riddiford, L. M. (1975). Juvenile hormone titers in the hemolymph during late larval development of the tobacco hornworm, *Manduca sexta* (L.). *Biological Bulletin*, 149, 506–521.

Fain, M. J., & Riddiford, L. M. (1976). Reassessment of the critical periods for prothoracicotropic hormone and juvenile hormone secretion in the larval molt of the tobacco hornworm *Manduca sexta. General and Comparative Endocrinology*, 30, 131–141.

Fain, M. J., & Riddiford, L. M. (1977). Requirements for molting of the crochet epidermis of the tobacco hornworm larva in vivo and in vitro. *Wilhelm Roux's Archives of Developmental Biology*, 181, 285–307.

Farris, J. S. (1970). Methods for computing Wagner trees. *Systematic Zoology*, 19, 83–92.

Fawcett, D. H., Lister, C. K., Kellett, E., & Finnegan, D. J. (1986). Transposable elements controlling I-R hybrid dysgenesis in *D. melanogaster* are similar to mammalian LINEs. *Cell*, 47, 1007–1015.

Faye, I. (1990). Acquired immunity in insects: the recognition of nonself and the subsequent onset of immune protein genes. *Research in Immunology,* 141, 927–932.

Faye, I., Pye, A., Rasmuson, T., Boman, H. G., & Boman, I. A. (1975). Simultaneous induction of antibacterial activity and selective synthesis of some hemolymph proteins in diapausing pupae of *Hyalophora cecropia* and *Samia cynthia. Infection and Immunity,* 12, 1426–1438.

Faye, I., & Wyatt, G. R. (1980). The synthesis of antibacterial proteins in isolated fat body from *Cecropia* silkmoth pupae. *Experientia,* 36, 1325–1326.

Fedoroff, N. V. (1989). Maize transposable elements. In *Mobile DNA,* ed. D. E. Berg & M. M. Howe, pp. 375–412. Washington, DC: American Society of Microbiology.

Fehrenbach, H. (1989). Fine structure of the eggshells of four primitive moths: *Hepialus hecta* (L.), *Wiseana umbraculata* (Guénée) (Hepialidae), *Mnesarchaea fusilella* Walker and *M. acuta* Philp. (Mnesarchaeidae) (Lepidoptera, Exoporia). *International Journal of Insect Morphology and Embryology,* 18, 261–274.

Fehrenbach, H. (in press). Eggshell fine structure in Exoporia–evolution and phylogenetic significance. *Invertebrate Taxonomy.*

Fehrenbach, H., Dittrich, V., & Zissler, D. (1987). Eggshell fine structure of three lepidopteran pests: *Cydia pomonella* (L.) (Tortricidae), *Heliothis virescens* (Fabr.), and *Spodoptera littoralis* (Boisd.) (Noctuidae). *International Journal of Insect Morphology and Embryology,* 16, 201–219.

Felsenfeld, G. (1992). Chromatin as an essential part of the transcriptional mechanism. *Nature,* 355, 219–224.

Fenerjian, M. G. (1991). Evolution and expression of chorion genes in *Drosophila* and the Chinese oak silkmoth. Ph.D. thesis, Harvard University.

Fenerjian, M. G., Martínez-Cruzado, J. C., Swimmer, C., King, D., & Kafatos, F. C. (1989). Evolution of the autosomal chorion cluster in *Drosophila.* II. Chorion gene expression and sequence comparisons of the *s16* and *s19* genes in evolutionarily distant species. *Journal of Molecular Evolution,* 29, 108–125.

Ferkovich, S. M., Mayer, M. S., & Rutter, R. R. (1973). Conversion of the sex pheromone of the cabbage looper. *Nature,* 242, 53–55.

Finlayson, L. H. (1956). Normal and induced degeneration of abdominal muscles during metamorphosis in the Lepidoptera. *Quarterly Journal of Microscopical Sciences,* 97, 215–233.

Finnerty, V. (1976). Genetic units of *Drosophila*–simple cistrons. In *The Genetics and Biology of* Drosophila, vol. 1b, ed. M. Ashburner & E. Novitski, pp. 721–765. New York: Academic Press.

Fitch, W. M. (1971). Toward defining the course of evolution: minimum change for a specific tree topology. *Systematic Zoology,* 20, 406–416.

Flyg, C., Dalhammar, G., Rasmuson, B., & Boman, H. G. (1987). Insect immunity: inducible antibacterial activity in *Drosophila. Insect Biochemistry,* 17, 153–160.

Fogel, S., Mortimer, R. K., Lusnak, K., & Travares, F. (1979). Meiotic gene conversion: a signal of the basic recombination event in yeast. *Cold Spring Harbor Symposia on Quantitative Biology,* 43, 1325–1341.

Fontana, P. G. (1976). Improved resolution of the meiotic chromosomes in both sexes of *Euxoa* species and their hybrids (Lepidoptera: Noctuidae). *Canadian Journal of Genetics and Cytology,* 18, 537–544.

Fotaki, M. E., & Iatrou, K. (1988). Identification of a transcriptionally active pseudogene in the chorion locus of the silkmoth *Bombyx mori.* Regional sequence conservation and biological function. *Journal of Molecular Biology,* 203, 849–860.

Fourcade-Peronnet, F., d'Auriol, L., Becker, J., Balibert, F., & Best-Belpomme, M. (1988). Primary structure and functional organization of *Drosophila 1731* retrotransposon. *Nucleic Acids Research,* 16, 6113–6125.

Fournier, A., Guerin, M.-A., Corlet, J., & Clarkson, S. G. (1984). Structure and in vitro transcription of a glycine tRNA gene from *Bombyx mori. The EMBO Journal,* 3, 1547–1552.

French, V. (1990). The development of segments in the invertebrates. *Seminars in Developmental Biology,* 1, 89–100.

Freund, R., & Meselson, M. (1984). Long terminal repeat nucleotide sequence and specific insertion of the *gypsy* transposon. *Proceedings of the National Academy of Sciences, U.S.A.,* 81, 4462–4464.

Fried, M., & Crothers, D. M. (1981). Equilibria and kinetics of lac repressor-operator interactions by polyacrylamide gel electrophoresis. *Nucleic Acids Research,* 9, 6505–6525.

Friedländer, M., & Hauschteck-Jungen, E. (1986). Regular and irregular divisions of Lepidoptera spermatocytes as related to the speed of meiotic prophase. *Chromosoma* (Berl.), 93, 227–230.

Friedländer, M., & Wahrman, J. (1970). The spindle as a basal body distributor. A study in the meiosis of the male silkworm moth, *Bombyx mori. Journal of Cell Science,* 7, 65–89.

Friesen, P. D., & Miller, L. K. (1986). The regulation of baculovirus gene expression. In *The Molecular Biology of Baculoviruses,* ed. W. Doerfler, pp. 31–49. Berlin, New York, and Tokyo: Springer-Verlag.

Friesen, P. D., & Miller, L. K. (1987). Divergent transcription of early 35- and 94-kilodalton protein genes encoded by the HindIII K genome fragment of the baculovirus *Autographa californica* nuclear polyhedrosis virus. *Journal of Virology,* 61, 2264–2272.

Friesen, P. D., & Nissen, M. S. (1990). Gene organization and transcription of TED, a lepidopteran retrotransposon integrated within the baculovirus genome. *Molecular and Cellular Biology,* 10, 3067–3077.

Fristrom, D., Wilcox, M., & Fristrom, J. (1993). The distribution of PS integrins, laminin A and F-actin during key stages in *Drosophila* wing development. *Development,* 117, 509–523.

Frohnhöfer, H. G., Lehmann, R., & Nüsslein-Volhard, C. (1986). Manipulating the anteroposterior pattern of the *Drosophila* embryo. *Journal of Embryological and Experimental Morphology,* 97, 169–179.

Fujii, T., Sakurai, H., Izumi, S., & Tomino, S. (1989). Structure of the gene for the arylphorin-type storage protein SP2 of *Bombyx mori. Journal of Biological Chemistry,* 264, 11020–11025.

Fujiwara, H., & Ishikawa, H. (1986). Molecular mechanism of introduction of the hidden break into the 28S rRNA of insects: implication based on structural studies. *Nucleic Acids Research,* 14, 6393–6401.

Fujiwara, H., & Ishikawa, H. (1987). Structure of the *Bombyx mori* rDNA: initiation site for transcription. *Nucleic Acids Research,* 15, 1245–1258.

Fujiwara, H., Ninaki, O., Kobayashi, K., Kusuda, J., & Maekawa, H. (1991). A chromosome fragment responsible for genetic mosaicism in larval body marking of the silkworm, *Bombyx mori. Genetical Research* (Camb.), 57, 11–16.

Fujiwara, H., Ogura, T., Takada, N., Miyajima, N., Ishikawa, H., & Maekawa, H. (1984). Introns and their flanking sequences of *Bombyx mori* rDNA. *Nucleic Acids Research,* 12, 6861–6869.

Fujiwara, Y., & Yamashita, O. (1990). Purification, characterization and developmental changes in the titer of a new larval serum protein of the silkworm, *Bombyx mori. Insect Biochemistry,* 20, 751–758.

Fujiwara, Y., & Yamashita, O. (1991). A larval serum protein of the silkworm, *Bombyx mori:* cDNA sequence and developmental specificity of the transcript. *Insect Biochemistry,* 21, 735–741.

Fukuda, S. (1940). Induction of pupation in silkworm by transplanting the prothoracic gland. *Proceedings of the Imperial Academy, Tokyo,* 16, 414–416.

Fukuda, S. (1944). The hormonal mechanism of larval molting and metamorphosis in the silkworm. *Journal of the Faculty of Sciences, Tokyo University, Section IV,* 6, 477–532.

Fukuda, S. (1951). The production of the diapause eggs by transplanting the suboesophageal ganglion in the silkworm. *Proceedings of the Japan Academy,* 27, 672–677.

Fukuta, M., Matsuno, K., Hui, C.-c., Nagata, T., Takiya, S., Xu, P.-X., Ueno, K., & Suzuki, Y. (1993). Molecular cloning of a POU domain-containing factor involved in the regulation of the *Bombyx* Sericin-1 gene. *Journal of Biological Chemistry,* 268, 19471–19475.

Furneaux, P. J. S., & Mackay, A. L. (1972). Crystalline protein in the chorion of insect egg shells. *Journal of Ultrastructural Research,* 38, 343–359.

Fuzeau-Braesch, S. (1985). Colour changes. In *Comprehensive Insect Physiology, Biochemistry and Pharmacology,* vol. 9, ed. G. A. Kerkut & L. I. Gilbert, pp. 549–589. Oxford: Pergamon.

Gabriel, A., Yen, T. J., Schwartz, D. C., Smith, C. L., Boeke, J. D., Sollner-Webb, B., & Cleveland, D. W. (1990). A rapidly rearranging retrotransposon within the miniexon gene locus of *Crithidia fasciculata. Molecular and Cellular Biology,* 10, 615–624.

Gabrielson, O. S., Marzouki, N., Ruet, A., Sentenac, A., & Fromageot, P. (1989). Two polypeptide chains in yeast transcription factor t interact with DNA. *Journal of Biological Chemistry,* 264, 7505–7511.

Gage, L. P. (1974a). The *Bombyx mori* genome: analysis by DNA reassociation kinetics. *Chromosoma* (Berl.), 45, 27–42.

Gage, L. P. (1974b). Polyploidization of the silk gland of *Bombyx mori. Journal of Molecular Biology,* 86, 97–108.

Gage, L. P., Friedlander, E., & Manning, R. F. (1975). Interspersion of inverted and middle repeated sequences within the genome of the silkworm, *Bombyx mori.* In *Molecular Mechanisms in the Control of Gene Expression,* ed. D. P. Nierlich, W. J. Rutter, & C. F. Fox, pp. 593–598. New York: Academic Press.

Galas, D. J., & Schmitz, A. (1978). DNAase footprinting: a simple method for the detection of DNA binding specificity. *Nucleic Acids Research,* 5, 3157–3170.

Galli, G., Hofstetter, H., & Birnstiel, M. L. (1981). Two conserved sequence blocks within eukaryotic tRNA genes are major promoter elements. *Nature,* 294, 626–631.

Gamo, T. (1982). Genetic variants of the *Bombyx mori* silkworm encoding sericin proteins of different lengths. *Biochemical Genetics,* 20, 165–177.

Gamo, T., Inokuchi, T., & Laufer, H. (1977). Polypeptides of fibroin and sericin secreted from the different sections of the silk gland in *Bombyx mori. Insect Biochemistry,* 7, 285–295.

Garcia-Bellido, A. (1977). Homeotic and atavic mutations in insects. *American Zoologist,* 17, 613–630.

Garcia-Bellido, A., Ripoll, P., & Morata, G. (1973). Developmental compartmentalization of the wing disc of *Drosophila. Nature,* 245, 251–253.

Garel, J. P. (1976). Quantitative adaptation of isoacceptor tRNAs to mRNA codons of alanine, glycine and serine. *Nature,* 260, 805–806.

Garel, J. P., Hentzen, D., & Daillie, J. (1974). Codon responses of $tRNA^{Ala}$, $tRNA^{Gly}$ and $tRNA^{Ser}$ from the posterior part of the silkgland of *Bombyx mori* L. *FEBS Letters,* 39, 359–363.

Garel, J. P., Mandel, P., Chavancy, G., & Daillie, J. (1970). Functional adaptation of tRNAs to fibroin biosynthesis in the silkgland of *Bombyx mori* L. *FEBS Letters,* 7, 327–329.

Garel, J. P., Mandel, P., Chavancy, G., & Daillie, J. (1971). Functional adaptation of tRNAs to protein biosynthesis in a highly differentiated cell system. III. Induction of isoacceptor tRNAs during the secretion of fibroin in the silkgland of *Bombyx mori* L. *FEBS Letters,* 12, 249–252.

Garrett, J. E., Knutzon, D. S., & Carroll, D. (1989). Composite transposable elements in the *Xenopus laevis* genome. *Molecular and Cellular Biology,* 9, 3018–3027.

Gauger, A., Glicksman, M. A., Salatino, R., Condie, J. M., Schubiger, G., & Brower, D. L. (1987). Segmentally repeated pattern of expression of a cell surface glycoprotein in *Drosophila* embryos. *Development,* 100, 237–244.

Gaumont, R. (1950). Etudes embryologiques sur l'oeuf de cheimatobie *Operophtera brumata* L., Lepidoptère Geometridae. *Annales de l'Institut National de la Recherche Agronomique, Paris* (C) 1, 253–273.

Gautreau, D., Zetlan, S. R., Mazur, G. D., & Goldsmith, M. R. (1993). A subtle defect underlies a major alteration in lamellar orientation in the Gr^{16} chorion mutant of *Bombyx mori. Developmental Biology,* 157, 60–72.

Gaw, Z.-Y., Lin, N. T., & Zia, T. U. (1959). Tissue culture methods for cultivation of virus *Grasserie. Acta Virologica,* 3, 55–60.

Gehring, W. J., & Hiromi, Y. (1986). Homeotic genes and the homeobox. *Annual Review of Genetics,* 20, 147–173.

Geiduschek, E. P., & Tocchini-Valentini, G. P. (1988). Transcription by RNA polymerase III. *Annual Review of Biochemistry,* 57, 873–914.

Geliebter, J., & Nathenson, S. G. (1988). Microrecombinations generate sequence diversity in the murine major histocompatibility complex: analysis of the K^{bm3}, K^{bm4}, K^{bm10}, and K^{bm11} mutants. *Molecular and Cellular Biology,* 8, 4342–4352.

Geng, C. (1990). Studies of hemolymph coagulation in *Manduca sexta.* Ph.D. thesis, Purdue University.

Geng, C., & Dunn, P. E. (1988). Hemostasis in larvae of *Manduca sexta:* formation of a fibrous coagulum by hemolymph proteins. *Biochemical and Biophysical Research Communications,* 155, 1060–1065.

Geng, C., & Dunn, P. E. (1989). Plasmatocyte depletion in larvae of *Manduca sexta* following injection with bacteria. *Developmental and Comparative Immunology,* 13, 17–23.

Giebultowicz, J. M., Zdarek, J., & Chroscikowska, U. (1980). Cocoon spinning behavior in *Ephestia kuehniella;* correlation with endocrine events. *Journal of Insect Physiology,* 26, 459–464.

Gilbert, L. I. (1989). The endocrine control of molting: the tobacco hornworm, *Manduca sexta,* as a model system. In *Ecdysone from Chemistry to Mode of Action,* ed. J. Koolman, pp. 448–471. Stuttgart: Georg Thieme Verlag.

Giraud, M. M., Castanet, J., Meunier, F. J., & Bouligand, Y. (1978). The fibrous structure of coelacanth scales: a twisted "plywood." *Tissue and Cell,* 10, 671–686.

Glass, C. K., Devary, O. V., & Rosenfeld, M. G. (1990). Multiple cell type-specific proteins differentially regulate target sequence recognition by the retinoic acid receptor. *Cell,* 63, 729–738.

Glass, C. K., Lipkin, S. M., Devary, O. V., & Rosenfeld, M. G. (1989). Positive and negative regulation of gene transcription by a retinoic acid–thyroid hormone receptor heterodimer. *Cell,* 59, 697–708.

Glen, D. M. (1982). Effects of natural enemies on a population of codling moth *Cydia pomonella. Annals of Applied Biology,* 101, 199–201.

Gloor, G. B., Nassif, N. A., Johnson-Schlitz, D. M., Preston, C. R., & Engels, W. R. (1991). Targetted gene replacement in *Drosophila* via P element-induced gap repair. *Science,* 253, 1110–1117.

Godward, M. B. E. (1985). The kinetochore. *International Review of Cytology,* 94, 77–105.

Gogos, J. A., Hsu, T., Bolton, J., & Kafatos, F. C. (1992). Sequence discrimination by alternatively spliced isoforms of a DNA binding zinc finger domain. *Science,* 257, 1951–1955.

Gogos, J. A., Tzertzinis, G., & Kafatos, F. C. (1991). Binding site selection analysis of protein-DNA interactions via solid phase sequencing of oligonucleotide mixtures. *Nucleic Acids Research,* 19, 1449–1453.

Goldschmidt, R. (1915). Some experiments on spermatogenesis in vitro. *Proceedings of the National Academy of Sciences, U.S.A.,* 1, 220–222.

Goldsmith, M. R. (1989). Organization and developmental timing of the *Bombyx mori* chorion gene clusters in strain C108. *Developmental Genetics,* 10, 16–23.

Goldsmith, M. R., & Basehoar, G. (1978). Organization of the chorion genes of *Bombyx mori,* a multigene family. I. Evidence for linkage to chromosome 2. *Genetics,* 90, 291–310.

Goldsmith, M. R., & Clermont-Rattner, E. (1979). Organization of the chorion genes of *Bombyx mori,* a multigene family. II. Partial localization of three gene clusters. *Genetics,* 92, 1173–1185.

Goldsmith, M. R., & Clermont-Rattner, E. (1980). Organization of the chorion genes of *Bombyx mori,* a multigene family. III. Detailed marker composition of three gene clusters. *Genetics,* 96, 201–212.

Goldsmith, M. R., & Kafatos, F. C. (1984). Developmentally regulated genes in silkmoths. *Annual Review of Genetics,* 18, 443–487.

Goldsmith, M. R., & Shi, J. (1994). A molecular map for the silkworm: constructing new links between basic and applied research. In *Silk Polymers: Materials Science and Biotechnology,* ed. D. Kaplan, W. W. Adams, B. Farmer, & C. Viney, *American Chemical Society Symposium Series,* 544, 45–58. Washington, DC: ACS.

Goodman, W. G., Tatham, G., Nesbit, D. J., Bultmann, H., & Sutton, R. D. (1987). The role of juvenile hormone in endocrine control of pigmentation in *Manduca sexta. Insect Biochemistry,* 17, 1065–1069.

Goodpasture, C. (1976). High-resolution chromosome analysis in Lepidoptera. *Annals of the Entomological Society of America,* 69, 764–771.

Gordon, D., Jover, E., Couraud, F., & Zlotkin, E. (1984). The binding of the insect selective neurotoxin (AaIT) from scorpion venom to locust synaptosomal membranes. *Biochimica et Biophysica Acta,* 778, 349–358.

Gorski, K., Carneiro, M., & Schibler, U. (1986). Tissue-specific in vitro transcription from the mouse albumin promoter. *Cell,* 47, 767–776.

Gotwals, P. J., & Fristrom, J. W. (1991). Three neighboring genes interact with the *Broad-Complex* and the *Stubble-stubbloid* locus to affect imaginal disc morphogenesis in *Drosophila. Genetics,* 127, 747–759.

Gould, A. P., Brookman, J. J., Strutt, D. I., & White, R. A. H. (1990). Targets of homeotic gene control in *Drosophila. Nature,* 348, 308–312.

Gould, S. J. (1977). *Ontogeny and Phylogeny.* Cambridge, MA: Harvard University Press.

Govind, S., & Steward, R. (1991). Dorsoventral pattern formation in *Drosophila:* signal transduction and nuclear targeting. *Trends in Genetics,* 7, 119–125.

Grace, T. D. H. (1962). Establishment of four strains of cells from insect tissues grown in vitro. *Nature,* 195, 788–789.

Graham, D. E., Neufeld, B. R., Davidson, E. H., & Britten, R. J. (1974). Interspersion of repetitive and non-repetitive DNA sequences in the sea urchin genome. *Cell,* 1, 127–138.

Granados, R. R., & Lawler, K. A. (1981). In vivo pathway of *Autographa californica* baculovirus invasion and infection. *Virology,* 108, 297–308.

Grandbastien, M.-A., Spielmann, A., & Caboche, M. (1989). *Tnt1,* a mobile retroviral-like transposable element of tobacco isolated by plant cell genetics. *Nature,* 337, 376–380.

Granger, N. A., Niemiec, S. M., Gilbert, L. I., & Bollenbacher, W. E. (1982). Juvenile hormone synthesis in vitro by larval and pupal corpora allata of *Manduca sexta. Molecular and Cellular Endocrinology,* 28, 587–604.

Granger, N. A., Whisenton, L. R., Janzen, W. P., & Bollenbacher, W. E. (1987). Interendocrine control by 20-hydroxyecdysone of the corpora allata of *Manduca sexta. Insect Biochemistry,* 17, 949–953.

Green, E. L. (1981). Breeding systems. In *The Mouse in Biomedical Research,* vol. 1, ed. H. L. Foster, J. D. Small, & J. G. Fox, pp. 91–104. New York: Academic Press.

Greenstein, M. E. (1972a). The ultrastructure of developing wings in the giant silkmoth, *Hyalophora cecropia.* I. Generalized epidermal cells. *Journal of Morphology,* 136, 1–22.

Greenstein, M. E. (1972b). The ultrastructure of developing wings in the giant silkmoth, *Hyalophora cecropia.* II. Scale-forming and socket-forming cells. *Journal of Morphology,* 136, 23–52.

Greer, C. A., Stewart, W. B., Teicher, M. H. & Shepherd, G. M. (1982). Functional development of the olfactory bulb and a unique glomerular complex in the neonatal rat. *Journal of Neuroscience*, 2, 1744–1759.

Gregoire, C. (1974). Hemolymph coagulation. In *The Physiology of Insecta*, 2nd ed., vol. 5, ed. M. Rockstein, pp. 309–360. New York: Academic Press.

Griffith, C. M., & Lai-Fook, J. (1986). Structure and formation of the chorion in the butterfly, *Calpodes. Tissue and Cell*, 18, 589–601.

Gröner, A., Granados, R. R., & Burand J. P. (1984). Interaction of *Autographa californica* nuclear polyhedrosis virus with two nonpermissive cell lines. *Intervirology*, 21, 203–209.

Gross, J. B., & Howland, R. B. (1940). The early embryology of *Prodenia eridania. Annals of the Entomological Society of America*, 33, 56–75.

Grossniklaus-Bürgin, C., & Lanzrein, B. (1990). Qualitative and quantitative analyses of juvenile hormone and ecdysteroids from the egg to the pupal molt in *Trichoplusia ni. Archives of Insect Biochemistry and Physiology*, 14, 13–30.

Grün, L., & Peter, M. G. (1984). Incorporation of radiolabelled tyrosine, N-acetyldopamine, N-β-alanyldopamine, and the arylphorin manducin into the sclerotized cuticle of tobacco hornworm (*Manduca sexta*) pupae. *Zeitschrift für Naturforschung*, 39C, 1066–1074.

Guarino, L. A., & Summers, M. D. (1986). Functional mapping of a transactivating gene required for expression of a baculovirus delayed-early gene. *Journal of Virology*, 57, 563–571.

Guarino, L. A., & Summers, M. D. (1987). Nucleotide sequence and temporal expression of a baculovirus regulatory gene. *Journal of Virology*, 61, 2091–2099.

Guay, P. S., & Guild, G. M. (1991). The ecdysone-induced puffing cascade in *Drosophila* salivary glands: A *Broad-Complex* early gene regulates intermolt and late gene transcription. *Genetics*, 129, 169–175.

Gudmundsson, G. H., Lidholm, D.-A., Åsling, B., Gan, R., & Boman, H. G. (1991). The cecropin locus: cloning and expression of a gene cluster encoding three antibacterial peptides in *Hyalophora cecropia. Journal of Biological Chemistry*, 266, 11510–11517.

Gupta, A. P. (1979). Hemocyte types: their structures, synonymies, interrelationships, and taxonomic significance. In *Insect Hemocytes: Development, Forms, Functions, and Techniques*, ed. A. P. Gupta, pp. 85–127. Cambridge: Cambridge University Press.

Gupta, A. P., ed. (1990). *Morphogenetic Hormones of Arthropods.*, vol. 1. New Brunswick, NJ: Rutgers University Press.

Gupta, M. L., & Narang, R. C. (1981). Karyotype and meiotic mechanism in Muga silkmoths, *Antheraea compta* (Roth.) and *A. assamensis* (Helf.) (Lepidoptera: Saturniidae). *Genetica*, 57, 21–27.

Guthrie, K. M., Anderson, A. J., Leon, M., & Gall, C. M. (1991). Spatially distributed increases in c-fos mRNA in odor-activated regions of the main olfactory bulb. *Society for Neuroscience, Abstracts*, 17, 141.

Gutz, H. (1971). Site specific induction of gene conversion in *Schizosaccharomyces pombe. Genetics*, 69, 317–337.

Gyorgyi, T. K., Roby-Shemkovitz, A. J., & Lerner, M. R. (1988). Characterization and cDNA cloning of the pheromone binding protein from the tobacco hornworm, *Manduca sexta:* a tissue specific, developmentally regulated protein. *Proceedings of the National Academy of Sciences, U.S.A.*, 85, 9851–9855.

Hagenbüchle, O., Larson, D., Hall, G. I., & Sprague, K. U. (1979). The primary transcription product of a silkworm alanine tRNA gene: Identification of in vitro sites of initiation, termination and processing. *Cell,* 18, 1217–1229.

Halpern, M. (1987). The organization and function of the vomeronasal system. *Annual Review of Neuroscience,* 10, 325–362.

Hamada, K., Gleason, S. L., Levi, B.-Z., Hirschfeld, S., Appella, E., & Ozato, K. (1989). H-2RIIBP, a member of the nuclear hormone receptor superfamily that binds to both the regulatory element of major histocompatibility class I genes and the estrogen response element. *Proceedings of the National Academy of Sciences, U.S.A.*, 86, 8289–8293.

Hamada, Y., Yamashita, O., & Suzuki, Y. (1987). Haemolymph control of sericin gene expression studied by organ transplantation. *Cell Differentiation,* 20, 65–76.

Hammock, B. D., Bonning, B. C., Possee, R. D., Hanzlik, T. N., & Maeda, S. (1990). Expression and effects of the juvenile hormone esterase in a baculovirus vector. *Nature,* 344, 458–461.

Hamodrakas, S. J., Bosshard, H. E., & Carlson, C. N. (1988). Structural models of the evolutionary conserved central domain of silkmoth chorion proteins. *Protein Engineering,* 2, 201–207.

Hamodrakas, S. J., Etmektzoglou, T., & Kafatos, F. C. (1985). Amino acid periodicities and their structural implications for the evolutionarily conservative central domain of some silkmoth chorion proteins. *Journal of Molecular Biology,* 186, 583–589.

Han, K., Levine, M. S., & Manley, J. L. (1989). Synergistic activation and repression of transcription by *Drosophila* homeobox proteins. *Cell,* 56, 573–583.

Hansson, B. S., Christensen, T. A., & Hildebrand, J. G. (1991). Functionally distinct subdivision of the macroglomerular complex in the antennal lobe of the male sphinx moth *Manduca sexta. Journal of Comparative Neurology,* 312, 264–278.

Hansson, B. S., Ljungberg, H., Hallberg, E., & Lofstedt, C. (1992). Functional specialization of olfactory glomeruli in a moth. *Science,* 256, 1313–1314.

Hanzlik, T. N., Abdel-Aal, Y. A. I., Harshman, L. G., & Hammock, B. D. (1989). Isolation and sequencing of cDNA clones coding for juvenile hormone esterase from *Heliothis virescens. Journal of Biological Chemistry,* 264, 12419–12425.

Harrap, K. A., & Robertson, J. S. (1968). A possible infection pathway in the development of a nuclear polyhedrosis virus. *Journal of General Virology,* 3, 221–225.

Hasegawa, K. (1951). Studies in voltinism in the silkworm, *Bombyx mori* L., with special reference to the organs concerning determination of voltinism (a preliminary note). *Proceedings of the Japanese Academy,* 27, 667–671.

Hashiguchi, T., Yoshitake, N., & Tsuchiya, Y. (1970). On the phenoloxidase system of the black pupae in the silkworm, *Bombyx mori* L. *Journal of Sericultural Science of Japan,* 39, 37–42. In Japanese with English summary.

Hashimoto, H. (1930). Hereditary superfluous legs in the silkworm. *Japanese Journal of Genetics,* 6, 45–54. In Japanese.

Hashimoto, H. (1941). Linkage studies in the silkworm. *Bulletin of the Sericultural Experiment Station of Japan,* 10, 328–363. In Japanese with English summary.

Hatamura, M. (1939). Genetical studies of the *d*-mottled silkworm. *Bulletin of the Sericultural Experiment Station of Japan,* 9, 353–375. In Japanese with English summary.

Hatfield, D., Matthews, C. R., & Rice, M. (1979). Aminoacyl-transfer RNA populations in mammalian cells. Chromatographic profiles and patterns of codon recognition. *Biochimica et Biophysica Acta,* 564, 414–423.

Hatfield, D., Varricchio, F., Rice, M., & Forget, B. G. (1982). The aminoacyl-tRNA population of human reticulocytes. *Journal of Biological Chemistry,* 257, 3183–3188.

Hattori, M., Kuhara, S., Takenaka, O., & Sakaki, Y. (1986). L1 family of repetitive sequences in primates may be derived from a sequence encoding a reverse transcriptase-related protein. *Nature,* 321, 625–627.

Hatzopoulos, A. K., & Regier, J. C. (1987). Evolutionary changes in the developmental expression of silkmoth chorion genes and their morphological consequences. *Proceedings of the National Academy of Sciences, U.S.A.,* 84, 479–483.

Hayashi, S., & Scott, M. P. (1990). What determines the specificity of action of *Drosophila* homeodomain proteins? *Cell,* 63, 883–894.

Hayes, P. H., Sato, T., & Denell, R. E. (1984). Homoeosis in *Drosophila:* the Ultrabithorax larval syndrome. *Proceedings of the National Academy of Science, U.S.A.,* 81, 545–549.

Hearne, C. M., Ghosh, S., & Todd, J. A. (1992). Microsatellites for linkage analysis of genetic traits. *Trends in Genetics,* 8, 288–294.

Heckel, D. G. (1991). Linkage mapping of insecticide resistance genes in the tobacco budworm, *Heliothis virescens. Sericologia,* 31, Supplement, 23.

Heckel, D. G. (1993). Comparative genetic linkage mapping in insects. *Annual Review of Entomology,* 38, 381–408.

Heckel, D. G., Abbott, A. G., & Brown, T. M. (1988). Genetic linkage mapping and insecticide resistance in *Heliothis virescens. Sericologia,* 28, Supplement, 49.

Hennig, W. (1966). *Phylogenetic Systematics.* Urbana: University of Illinois Press.

Hennig, W. (1981). *Insect Phylogeny.* New York: Wiley.

Henrich, V. C., Sliter, T. J., Lubahn, D. B., MacIntyre A. L., & Gilbert, L. I. (1990). A steroid/thyroid hormone receptor superfamily member in *Drosophila melanogaster* that shares extensive sequence similarity with a mammalian homologue. *Nucleic Acids Research,* 18, 4143–4148.

Hentzen, D., Chevallier, A., & Garel, J. P. (1981). Differential usage of isoaccepting $tRNA^{Ser}$ species in silkglands of *Bombyx mori. Nature,* 290, 267–269.

Herr, W., Sturm, R. A., Clerc, R. G., Corcoran, L. M., Baltimore, D., Sharp, P. A., Ingraham, H. A., Rosenfeld, M. G., Finney, M., Ruvkun, G., & Horvitz, H. R. (1988). The POU domain: a large conserved region in the mammalian *pit-1, oct-1, oct-2* and *Caenorhabditis elegans unc-86* gene products. *Genes and Development,* 2, 1513–1516.

Herrera, R. J., & Wang, J. (1991). Evidence for a relationship between the *Bombyx mori* middle repetitive Bm1 sequence family and U1 snRNA. *Genetica,* 84, 31–37.

Hewes, R. S., & Truman, J. W. (1991). The roles of central and peripheral eclosion hormone release in the control of ecdysis behavior in *Manduca sexta. Journal of Comparative Physiology,* A, 168, 697–707.

Hibner, B. L. (1989). Organization and evolution of the early chorion locus of the silkmoth, *Bombyx mori.* Ph.D. thesis, University of Rochester.

Hibner, B. L., Burke, W. D., & Eickbush, T. H. (1991). Sequence identity in an early chorion multigene family is the result of localized gene conversion. *Genetics,* 128, 595–606.

Hibner, B. L., Burke, W. D., Lecanidou, R., Rodakis, G. C., & Eickbush, T. H. (1988). Organization and expression of three genes from the silkmoth early chorion locus. *Developmental Biology,* 125, 423–431.

Hildebrand, J. G. (1985). Metamorphosis of the insect nervous system. Influences of the periphery on the postembryonic development of the antennal sensory pathway in the brain of *Manduca sexta.* In *Model Neural Networks and Behavior,* ed. A. Selverston, pp. 129–148. New York: Plenum.

Hill-Perkins, M. S., & Possee, R. D. (1990). A baculovirus expression vector derived from the basic protein promoter of *Autographa californica* nuclear polyhedrosis virus. *Journal of General Virology,* 71, 971–976.

Hink, W. F., & Briggs, J. D. (1968). Bactericidal factors in hemolymph from normal and immune wax moth larvae, *Galleria mellonella. Journal of Insect Physiology,* 14, 1025–1034.

Hinton, H. E. (1981). *Biology of Insect Eggs,* vol. I. Oxford: Pergamon.

Hirose, S., & Suzuki, Y. (1988). In vitro transcription of eukaryotic genes is affected differently by the degree of DNA supercoiling. *Proceedings of the National Academy of Sciences, U.S.A.,* 85, 718–722.

Hirschler, J. (1905). Badania embryologiczne nad motylem *Catocala nupta* L. (Embryologishche Untersuchungen an *Catocla nupta* L.). *Bulletin of the International Academy of Science, Cracovie,* 802–810.

Hiruma, K. (1980). Possible roles of juvenile hormone in the prepupal stage of *Mamestra brassicae. General and Comparative Endocrinology,* 41, 392–399.

Hiruma, K., Hardie, J., & Riddiford, L. M. (1991). Hormonal regulation of epidermal metamorphosis in vitro. Control of expression of a larval-specific cuticle gene. *Developmental Biology,* 144, 369–378.

Hiruma, K, & Riddiford, L. M. (1984). Regulation of melanization of tobacco hornworm larval cuticle in vitro. *Journal of Experimental Zoology,* 230, 393–403.

Hiruma, K., & Riddiford, L. M. (1985). Hormonal regulation of dopa decarboxylase during a larval molt. *Developmental Biology,* 110, 509–513.

Hiruma, K., & Riddiford, L. M. (1990). Regulation of dopa decarboxylase gene expression in the larval epidermis of the tobacco hornworm by 20-hydroxyecdysone and juvenile hormone. *Developmental Biology,* 138, 214–224.

Hiruma, K., & Riddiford, L. M. (1993). Molecular mechanisms of cuticular melanization in the tobacco hornworm, *Manduca sexta* (L) (Lepidoptera: Sphingidae). *International Journal of Insect Morphology and Embryology,* 22, 103–117.

Hiruma, K., Riddiford, L. M., Hopkins, T. L., & Morgan, T. D. (1985). Roles of dopa decarboxylase and phenoloxidase in the melanization of the tobacco hornworm and their control by 20-hydroxyecdysone. *Journal of Comparative Physiology,* B, 155, 659–669.

Hishinuma, A., Hockfield, S., McKay, R., & Hildebrand, J. G. (1988). Monoclonal antibodies reveal cell-type-specific antigens in the sexually dimorphic olfactory system of *Manduca sexta.* II. Expression of antigens during postembryonic development. *Journal of Neuroscience,* 8, 308–315.

Hodges, R. W., Dominick, T., Davis, D. R., Ferguson, D. C., Franclemont, J. G., Munroe, E. G., & Powell, J. A., eds. (1983). *Check List of the Lepidoptera of America North of Mexico including Greenland.* London: E. W. Classey and Wedge Entomological Research Foundation.

Hoffmann, D., Hultmark, D., & Boman, H. G. (1981). Insect immunity: *Galleria mellonella* and other Lepidoptera have cecropia-P9-like factors active against gram negative bacteria. *Insect Biochemistry,* 11, 537–548.

Höfte, H., & Whiteley, H. R. (1989). Insecticidal crystal proteins of *Bacillus thuringiensis. Microbiological Review,* 53, 242–255.

Holman, G. M., Cook, B. J., & Nachman, R. J. (1986a). Primary structure and synthesis of a blocked myotropic neuropeptide isolated from the cockroach, *Leucophea maderae. Comparative Biochemistry and Physiology,* C, 85, 219–224.

Holman, G. M., Cook, B. J., & Nachman, R. J. (1986b). Isolation, primary structure and synthesis of leucomyosuppressin, an insect neuropeptide that inhibits the spontaneous contractions of the cockroach hindgut. *Comparative Biochemistry and Physiology,* C, 85, 329–333.

Holman G. M., Nachman, R. J., & Wright, M. S. (1990a). Comparative aspects of insect myotropic peptides. *Progress in Clinical and Biological Research,* 342, 35–39.

Holman G. M., Nachman, R. J., & Wright, M. S. (1990b). Insect neuropeptides. *Annual Review of Entomology,* 35, 201–217.

Hopkins, T. R., & Kramer, K. J. (1992). Insect cuticle sclerotization. *Annual Review of Entomology,* 37, 273–302.

Hori, M., Hiruma, K., & Riddiford, L. M. (1984). Cuticular melanization in the tobacco hornworm larva. *Insect Biochemistry,* 14, 267–274.

Hori, M., & Riddiford, L. M. (1982). Regulation of ommochrome biosynthesis in the tobacco hornworm, *Manduca sexta,* by juvenile hormone. *Journal of Comparative Physiology,* B, 147, 1–9.

Horie, Y., & Watanabe, K. (1983). Design of the composition of the artificial diet for the silkworm, *Bombyx mori,* by linear programming method: Application of the ingredients of feeds for domesticated animals and fowls. *Bulletin of the Sericultural Experiment Station of Japan,* 29, 259–283.

Horn, D. H. S. (1989). Historical introduction. In *Ecdysone from Chemistry to Mode of Action,* ed. J. Koolman, pp. 8–19. Stuttgart: Georg Thieme Verlag.

Horodyski, F. M., & Riddiford, L. M. (1989). Expression and hormonal control of a new larval cuticular multigene family at the onset of metamorphosis of the tobacco hornworm. *Developmental Biology,* 132, 292–303.

Horodyski, F. M., Riddiford, L. M., & Truman, J. W. (1989). Isolation and expression of the eclosion hormone from the tobacco hornworm, *Manduca sexta. Proceedings of the National Academy of Sciences, U.S.A.,* 86, 8123–8127.

Horohov, D. W., & Dunn, P. E. (1982). Changes in the circulating hemocyte population of *Manduca sexta* larvae following injection of bacteria. *Journal of Invertebrate Pathology,* 40, 327–339.

Horohov, D. W., & Dunn, P. E. (1983). Phagocytosis and nodule formation by hemocytes of *Manduca sexta* larvae following injection of *Pseudomonas aeruginosa*. *Journal of Invertebrate Pathology,* 41, 203–213.

Hospital, F., Chevalet, C., & Mulsant, P. (1992). Using markers in gene introgression breeding programs. *Genetics,* 132, 1199–1210.

Hotta, Y., Bennett, M. D., Toledo, L. A., & Stern, H. (1979). Regulation of R-protein and endonuclease activities in meiocytes by homologous chromosome pairing. *Chromosoma* (Berl.), 72, 191–201.

Hotta, Y., & Stern, H. (1981). Small nuclear RNA molecules that regulate nuclease accessibility in specific chromatin regions of meiotic cells. *Cell,* 27, 309–319.

Hsaio, T. H. (1985). Feeding behavior. In *Comprehensive Insect Physiology, Biochemistry and Pharmacology,* vol. 9, ed. G. A. Kerkut & L. I. Gilbert, pp. 471–512. Oxford: Pergamon.

Hsu, T., Gogos, J. A., Kirsh, S. A., & Kafatos, F. C. (1992). Multiple zinc finger forms resulting from developmentally regulated alternative splicing of a transcription factor gene. *Science,* 257, 1946–1950.

Hu, W., Shimada, T., & Kobayashi, M. (1992). Developmental changes in electrophoretic patterns of sericin in the silkgland of *Bombyx mori. Journal of Sericultural Science of Japan,* 61, 236–240.

Huber, R., & Hoppe, W. (1965). Zur Chemie des Ecdysons. VII. Die Kristall- und Molekülstrukturanalyse des Insektenverpuppungshormons Ecdyson mit der automatisierten Faltmolekülmethode. *Chemische Berichte,* 98, 2403–2424.

Huebers, H. A., Huebers, E., Finch, C. A., Webb, B. A., Truman, J. W., Riddiford, L. M., Martin, A. W., & Massover, W. H. (1988). Iron binding proteins and their roles in the tobacco hornworm, *Manduca sexta* (L.). *Journal of Comparative Physiology,* B, 158, 291–300.

Hui, C.-c. (1990). A possible involvement of homeobox genes in silk gene regulation. Ph.D. thesis, Nagoya University.

Hui, C.-c., Matsuno, K., & Suzuki, Y. (1990). Fibroin gene promoter contains a cluster of homeodomain binding sites that interact with three silk gland factors. *Journal of Molecular Biology,* 213, 651–670.

Hui, C.-c., Matsuno, K., Ueno, K., & Suzuki, Y. (1992). Molecular characterization and silk gland expression of *Bombyx* engrailed and invected genes. *Proceedings of the National Academy of Sciences, U.S.A.,* 89, 167–171.

Hui, C-c., & Suzuki, Y. (1989). Enhancement of transcription from the Ad2 major late promoter by upstream elements of the fibroin- and sericin-1 genes in silk gland extracts. *Gene,* 85, 403–411.

Hui, C-c., & Suzuki, Y. (1990). Homeodomain binding sites in the promoter region of silk protein genes. *Development, Growth and Differentiation,* 32, 263–273.

Hui, C-c., Suzuki, Y., Kikuchi, Y., & Mizuno, S. (1990). Homeodomain binding sites in the 5′ flanking region of the *Bombyx mori* silk fibroin light-chain gene. *Journal of Molecular Biology,* 213, 395–398.

Huie, L. H. (1918). The formation of the germ band in the egg of the Holly Tortrix moth, *Eudemis naevana* (HB). *Proceedings of the Royal Society, Edinburgh,* 38, 154–165.

Huijser, P., Hennig, W., & Dijkhof, R. (1987). Poly (dC-dA/dG-dT) repeats in the *Drosophila* genome: a key function for dosage compensation and position effects? *Chromosoma* (Berl.), 95, 209–215.

Hülskamp, M., & Tautz, D. (1991). Gap genes and gradients–the logic behind the gaps. *BioEssays,* 13, 261–268.

Hultmark, D., Engström, Å., Andersson, K., Steiner, H., Bennich, H., & Boman, H. G. (1983). Insect immunity: attacins, a family of antibacterial proteins from *Hyalophora cecropia. The EMBO Journal,* 2, 571–576.

Hultmark, D., Engström, Å., Bennich, H., Kapur, R., & Boman, H. G. (1982). Insect immunity: isolation and structure of cecropin D and four minor antibacterial components from *cecropia* pupae. *European Journal of Biochemistry,* 127, 207–217.

Huque, T., & Bruch, R. C. (1986). Odorant- and guanine nucleotide-stimulated phosphoinositide turnover in olfactory cilia. *Biochemical and Biophysical Research Communications,* 137, 36–42.

Hurlbert, R. E., Karlinsey, J. E., & Spence, K. D. (1985). Differential synthesis of bacteria-induced proteins of *Manduca sexta* larvae and pupae. *Journal of Insect Physiology,* 31, 205–215.

Hutchison, C. A., III, Hardies, S. C., Loeb, D. D., Shehee, W. R., & Edgell, M. H. (1989). LINEs and related retroposons: long interspersed repeated sequences in the eucaryotic genome. In *Mobile DNA,* ed. D. E. Berg & M. M. Howe, pp. 593–617. Washington, DC: American Society of Microbiology.

Iatrou, K., & Meidinger, R. G. (1989). *Bombyx mori* nuclear polyhedrosis virus-based vectors for expressing passenger genes in silkmoth cells under viral or cellular promoter control. *Gene,* 75, 59–71.

Iatrou, K., & Meidinger, R. G. (1990). Tissue-specific expression of silkmoth chorion genes in vivo using *Bombyx mori* nuclear polyhedrosis virus as a transducing vector. *Proceedings of the National Academy of Sciences, U.S.A.,* 87, 3650–3654.

Iatrou, K., Meidinger, R. G., & Goldsmith, M. R. (1989). Recombinant baculoviruses as vectors for identifying proteins encoded by intron-containing members of complex multigene families. *Proceedings of the National Academy Sciences, U.S.A.,* 86, 9129–9133.

Iatrou, K., & Tsitilou, S. G. (1983). Coordinately expressed chorion genes of *Bombyx mori:* is developmental specificity determined by secondary structure recognition? *The EMBO Journal,* 2, 1431–1440.

Iatrou, K., Tsitilou, S. G., Goldsmith, M. R., & Kafatos, F. C. (1980). Molecular analysis of the *Gr*^{B} mutation in *Bombyx mori* through the use of a chorion cDNA library. *Cell,* 20, 659–669.

Iatrou, K., Tsitlou, S. G., & Kafatos, F. C. (1984). DNA sequence transfer between two high-cysteine chorion gene families in the silkmoth *Bombyx mori. Proceedings of the National Academy of Sciences, U.S.A.,* 81, 4452–4456.

Ibotombi, N., Bhagirath, T., & Kundu, S. C. (1988). Pachytene synaptonemal complex analysis in the Indian eri silkworm *Philosamia cynthia. Secricologia,* 28, 195–200.

Iggo, R. D., & Lane, D. P. (1989). Nuclear protein p68 is an RNA-dependent ATPase. *The EMBO Journal,* 8, 1827–1831.

Ikenaga, H., & Saigo, K. (1982). Insertion of a movable genetic element, 297, into the T-A-T-A box for the H3 histone gene in *Drosophila melanogaster. Proceedings of the National Academy of Sciences, U.S.A.,* 79, 4143–4147.

Ilenchuk, T. T., & Davey, K. G. (1987). Effects of various compounds on Na/K-ATPase activity, JH I binding capacity and patency response in follicles of *Rhodnius prolixus. Insect Biochemistry,* 17, 1085–1088.

Imai, K., Konno, T., Nakazawa, Y., Komiya, T., Isobe, M., Koga, K., Goto, T., Yaginuma, T., Sakakibara, K., Hasegawa, K., & Yamashita, Y. (1991). Isolation and structure of diapause hormone of the silkworm, *Bombyx mori*. *Proceedings of the Japan Academy*, B, 67, 98–101.

Imboden, H., & Law, J. H. (1983). Heterogeneity of vitellins and vitellogenins of the tobacco hornworm, *Manduca sexta* L. Time course of vitellogenin appearance in the haemolymph of the adult female. *Insect Biochemistry*, 13, 151–162.

Ingham, P. W. (1988). The molecular genetics of embryonic pattern formation in *Drosophila*. *Nature*, 335, 25–34.

Ingham, P. W., & Martinez Arias, A. (1992). Boundaries and fields in early embryos. *Cell*, 68, 221–235.

Inouye, S., Hsu, M., Eagle, S., & Inouye, M. (1989). Reverse transcriptase associated with the biosynthesis of the branched RNA-linked msDNA in *Myxococcus xanthus*. *Cell*, 56, 709–717.

Inouye, S., Yuki, S., & Saigo, K. (1984). Sequence-specific insertion of the *Drosophila* transposable element *17.6*. *Nature*, 310, 332–333.

Ishikawa, E., & Suzuki, Y. (1985). Tissue- and stage-specific expression of sericin genes in the middle silk gland of *Bombyx mori*. *Development, Growth and Differentiation*, 27, 73–82.

Itikawa, N. (1943). Genetical and embryological studies of a dominant mutant, "new additional crescent," of the silkworm. *Japanese Journal of Genetics*, 19, 182–188. In Japanese with English summary.

Itikawa, N. (1944). Anatomical observation of the abnormal embryos of "additional crescent" and "new additional crescent" silkworms. *Japanese Journal of Genetics*, 20, 83. In Japanese.

Itikawa, N. (1951). Genetical studies on the $E^{A}E^{Nc}/+$ silkworm. *Japanese Journal of Genetics*, 26, 244. In Japanese.

Ito, S. (1977). Cytogenetical studies on the chromosomes of silk gland cells of the silkworm with special reference to the structure and behavior of the sex chromosomes. *Japanese Journal of Genetics*, 52, 327–340.

Iwami, M. (1990). The genes encoding bombyxin, a brain secretory peptide of *Bombyx mori:* structure and expression. In *Molting and Metamorphosis*, ed. E. Ohnishi & H. Ishizaki, pp. 49–66. Tokyo and Berlin: Japan Scientific Society Press and Springer-Verlag.

Iwami, M., Adachi, T., Kondo, H., Kawakami, A., Suzuki, Y., Nagasawa, H., Suzuki, A., & Ishizaki, H. (1990). A novel family C of the genes that encode bombyxin, an insulin-related brain secretory peptide of the silkmoth *Bombyx mori:* isolation and characterization of gene C-1. *Insect Biochemistry*, 20, 295–303.

Iwami, M., Kawakami, A., Ishizaki, H., Takahashi, S. Y., Adachi, T., Suzuki, Y., Nagasawa, H., & Suzuki, A. (1989). Cloning of a gene encoding bombyxin, an insulin-like brain secretory peptide of the silkmoth *Bombyx mori* with prothoracicotropic activity. *Development, Growth and Differentiation*, 31, 31–37.

Izumi, S., Sakurai, H., Fujii, T., Ikeda, W., & Tomino, S. (1988). Cloning of mRNA sequence coding for sex-specific storage protein of *Bombyx mori*. *Biochimica et Biophysica Acta*, 949, 181–188.

Izzo, J. (1991). Organization, chromatin structure and molecular evolution of the *Bombyx mori* chorion gene complex. Ph.D. thesis, University of Rochester.

Jackson, J. A., & Fink, G. R. (1981). Gene conversion between duplicated genetic elements in yeast. *Nature,* 292, 306–307.

Jagadeeswaran, P., Forget, B. G., & Weissman, S. M. (1981). Short, interspersed repetitive DNA elements in eukaryotes: transposable DNA elements generated by reverse transcription? *Cell,* 26, 141–142.

Jakubczak, J. L., Burke, W. D., & Eickbush, T. H. (1991). Retrotransposable elements R1 and R2 interrupt the rRNA genes of most insects. *Proceedings of the National Academy of Sciences, U.S.A.,* 88, 3295–3299.

Jakubczak, J. L., Xiong, Y., & Eickbush, T. H. (1990). *Type I (R1)* and *Type II (R2)* ribosomal DNA insertions of *Drosophila melanogaster* are retrotransposable elements closely related to those of *Bombyx mori. Journal of Molecular Biology,* 212, 37–52.

Jamrich, M., & Miller, O. L. (1984). The rare transcripts of interrupted rDNA genes in *Drosophila melanogaster* are processed or degraded during synthesis. *The EMBO Journal,* 3, 1541–1545.

Jan, N. J., & Jan, L. Y. (1990). Genes required for specifying cell fates in *Drosophila* embryonic sensory nervous system. *Trends in Neuroscience,* 13, 493–498.

Janeway, C. A., Jr. (1989). Approaching the asymptote? Evolution and revolution in immunology. *Cold Spring Harbor Symposia on Quantitative Biology,* 54, 1–13.

Jans, P., Benz, G., & Friedländer, M. (1984). A pyrene spermatogenesis-inducing factor is present in the haemolymph of male and female pupae of the codling moth. *Journal of Insect Physiology,* 30, 495–497.

Jantzen, H.-M., Admon, A., Bell, S. P., & Tjian, R. (1990). Nucleolar transcription factor hUBF contains a DNA-binding motif with homology to HMG proteins. *Nature,* 344, 830–836.

Jarnot, B., Watson, C., Laffan, E., Nichols, L., Geysen, J., & Berry, S. J. (1988). Cortical cytoskeleton of giant moth eggs. *Molecular Reproduction and Development,* 1, 35–48.

Jastreboff, P. J., Pedersen, P. E., Greer, C. A., Stewart, W. B., Kauer, J. S., Benson, T. E., & Shepherd, G. M. (1984). Specific olfactory receptor population projecting to identified glomeruli in the rat olfactory bulb. *Proceedings of the National Academy of Sciences, U.S.A.,* 81, 5250–5254.

Jaynes, J. M., Burton, C. A., Barr, S. B., Jeffers, G. W., Julian, G. R., White, K. L., Enright, F. M., Klei, T. R., & Laine, R. A. (1988). In vitro cytocidal effect of novel lytic peptides on *Plasmodium falciparum* and *Trypanosoma cruzi. The FASEB Journal,* 2, 2878–2883.

Jaynes, J. M., Xanthopoulos, K. G., Destefano-Beltran, L., & Dodds, J. H. (1987). Increasing bacterial disease resistance in plants utilizing antibacterial genes from insects. *BioEssays,* 6, 263–270.

Jeffreys, A. J., Wilson, V., & Thein, S. L. (1985). Hypervariable "minisatellite" regions in human DNA. *Nature,* 314, 67–73.

Jesudason, P., Venkatesh, K., & Roe, R. M. (1990). Haemolymph juvenile hormone esterase during the life cycle of the tobacco hornworm, *Manduca sexta* (L.). *Insect Biochemistry,* 20, 593–603.

Jindra, M., Sehnal, F., & Riddiford, L. M. (1994). Isolation, characterization and developmental expression of the ecdysteroid-induced *E75* gene of the wax moth, *Galleria mellonella. European Journal of Biochemistry,* 221, 665–675.

Johannsen, O. A. (1929). Some phases in the embryonic development of *Diacrisia virginica* FABR (Lepidoptera). *Journal of Morphology and Physiology,* 48, 493–541.

Johannsen, O. A., & Butt, F. H. (1941). *Embryology of Insects and Myriapods.* New York: McGraw-Hill.

John, B., & Miklos, G. L. G. (1979). Functional aspects of satellite DNA and heterochromatin. *International Review of Cytology,* 58, 1–113.

John, B., & Miklos, G. L. G. (1988). *The Eukaryotic Genome in Development and Evolution.* New York: Allen and Unwin.

Johnson, M. S., & Turner, J. R. G. (1979). Absence of dosage compensation for a sex-linked enzyme in butterflies (*Heliconius*). *Heredity,* 43, 71–77.

Johnson, P. F. (1990). Transcriptional activators in hepatocytes. *Cell Growth and Differentiation,* 1, 47–52.

Johnson, P. F., & McKnight, S. L. (1989). Eukaryotic transcriptional regulatory proteins. *Annual Review of Biochemistry,* 58, 799–839.

Johnson, W. A., & Hirsh, J. (1990). A *Drosophila* "POU-protein" binds to a sequence element regulating gene expression in specific dopaminergic neurons. *Nature,* 343, 467–470.

Jolly, M. S., Datta, R. K., Noamani, M. K. R., Iyengar, M. N. S., Nagaraj, C. S., Basavaraj, H. K., Kshama, G., & Rao, P. R. M. (1989). Studies on genetic divergence in mulberry silkworm, *Bombyx mori* L. *Sericologia,* 29, 545–553.

Jones, C. W., & Kafatos, F. C. (1980). Structure, organization and evolution of developmentally regulated chorion genes in a silkmoth. *Cell,* 22, 855–867.

Jones, D. T., & Reed, R. R. (1989). G_{olf}: an olfactory neuron specific-G protein involved in odorant signal transduction. *Science,* 244, 790–795.

Jones, G., Brown, N., Manczak, M., Hiremath, S., & Kafatos, F. C. (1990). Molecular cloning, regulation, and complete sequence of a hemocyanin-related, juvenile hormone-suppressible protein from insect hemolymph. *Journal of Biological Chemistry,* 265, 8596–8602.

Jones, G., Hiremath, S. T., Hellmann, G. M., & Rhoads, R. E. (1988). Juvenile hormone regulation of mRNA levels for a highly abundant hemolymph protein in larval *Trichoplusia ni. Journal of Biological Chemistry,* 263, 1089–1092.

Judy, K. J., Schooley, D. A., Dunham, L. L., Hall, M. S., Bergot, B. J., & Siddall, J. B. (1973). Isolation, structure, and absolute configuration of a new natural insect juvenile hormone from *Manduca sexta. Proceedings of the National Academy of Sciences, U.S.A.,* 70, 1509–1513.

Jukes, T. H., & Canter, C. R. (1969). Evolution of protein molecules. In *Mammalian Protein Metabolism,* ed. M. N. Munro, pp. 21–132. New York: Academic Press.

Jürgens, G., Wieschaus, E., Nüsslein-Volhard, C., & Kluding, H. (1984). Mutations affecting the pattern of the larval cuticle in *Drosophila melanogaster.* II. Zygotic loci on the third chromosome. *Wilhelm Roux's Archives of Developmental Biology,* 193, 283–295.

Jurka, J., & Smith, T. (1988). A fundamental division in the *Alu* family of repeated sequences. *Proceedings of the National Academy of Sciences, U.S.A.,* 85, 4775–4778.

Kadonaga, J. T. (1990). Gene transcription: basal and regulated transcription by RNA polymerase. II. *Current Opinion in Cell Biology,* 2, 496–501.

Kafatos, F. C. (1972). The cocoonase zymogen cells of silk moths: a model of terminal cell differentiation for specific protein synthesis. *Current Topics in Developmental Biology,* 7, 125–191.

Kafatos, F. C., Mitsialis, S. A., Nguyen, H. T., Spoerel, N., Tsitilou, S. G., & Mazur, G. D. (1987). Evolution of structural genes and regulatory elements for the insect chorion. In *Development as an Evolutionary Process,* vol. 8, ed. R. Raff & E. C. Raff, pp. 161–178. New York: Alan R. Liss.

Kafatos, F. C., Mitsialis, S. A., Spoerel, N., Mariani, B., Lingappa, J. R., & Delidakis, C. (1985). Studies of developmentally regulated expression and amplification of insect chorion genes. *Cold Spring Harbor Symposia on Quantitative Biology,* 50, 537–547.

Kafatos, F. C., Regier, J. C., Mazur, G. D., Nadel, M. R., Blau, H. M., Petri, W. H., Wyman, A. R., Gelinas, R. E., Moore, P. B., Paul, M., Efstratiadis, A., Vournakis, J. M., Goldsmith, M. R., Hunsley, J. R., Baker, B., Nardi, M., & Koehler, M. (1977). The eggshell of insects: differentiation-specific proteins and the control of their synthesis and accumulation during development. In *Results and Problems in Cell Differentiation,* vol. 8, ed. W. Beermann, pp. 45–145. Berlin: Springer-Verlag.

Kafatos, F. C., & Williams, C. M. (1964). Enzymatic mechanism for the escape of certain moths from their cocoons. *Science,* 146, 538–540.

Kaissling, K.-E. (1974). Sensory transduction in insect olfactory receptors. In *Biochemistry of Sensory Functions,* ed. L. Jaenicke, pp. 243–273. Berlin: Springer-Verlag.

Kaissling, K.-E. (1986). Chemo-electrical transduction in insect olfactory receptors. *Annual Review of Neuroscience,* 9, 121–145.

Kaissling, K.-E., Kasang, G., Bestmann, H. J., Stransky, W., & Vostrowsky, O. (1978). A new pheromone of the silkworm moth *Bombyx mori,* sensory pathway and behavioral effect. *Naturwissenschaften,* 65, 382–384.

Kaissling, K.-E., Keil, T. A., & Williams, J. L. D. (1991). Pheromone stimulation in perfused sensory hairs of the moth *Antheraea polyphemus. Journal of Insect Physiology,* 37, 71–78.

Kajiura, Z., & Yamashita, O. (1989). Stimulated synthesis of the female-specific storage protein in male larvae of the silkworm *Bombyx mori* treated with juvenile hormone analog. *Archives of Insect Biochemistry and Physiology,* 12, 99–109.

Kalthoff, K. (1983). Cytoplasmic determinants in dipteran eggs. In *Space, Time, and Pattern in Embryonic Development,* ed. W. R. Jeffery & R. A. Raff, pp. 313–348. New York: Alan R. Liss.

Kanaratanakul, S., Tharvornanukulkit, C., Wongthong, S., Chareonying, S., Campiranon, A., & Saksoong, P. (1987a). Heterosis in F1 hybrid between polyvoltine and bivoltine silkworm (*Bombyx mori* Linn.). *Sericologia,* 27, 373–380.

Kanaratanakul, S., Tharvornanukulkit, C., Wongthong, S., Chareonying, S., Campiranon, A., & Saksoong, P. (1987b). Combining ability of single cocoon filament length in F1 multivoltine and bivoltine silkworm hybrids. *Sericologia,* 27, 463–470.

Kanda, T. (1992). Genetical study of feeding habits of the silkworm, *Bombyx mori,* to new-low cost diet designed by linear programming method and application to the breeding. *Bulletin of the National Institute of Sericultural and Entomological Science,* 5, 1–89. In Japanese with English summary.

Kanda, T., Tamura, T., & Inoue, H. (1988). Feeding response of the silkworm larva to the LP-1 artificial diet designed by a linear programming method and its inheritance. *Journal of Sericultural Science of Japan,* 57, 489–494. In Japanese with English summary.
Kania, A., & Natori, S. (1989). Cloning of gene cluster for sarcotoxin I, antibacterial proteins of *Sarcophaga peregrina. FEBS Letters,* 258, 199–202.
Kanost, M. R. (1983). The induction of lysozyme and other antibacterial hemolymph proteins in the tobacco hornworm, *Manduca sexta.* Ph.D. thesis, Purdue University.
Kanost, M. P., Dai, W., & Dunn, P. E. (1988). Peptidoglycan fragments elicit antibacterial protein synthesis in larvae of *Manduca sexta. Archives of Insect Biochemistry and Physiology,* 8, 147–164.
Kanost, M. R., Kawooya, J. K., Law, J. H., Ryan, R. O., Van Heusden, M. C., & Ziegler, R. (1990). Insect haemolymph proteins. *Advances in Insect Physiology,* 22, 299–396.
Kanost, M. R., Prasad, S. V., & Wells, M. A. (1989). Primary structure of a member of the serpin superfamily of proteinase inhibitors from an insect, *Manduca sexta. Journal of Biological Chemistry,* 264, 965–972.
Kaomini, M. (1993). Chromosomal arrangement and fine structure mapping of the chorion genes of the domesticated silkmoth, *Bombyx mori.* Ph.D. thesis, University of Rhode Island.
Karch, F, Bender, W., & Weiffenbach, B. (1990). *abdA* expression in *Drosophila* embryos. *Genes and Development,* 4, 1573–1587.
Karim, F. D., & Thummel, C. S. (1991). Ecdysone coordinates the timing and amounts of E74A and E74B transcription in *Drosophila. Genes and Development,* 5, 1067–1079.
Karin, M. (1991). Signal transduction and gene control. *Current Opinion in Cell Biology,* 3, 467–473.
Karlson, P., Hoffmeister, H., Hummel, H., Hocks, P., & Spiteller, G. (1965). Zur Chemie des Ecdysons. VI. Reaktionen des Ecdysonmoleküls. *Chemische Berichte,* 98, 2394–2402.
Karp, R. D., & Duwel-Eby, L. E. (1991). Adaptive immune responses in insects. In *Phylogenesis of Immune Functions,* ed. G. W. Warr & N. Cohen, pp. 1–18. Boca Raton FL: CRC Press.
Kasang, G. (1971). Bombykol reception and metabolism on the antennae of the silkmoth *Bombyx mori.* In *Gustation and Olfaction,* ed. G. Ohloff & A. F. Thomas, pp. 245–53. New York: Academic Press.
Kassavetis, G. A., Riggs, D. L., Negri, R., Nguyen, L. H., & Geiduschek, E. P. (1989). Transcription factor IIIB generates extended DNA interactions in RNA polymerase III transcription complexes on tRNA genes. *Molecular and Cellular Biology,* 9, 2551–2566.
Kastern, W. H., Watson, C. A., & Berry, S. J. (1990). Maternal messenger RNA distribution in silkmoth eggs. I. Clone Ec4B is associated with the cortical cytoskeleton. *Development,* 108, 497–505.
Kataoka, H., Toschi, A., Li, J. P., Carney, R. L., Schooley, D. A., & Kramer, S. J. (1989). Identification of an allatotropin from adult *Manduca sexta. Science,* 243, 1481–1483.
Kataoka, H., Troetschler, R. G., Kramer, S. J., Cesarin, B. J., & Schooley, D. A. (1987). Isolation and primary structure of the eclosion hormone of the to-

bacco hornworm, *Manduca sexta. Biochemical and Biophysical Research Communications,* 146, 746–750.

Kataoka, H., Troetschler, R. G., Li, J. P., Kramer, S. J., Carney, R. L., & Schooley, D. A. (1989). Isolation and identification of a diuretic hormone from the tobacco hornworm, *Manduca sexta. Proceedings of the National Academy of Sciences, U.S.A.,* 86, 2976–2980.

Katsuki, M., Murakami, A., & Watanabe, I. (1980). Fate mapping of some tissues in the genetic mosaics of the silkworm, *Bombyx mori. Zoological Magazine,* 89, 269–276. In Japanese with English summary.

Kawaguchi, Y., Banno, Y., Koga, K., Doira, H., & Fujii, H. (1991). Manifestation of "Giant egg" mutant of *Bombyx mori* (Lepidoptera: Bombycidae). 3. Reciprocal transplant of ovaries. *Japanese Journal of Applied Entomology and Zoology,* 35, 109–113.

Kawaguchi, Y., & Doira, H. (1973). Gene-controlled incorporation of haemolymph protein into the ovaries of *Bombyx mori. Journal of Insect Physiology,* 19, 2083–2096.

Kawaguchi, Y., Doira, H., Banno, Y., & Fujii, H. (1985). Biological effects of N-methyl-N-nitrosourea as revealed by soaking of newly laid eggs of *Bombyx mori. Journal of Sericultural Science of Japan,* 54, 213–222.

Kawaguchi, Y., Doira, H., Banno, Y., & Fujii, H. (1990). Ovary-dependent genetic determination of the egg shape and the yolk protein in the small-egg 2 mutant of *Bombyx mori. Sericologia,* 30, 489–498.

Kawaguchi, Y., & Fujii, H. (1983). Synthesis and accumulation of RNA during oogenesis of *sm*^{n} mutant in *Bombyx mori. Journal of Sericultural Science of Japan,* 52, 233–241.

Kawaguchi, Y., & Fujii, H. (1984). Ovarian protein of *sm*^{n} mutant in *Bombyx mori. Journal of Sericultural Science of Japan,* 53, 448–455. In Japanese with English summary.

Kawaguchi, Y., Miyaji, Y., Doira, H., & Fujii, H. (1988a). Changes of proteins during development of ovary in the small egg 2 mutant of *Bombyx mori. Journal of Sericultural Science of Japan,* 57, 289–297.

Kawaguchi, Y., Nho, S. K., Miyaji, Y., & Fujii, H. (1988b). Characteristics of the *sm-2* egg in *Bombyx mori. Journal of Sericultural Science of Japan,* 57, 157–164.

Kawaguchi, Y., Shito, K., Fujii, H., & Doira, H. (1987). Manifestation of "Giant egg" mutant of *Bombyx mori* (Lepidoptera: Bombycidae). 1. Characteristics of the *Ge egg. Japanese Journal of Applied Entomology and Zoology,* 31, 344–349.

Kawahara, A., Baker, B. S., & Tata, J. R. (1991). Developmental and regional expression of thyroid hormone receptor genes during *Xenopus* metamorphosis. *Development,* 112, 933–943.

Kawakami, A., Iwami, M., Nagasawa, H., Suzuki, A., & Ishizaki, H. (1989). Structure and organization of four clustered genes that encode bombyxin, an insulin-related brain secretory peptide of the silkmoth *Bombyx mori. Proceedings of the National Academy of Sciences, U.S.A.,* 86, 6843–6847.

Kawakami, A., Kataoka, H., Oka, T., Mizoguchi, A., Kimura-Kawakami, M., Adachi, T., Iwami, M., Nagasawa, H., Suzuki, A., & Ishizaki, H. (1990). Molecular cloning of the *Bombyx mori* prothoracicotropic hormone. *Science,* 247, 1333–1335.

Kawamura, N. (1988). The egg size determining gene, *Esd,* is a unique morphological marker on the W chromosome of *Bombyx mori. Genetica,* 76, 195–201.

Kawamura, N. (1990). Is the egg size determining gene, *Esd,* on the W chromosome identical with the sex-linked giant egg gene, *Ge,* in the silkworm? *Genetica,* 81, 205–210.

Kawamura, N., & Nakada, T. (1981). Studies on the increase in egg size in tetraploid silkworms induced from a normal and a giant-egg strains. *Japanese Journal of Genetics,* 56, 249–256.

Kawamura, N., & Niino, T. (1991). Identification of the Z-W bivalent in the silkworm, *Bombyx mori. Genetica,* 83, 121–123.

Kawasaki, H., Sato, H., & Suzuki, M. (1971). Structural proteins in the eggshell of the oriental garden cricket, *Gryllus mitratus. Biochemical Journal,* 125, 495–505.

Kawase, S. (1955). Tyrosinase activity in the integument of a marking mutant in the silkworm, *Bombyx mori. Japanese Journal of Genetics,* 30, 1–6.

Kawazoé, A. (1987). Comparative karyotype analysis of the silkworms, *Bombyx mori* Linnaeus and *B. mandarina* Moore (Lepidoptera: Bombycidae). *La Kromosomo II,* 46, 1521–1532.

Kawooya, J. K., Keim, P. S., Law, J. H., Riley, C. T., Ryan, R. O., & Shapiro, J. P. (1985). Why are green caterpillars green? In *Bioregulators for Pest Control,* ed. P. A. Hedin, *American Chemical Society Symposium Series,* vol. 276, pp. 511–521. Washington, DC: American Chemical Society.

Kawooya, J. K., & Law, J. H. (1983). Purification and properties of microvitellogenin of *Manduca sexta.* Role of juvenile hormone in appearance and uptake. *Biochemical and Biophysical Research Communications,* 117, 643–650.

Kawooya, J. K., Osir, E. O., & Law, J. H. (1987). Microvitellogenin, a 31,000 dalton female specific protein. In *Molecular Entomology,* ed. J. H. Law, pp. 425–532. New York: Alan R. Liss.

Kay, I., Wheeler, C. H., Coast, G. M., Totty, N. F., Cusinato, O., Patel, M., & Goldsworthy, G. J. (1991). Characterization of a diuretic peptide from *Locusta migratoria. Biological Chemistry Hoppe-Seyler,* 372, 929–934.

Kaya, H. K. (1977). Transmission of a nuclear polyhedrosis virus isolated from *Autographa californica* to *Alsophila pometaria, Hyphantria cunea* and other forest defoliators. *Journal of Economic Entomology,* 70, 9–12.

Kayser, H. (1985). Pigments. In *Comprehensive Insect Physiology, Biochemistry and Pharmacology,* vol. 10, ed. G. A. Kerkut & L. I. Gilbert, pp. 368–415. Oxford: Pergamon.

Keddie, B. A., Aponte, G. W., & Volkman, L. E. (1989). The pathway of infection of *Autographa californica* nuclear polyhedrosis virus in an insect host. *Science,* 243, 1728–1730.

Keeley, L. L. (1985). Physiology and biochemistry of the fat body. In *Comprehensive Insect Physiology, Biochemistry and Pharmacology,* vol. 3, ed. G. A. Kerkut & L. I. Gilbert, pp. 211–248. Oxford: Pergamon.

Keil, T. A. (1989). Fine structure of the pheromone sensitive sensilla on the antenna of the hawkmoth, *Manduca sexta. Tissue and Cell,* 21, 139–151.

Keil, T. A., & Steiner, C. (1990a). Morphogenesis of the antenna of the male silkmoth, *Antheraea polyphemus.* I. The leaf-shaped antenna of the pupa from diapause to apolysis. *Tissue and Cell,* 22, 319–336.

Keil, T. A., & Steiner, C. (1990b). Morphogenesis of the antenna of the male silkmoth, *Antheraea polyphemus.* II. Differential mitoses of "dark" precursor cells create the anlagen of sensilla. *Tissue and Cell,* 22, 705–720.

Keil, T. A., & Steiner, C. (1990c). Development of an insect olfactory organ, morphogenesis of the antenna of the silkmoth, *Antheraea polyphemus. Olfaction and Taste,* 10, 226–235.

Keil, T. A., & Steiner, C. (1991). Morphogenesis of the antenna of the male silkmoth, *Antheraea polyphemus.* III. Development of olfactory sensilla and the properties of hair-forming cells. *Tissue and Cell,* 23, 821–851.

Keino, H., & Takesue, S. (1982). Scanning electron microscopic study on the early development of silkworm eggs (*Bombyx mori* L.). *Development, Growth and Differentiation,* 24, 287–294.

Keller, R. E., Danilchik, M., Gimlich, R., & Shih, J. (1985a). Convergent extension by cell intercalation during gastrulation of *Xenopus laevis.* In *Molecular Determinants of Animal Form,* ed. G. M. Edelman, *UCLA Symposium on Molecular and Cell Biology,* pp. 111–141. New York: Alan R. Liss.

Keller, R. E., Danilchik, M., Gimlich, R., & Shih, J. (1985b). The function of convergent extension during gastrulation of *Xenopus laevis. Journal of Embryology and Experimental Morphology,* 89, Supplement, 185–209.

Kent, K. S., & Levine, R. B. (1988). Neural control of leg movements in a metamorphic insect: persistence of larval leg motor neurons to innervate the adult legs of *Manduca sexta. Journal of Comparative Neurology,* 276, 30–43.

Keppi, E., Zachary, D., Robertson, M., Hoffmann, D., & Hoffmann, J. A. (1986). Induced antibacterial proteins in the haemolymph of *Phormia terranovae* (Diptera): purification and possible origin of one protein. *Insect Biochemistry,* 16, 395–402.

Kessel, M., & Gruss, P. (1990). Murine developmental control genes. *Science,* 249, 375–379.

Kidd, S. J., & Glover, D. M. (1981). *Drosophila melanogaster* ribosomal DNA containing Type II insertions is variably transcribed in different strains and tissues. *Journal of Molecular Biology,* 151, 645–662.

Kiely, M. L., & Riddiford, L. M. (1985). Temporal programming of epidermal cell protein synthesis during the larval-pupal transformation of *Manduca sexta. Roux's Archives of Developmental Biology,* 194, 325–335.

Kiguchi, K., & Riddiford, L. M. (1978). The role of juvenile hormone in pupal development of the tobacco hornworm, *Manduca sexta. Journal of Insect Physiology,* 24, 673–680.

Kikkawa, H. (1953). Biochemical genetics of *Bombyx mori* (silkworm). *Advances in Genetics,* 5, 89–140.

Kim, J. H., Moon, M. J., Kim, S. H., Kim, W. K., & Kim, C. W. (1983). A study of egg maturation in the ovary of *Pieris rapae* L. *Korean Journal of Electron Microscopy,* 13, 23–31.

Kimmel, B. E., ole-Moiyoi, O. K., & Young, J. R. (1987). Ingi, a 5.2-kb dispersed sequence element from *Trypanosoma brucei* that carries half of a smaller mobile element at either end and has homology with mammalian LINEs. *Molecular and Cellular Biology,* 7, 1465–1475.

Kimura, K., Oyama, F., Ueda, H., Mizuno, S., & Shimura, K. (1985). Molecular cloning of the fibroin light chain complementary DNA and its use in the study of the expression of the light chain gene in the posterior silk gland of *Bombyx mori. Experientia,* 41, 1167–1171.

Kimura, K.-i., & Truman, J. W. (1990). Postmetamorphic cell death in the nervous and muscular systems of *Drosophila melanogaster. Journal of Neuroscience,* 10, 403–411.

King, R. C. (1970). *Ovarian Development in* Drosophila melanogaster. New York: Academic Press.

King, R. C., & Büning, J. (1985). The origin and functioning of insect oocytes and nurse cells. In *Comprehensive Insect Physiology, Biochemistry and Pharmacology,* vol. 1, ed. G. A. Kerkutt & L. I. Gilbert, pp. 37–82. Oxford: Pergamon.

Klein, H. L., & Petes, T. D. (1981). Intrachromosomal gene conversion in yeast. *Nature,* 289, 144–148.

Klein, U. (1987). Sensillum-lymph proteins from antennal olfactory hairs of the moth *Antheraea polyphemus* (Saturniidae). *Insect Biochemistry,* 17, 1193–1204.

Klug, A., & Rhodes, D. (1987). "Zinc-fingers": a novel protein motif for nucleic acid recognition. *Trends in Biochemical Sciences,* 12, 464–469.

Knowles, B. H., & Ellar, D. J. (1987). Colloid-osmotic lysis is a general feature of the mechanism of action of *Bacillus thuringiensis* k-endotoxins with different insect specificity. *Biochimica et Biophysica Acta,* 924, 509–518.

Kobayashi, Y., & Ando, H. (1981). Embryonic development of the primitive moth, *Neomicropteryx nipponensis* (Lepidoptera, Micropterygidae): Morphogenesis of the embryo by external observation. *Journal of Morphology,* 169, 49–60.

Kobayashi, Y., & Ando, H. (1982). The early embryonic development of the primitive moth, *Neomicropteryx nipponensis* Issiki (Lepidoptera, Micropterygidae). *Journal of Morphology,* 172, 259–269.

Kobayashi, Y., & Ando, H. (1987). Early embryonic development and external features of developing embryos in the primitive moth, *Eriocrania* sp. (Lepidoptera, Eriocraniidae). In *Recent Advances in Insect Embryology in Japan and Poland,* ed. H. Ando & Cz. Jura, pp. 159–194. Tsukuba, Japan: ISEBU.

Kobayashi, Y., & Ando, H. (1988). Phylogenetic relationships among the lepidopteran and trichopteran suborders (Insecta) from the embryological standpoint. *Zeitschrift für Zoologische Systematik und Evolutionsforschung,* 26, 186–210.

Kobayashi, Y., & Miya, K. (1987). Structure of egg cortex relating to presumptive embryonic and extraembryonic regions in silkworm, *Bombyx mori* (Bombycidae: Lepidoptera). In *Recent Advances in Insect Embryology in Japan and Poland,* ed. H. Ando & Cz. Jura, pp. 181–194. Tsukuba, Japan: ISEBU.

Kobori, J. A., Strauss, E., Minard, K., & Hood, L. (1986). Molecular analysis of the hotspot of recombination in the murine major histocompatibility complex. *Science,* 234, 173–179.

Kochansky, J., Tette, J., Taschenberg, E. F., Carde, R. T., Kaissling, K.-E., & Roelofs, W. L. (1975). Sex pheromone of the moth *Antheraea polyphemus* (Saturniidae). *Insect Biochemistry,* 17, 1193–1204.

Kockum, K., Faye, I., Hofsten, P. V., Lee, J.-Y., Xanthopoulos, K. G., & Boman, H. G. (1984). Insect immunity: isolation and sequence of two cDNA clones corresponding to acidic and basic attacins from *Hyalophora cecropia. The EMBO Journal,* 3, 2071–2075.

Koelle, M. R., Segraves, W. A., & Hogness, D. S. (1992). DHR3: a *Drosophila* steroid receptor homolog. *Proceedings of the National Academy of Sciences, U.S.A.*, 89, 6167–6171.

Koelle, M. R., Talbot, W. S., Segraves, W. A., Bender, M. T., Cherbas, P., & Hogness, D. S. (1991). The *Drosophila* EcR gene encodes an ecdysone receptor, a new member of the steroid receptor superfamily. *Cell*, 67, 59–77.

Kollberg, U., Kelber, G., Obermaier, B., & Wolbert, P. (1991). cDNA-cloning, characterization, and expression of a pupal cuticular protein gene in *Galleria mellonella* L. *Sericologia*, 31, Supplement, 12.

Kolodrubetz, D. (1990). Consensus sequence for HMG I-like DNA binding domains. *Nucleic Acids Research*, 18, 5565.

Komano, H., Mizuno, D., & Natori, S. (1980). Purification of lectin induced in the hemolymph of *Sarcophaga peregrina* larvae by injury. *Journal of Biological Chemistry*, 255, 2919–2924.

Kondo, K., Hodgkin, J., & Waterston, R. H. (1988). Differential expression of five tRNA''PUAG amber suppressors in *Caenorhabditis elegans*. *Molecular and Cellular Biology*, 8, 3627–3635.

Kono, T., Nagasawa, H., Isogai, A., Fugo, H., & Suzuki, A. (1987). Amino acid sequence of eclosion hormone of the silkworm, *Bombyx mori*. *Agricultural and Biological Chemistry*, 51, 2307–2308.

Kono, T., Nagasawa, H., Kataoka, H., Isogai, A., Fugo, H., & Suzuki, A. (1990). Eclosion hormone of the silkworm *Bombyx mori*. Expression in *Escherichia coli* and location of disulfide bonds. *FEBS Letters*, 263, 358–360.

Konsolaki, M., Komitopoulou, K., Tolias, P. P., King, D. L., Swimmer, C., & Kafatos, F. C. (1990). The chorion genes of the medfly, *Ceratitis capitata*. I. Structural and regulatory conservation of the *s36* gene relative to two *Drosophila* species. *Nucleic Acids Research*, 18, 1731–1737.

Kopec, S. (1922). Studies on the necessity of the brain for the inception of insect metamorphosis. *Biological Bulletin*, 42, 323–342.

Korenberg, J. R., & Rykowski, M. C. (1988). Human genome organization: Alu, lines, and the molecular structure of metaphase chromosome bands. *Cell*, 53, 391–400.

Kornberg, T. (1981). *engrailed:* a gene controlling compartment and segment formation in *Drosophila*. *Proceedings of the National Academy of Sciences, U.S.A.*, 78, 1095–1099.

Kornberg, T., Siden, I., O'Farrell, P., & Simon, P. (1985). The *engrailed* locus of *Drosophila:* in situ localization of transcripts reveals compartment-specific expression. *Cell*, 40, 45–53.

Kostriken, R., & Heffron, F. (1984). The product of the HO gene is a nuclease: purification and characterization of the enzyme. *Cold Spring Harbor Symposia on Quantitative Biology*, 49, 89–96.

Kraft, R., & Jäckle, H. (in press). *Drosophila* mode of metamerization in the embryogenesis of the lepidopteran insect *Manduca sexta*. *Proceedings of the National Academy of Sciences, U.S.A.*

Kramer, S. J., Toschi, A., Miller, C. A., Kataoka, H., Quistad, G. B., Li, J. P., Carney, R. L., & Schooley, D. A. (1991). Identification of an allatostatin from the tobacco hornworm *Manduca sexta*. *Proceedings of the National Academy of Sciences, U.S.A.*, 88, 9458–9462.

Krappa, R., & Knebel-Mörsdorf, D. (1991). Identification of the very early transcribed baculovirus gene PE-38. *Journal of Virology*, 65, 805–812.

Krause, G. (1939). Die eitypen der insekten. *Biologisches Zentralblatt,* 59, 495–536.

Krause, G., & Krause, J. H. (1964). Schichtenbau und Segmentierung junger Keimanlagen von *Bombyx mori* L. (Lepidoptera) in vitro ohne Dottersystem. *Wilhelm Roux' Archive für Entwicklungsmechanik der Organismen,* 155, 451–510.

Krause, G., & Krause, J. H. (1965). Über das Vermögen median durchschnittener Keimanlagen von *Bombyx mori* L. sich in ovo und sich ohne Dottersystem in vitro zwillingartig zu entwickeln. *Zeitschrift für Naturforschung,* 20b, 334–339.

Kreil, G. (1990). Processing of precursors by dipeptidylaminopeptidases: a case of molecular ticketing. *Trends in Biochemical Sciences,* 15, 23–26.

Kremky, T., & Michalska, E. (1988). Resistance of some mulberry silkworm lines to polyhedrosis. *Sericologia,* 28, 61–70.

Kress, H., & Swida, U. (1990). *Drosophila* glue protein gene expression. *Naturwissenschaften,* 77, 317–324.

Krieger, J., Raming, K., & Breer, H. (1991). Cloning of genomic and complementary DNA encoding insect pheromone binding proteins: evidence for microdiversity. *Biochimica et Biophysica Acta,* 1088, 277–284.

Kristensen, N. P. (1984). Studies on the morphology and systematics of primitive Lepidoptera (Insecta). *Steenstrupia,* 10, 141–191.

Kristensen, N. P. (1989). Insect phylogeny based on morphological evidence. In *The Hierarchy of life,* ed. B. Fernholm, K. Bremer, & H. Jörnvall, pp. 295–306. Amsterdam: Elsevier.

Kuhn, D. (1988). *Textile Technology: Spinning and Reeling,* vol. 5., *Science and Civilisation in China,* part IX, ed. J. Needham. Cambridge: Cambridge University Press.

Kumaran, A. K. (1991). Modes of action of juvenile hormones at cellular and molecular levels. *Morphogenetic Hormones of Arthropods,* vol. 1, ed. A. Gupta, pp. 181–228. New Brunswick, NJ: Rutgers University Press.

Kumaran, A. K., Memmel, N. A., Wang, C., & Trewitt, P. M. (1993). Developmental regulation of arylphorin gene activity in fat body cells and gonadal sheath cells of *Galleria mellonella. Insect Biochemistry and Molecular Biology,* 23, 145–151.

Kumaran, A. K., Ray, A., Tertadian, J. A., & Memmel, N. A. (1987). Effects of juvenile hormone, ecdysteroids and nutrition on larval hemolymph protein gene expression in *Galleria mellonella. Insect Biochemistry,* 17, 1053–1058.

Kurihara, K., & Koyama, N. (1972). High activity of adenyl cyclase in olfactory and gustatory organs. *Biochemical and Biophysical Research Communications,* 48, 30–34.

Kusuda, J., Tazima, Y., Onimaru, K., Ninaki, O., & Suzuki, Y. (1986). The sequence around the 5′ end of the fibroin gene from the wild silkworm, *Bombyx mandarina,* and comparison with that of the domesticated species, *B. mori. Molecular and General Genetics,* 203, 359–364.

Kuwana, J., & Takami, T. (1968). Insecta. In *Invertebrate Embryology,* ed. M. Kume & K. Dan, pp. 405–484. New York: Garland.

Kylsten, P., Samakovlis, C., & Hultmark D. (1990). The cecropin locus in *Drosophila;* a compact gene cluster involved in the response to infection. *The EMBO Journal,* 9, 217–224.

Kyrki, J. (1984). The Yponomeutoidea: a reassessment of the superfamily and its suprageneric groups (Lepidoptera). *Entomologica Scandinavica,* 15, 71–84.

Ladendorff, N., & Kanost, M. R. (1990). Isolation and characterization of bacteria-induced protein P4 from hemolymph of *Manduca sexta. Archives of Insect Biochemistry and Physiology,* 15, 33–42.

Ladendorff, N., & Kanost, M. R. (1991). Bacteria-induced protein P4 (hemolin) from *Manduca sexta:* a member of the immunoglobulin superfamily which can inhibit hemocyte aggregation. *Archives of Insect Biochemistry and Physiology,* 18, 285–300.

Laing, D. G., Panhuber, H., Pittman, E. A., Willcox, M. E., & Eagleson, G. K. (1985). Prolonged exposure to an odor or deodorized air alters the size of mitral cells in the olfactory bulb. *Brain Research,* 336, 81–87.

Lamb, T. D., & Pugh, E. N., Jr. (1992). G-protein cascades: gain and kinetics. *Trends in Neuroscience,* 15, 291–298.

Lambert, J., Keppi, E., Dimarcq, J.-L., Wicker, C., Reichhart, J.-M., Dunbar, B., Lepage, P., Van Dorsselaer, A., Hoffmann, J., Fothergill, J., & Hoffmann, D. (1989). Insect immunity: isolation from immune blood of the dipteran *Phormia terranovae* of two insect antibacterial peptides with sequence homology to rabbit lung macrophage bactericidal peptides. *Proceedings of the National Academy of Sciences, U.S.A.,* 86, 262–266.

Lampe, D. J., & Willis, J. H. (1994). Characterization of a cDNA and gene encoding a cuticular protein from rigid cuticles of the giant silk moth, *Hyalophora cecropia. Insect Biochemistry and Molecular Biology,* 24, 419–435.

Lancet, D., Greer, C. A., Kauer, J. S., & Shepherd, G. M. (1982). Mapping of odor-related neuronal activity in the olfactory bulb by high-resolution 2-deoxyglucose autoradiography. *Proceedings of the National Academy of Sciences, U.S.A.,* 79, 670–674.

Lande, R., & Thompson, R. (1990). Efficiency of marker-assisted selection in the improvement of quantitative traits. *Genetics,* 124, 743–756.

Lander, E. S., & Botstein, D. (1989). Mapping Mendelian factors underlying quantitative traits using RFLP linkage maps. *Genetics,* 121, 185–199.

Lange, A. B., Orchard, I., & Te Brugge, V. A. (1991). Evidence for the involvement of a SchistoFLRF-amide-like peptide in the neural control of locust oviduct. *Journal of Comparative Physiology, A,* 168, 383–391.

Larson, D., Bradford-Wilcox, J., Young, L. S., & Sprague, K. U. (1983). A short 5′ flanking region containing conserved sequences is required for silkworm tRNA gene activity. *Proceedings of the National Academy of Sciences, U.S.A.,* 80, 3416–3420.

Lautenschlager, F. (1932). Die Embryonalentwicklung der weiblichen Keimdruse bei der Psychide *Solenobia triquetrella. Zoologische Jahrbücher Anatomie und Ontogenie der Tiere,* 56, 121–162.

Law, J. H., & Wells, M. A. (1989). Insects as biochemical models. *Journal of Biological Chemistry,* 264, 16335–16338.

Lawrence, P. A. (1992). *The Making of a Fly. The Genetics of Animal Design.* London: Blackwell Scientific.

Lawrence, P. A., & Johnston, P. (1984). On the role of the *engrailed* gene in the internal organs of *Drosophila. The EMBO Journal,* 3, 2839–2844.

Lawrence, P. A., & Morata, G. (1977). The early development of mesothoracic compartments in *Drosophila. Developmental Biology,* 56, 40–51.

Le Blancq, S. M., Swinkels, B. W., Gibson, W. C., & Borst, P. (1988). Evidence for gene conversion between the phosphoglycerate kinase genes of *Trypanosoma brucei*. *Journal of Molecular Biology*, 200, 439–447.

Lecanidou, R., Eickbush, T. H., & Kafatos, F. C. (1984). Ribosomal DNA genes of *Bombyx mori:* a minor fraction of the repeating units contain insertions. *Nucleic Acids Research*, 12, 4703–4713.

Lecanidou, R., Eickbush, T. H., Rodakis, G. C., & Kafatos, F. C. (1983). Novel B family sequence from an early chorion cDNA library of *Bombyx mori*. *Proceedings of the National Academy of Sciences, U.S.A.*, 80, 1955–1959.

Lecanidou, R., Rodakis, G. C., Eickbush, T. H., & Kafatos, F. C. (1986). Evolution of the silkmoth chorion gene superfamily: gene families CA and CB. *Proceedings of the National Academy of Sciences, U.S.A.*, 83, 6514–6518.

Leclerc, R. F., & Miller, S. G. (1990). Identification and molecular analysis of storage proteins from *Heliothis virescens*. *Archives of Insect Biochemistry and Physiology*, 14, 131–150.

Leclerc, R. F., & Regier, J. C. (1990). Heterochrony in insect development and evolution. *Seminars in Developmental Biology*, 1, 271–279.

Lee, J.-K., & Strausfeld, N. J. (1990). Structure, distribution and number of surface sensilla and their receptor cells on the olfactory appendage of the male moth *Manduca sexta*. *Journal of Neurocytology*, 19, 519–538.

Lee, J.-Y., Edlund, T., Ny, T., Faye, I., & Boman, H. G. (1983). Insect immunity: isolation of cDNA clones corresponding to attacins and immune protein P4 from *Hyalophora cecropia*. *The EMBO Journal*, 2, 577–581.

Lees, A. D. (1955). *The Physiology of Diapause in Arthropods*. Cambridge: Cambridge University Press.

Lehmann, R., & Nüsslein-Volhard, C. (1991). The maternal gene *nanos* has a central role in posterior pattern formation of the *Drosophila* embryo. *Development*, 112, 679–691.

LeMotte, P. K., Kuroiwa, A., Fessler, L., & Gehring, W. J. (1989). The homeotic *Sex combs reduced* of *Drosophila:* gene structure and embryonic expression. *The EMBO Journal*, 8, 219–227.

Lenardo, M. J., & Baltimore, D. (1989). NF-κB: a pleiotropic mediator of inducible and tissue-specific gene control. *Cell*, 58, 227–229.

Levenbook, L. (1985). Insect storage proteins. In *Comprehensive Insect Physiology, Biochemistry and Pharmacology*, vol. 10, ed. G. A. Kerkut & L. I. Gilbert, pp. 307–346. Oxford: Pergamon.

Levine, J., & Spradling, A. C. (1985). DNA sequence of a 3.8 kilobase pair region controlling *Drosophila* chorion gene amplification. *Chromosoma* (Berl.), 92, 136–142.

Levine, R. B. (1990). Expansion of the central arborizations of persistent sensory neurons during insect metamorphosis: the role of the steroid hormone, 20-hydroxyecdysone. *Journal of Neuroscience*, 9, 1045–1054.

Levine, R. B., Pak, C., & Linn, D. (1985). The structure, function and metamorphic reorganization of somatotopically projecting sensory neurons in *Manduca sexta* larvae. *Journal of Comparative Physiology*, 157, 1–13.

Levine, R. B., & Truman, J. W. (1982). Metamorphosis of the insect nervous system: changes in the morphology and synaptic interactions of identified cells. *Nature*, 299, 250–252.

Levine, R. B., & Truman, J. W. (1985). Dendritic reorganization of abdominal motoneurons during metamorphosis of the moth *Manduca sexta. Journal of Neuroscience,* 5, 2424–2431.

Levine, R. B., Truman, J. W., Linn, D., & Bate, C. M. (1986). Endocrine regulation of the form and function of axonal arbors during insect metamorphosis. *Journal of Neuroscience,* 6, 293–299.

Levine, R. B., & Weeks, J. C. (1990). Hormonally mediated changes in simple reflex circuits during metamorphosis in *Manduca. Journal of Neurobiology,* 21, 1022–1036.

Lewin, B. (1990a). Commitment and activation at Pol II promoters: a tail of protein-protein interactions. *Cell,* 61, 1161–1164.

Lewin, B. (1990b). RNA polymerase III has a downstream promoter. In *Genes IV,* pp. 574–576. Oxford and Cambridge, MA: Oxford University Press and Cell Press.

Lewis, E. (1963). Genes and developmental pathways. *American Zoologist,* 3, 33–56.

Lewis, E. B. (1978). A gene complex controlling segmentation in *Drosophila. Nature,* 276, 565–570.

Li, W.-c. (1992). Molecular cloning, structure, developmental expression and hormonal control of the two insecticyanin genes in the tobacco hornworm, *Manduca sexta.* Ph.D. thesis, University of Washington.

Li, W.-c., & Riddiford, L. M. (1992). Two distinct genes encode two major isoelectric forms of insecticyanin in the tobacco hornworm, *Manduca sexta. European Journal of Biochemistry,* 205, 491–499.

Li, X.-J., Wolfgang, W., Wu, Y.-N., North, R. A., & Forte, M. (1991). Cloning, heterologous expression and developmental regulation of a *Drosophila* receptor for tachykinin-like peptides. *The EMBO Journal,* 10, 3221–3229.

Lidholm, D. A., Gudmundsson, G. H., Xanthopoulos, K. G., & Boman, H. G. (1987). Insect immunity: cDNA clones coding for the precursor forms of cecropins A and D, antibacterial proteins from *Hyalophora cecropia. FEBS Letters,* 226, 8–12.

Lim, D., & Maas, W. K. (1989). Reverse transcriptase-dependent synthesis of a covalently linked, branched DNA-RNA compound in *E. coli* B. *Cell,* 56, 891–904.

Lin, F.-K., Furr, T. D., Chang, S. H., Horwitz, J., Agris, P. F., & Ortwerth, B. J. (1980). The nucleotide sequence of two bovine lens phenylalanine tRNAs. *Journal of Biological Chemistry,* 255, 6020–6023.

Lindsley, D. L., Sandler, L., Baker, B. S., Carpenter, A. T. C., Denell, R. E., Davis, B. K., Gethmann, R. C., Hardy, R. W., Hessler, A., Miller, S. M., Nozawa, H., Parry, D. M., & Gould-Somero, M. (1972). Segmental aneuploidy and the genetic gross structure of the *Drosophila* genome. *Genetics,* 71, 157–184.

Liu, K. (1992). Peptidoglycan regulation of antibacterial gene expression in MRRL-CH-1, an embryonic cell line derived from *Manduca sexta.* M.S. thesis, Purdue University.

Lizardi, P. M., Mahdavi, V., Shields, D., & Candelas, G. (1979). Discontinuous translation of silk fibroin in a reticulocyte cell-free system and in intact silkgland cells. *Proceedings of the National Academy of Sciences, U.S.A.,* 76, 6211–6215.

Lobo, S. M., & Hernandez, N. (1989). A 7 bp mutation converts a human RNA polymerase II snRNA promoter into an RNA polymerase III promoter. *Cell,* 58, 55–67.

Lobo, S. M., Lister, J., Sullivan, M. L., & Hernandez, N. (1991). The cloned RNA polymerase II transcription factor IID selects RNA polymerase III to transcribe the human U6 gene in vitro. *Genes and Development,* 5, 1477–1489.

Locke, M. (1980). The cell biology of fat body development. In *Insect Biology in the Future,* ed. M. Locke & D. S. Smith, pp. 227–252. New York: Academic Press.

Locke, M. (1984). Epidermal cells. In *Biology of the Integument: Invertebrates,* vol. 1, ed. J. Bereiter-Hahn, A. G. Hatoltsy, & K. S. Richards, pp. 503–522. Heidelberg: Springer-Verlag.

Lockshin, R. A. (1969). Programmed cell death. Activation of lysis by a mechanism involving synthesis of protein. *Journal of Insect Physiology,* 15, 1505–1516.

Lockshin, R. A. (1985). Programmed cell death. In *Comprehensive Insect Physiology, Biochemistry and Pharmacology,* vol. 2, ed. G. A. Kerkut & L. I. Gilbert, pp. 301–317. Oxford: Pergamon.

Loeb, D. D., Padgett, R. W., Hardies, S. C., Shehee, W. R., Comer, M. B., Edgell, M. H., & Hutchison, C. A. III. (1986). The sequence of a large L1Md element reveals a tandemly repeated 5′ end and several features found in retrotransposons. *Molecular and Cellular Biology,* 6, 168–182.

Lohs-Schardin, M., Cremer, C., & Nüsslein-Volhard, C. (1979). A fate map for the larval epidermis of *Drosophila melanogaster:* localized cuticle defects following irradiation of the blastoderm with an ultraviolet laser microbeam. *Developmental Biology,* 73, 239–255.

Long, E. O., & Dawid, I. B. (1979). Expression of ribosomal DNA insertions in *Drosophila melanogaster. Cell,* 18, 1185–1196.

Lounibos, L. P. (1976). Initiation and maintenance of cocoon spinning behaviour by saturniid silkmoths. *Physiological Entomology,* 1, 195–206.

Lucas, F., Shaw, J. T. B., & Smith, S. G. (1958). The silk fibroins. *Advances in Protein Chemistry,* 13, 107–242.

Ludwig, J., Margalit, T., Eismann, E., Lancet, D., & Kaupp, U. B. (1990). Primary structure of cAMP-gated channel from bovine olfactory epithelium. *FEBS Letters,* 270, 24–29.

Luo, Y., Amin, J., & Voellmy, R. (1991). Ecdysterone receptor is a sequence-specific transcription factor involved in the developmental regulation of heat shock genes. *Molecular and Cellular Biology,* 11, 3660–3675.

Lüscher, M. (1944). Experimentelle Untersuchungen über die larvale und die imaginale Determination im Ei der Kleidermotte (*Tineola biselliella* Hum.). *Revue suisse de Zoologie,* 51, 531–627.

Lynn, D. E., Miller, S. G., & Oberlander, H. (1982). Establishment of a cell line for lepidopteran wing imaginal discs: induction of newly synthesized proteins by 20-hydroxyecdysone. *Proceedings of the National Academy of Sciences, U.S.A.,* 79, 2589–2593.

Lyon, M. F. (1992). Some milestones in the history of X-chromosome inactivation. *Annual Review of Genetics,* 26, 17–28.

Lyons, K. M., Stein, J. H., & Smithies, O. (1988). Length polymorphisms in human proline-rich protein genes generated by intragenic unequal crossing over. *Genetics,* 120, 267–278.

McCutchen, B. F., Choudary, P. V., Crenshaw, R., Maddox, D., Kamita, S. G., Palekar, N., Volrath, S., Fowler, E., Hammock, B. D., & Maeda, S. (1991). Development of a recombinant baculovirus expressing an insect-selective neurotoxin: potential for pest control. *Bio/Technology,* 9, 848–852.

McEwen, B. S. (1991). Non-genomic and genomic effects of steroids on neural activity. *Trends in Pharmacological Science,* 12, 141–147.

McGinnis, W., Garber, R. L., Wirz, J., Kuroiwa, A., & Gehring, W. J. (1984). A homologous protein-coding sequence in *Drosophila* homeotic genes and its conservation in other metazoans. *Cell,* 37, 403–408.

McGinnis, W., & Krumlauf, R. (1992). Homeobox genes and axial patterning. *Cell,* 68, 283–302.

McIntosh, A. H., & Ignoffo, C. M. (1981). Replication and infectivity of the single-embedded nuclear polyhedrosis virus, *Baculovirus heliothis,* in homologous cell lines. *Journal of Invertebrate Pathology,* 37, 258–264.

McIntosh, A. H., & Ignoffo, C. M. (1989). Replication of *Autographa californica* nuclear polyhedrosis virus in five lepidopteran cell lines. *Journal of Invertebrate Pathology,* 54, 97–102.

McIntosh, A. H., Ignoffo, C. M., & Andrews, P. L. (1985). In vitro host range of five baculoviruses in lepidopteran cell lines. *Intervirology,* 23, 150–156.

Maeda, S. (1989a). Expression of foreign genes in insects using baculovirus vectors. *Annual Review of Entomology,* 34, 351–372.

Maeda, S. (1989b). Increased insecticidal effect by a recombinant baculovirus carrying a synthetic diuretic hormone gene. *Biochemical and Biophysical Research Communications,* 165, 1177–1183.

Maeda, S., Kamita, S. G., & Kataoka, H. (1991). The basic protein of *Bombyx mori* nuclear polyhedrosis virus: the existence of an additional arginine repeat. *Virology,* 180, 807–810.

Maeda, S., Volrath, S. L., Hanzlik, T. N., Harper, S. A., Majima, K., Maddox, D. W., Hammock, B. D., & Fowler, E. (1991). Insecticidal effects of an insect-specific neurotoxin expressed by a recombinant baculovirus. *Virology,* 184, 777–780.

Maekawa, H., Doira, H., & Sakaguchi, B. (1980). Flimsy cocoon mutant of *Bombyx mori* larva produces a reduced amount of fibroin mRNA. *Cell Structure and Function,* 5, 233–238.

Maekawa, H., & Suzuki, Y. (1980). Repeated turn-off and turn-on of fibroin gene transcription during silk gland development of *Bombyx mori. Developmental Biology,* 78, 394–406.

Maeki, K. (1981a). Notes on the W-chromosome of the butterfly, *Graphium sarpedon* (Papilionidae, Lepidoptera). *Proceedings of the Japan Academy,* 57, Series B, 371–373.

Maeki, K. (1981b). The chromosome of the Lepidoptera. *Tyô to Ga,* 32, 13–28. In Japanese with English summary.

Maeki, K. (1982). Maturation, fertilization, and the chromosomes of early developmental stages in butterfly eggs. *La Kromosomo II,* 27–28, 839–848. In Japanese with English summary.

Mahowald, A. P. (1971). Origin and continuity of cell organelles. In *Results and Problems in Cell Differentiation,* vol. 2, ed. J. Reinert & H. Ursprung, pp. 158–169. New York and Berlin: Springer-Verlag.

Majima, R., Kawakami, M., & Shimura, K. (1975). The biosynthesis of transfer RNA in insects. I. Increase of amino acid acceptor activity of specific tRNAs utilized for silk protein biosynthesis in the silkgland of *Bombyx mori. Journal of Biochemistry,* 78, 391–400.

Malter, J. S. (1989). Identification of an AUUUA-specific messenger RNA binding protein. *Science,* 246, 664–666.

Malter, J. S., & Hong, Y. (1991). A redox switch and phosphorylation are involved in the post-translational up-regulation of the adenosine-uridine binding factor by phorbol ester and ionophore. *Journal of Biological Chemistry,* 266, 3167–3171.

Manning, J. E., Schmid, C. W., & Davidson, N. (1975). Interspersion of repetitive and non-repetitive DNA sequences in the *Drosophila melanogaster* genome. *Cell,* 4, 141–155.

Margaritis, L. H. (1985). Comparative study of the eggshell of the fruit flies *Dacus oleae* and *Ceratitis capitata* (Diptera: Trypetidae). *Canadian Journal of Zoology,* 63, 2194–2206.

Margaritis, L. H., Kafatos, F. C., & Petri, W. H. (1980). The eggshell of *Drosophila melanogaster.* I. Fine structure of the layers and regions of the wild-type eggshell. *Journal of Cell Science,* 43, 1–35.

Margottin, F., Dujardin, G., Gerard, M., Egly, J.-M., Huet, J., & Sentenac, A. (1991). Participation of the TATA factor in transcription of the yeast U6 gene by RNA polymerase C. *Science,* 251, 424–426.

Mariani, B. D., Lingappa, J. R., & Kafatos, F. C. (1988). Temporal regulation in development: negative and positive cis regulators dictate the precise timing of expression of a *Drosophila* chorion gene. *Proceedings of the National Academy of Sciences, U.S.A.,* 85, 3029–3033.

Marlor, R., Parkhurst, S., & Corces, V. (1986). The *Drosophila melanogaster gypsy* transposable element encodes putative gene products homologous to retroviral proteins. *Molecular and Cellular Biology,* 6, 1129–1134.

Marti, T., Takio, K., Walsh, K. A., Terzi, G., & Truman, J. W. (1987). Microanalysis of the amino acid sequence of the eclosion hormone from the tobacco hornworm *Manduca sexta. FEBS Letters,* 219, 415–418.

Martin, J. A., & Pashley, D. P. (1992). Molecular systematic analysis of butterfly family and some subfamily relationships (Lepidoptera: Papilionoidea). *Annals of the Entomological Society of America,* 85, 127–139.

Martinez-Arias, A., & Lawrence, P. A. (1985). Parasegments and compartments in the *Drosophila* embryo. *Nature,* 313, 639–642.

Martínez-Cruzado, J. C., Swimmer, C., Fenerjian, M. G., & Kafatos, F. C. (1988). Evolution of the autosomal chorion locus in *Drosophila.* I. General organization of the locus and sequence comparisons of genes *s15* and *s19* in evolutionarily distant species. *Genetics,* 119, 663–667.

Maschlanka, H. (1938). Physiologische Untersuchungen am Ei der Mehlmotte *Ephestia kuhniella. Wilhelm Roux' Archiv für Entwicklungsmechanik der Organismen,* 137, 714–772.

Matsubara, M., Tsusué, M., & Akino, M. (1963). Occurrence of two different enzymes in the silkworm, *Bombyx mori,* to reduce folate and sepiapterin. *Nature,* 199, 908–909.

Matsumoto, K., Murakami, K., & Okada, N. (1986). Gene for lysine tRNA1 may be a progenitor of the highly repetitive and transcribable sequences present in the salmon genome. *Proceedings of the National Academy of Sciences, U.S.A.,* 83, 3156–3160.

Matsumoto, N., Okada, M., Takahashi, H., Ming, Q. X., Nakajima, Y., Nakanishi, Y., Komano, H., & Natori, S. (1986). Molecular cloning of a cDNA and assignment of the C-terminal of sarcotoxin IA, a potent antibacterial protein of *Sarcophaga peregrina. Biochemical Journal,* 239, 717–722.

Matsuno, K., Hui, C.-c., Takiya, S., Suzuki, T., Ueno, K., & Suzuki, Y. (1989). Transcription signals and protein binding sites for sericin gene transcription in vitro. *Journal of Biological Chemistry,* 264, 18707–18713.

Matsuno, K., Takiya, S., Hui, C.-c., Suzuki, T., Fukuta, M., Ueno, K., & Suzuki, Y. (1990). Transcriptional stimulation via SC site of *Bombyx* sericin-1 gene through an interaction with a DNA binding protein SGF-3. *Nucleic Acids Research,* 18, 1853–1858.

Matsuyama, K., & Natori, S. (1988a). Purification of three antibacterial proteins from the culture medium of NIH-SAPE-4, an embryonic cell line of *Sarcophaga peregrina. Journal of Biological Chemistry,* 263, 17112–17116.

Matsuyama, K., & Natori, S. (1988b). Molecular cloning of cDNA for sapecin and unique expression of the sapecin gene during the development of *Sarcophaga peregrina. Journal of Biological Chemistry,* 263, 17117–17121.

Matsuzaki, K. (1966). Fractionation of amino acid-specific t-RNA from silkgland by methylated albumin column chromatography. *Biochimica et Biophysica Acta,* 114, 222–226.

Matsuzaki, M. (1972). Oogenesis in adult net-spinning caddisfly: *Parastenopsyche sauteri* (Trichoptera, Stenopsychidae), as revealed by electron microscopic observation. *Scientific Reports of Fukushima University,* 22, 27–40.

Mayer, A. G. (1896). The development of the wing scales and their pigment in butterflies and moths. *Bulletin of the Museum of Comparative Zoology,* 29, 209–236.

Mayer, M. S. (1974). Hydrolysis of sex pheromone by the antennae of *Trichoplusia ni. Experientia,* 31, 452–454.

Mazda, T., Tsusué, M., & Sakate, S. (1980). Purification and identification of a yellow pteridine characteristic of the larval colour of the Kiuki mutant of the silkworm, *Bombyx mori. Insect Biochemistry,* 10, 357–362.

Mazda, T., Tsusué, M., Sakate, S., & Doira, H. (1981). Studies on the genetical and biochemical properties of the Kiuki mutant of the silkworm, *Bombyx mori. Japanese Journal of Genetics,* 56, 19–26.

Mazo, A. M., Huang, D.-H., Mozer, B. A., & Dawid, I. B. (1990). The *trithorax* gene, a trans-acting regulator of the bithorax complex in *Drosophila,* encodes a protein with zinc-binding domains. *Proceedings of the National Academy of Sciences, U.S.A.,* 87, 2112–2116.

Mazur, G. D., Regier, J. C., & Kafatos, F. C. (1980). The silkmoth chorion: morphogenesis of surface structures and its relation to synthesis of specific proteins. *Developmental Biology,* 76, 305–321.

Mazur, G. D., Regier, J. C., & Kafatos, F. C. (1982). Order and defects in the silkmoth chorion–a biological analogue of a cholesteric liquid crystal. In *Insect Ultrastructure,* vol. 1, ed. R. C. King & H. Akai, pp. 150–185. New York: Plenum.

Mazur, G. D., Regier, J. C., & Kafatos, F. C. (1989). Morphogenesis of silkmoth chorion: sequential modification of an early helicoidal framework through expansion and densification. *Tissue and Cell,* 21, 227–242.

Memmel, N. A., & Kumaran, A. K. (1988). Role of ecdysteroids and juvenile hormone in regulation of larval haemolymph protein gene expression in *Galleria mellonella*. *Journal of Insect Physiology*, 34, 585–591.

Memmel, N. A., Ray, A., & Kumaran, A. K. (1988). Role of hormones in starvation-induced delay in larval hemolymph protein gene expression in *Galleria mellonella*. *Roux's Archives of Developmental Biology*, 197, 496–502.

Memmel, N. A., Trewitt, P. M., Silhacek, D. L., & Kumaran, A. K. (1992). Nucleotide sequence and structure of the arylphorin gene from *Galleria mellonella*. *Insect Biochemistry and Molecular Biology*, 22, 333–342.

Merrifield, R. B., Vizioli, L. D., & Boman, H. G. (1982). Synthesis of the antibacterial peptide cecropin A(1-33). *Biochemistry*, 21, 5020–5030.

Merryweather, A. T., Weyer, U., Harris, M. P. G., Hirst, M., Booth, T., & Possee, R. D. (1990). Construction of genetically engineered baculovirus insecticides containing the *Bacillus thuringiensis* subsp. *kurstaki* HD-73 delta endotoxin. *Journal of General Virology*, 71, 1535–1544.

Meyer, A. S., Schneiderman, H. A., Hanzmann, E., & Ko, J. H. (1968). The two juvenile hormones from the *Cecropia* silk moth. *Proceedings of the National Academy of Sciences, U.S.A.*, 60, 853–860.

Meyerowitz, E. M., Crosby, M. A., Garfinkel, M. D., Martin, C. H., Mathers, P. H., & Vijayraghavan, K. (1985). The 68 glue puff of *Drosophila*. *Cold Spring Harbor Symposia on Quantitative Biology*, 50, 347–353.

Meyerowitz, E. M., Raghavan, K. V., Mathers, P. H., & Roark, M. (1987). How *Drosophila* larvae make glue: control of Sgs-3 gene expression. *Trends in Genetics*, 3, 288–293.

Meyrand, P., & Marder, E. (1991). Matching neural and muscle oscillators: control by FMRFamide-like peptides. *Journal of Neuroscience*, 11, 1150–1161.

Meza, L., Araya, A., Leon, G., Krauskopf, M., Siddiqui, M. A. Q., & Garel, J. P. (1977). Specific alanine tRNA species associated with fibroin biosynthesis in the posterior silk-gland of *Bombyx mori* L. *FEBS Letters*, 77, 255–260.

Michaille, J.-J., Couble, P., Prudhomme, J.-C., & Garel, A. (1986). A single gene produces multiple sericin messenger RNAs in the silk gland of *Bombyx mori*. *Biochimie*, 68, 1165–1173.

Michaille, J.-J., Garel, A., & Prudhomme, J.-C. (1989). The expression of five middle silk gland specific genes is transcriptionally regulated during the larval development of *Bombyx mori*. *Insect Biochemistry*, 19, 19–27.

Michaille, J.-J., Garel, A., & Prudhomme, J.-C. (1990a). Expression of Ser1 and Ser2 genes in the middle silkgland of *Bombyx mori* during the fifth instar. *Sericologia*, 30, 49–60.

Michaille, J.-J., Garel, A., & Prudhomme, J.-C. (1990b). Cloning and characterization of the highly polymorphic Ser2 gene of *Bombyx mori*. *Gene*, 86, 177–184.

Michaille, J.-J., Mathavan, S., Gaillard, J., & Garel, A. (1990). The complete sequence of *Mag*, a new retrotransposon in *Bombyx mori*. *Nucleic Acids Research*, 18, 674.

Michel, F., & Lang, B. F. (1985). Mitochondrial class II introns encode proteins related to the reverse transcriptases of retroviruses. *Nature*, 316, 641–643.

Michel, W. C., Fadool, D. A., & Ache, B. W. (1992). Cyclic nucleotides mediate an odor-evoked potassium conductance in lobster olfactory receptor cells. *Journal of Neuroscience*, 12, 3979–3984.

Miller, D. W., & Miller, L. K. (1982). A virus mutant with an insertion of a *copia*-like transposable element. *Nature,* 299, 562–564.

Miller, L. K. (1988). Baculoviruses as gene expression vectors. *Annual Review of Microbiology,* 42, 177–199.

Miller, L. K. (1989). Insect baculoviruses: powerful gene expression vectors. *BioEssays,* 11, 91–95.

Miller, L. K. (1991). Molecular baculovirology: from genes to strategies. In *Entomology Serving Society: Emerging Technologies and Challenges,* ed. S. B. Vinson & R. Metcalf, pp. 58–85. Lanham: Entomological Society of America.

Miller, S. G., Leclerc, R. F., Seo, S-J., & Malone, C. (1990). Synthesis and transport of storage proteins by testes in *Heliothis virescens. Archives of Insect Biochemistry and Physiology,* 14, 151–170.

Mine, E., Izumi, S., Katsuki, M., & Tomino, S. (1983). Developmental and sex-dependent regulation of storage protein synthesis in the silkworm, *Bombyx mori. Developmental Biology,* 97, 329–337.

Miner, J. N., & Yamamoto, K. R. (1991). Regulatory cross-talk at composite response elements. *Trends in Biochemical Sciences,* 16, 423–426.

Minet, J. (1986). Ebauche d'une classification moderne de l'ordre des Lépidoptères. *Alexanor,* 14, 291–313.

Minnick, M. F., Rupp, R. A., & Spence, K. D. (1986). A bacteria-induced lectin which triggers hemocyte coagulation in *Manduca sexta. Biochemical and Biophysical Research Communications,* 137, 729–735.

Mitsialis, S. A., & Kafatos, F. C. (1985). Regulatory elements controlling chorion gene expression are conserved between flies and moths. *Nature,* 317, 453–456.

Mitsialis, S. A., Spoerel, N., Leviten, M., & Kafatos, F. C. (1987). A short 5′-flanking DNA region is sufficient for developmentally correct expression of moth chorion genes in *Drosophila. Proceedings of the National Academy of Sciences, U.S.A.,* 84, 7987–7991.

Mitsialis, S. A., Veletza, S., & Kafatos, F. C. (1989). Transgenic regulation of moth chorion gene promoters in *Drosophila:* tissue, temporal and quantitative control of four bidirectional promoters. *Journal of Molecular Evolution,* 29, 486–495.

Mitsui, T., & Riddiford, L. M. (1978). Hormonal requirements for the larval-pupal transformation of the epidermis of *Manduca sexta* in vitro. *Developmental Biology,* 62, 193–205.

Miya, K. (1958). Studies on the embryonic development of the gonad in the silkworm, *Bombyx mori* L. 1. Differentiation of germ cells. *Journal of the Faculty of Agriculture, Iwate University,* 3, 436–467.

Miya, K. (1973). Analyses of early embryonic development of the silkworm, *Bombyx mori,* by centrifugation. 1. Effects of centrifugation on the different developmental stages. *Journal of the Faculty of Agriculture, Iwate University,* 11, 213–229.

Miya, K. (1978). Electron microscope studies on the early embryonic development of the silkworm, *Bombyx mori.* I. Architecture of the newly laid egg and the changes by sperm entry. *Journal of the Faculty of Agriculture, Iwate University,* 3, 436–467.

Miya, K. (1984). Early embryogenesis of "kidney-shaped egg" in the silkmoth, *Bombyx mori* L. *Journal of the Faculty of Agriculture, Iwate University,* 17, 131–149.

Miya, K. (1985a). Determination and formation of the basic body pattern in embryo of the domesticated silkmoth, *Bombyx mori* (Lepidoptera, Bombycidae). In *Recent Advances in Insect Embryology in Japan,* ed. H. Ando & K. Miya, pp. 107–123. Tsukuba, Japan: ISEBU.

Miya, K. (1985b). Embryogenesis of an embryonic lethal, "kidney-shaped egg" in *Bombyx mori,* with special reference to mesoderm differentiation. *International Journal of Invertebrate Reproduction and Development,* 8, 263–267.

Miya, K., & Kobayashi, Y. (1974). The embryonic development of *Atrachya menetriesi* Faldermann (Coleoptera, Chrysomelidae). II. Analyses of early development by ligation and low temperature treatment. *Journal of the Faculty of Agriculture, Iwate University,* 12, 39–55.

Mizoguchi, A., Oka, T., Kataoka, H., Nagasawa, H., Suzuki, A., & Ishizaki, H. (1990). Immunohistochemical localization of prothoracicotropic hormone-producing neurosecretory cells in the brain of *Bombyx mori. Development, Growth and Differentiation,* 32, 591–598.

Moerman, D. G., & Waterston, R. H. (1989). Mobile elements in *Caenorhabditis elegans* and other nematodes. In *Mobile DNA,* ed. D. E. Berg & M. M. Howe, pp. 537–556. Washington, DC: American Society of Microbiology.

Monnier, D., Colas, J. F., Rosay, P., Hen, R., Borrelli, E., & Maroteaux, L. (1992). NKD, a developmentally regulated tachykinin receptor in *Drosophila. Journal of Biological Chemistry,* 267, 1298–1302.

Morata, G., & Kerridge, S. (1981). Sequential functions of the bithorax complex of *Drosophila. Nature,* 290, 778–781.

Morata, G., & Lawrence, P. A. (1975). Control of compartment development by the *engrailed* gene in *Drosophila. Nature,* 255, 614–617.

Morata, G., Macias, A., Urquia, N., & Gonzalez-Reyes, A. (1990). Homeotic genes. *Seminars in Cell Biology,* 1, 219–227.

Morgan, B. A., Johnson, W. A., & Hirsh, J. (1986). Regulation of splicing produces different forms of dopa decarboxylase in the CNS and hypoderm of *Drosophila melanogaster. The EMBO Journal,* 5, 3335–3342.

Morishima, I., Suginaka, S., Ueno, T., & Hirano, H. (1990). Isolation and structure of cecropins, inducible antibacterial peptides, from the silkworm, *Bombyx mori. Comparative Biochemistry and Physiology,* B, 95, 551–554.

Morishima, I., Yamada, K., & Ueno, T. (1992). Bacterial peptidoglycan as elicitor of antibacterial protein synthesis in larvae of the silkworm, *Bombyx mori. Insect Biochemistry,* 22, 363–367.

Morton, D. B., & Truman, J. W. (1985). Steroid regulation of the peptide-mediated increase in cyclic GMP in the nervous system of the hawkmoth, *Manduca sexta. Journal of Comparative Physiology,* A, 157, 423–432.

Morton, D. B., & Truman, J. W. (1986). Substrate phosphoprotein availability regulates eclosion hormone sensitivity in an insect CNS. *Nature,* 323, 264–267.

Morton, D. B., & Truman, J. W. (1988a). The EGPs–the eclosion hormone and cyclic GMP regulated phosphoproteins. I. Appearance and partial characterization in the CNS of *Manduca sexta. Journal of Neuroscience,* 8, 1326–1337.

Morton, D. B., & Truman, J. W. (1988b). The EGPs–the eclosion hormone and cyclic GMP regulated phosphoproteins. II. Regulation of appearance by the steroid hormone 20-hydroxyecdysone in *Manduca sexta. Journal of Neuroscience,* 8, 1338–1345.

Morton, D. G., & Sprague, K. U. (1984). In vitro transcription of a silkworm 5S RNA gene requires an upstream signal. *Proceedings of the National Academy of Sciences, U.S.A.,* 81, 5519–5522.

Moschonas, N. R., Thireos, G., & Kafatos, F. C. (1988). Evolution of chorion structural genes and regulatory mechanisms in two wild silkmoths: a preliminary analysis. *Journal of Molecular Evolution,* 27, 187–193.

Mounier, N., & Prudhomme, J.-C. (1986). Isolation of actin genes in *Bombyx mori:* the coding sequence of a cytoplasmic actin gene expressed in the silk gland is interrupted by a single intron in an unusual position. *Biochimie,* 68, 1053–1061.

Mounier, N., & Prudhomme, J.-C. (1991). Differential expression of muscle and cytoplasmic actin genes during development of *Bombyx mori. Insect Biochemistry,* 21, 523–533.

Mount, S. M., & Rubin, G. M. (1985). Complete nucleotide sequence of the *Drosophila* transposable element *copia:* homology between *copia* and retroviral proteins. *Molecular and Cellular Biology,* 5, 1630–1638.

Muller, G., & Zimmermann, R. (1987). Import of honeybee prepromelittin into the endoplasmic reticulum: structural basis for independence of SRP and docking protein. *The EMBO Journal,* 6, 2099–2107.

Mulnix, A. B. (1991). Structure and regulation of expression of the lysozyme gene from the tobacco hornworm, *Manduca sexta.* Ph.D. thesis, Purdue University.

Munz, P., Amstutz, H., Kohli, J., & Leupold, U. (1982). Recombination between dispersed serine tRNA genes in *Schizosaccharomyces pombe. Nature,* 300, 225–231.

Murakami, A., & Imai, H. (1974). Cytological evidence for holocentric chromosomes of the silkworms, *Bombyx mori* and *B. mandarina,* (Bombycidae, Lepidoptera). *Chromosoma* (Berl.), 47, 167–178.

Murlis, J. (1986). The structure of odour plumes. In *Mechanisms in Insect Olfaction,* ed. T. L. Payne, M. C. Birch, & C. E. J. Kennedy, pp. 27–38. London and New York: Oxford University Press and Clarendon Press.

Murlis, J., & Jones, C. D. (1981). Fine-scale structure of odour plumes in relation to insect orientation to distant pheromone and other attractant sources. *Physiological Entomology,* 6, 71–86.

Murphy, S., Moorefield, B., & Pieler, T. (1989). Common mechanisms of promoter recognition by RNA polymerases II and III. *Trends in Genetics,* 5, 122–126.

Murre, C., McCaw, P. S., & Baltimore, D. (1989). A new DNA binding and dimerization motif in immunoglobulin enhancer binding, *daughterless, MyoD,* and *myc* proteins. *Cell,* 56, 777–783.

Murtha, M. T., Leckman, J. F., & Ruddle, F. H. (1991). Detection of homeobox genes in development and evolution. *Proceedings of the National Academy of Sciences, U.S.A.,* 88, 10711–10715.

Myohara, M., & Kiguchi, K. (1990). Induction of localized cuticle defects by UV laser irradiation of early embryos in *Bombyx mori* Linne. *Proceedings of the Arthropod Embryology Society, Japan,* 25, 13–14.

Nachman, R. J., Holman, G. M., Cook, B. J., Haddon, W. F., & Ling, N. (1986). Leucosulfakinin-II, a blocked sulfated insect neuropeptide with homology to cholecystokinin and gastrin. *Biochemical and Biophysical Research Communications,* 140, 357–364.

Nachman, R. J., Holman, G. M., Haddon, W. F., & Ling, N. (1986). Leucosulfakinin, a sulfated insect neuropeptide with homology to gastrin and cholecystokinin. *Science,* 234, 71–73.

Nadel, M. R., Goldsmith, M. R., Goplerud, J., & Kafatos, F. C. (1980). Specific protein synthesis in cellular differentiation. V. A secretory defect of chorion formation in the *Gr^col^* mutant of *Bombyx mori. Developmental Biology,* 75, 41–58.

Nadel, M. R., Thireos, G., & Kafatos, F. C. (1980). Effect of the pleiotropic Gr^B mutation of *Bombyx mori* on chorion protein synthesis. *Cell,* 20, 649–658.

Nagata, M., Tsuchida, K., Shimizu, K., & Yoshitake, N. (1987). Physiological aspects of *nm-g* mutant: an ecdysteroid-deficient mutant of the silkworm, *Bombyx mori. Journal of Insect Physiology,* 33, 723–727.

Nagl, W. (1978). *Endopolyploidy and Polyteney in Differentiation and Evolution.* Amsterdam, New York, and Oxford: North-Holland.

Nagy, L. M., Booker, R., & Riddiford, L. M. (1991). Isolation and embryonic expression of an abdominal-A-like gene from the lepidopteran, *Manduca sexta. Development,* 112, 119–129.

Nagy, L., Riddiford, L. M., & Kiguchi, K. (in press). Morphogenesis in the early embryo of the lepidopteran *Bombyx mori. Developmental Biology.*

Nakajima, Y., Qu, X.-M., & Natori, S. (1987). Interaction between liposomes and sarcotoxin IA, a potent antibacterial protein of *Sarcophaga peregrina* (flesh fly). *Journal of Biological Chemistry,* 262, 1665–1669.

Nakamura, T., & Gold, G. H. (1987). A cyclic nucleotide-gated conductance in olfactory receptor cilia. *Nature,* 325, 442–444.

Nakamura, Y., Leppert, M., O'Connell, P., Wolff, R., Holm, T., Culver, M., Martin, C., Fujimoto, E., Hoff, M., Kumlin, E., & White, R. (1987). Variable number of tandem repeat (VNTR) markers for human gene mapping. *Science,* 235, 1616–1622.

Nakato, H., Toriyama, M., Izumi, S., & Tomino, S. (1990). Structure and expression of mRNA for a pupal cuticle protein of the silkworm, *Bombyx mori. Insect Biochemistry,* 20, 667–677.

Nambu, J. R., Murphy-Erdosh, C., Andrews, P. C., Feistner, G. J., & Scheller, R. H. (1988). Isolation and characterization of a *Drosophila* neuropeptide gene. *Neuron,* 1, 55–61.

Nanbu, R., Nakajima, Y., Ando, K., & Natori, S. (1988). Novel feature of expression of the sarcotoxin IA gene in development of *Sarcophaga peregrina. Biochemical and Biophysical Research Communications,* 150, 540–544.

Nardi, J. (1993). Modulated expression of a surface epitope on migrating germ cells of *Manduca sexta* embryos. *Development,* 118, 967–975.

Nardi, J., & Magee-Adams, S. M. (1986). Formation of scale spacing patterns in a moth wing. I. Epithelial feet may mediate cell rearrangement. *Developmental Biology,* 116, 278–290.

Natori, S. (1977). Bactericidal substance induced in the haemolymph of *Sarcophaga peregrina* larvae. *Journal of Insect Physiology,* 23, 1169–1173.

Natori, S. (1989). Activation of arylphorin receptor and induction of phosphorylation of ribosomal protein S6 by 20-hydroxyecdysone. In *Ecdysone: From Chemistry to Mode of Action,* ed. J. A. Koolman, pp. 426–431. Stuttgart: Georg Thieme.

Neville, A. C. (1975). *Biology of Arthropod Cuticle.* New York: Springer-Verlag.

Neville, A. C. (1988). The need for a constraining layer in the formation of monodomain helicoids in a wide range of biological structures. *Tissue and Cell,* 18, 603–620.

Neville, A. C., & Luke, B. M. (1971). A biological system producing a self-assembling cholesteric protein liquid crystal. *Journal of Cell Science,* 8, 93–109.

Nichols, R., Schneuwly, S. A., & Dixon, J. E. (1988). Identification and characterization of a *Drosophila* homologue to the vertebrate neuropeptide cholecystokinin. *Journal of Biological Chemistry,* 263, 12167–12170.

Nicolas, A., Treco, D., Schultes, N. P., & Szostak, J. W. (1989). An initiation site for meiotic gene conversion in the yeast *Saccharomyces cerevisiae. Nature,* 338, 35–39.

Nielsen, E. S. (1989). Phylogeny of major lepidopteran groups. In *The Hierarchy of Life,* ed. B. Fernholm, K. Bremer, & H. Jörnvall, pp. 281–294. Amsterdam: Elsevier.

Nijhout, H. F. (1975). A threshold size for metamorphosis in the tobacco hornworm, *Manduca sexta. Biological Bulletin,* 149, 214–225.

Nijhout, H. F. (1991). *The Development and Evolution of Butterfly Wing Patterns.* Washington and London: Smithsonian Institution Press.

Nijhout, M. M., & Riddiford, L. M. (1974). The control of egg maturation by juvenile hormone in the tobacco hornworm moth, *Manduca sexta. Biological Bulletin,* 146, 377–392.

Nijhout, M. M., & Riddiford, L. M. (1979). Juvenile hormone and ovarian growth in *Manduca sexta. Journal of Invertebrate Reproduction,* 1, 209–219.

Nishio, K., & Kawakami, M. (1984). Purification and properties of alanyl-tRNA synthetase from *Bombyx mori:* a monomeric enzyme. *Journal of Biochemistry,* 96, 1867–1874.

Nissen, M. S., & Friesen, P. D. (1989). Molecular analysis of the transcriptional regulatory region of an early baculovirus gene. *Journal of Virology,* 63, 493–503.

Norbury, C., & Nurse, P. (1992). Animal cell cycles and their control. *Annual Review of Biochemistry,* 61, 441–470.

Numata, H., Numata, A., Takahashi, C., Nakagawa, Y., Iwatani, K., Takahashi, S., Miura, K., & Chinzei, Y. (1992). Juvenile hormone I is the principal juvenile hormone in a hemipteran insect, *Riptortus clavatus. Experientia,* 48, 606–610.

Nunome, J. (1937). The silk gland development of *Bombyx mori. Bulletin of Applied Zoology,* 9, 68–92. In Japanese.

Nüsslein-Volhard, C., Frohnhöfer, H. G., & Lehman, R. (1987). Determination of anteroposterior polarity in *Drosophila. Science,* 238, 1675–1681.

Nüsslein-Volhard, C., & Wieschaus, E. (1980). Mutations affecting segment number and polarity in *Drosophila. Nature,* 287, 795–801.

Nüsslein-Volhard, C., Wieschaus, E., & Kluding, H. (1984). Mutations affecting the pattern of the larval cuticle in *Drosophila melanogaster.* I. Zygotic loci on the second chromosome. *Wilhelm Roux's Archives of Developmental Biology,* 193, 267–282.

Obara, T., & Suzuki, Y. (1988). Temporal and spatial control of silk gene transcription analyzed by nuclear run-on assays. *Developmental Biology,* 127, 384–391.

Oberlander, H., & Fulco, L. (1967). Growth and partial metamorphosis of imaginal disks of the greater wax moth, *Galleria mellonella,* in vitro. *Nature,* 216, 1140–1141.

Oberlander, H., & Miller, S. (1987). Lepidopteran cell lines: tools for research in physiology, development and genetics. *Advances in Cell Culture,* 5, 187–208.

O'Brien, M. A., Katahira, E. J., Flanagan, T. R., Arnold, L. W., Haughton, G., & Bollenbacher, W. E. (1988). A monoclonal antibody to the insect prothoracicotropic hormone. *Journal of Neuroscience,* 8, 3247–3257.

Ogura, T., Okano, K., Tsuchida, K., Miyajima, N., Tanaka, H., Takada, N., Izumi, S., Tomino, S., & Maekawa, H. (in press). A defective non-LTR retrotransposon is dispersed throughout the genome of the silkworm, *Bombyx mori. Chromosoma,* 103.

Ohashi, M., Tsusué, M., & Kiguchi, K. (1983). Juvenile hormone control of larval colouration in the silkworm, *Bombyx mori:* characterization and determination of epidermal brown colour induced by the hormone. *Insect Biochemistry,* 13, 123–127.

Ohnishi, E., & Ishizaki, H., eds. (1990). *Molting and Metamorphosis.* Tokyo: Japan Scientific Press.

Ohshima, Y., & Suzuki, Y. (1977). Cloning of the silk fibroin gene and its flanking sequences. *Proceedings of the National Academy of Sciences, U.S.A.,* 74, 5363–5367.

Ohta, S., Suzuki, Y., Hara, W., Takiya, S., & Suzuki, T. (1988). Fibroin gene transcription in the embryonic stages of the silkworm, *Bombyx mori. Development, Growth and Differentiation,* 30, 293–299.

Ohta, T. (1980). *Evolution and Variation of Multigene Families.* Berlin: Springer-Verlag.

Ohta, T., & Hirose, S. (1990). Purification of a DNA supercoiling factor from the posterior silk gland of *Bombyx mori. Proceedings of the National Academy of Sciences, U.S.A.,* 87, 5307–5311.

Okada, M. (1960). Embryonic development of the rice stem-borer, *Chilo suppressallis. Scientific Reports, Tokyo Kyoiku Daigaku, Section B,* 9, 243–296.

Okada, M., & Natori, S. (1984). Mode of action of a bactericidal protein induced in the haemolymph of *Sarcophaga peregrina* (flesh fly) larvae. *Biochemical Journal,* 222, 119–124.

Okada, M., & Natori, S. (1985a). Primary structure of sarcotoxin I, an antibacterial protein induced in the hemolymph of *Sarcophaga peregrina* (flesh fly) larvae. *Journal of Biological Chemistry,* 260, 7174–7177.

Okada, M., & Natori, S. (1985b). Ionophore activity of sarcotoxin I, a bactericidal protein of *Sarcophaga peregrina. Biochemical Journal,* 229, 453–458.

Okamoto, H., Ishikawa, E., & Suzuki, Y. (1982). Structural analysis of sericin genes. Homologies with fibroin gene in the 5′ flanking nucleotide sequences. *Journal of Biological Chemistry,* 257, 15192–15199.

Okazaki, S., Tsuchida, K., Maekawa, H., Ishikawa, H., & Fujiwara, H. (1993). Identification of a pentanucleotide telomeric sequence, (TTAGG)n, in the silkworm, *Bombyx mori* and other insects. *Molecular and Cellular Biology,* 13, 1424–1432.

Oland, L. A., Orr, G., & Tolbert, L. P. (1990). Construction of a protoglomerular template by olfactory axons initiates the formation of olfactory glomeruli in the insect brain. *Journal of Neuroscience,* 10, 2096–2112.

Oland, L. A., & Tolbert, L. P. (1987). Glial patterns during early development of antennal lobes of *Manduca sexta:* a comparison between normal lobes and lobes deprived of antennal axons. *Journal of Comparative Neurology,* 255, 196–207.
Oland, L. A., Tolbert, L. P., & Mossman, K. L. (1988). Radiation-induced reduction of the glial population during development disrupts the formation of olfactory glomeruli in an insect. *Journal of Neuroscience,* 8, 353–367.
Oquendo, P., Hundt, E., Lawler, J., & Seed, B. (1989). CD36 directly mediates cytoadherence of *Plasmodium falciparum* parasitized erythrocytes. *Cell,* 58, 95–101.
O'Reilly, D. R., Crawford, A. M., & Miller, L. K. (1989). Viral proliferating cell nuclear antigen. *Nature,* 337, 606.
O'Reilly, D. R., & Miller, L. K. (1989). A baculovirus blocks insect molting by producing ecdysteroid UDP-glycosyl transferase. *Science,* 245, 1110–1112.
O'Reilly, D. R., & Miller, L. K. (1990). Regulation of expression of a baculovirus ecdysteriod UDPglycosyltransferase gene. *Journal of Virology,* 64, 1321–1328.
Orgel, L. E., & Crick, F. H. C. (1980). Selfish DNA: the ultimate parasite. *Nature,* 284, 604–608.
Orkin, S. H. (1990). Globin gene regulation and switching: circa 1990. *Cell,* 63, 665–672.
Oro, A. E., McKeown, M., & Evans, R. M. (1990). Relationship between the product of the *Drosophila ultraspiracle* locus and the vertebrate retinoid X receptor. *Nature,* 347, 298–301.
Orr, W. C., Galanopoulos, V. R., Romano, C. P., & Kafatos, F. C. (1989). A female sterile screen of the *Drosophila melanogaster* X chromosome using hybrid dysgenesis: Identification and characterization of egg morphology mutants. *Genetics,* 122, 847–858.
O'Shea, M., Witten, J. L., & Schaffer, M. H. (1984). Isolation and characterization of two myoactive neuropeptides: further evidence of an invertebrate neuropeptide family. *Journal of Neuroscience,* 4, 521–529.
Osir, E. O., & Riddiford, L. M. (1988). Nuclear binding of juvenile hormone and its analogs in the epidermis of the tobacco hornworm. *Journal of Biological Chemistry,* 263, 13812–13818.
Ottonello, S., Rivier, D. H., Doolittle, G. M., Young, L. S., & Sprague, K. U. (1987). The properties of a new polymerase III transcription factor reveal that transcription complexes can assemble by more than one pathway. *The EMBO Journal,* 6, 1921–1927.
Ouweneel, W. J. (1976). Developmental genetics of homeosis. *Advances in Genetics,* 18, 179–248.
Ozyhar, A., Strangmann-Diekmann, M., Kiltz, H. H., & Pongs, O. (1991). Characterization of a specific ecdysteroid receptor-DNA complex reveals common properties for invertebrate and vertebrate hormone-receptor/DNA interactions. *European Journal of Biochemistry,* 200, 329–335.
Palida, F. A., Hale, C., & Sprague, K. U. (1993). Transcription of a silkworm $tRNA^{Ala}C$ gene is directed by two AT-rich upstream sequence elements. *Nucleic Acids Research,* 21, 5875–5881.
Palli, S. R., Hiruma, K., & Riddiford, L. M. (1992). An ecdysteroid-inducible *Manduca* gene similar to the *Drosophila* DHR3 gene, a member of the steroid hormone receptor superfamily. *Developmental Biology,* 150, 306–318.

Palli, S. R., & Locke, M. (1987). The synthesis of hemolymph proteins by the larval midgut of an insect *Calpodes ethlius* (Lepidoptera: Hesperiidae). *Insect Biochemistry,* 17, 561–572.

Palli, S. R., & Locke, M. (1988). The synthesis of hemolymph proteins by the larval fat body of an insect *Calpodes ethlius* (Lepidoptera: Hesperiidae). *Insect Biochemistry,* 18, 405–413.

Palli, S. R., McClelland, S., Hiruma, K., Latli, B., & Riddiford, L. M. (1991). Developmental expression and hormonal regulation of the nuclear 29 kDa juvenile hormone-binding protein in *Manduca sexta* larval epidermis. *Journal of Experimental Zoology,* 260, 337–344.

Palli, S. R., Osir, E. O., Eng, W.-S., Boehm, M. F., Edwards, M., Kulcsar, P., Ujvari, I., Hiruma, K., Prestwich, G. D., & Riddiford, L. M. (1990). Juvenile hormone receptors in insect larval epidermis: identification by photoaffinity labeling. *Proceedings of the National Academy of Sciences, U.S.A.,* 87, 796–800.

Palli, S. R., Riddiford, L. M., & Hiruma, K. (1991). Juvenile hormone and "retinoic acid" receptors in *Manduca* epidermis. *Insect Biochemistry,* 21, 7–15.

Palli, S. R., Touhara, K., Charles, J-P., Bonning, B. C., Atkinson, J. K., Trowell, S. C., Hiruma, K., Goodman, W. G., Kyriakides, T., Prestwich, G. D., Hammock, B. D., & Riddiford, L. M. (1994). A nuclear juvenile hormone binding protein from larvae of *Manduca sexta:* a putative receptor for the metamorphic action of juvenile hormone. *Proceedings of the National Academy of Sciences, U.S.A,* 91, 6191–6195.

Panhuber, H., & Laing, D. G. (1987). The size of mitral cells is altered when rats are exposed to an odor from their day of birth. *Developmental Brain Research,* 34, 133–140.

Papanikolaou, A. M., Margaritis, L. H., & Hamodrakas, S. J. (1985). Ultrastructural analysis of chorion formation in the silkmoth *Bombyx mori. Canadian Journal of Zoology,* 64, 1158–1173.

Pardue, M. L., Lowenhaupt, K., Rich, A., & Nordheim, A. (1987). $(dC\text{-}dA)_n(dG\text{-}dT)_n$ sequences have evolutionarily conserved chromosome locations in *Drosophila* with implications for roles in chromosome structure and function. *The EMBO Journal,* 6, 1781–1789.

Parham, P., Lomen, C. E., Lawlor, D. A., Ways, J. P., Holmes, N., Coppin, H. L., Salter, R. D., Wan, A. M., & Ennis, P. D. (1988). Nature of polymorphism in HLA-A, -B, and -C molecules. *Proceedings of the National Academy of Sciences, U.S.A.,* 85, 4005–4009.

Park, Y.-S., & Kramer, J. M. (1990). Tandemly duplicated *Caenorhabditis elegans* collagen genes differ in their modes of splicing. *Journal of Molecular Biology,* 211, 395–406.

Parks, S. & Spradling, A. (1987). Spatially regulated expression of chorion genes during *Drosophila* oogenesis. *Genes and Development,* 1, 497–509.

Parsons, M. C., & Weil, P. A. (1990). Purification and characterization of *Saccharomyces cerevisiae* transcription factor TFIIIC. *Journal of Biological Chemistry,* 265, 5095–5103.

Pasteur, L. (1870). *Études sur la maladie des vers á soie,* vol. I. *La pebrine et la flacherie.* Paris: Gauthier-Villars.

Patel, N., Ball, E., & Goodman, C. S. (1992). Changing role of *even-skipped* during the evolution of insect pattern formation. *Nature,* 357, 339–342.

Patel, N. H., Martin-Blanco, E., Colemen, K. G., Poule, S. J., Ellis, M. C., Kornberg, T. B., & Goodman, C. G. (1989). Expression of *engrailed* proteins in arthropods, annelids and chordates. *Cell,* 58, 955–968.

Paterson, A. H., Damon, S., Hewitt, J. D., Zamir, D., Rabinowitch, H. D., Lincoln, S. E., Lander, E. S., & Tanksley, S. D. (1991). Mendelian factors underlying quantitative traits in tomato: comparison across species, generations, and environments. *Genetics,* 127, 181–197.

Paterson, A. H., Lander, E. S., Hewitt, J. D., Peterson, S., Lincoln, S. E., & Tanksley, S. D. (1988). Resolution of quantitative traits into Mendelian factors by using a complete linkage map of restriction fragment length polymorphisms. *Nature,* 335, 721–726.

Pau, R. N., Weaver, R. J., & Edwards-Jones, K. (1986). Regulation of cockroach oothecin synthesis by juvenile hormone. *Archives of Insect Biochemistry and Physiology,* 1, Supplement, 59–73.

Paul, M., Goldsmith, M. R., Hunsley, J. R., & Kafatos, F. C. (1972). Specific protein synthesis in cellular differentiation: production of eggshell proteins by silkmoth follicular cells. *Journal of Cell Biology,* 55, 653–680.

Pedersen, P. E., Jastreboff, P. J., Stewart, W. B., & Shepherd, G. M. (1986). Mapping of an olfactory receptor population that projects to a specific region in the rat olfactory bulb. *Journal of Comparative Neurology,* 250, 93–108.

Perdrix-Gillot, S. (1979). DNA synthesis and endomitoses in the giant nuclei of the silk gland of *Bombyx mori. Biochimie,* 61, 171–204.

Peter, M. G., & Scheller, K. (1991). Arylphorins and the integument. In *Physiology of the Insect Epidermis,* ed. K. Binnington & A. Retnakaran, pp. 113–122. East Melbourne: CSIRO Publications.

Petes, T. D. (1980). Unequal meiotic recombination within tandem arrays of yeast ribosomal RNA genes. *Cell,* 19, 765–774.

Petkovich, M., Brand, N. J., Krust, A., & Chambon, P. (1987). A human retinoic acid receptor which belongs to the family of nuclear receptors. *Nature,* 330, 444–450.

Philpott, M. L., & Hammock, B. D. (1990). Juvenile hormone esterase is a biochemical anti-juvenile hormone agent. *Insect Biochemistry,* 20, 451–459.

Piepho, H. (1942). Untersuchungen zur Entwicklungsphysiologie der Insektenmetamorphose. Uber bei die Puppenhäutung der Wachsmotte *Galleria mellonella* L. *Wilhelm Roux' Archiv für Entwicklungsmechanik der Organismen,* 141, 500–583.

Piepho, H. (1950). Hormonale Grundlagen der Spinntatigkeit bei Schmetterlingsraupen. *Zeitschrift für Tierpsychologie,* 7, 424–434.

Plantevin, G., Bosquet, G., Calvez, B., & Nardon, C. (1987). Relationships between juvenile hormone levels and synthesis of major hemolymph proteins in *Bombyx mori* larvae. *Comparative Biochemistry and Physiology,* B, 86, 501–507.

Poch, O., Sauvaget, I., Delarue, M., & Tordo, N. (1989). Identification of four conserved motifs among the RNA-dependent polymerase encoding elements. *The EMBO Journal,* 8, 3867–3874.

Powers, P. A., & Smithies, O. (1986). Short gene conversions in the human fetal globin gene region: a by-product of chromosome pairing during meiosis. *Genetics,* 112, 343–358.

Pratt, G. E., Farnsworth, D. E., & Feyereisen, R. (1990). Changes in the sensitivity of adult cockroach corpora allata to a brain allatostatin. *Molecular and Cellular Endocrinology,* 70, 185–195.

Pratt, G. E., Farnsworth, D. E., Fok, K. F., Siegel, N. R., McCormack, A. L., Shabanowitz, J., Hunt, D. F., & Feyereisen, R. (1991). Identity of a second type of allatostatin from cockroach brains: an octadecapeptide amide with a tyrosine-rich address sequence. *Proceedings of the National Academy of Sciences, U.S.A.,* 88, 2412–2416.

Presser, B. D., & Rutschky, C. W. (1956). The embryonic development of the corn earworm, *Heliothis zea* (Boddie) (Lepidoptera, Phalaenidae). *Annals of the Entomological Society of America,* 50, 133–164.

Prestwich, G. D. (1987). Chemistry of pheromone and hormone metabolism in insects. *Science,* 237, 999–1006.

Prestwich, G. D. (1991). Photoaffinity labeling and biochemical characterization of binding proteins for pheromones, juvenile hormones, and peptides. *Insect Biochemistry,* 21, 27–40.

Prestwich, G. D., Graham, S. M., Handley, M., Latli, B., Streinz, L., & Tasayco, M. J. (1989). Enzymatic processing of pheromones and pheromone analogs. *Experientia,* 45, 263–270.

Prestwich, G. D., Graham, S. M., Kuo, J.-W., & Vogt, R. G. (1989). Tritium-labeled enantiomers of disparlure: synthesis and in vitro metabolism. *Journal of the American Chemical Society,* 111, 636–642.

Prudhomme, J.-C., Couble, P., Garel, J.-P., & Daillie, J. (1985). Silk synthesis. In *Comprehensive Insect Physiology, Biochemistry and Pharmacology,* vol. 10, ed. G. A. Kerkut & L. I. Gilbert, pp. 571–594. New York: Pergamon.

Prugh, J., Della Croce, K., & Levine, R. B. (1992). Effects of the steroid hormone, 20-hydroxyecdysone, on the growth of neurites by identified insect motoneurons in vitro. *Developmental Biology,* 154, 331–347.

Pye, A. E., & Boman, H. G. (1977). Insect immunity. III. Purification and partial characterization of immune protein P5 from hemolymph of *Hyalophora cecropia* pupae. *Infection and Immunity,* 17, 408–414.

Qu, X.-M., Steiner, H., Engström, A., Bennich, H., & Boman, H. G. (1982). Insect immunity: isolation and structure of cecropins B and D from pupae of the Chinese oak silk moth, *Antheraea pernyi. European Journal of Biochemistry,* 127, 219–224.

Rae, P. M. M. (1981). Coding region deletions associated with the major form of rDNA interruption in *Drosophila melanogaster. Nucleic Acids Research,* 9, 4997–5010.

Raikhel, A. S., & Dhadialla, T. S. (1992). Accumulation of yolk proteins in insect oocytes. *Annual Review of Entomology,* 37, 217–251.

Raming, K., Krieger, J., & Breer, H. (1989). Molecular cloning of an insect pheromone-binding protein. *FEBS Letters,* 256, 215–218.

Raming, K., Krieger, J., & Breer, H. (1990). Primary structure of a pheromone-binding protein from *Antheraea pernyi:* homologies with other ligand-carrying proteins. *Journal of Comparative Physiology,* B, 160, 503–509.

Rappolee, D. A., Wang, A., Mark, D., & Werb, Z. (1989). A novel method for studying mRNA phenotypes in single or small numbers of cells. *Journal of Cell Biochemistry,* 39, 1–11.

Rasch, E. M. (1974). The DNA content of sperm and hemocyte nuclei of the silkworm, *Bombyx mori* L. *Chromosoma* (Berl.), 45, 1–26.

Rasch, E. M., Barr, H. J., & Rasch, R. W. (1971). The DNA content of sperm of *Drosophila melanogaster. Chromosoma* (Berl.), 33, 1–18.

Rasmuson, T., & Boman, H. G. (1979). Insect immunity. V. Purification and some properties of immune protein P4 from hemolymph of *Hyalophora cecropia* pupae. *Insect Biochemistry,* 9, 259–264.

Rasmussen, S. W. (1973). Ultrastructural studies of spermatogenesis in *Drosophila melanogaster* Meigen. *Zeitschrift für Zellforschung,* 140, 125–144.

Rasmussen, S. W. (1977). The transformation of the synaptonemal complex into the "elimination chromatin" in *Bombyx mori* oocytes. *Chromosoma* (Berl.), 60, 205–221.

Rasmussen, S. W., & Holm, P. B. (1982). The meiotic prophase in *Bombyx mori.* In *Insect Ultrastructure,* vol. 1, ed. R. C. King & H. Akai, pp. 61–85. New York and London: Plenum.

Ratcliffe, N. A., & Gagen, S. J. (1977). Studies on the in vitro cellular reactions of insects: an ultrastructural analysis of nodule formation in *Galleria mellonella. Tissue and Cell,* 9, 73–85.

Ratcliffe, N. A., Leonard, C., & Rowley, A. F. (1984). Prophenoloxidase activation: nonself recognition and cell cooperation in insect immunity. *Science,* 226, 557–559.

Ratcliffe, N. A., & Rowley, A. F. (1979). Role of hemocytes in defense against biological agents. In *Insect Hemocytes: Development, Forms, Functions, and Techniques,* ed. A. P. Gupta, pp. 331–414. Cambridge: Cambridge University Press.

Rattner, J. B. (1991). The structure of the mammalian centromere. *BioEssays,* 13, 51–56.

Rattner, J. B., Goldsmith, M., & Hamkalo, B. A. (1980). Chromatin organization during meiotic prophase of *Bombyx mori. Chromosoma* (Berl.), 79, 215–224.

Rattner, J. B., Goldsmith, M., & Hamkalo, B. A. (1981). Chromosome organization during male meiosis in *Bombyx mori. Chromosoma* (Berl.), 82, 341–351.

Ray, A., Memmel, N. A., & Kumaran, A. K. (1987). Developmental regulation of the larval hemolymph protein genes in *Galleria mellonella. Roux's Archives of Developmental Biology,* 196, 414–420.

Rebers, J. E., & Riddiford, L. M. (1988). Structure and expression of a *Manduca sexta* larval cuticle gene homologous to *Drosophila* cuticle genes. *Journal of Molecular Biology,* 203, 411–423.

Reed, E. M., & Day, M. F. (1966). Embryonic movements during the development of the light brown apple moth. *Australian Journal of Zoology,* 14, 253–263.

Reed, R. R. (1992). Signaling pathways in odorant detection. *Neuron,* 8, 205–209.

Regier, J. C. (1986). Evolution and higher-order structure of architectural proteins in silkmoth chorion. *The EMBO Journal,* 5, 1981–1989.

Regier, J. C., & Hatzopoulos, A. K. (1988). Evolution in steps: the role of regulatory alterations in the diversification of the moth chorion morphogenetic pathway. In *Self-Assembling Architecture, 46th Symposium of the Society for Developmental Biology,* ed. J. E. Varner, pp. 179–202. New York: Alan R. Liss.

Regier, J. C., Hatzopoulos, A. K., & Durot, A. R. (1986). Patterns of region-specific chorion gene expression in the silkmoth and identification of shared 5′ flanking genomic elements. *Developmental Biology,* 118, 432–441.

Regier, J. C., & Kafatos, F. C. (1985). Molecular aspects of chorion formation. In *Comprehensive Insect Physiology, Biochemistry and Pharmacology,* vol. 1, ed. G. A. Kerkut & L. I. Gilbert, pp. 113–151. Oxford: Pergamon.

Regier, J. C., Mazur, G. D., & Kafatos, F. C. (1980). The silkmoth chorion: morphological and biochemical characterization of four surface regions. *Developmental Biology,* 76, 296–304.

Regier, J. C., Mazur, G. D., Kafatos, F. C., & Paul, M. (1982). Morphogenesis of silkmoth chorion: initial framework formation and its relation to synthesis of specific proteins. *Developmental Biology,* 92, 159–174.

Regier, J. C., & Vlahos, N. S. (1988). Heterochrony and the introduction of novel modes of morphogenesis during the evolution of moth choriogenesis. *Journal of Molecular Evolution,* 28, 19–31.

Reichhart, J.-M., Essrich, M., Dimarcq, J. L., Hoffmann, D., Hoffmann, J. A., & Lagueux, M. (1989). Insect immunity: isolation of cDNA clones corresponding to diptericin, an inducible antibacterial peptide from *Phormia terranovae* (Diptera) and transcriptional profiles during immunization. *European Journal of Biochemistry,* 182, 423–427.

Reiter, R. S., Williams, J. G. K., Feldmann, K. A., Rafalski, J. A., Tingey, S. V., & Scolnik, P. A. (1992). Global and local genome mapping in *Arabidopsis thaliana* by using recombinant inbred lines and random amplified polymorphic DNAs. *Proceedings of the National Academy of Sciences, U.S.A.,* 89, 1477–1481.

Rembold, H., & Sehnal, F. (1987). Juvenile hormones and their titer regulation in *Galleria mellonella. Insect Biochemistry,* 7, 997–1001.

Rempel, J. G. (1951). A study of the embryology of *Mamestra configurata* (Walker) (Lepidoptera, Phalaenidae). *Canadian Entomologist,* 83, 1–19.

Reynaud, C.-A., Auquez, V., Grimal, H., & Weill, J.-C. (1987). A hyperconversion mechanism generates the chicken light chain preimmune repertoire. *Cell,* 48, 379–388.

Reynolds, S. E. (1977). Control of cuticle extensibility in the wings of adult *Manduca* at the time of eclosion: effects of eclosion hormone and bursicon. *Journal of Experimental Biology,* 70, 27–37.

Richard, D. S., Applebaum, S. W., Sliter, T. J., Baker, F. C., Schooley, D. A., Reuter, C. C., Henrich V. C., & Gilbert, L. I. (1989). Juvenile hormone bisepoxide biosynthesis in vitro by the ring gland of *Drosophila melanogaster:* a putative juvenile hormone in the higher Diptera. *Proceedings of the National Academy of Sciences, U.S.A.,* 86, 1421–1425.

Richards, G. (1978). Sequential gene activation by ecdysone in polytene chromosomes of *Drosophila melanogaster.* VI. Inhibition by juvenile hormones. *Developmental Biology,* 66, 32–42.

Richardson, J. C., Jorgensen, C. D., & Croft, B. A. (1982). Embryogenesis of the codling moth *Laspeyresia pomonella:* use in validating phenology models. *Annals of the Entomological Society of America,* 75, 201–209.

Riddiford, L. M. (1970). Effects of juvenile hormone on the programming of postembryonic development in eggs of the silkworm *Hyalophora cecropia. Developmental Biology,* 22, 249–263.

Riddiford, L. M. (1976). Hormonal control of insect epidermal cell commitment in vitro. *Nature,* 259, 115–117.

Riddiford, L. M. (1978). Ecdysone-induced change in cellular commitment of the epidermis of the tobacco hornworm, *Manduca sexta,* at the initiation of metamorphosis. *General and Comparative Endocrinology,* 34, 438–446.

Riddiford, L. M. (1985). Hormone action at the cellular level. In *Comprehensive Insect Physiology, Biochemistry and Pharmacology,* vol. 8, ed. G. A. Kerkut & L. I. Gilbert, pp. 37–84. Oxford: Pergamon.

Riddiford, L. M. (1986). Hormonal regulation of sequential larval cuticular gene expression. *Archives of Insect Biochemistry and Physiology,* Supplement 1, 75–86.

Riddiford, L. M. (1987). Hormonal control of sequential gene expression in insect epidermis. *UCLA Symposia on Molecular Cellular Biology,* 49, 211–222.

Riddiford, L. M. (1991). Hormonal control of sequential gene expression in insect epidermis. In *Physiology of the Insect Epidermis,* ed. K. Binnington & A. Retnakaran, pp. 46–54. East Melbourne: CSIRO Publications.

Riddiford, L. M., & Ajami, A. M. (1973). Juvenile hormone: its assay and effects on pupae of *Manduca sexta. Journal of Insect Physiology,* 19, 749–762.

Riddiford, L. M., Baeckmann, A., Hice, R. H., & Rebers, J. (1986). Developmental expression of three genes for larval cuticular proteins of the tobacco hornworm, *Manduca sexta. Developmental Biology,* 118, 82–94.

Riddiford, L. M., Curtis, A. T., & Kiguchi, K. (1979). Culture of the epidermis of the tobacco hornworm, *Manduca sexta. Tissue Culture Association Manual,* 5, 975–985.

Riddiford, L. M., & Hice, R. H. (1985). Developmental profiles of the mRNAs for *Manduca* arylphorin and two other storage proteins during the final larval instar of *Manduca sexta. Insect Biochemistry,* 15, 489–502.

Riddiford, L. M., & Hiruma, K. (1990). Hormonal control of sequential gene expression in lepidopteran epidermis. In *Molting and Metamorphosis,* ed. E. Ohnishi & H. Ishizaki, pp. 207–222. Tokyo: Japan Scientific Societies Press.

Riddiford, L. M., Osir, E. O., Fittinghoff, C. M., & Green, J. M. (1987). Juvenile hormone analogue binding in *Manduca* epidermis. *Insect Biochemistry,* 17, 1039–1043.

Riddiford, L. M., Palli, S. R., & Hiruma, K. (1990). Hormonal control of sequential gene expression in *Manduca* epidermis. In *Progress in Comparative Endocrinology,* ed. A. Epple, C. G. Scanes, & M. H. Stetson, pp. 226–231. New York: Wiley-Liss.

Riddiford, L. M., Palli, S. R., Hiruma, K., Li, W.-c., Green, J., Hice, R. H., Wolfgang, W. J., & Webb, B. A. (1990). Developmental expression, synthesis and secretion of insecticyanin by the epidermis of the tobacco hornworm, *Manduca sexta. Archives of Insect Biochemistry and Physiology,* 14, 171–190.

Riddiford, L. M., & Truman, J. W. (1993). Hormone receptors and the regulation of insect metamorphosis. *American Zoologist,* 33, 340–347.

Ridley, M. (1986). *Evolution and Classification: The Reformation of Cladism.* London: Longman.

Rieder, C. L. (1982). The formation, structure, and composition of the mammalian kinetochore and kinetochore fiber. *International Review of Cytology,* 79, 1–58.

Riggs, A. D., & Pfeifer, G. P. (1992). X-chromosome inactivation and cell memory. *Trends in Genetics,* 8, 169–174.

Rinehart, F. P., Ritch, T. G., Deininger, P. L., & Schmid, C. W. (1981). Renaturation rate studies of a single family of interspersed repeated DNA sequences in human deoxyribonucleic acid. *Biochemistry,* 20, 3003–3010.

Risley, M. S. (1986). The organization of meiotic chromosomes and synaptonemal complexes. In *Chromosome Structure and Function,* ed. M. S. Risley, pp. 126–151. New York: Van Nostrand Reinhold.

Robb, S., Packman, L. C., & Evans, P. D. (1989). Isolation, primary structure and bioactivity of Schisto FLRF-amide, a FMRF-amide-like neuropeptide from the locust *Schistocerca gregaria. Biochemical and Biophysical Research Communications,* 160, 850–856.

Robertson, M., & Postlethwait, J. H. (1986). The humoral antibacterial response of *Drosophila* adults. *Developmental and Comparative Immunology,* 10, 167–179.

Robinow, S., Talbot, W. S., Hogness, D. S., & Truman, J. W. (1993). Programmed cell death in the *Drosophila* CNS is ecdysone-regulated and coupled with a specific ecdysone receptor isoform. *Development,* 119, 1251–1259.

Robinson, G. (1988). A phylogeny for the Tineoidea (Lepidoptera). *Entomologica Scandinavica,* 19, 117–129.

Robinson, R. (1971). *Lepidoptera Genetics.* Oxford: Pergamon.

Rodakis, G. C., Lecanidou, R., & Eickbush, T. H. (1984). Diversity in a chorion multigene family created by tandem duplications and a putative gene-conversion event. *Journal of Molecular Evolution,* 20, 265–273.

Rodakis, G. C., Moschonas, N. K., & Kafatos, F. C. (1982). Evolution of a multigene family of chorion proteins in silkmoths. *Molecular and Cellular Biology,* 2, 554–563.

Rogers, J. (1985). The origin and evolution of retroposons. *International Review of Cytology,* 93, 187–279.

Roiha, H., & Glover, D. M. (1981). Duplication rDNA sequences of variable lengths flanking the short Type I insertions in the rDNA of *Drosophila melanogaster. Nucleic Acids Research,* 9, 5521–5532.

Roiha, H., Miller, J. R., Woods, L. C., & Glover, D. M. (1981). Arrangements and rearrangements of sequences flanking the two types of rDNA insertion in *D. melanogaster. Nature,* 290, 749–753.

Röller, H., & Bjerke, J. S. (1965). Purification and isolation of juvenile hormone and its action in lepidopteran larvae. *Life Sciences,* 4, 1617–1624.

Röller, H., Dahm, K. H., Sweely, C. C., & Trost, B. M. (1967). The structure of the juvenile hormone. *Angewandte Chemie,* 6, 179–180.

Romano, C. P., Martínez-Cruzado, J. C., & Kafatos, F. C. (1991). The relative importance of transcriptional and post transcriptional regulation of *Drosophila* chorion gene expression during oogenesis. *Developmental Genetics,* 12, 196–205.

Rories, C., & Spelsberg, T. C. (1989). Ovarian steroid hormone action on gene expression: mechanisms and models. *Annual Review of Physiology,* 51, 653–682.

Rosenthal, G. A., & Dahlman, D. L. (1991). Studies of L-canavanine incorporation into insectan lysozyme. *Journal of Biological Chemistry,* 266, 15684–15687.

Rountree, D. B., & Bollenbacher, W. E. (1986). The release of the prothoracicotropic hormone in the tobacco hornworm, *Manduca sexta,* is controlled intrinsically by juvenile hormone. *Journal of Experimental Biology,* 120, 41–58.

Royet, J. P., Jourdan, F., Ploye, H., & Souchier, C. (1989). Morphometric modification associated with early sensory experience in the rat olfactory bulb. II. Stereological study of the population of olfactory glomeruli. *Journal of Comparative Neurology,* 289, 594–609.

Royet, J. P., Sicard, G., Souchier, C., & Jourdan, F. (1987). Specificity of spatial patterns of glomerular activation in the mouse olfactory bulb: computer-assisted image analysis of 2-deoxyglucose autoradiograms. *Brain Research,* 417, 1–11.

Rubin, G. M., & Spradling, A. C. (1982). Genetic transformation of *Drosophila* with transposable element vectors. *Science,* 218, 348–353.

Ruiz i Altaba, A., & Melton, D. A. (1989). Bimodal and graded expression of the *Xenopus* homeobox gene *Xhox3* during embryonic development. *Development,* 106, 173–183.

Ruppert, S., Scherer, G., & Schütz, G. (1984). Recent gene conversion involving bovine vasopressin and oxytocin precursor genes suggested by nucleotide sequence. *Nature,* 308, 554–557.

Russell, V. W., & Dunn, P. E. (1990). Lysozyme in the pericardial complex of *Manduca sexta. Insect Biochemistry,* 20, 501–509.

Russell, V. W., & Dunn, P. E. (1991). Lysozyme in the midgut of *Manduca sexta* during metamorphosis. *Archives of Insect Biochemistry and Physiology,* 17, 67–80.

Ryan, R. O., Cole, K. D., Kawooya, J. K., Wells, M. A., & Law, J. H. (1988). Identification and characterization of a novel postlarval hemolymph protein from *Manduca sexta. Archives of Insect Biochemistry and Physiology,* 9, 81–90.

Ryan, R. O., Keim, P. S., Wells, M. A., & Law, J. H. (1985). Purification and properties of a predominantly female-specific protein from the hemolymph of the larva of the tobacco hornworm, *Manduca sexta. Journal of Biological Chemistry,* 260, 782–787.

Rybczynski, R., Reagan, J., & Lerner, M. R. (1989). A pheromone-degrading aldehyde oxidase in the antennae of the moth *Manduca sexta. Journal of Neuroscience,* 9, 1341–1353.

Rybczynski, R., Vogt, R. G., & Lerner, M. R. (1990). Antennal-specific pheromone-degrading aldehyde oxidases from the moths *Antheraea polyphemus* and *Bombyx mori. Journal of Biological Chemistry,* 265, 19712–19715.

Sahara, K., Kawamura, N., & Iizuka, T. (1990). Effect of environmental factors on the egg size-determining gene, *Esd,* as revealed by ovary transplantation in the silkworm, *Bombyx mori. Journal of Sericultural Science of Japan,* 59, 293–303.

Saigo, K., Kugimiya, W., Matsuo, Y., Inouye, S., Yoshioka, K., & Yuki, S. (1984). Identification of the coding sequence for reverse transcriptase-like enzyme in a transposable genetic element in *Drosophila melanogaster. Nature,* 312, 659–661.

Saiki, R. K., Gelfand, D. H., Stoffel, S., Scharf, S. J., Higuchi, R., Horn, G. T., Mullis, K. B., & Erlich, H. A. (1988). Primer-directed enzymatic amplification of DNA with a thermostable DNA polymerase. *Science,* 239, 487–491.

Saito, S. (1934). A study on the development of the tussar worm, *Antheraea pernyi* Guer. *Journal of the Faculty of Agriculture, Hokkaido Imperial University,* 33, 249–266.

Saito, S. (1937). On the development of the tusser, *Antheraea pernyi,* with special reference to the comparative embryology of insects. *Journal of the Faculty of Agriculture, Hokkaido Imperial University,* 60, 35–109.

Saitou, N., & Nei, M. (1987). The neighbor-joining method: a new method for reconstructing phylogenetic trees. *Molecular Biology and Evolution,* 4, 406–425.

Sakaguchi, B. (1952). The effect of centrifugal force upon the development of silkworm and eri-silkworm eggs. *Annual Report of the National Institute of Genetics, Japan,* 2, 21–22.

Sakaguchi, B. (1982). Some functional aspects of egg cell and blastoderm in dipteran insects. In *The Ultrastructure and Functioning of Insect Cells,* ed. H. Akai, R. C. King, & S. Morohoshi, pp. 45–48. Tokyo: Society for Insect Cells.

Sakaguchi, B., Sugahara, K., Nakajima, Y., Koga, K., Ninaki, O., Sugimoto, Y., Goldsmith, M. R., Mazur, G. D., & Kafatos, F. C. (1990). Expression of chorion multigene families of *Bombyx mandarina* in comparison with those of *Bombyx mori.* In *Advances in Invertebrate Reproduction,* vol. 5, ed. M. Hoshi & O. Yamashita, pp. 517–522. Amsterdam: Elsevier (Biomedical Division).

Sakaguchi, B., Sugawara, K., Koga, K., Goldsmith, M. R., Spoerel, N., Nguyen, H. T., Mazur, G. D., & Kafatos, F. C. (1988). Expression of chorion gene family in progenies between *Bombyx mori* and *Bombyx mandarina. Genome,* 30, S182.

Sakamoto, K., & Okada, N. (1985). Rodent type 2 *Alu* family, rat identifier sequence, rabbit C family, and bovine or goat 73 bp repeat may have evolved from tRNA genes. *Journal of Molecular Evolution,* 22, 134–140.

Sakata, T. (1938). Genetical studies on deformed gonads in the silkworm. *Journal of Sericultural Science of Japan,* 9, 284. In Japanese.

Sakonju, S., Bogenhagen, D. F., & Brown, D. D. (1980). A control region in the center of the 5S RNA gene directs specific initiation of transcription: I. The 5′ border of the region. *Cell,* 19, 13–25.

Sakurai, H., Fujii, T., Izumi, S., & Tomino, S. (1988). Structure and expression of gene coding for sex-specific storage protein of *Bombyx mori. Journal of Biological Chemistry,* 263, 7876–7880.

Sakurai, S., & Gilbert, L. I. (1990). Biosynthesis and secretion of ecdysteroids by the prothoracic glands. In *Molting and Metamorphosis,* ed. E. Ohnishi & H. Ishizaki, pp. 83–106. Tokyo: Japan Science Societies Press.

Sakurai, S., Okuda, M., & Ohtaki, T. (1989). Juvenile hormone inhibits ecdysone secretion and responsiveness to prothoracicotropic hormone in prothoracic glands of *Bombyx mori. General and Comparative Endocrinology,* 75, 222–230.

Sakurai, S., & Tsujita, M. (1976a). Genetical and biochemical studies of pteridine granule membrane in larval hypodermal cells of the silkworm. I. Purification and characterization of the membrane protein from pteridine granules of a normal strain. *Japanese Journal of Genetics,* 51, 39–52.

Sakurai, S., & Tsujita, M. (1976b). Genetical and biochemical studies of pteridine granule membrane in larval hypodermal cells of the silkworm. II. Genetic variations in membrane proteins of pteridine granules isolated from several mutants with transparent larval skin. *Japanese Journal of Genetics,* 51, 79–89.

Salkeld, E. H. (1973). The chorionic architecture and shell structure of *Amathes c-nigrum* (Lepidoptera: Noctuidae). *Canadian Entomologist,* 105, 1–10.

Salkeld, E. H. (1975). Biosystematics of the genus *Euxoa* (Lepidoptera: Noctuidae). IV. Eggs of the subgenus *Euxoa* Hbn. *Canadian Entomologist,* 107, 1137–1152.

Salt, G. (1970). *The Cellular Defense Reactions of Insects.* Cambridge: Cambridge University Press.

Samakovlis, C., Kimbrell, D. A., Kylsten, P., Engström, Å., & Hultmark, D. (1990). The immune response in *Drosophila:* pattern of cecropin expression and biological activity. *The EMBO Journal,* 9, 2969–2976.

Sander, K. (1960). Analyse des ooplasmatischen Reaktionssystems von *Euscelis plebejus* Fall. (Cicadina) durch Isolieren und Kombinieren von Keimteilen. II. *Wilhelm Roux' Archiv für Entwicklungsmechanik der Organismen,* 151, 660–707.

Sander, K. (1976). Specification of the basic body pattern in insect embryogenesis. *Advances in Insect Physiology,* 12, 125–238.

Sander, K. (1983). The evolution of patterning mechanisms: gleanings from insect embryogenesis and spermatogenesis. In *Development and Evolution,* ed. B. B. Goodwin, N. Holder, & C. C. Wylie, pp. 137–159. Cambridge: Cambridge University Press.

Sander, K. (1988). Studies in insect segmentation: from teratology to phenogenetics. *Development,* 104, Supplement, 112–121.

Sander, K., Gutzeit, H. O., & Jäckle, H. (1985). Insect embryogenesis: morphology, physiology, genetical and molecular aspects. In *Comprehensive Insect Physiology, Biochemistry and Pharmacology,* vol. 1, ed. G. A. Kerkut & L. I. Gilbert, pp. 319–386. Oxford: Pergamon.

Sanes, J. R., & Hildebrand, J. G. (1976a). Structure and development of antennae in a moth, *Manduca sexta. Developmental Biology,* 51, 282–299.

Sanes, J. R., & Hildebrand, J. G. (1976b). Origin and morphogenesis of sensory neurons in an insect antenna. *Developmental Biology,* 51, 300–319.

Sanes, J. R., & Hildebrand, J. G. (1976c). Acetylcholine and its metabolic enzymes in developing antennae of the moth, *Manduca sexta. Developmental Biology,* 52, 105–120.

Santamaria, P., & Nüsslein-Volhard, C. (1983). Partial rescue of *dorsal,* a maternal effect mutation affecting the dorsoventral pattern of the *Drosophila* embryo, by the injection of wild-type cytoplasm. *European Journal of Molecular Biology,* 2, 1695–1699.

Sasaki, M., & Riddiford, L. M. (1984). Regulation of reproductive behavior and egg maturation in the tobacco hawk moth, *Manduca sexta. Physiological Entomology,* 9, 315–327.

Sasaki, S. (1940). Inheritance of a new mutant "no-crescents supernumerary legs" in the silkworm. *Journal of Sericultural Science of Japan,* 11, 1–13. In Japanese.

Sass, M., Kiss, A., & Locke, M. (1993). Classes of integument peptides. *Insect Biochemistry and Molecular Biology,* 23, 845–857.

Sass, M., Kiss, A., & Locke, M. (1994). Integument and hemolymph peptides. *Journal of Insect Physiology,* 40, 407–421.

Sato, Y., & Yamashita, O. (1991a). Synthesis and secretion of egg-specific protein from follicle cells of the silkworm, *Bombyx mori. Insect Biochemistry,* 21, 233–237.

Sato, Y., & Yamashita, O. (1991b). Structure and expression of a gene coding for egg-specific protein in the silkworm, *Bombyx mori. Insect Biochemistry,* 21, 495–503.

Sawada, H., Tsusué, M., Yamamoto, T., & Sakurai, S. (1990). Occurrence of xanthommatin containing pigment graules in the epidermal cells of the silkworm, *Bombyx mori. Insect Biochemistry,* 20, 785–792.

Scarborough, R. M., Jamieson, G. C., Kalish, F., Kramer, S. J., Miller, C. A., & Schooley, D. A. (1984). Isolation and primary structure of two peptides with cardioacceleratory and hyperglycemic activity from the corpora cardiaca of *Periplaneta americana. Proceedings of the National Academy of Sciences, U.S.A.,* 81, 5575–5579.

Schaack, J., Sharp, S., Dingermann, T., Burke, D. J., Cooley, L., & Soll, D. (1984). The extent of a eukaryotic tRNA gene. 5′- and 3′-flanking sequence dependence for transcription and stable complex formation. *Journal of Biological Chemistry,* 259, 1461–1467.

Schaack, J., Sharp, S., Dingermann, T., & Soll, D. (1983). Transcription of eukaryotic tRNA genes in vitro. II. Formation of stable complexes. *Journal of Biological Chemistry,* 258, 2447–2453.

Scharrer, B. (1987). Insects as models in neuroendocrine research. *Annual Review of Entomology,* 32, 1–16.

Scheepens, M. H. M., & Wysoki, M. (1989). Pathogenicity of AcNPV for larvae of *Boarmia selenaria, Heliothis armigera, Heliothis peltigera, Spodoptera littoralis* and *Ephestia cautella. Journal of Invertebrate Pathology,* 53, 183–189.

Scherer, S., & Davis, R. D. (1980). Recombination of dispersed repeated DNA sequences in yeast. *Science,* 209, 1380–1384.

Schlenstedt, G., Gudmundsson, G. H., Boman, H. G., & Zimmerman, R. (1990). A large presecretory protein translocates both cotranslationally, using signal recognition particle and ribosome, and post-translationally, without these ribonucleoparticles, when synthesized in the presence of mammalian ribosomes. *Journal of Biological Chemistry,* 265, 13960–13968.

Schneider, L. E., & Taghert, P. H. (1988). Isolation and characterization of a *Drosophila* gene that encodes multiple neuropeptides related to Phe-Met-Arg-Phe-NH_2 (FMRFamide). *Proceedings of the National Academy of Sciences, U.S.A.,* 85, 1993–1997.

Schneider, L. E., & Taghert, P. H. (1990). Organization and expression of the *Drosophila* Phe-Met-Arg-Phe-NH_2 neuropeptide gene. *Journal of Biological Chemistry,* 265, 6890–6895.

Schneiderman, A. M., Matsumoto, S. G., & Hildebrand, J. G. (1986). Trans-sexually grafted antennae influence development of sexually dimorphic neurons in moth brain. *Nature,* 298, 844–846.

Schoofs, L., Holman, G. M., Hayes, T. K., & De Loof, A. (1990a). Isolation and identification of a sulfakinin-like peptide with sequence homology to vertebrate gastrin and cholecystokinin, from the brain of *Locusta migratoria.* In *Chromatography and Isolation of Insect Hormones and Pheromones,* ed. A. McCaffery & I. Wilson, pp. 231–241. London: Plenum.

Schoofs, L., Holman, G. M., Hayes, T. K., Nachman, R. J., & De Loof, A. (1990b). Isolation, identification and synthesis of locustamyotropin II, an additional neuropeptide of *Locusta migratoria:* member of the cephalomyotropic peptide family. *Insect Biochemistry,* 20, 479–484.

Schoofs, L., Holman, G. M., Hayes, T. K., Nachman, R. J., & De Loof, A. (1990c). Locustatachykinin I and II, two novel insect neuropeptides with homology to peptides of the vertebrate tachykinin family. *FEBS Letters,* 261, 397–401.

Schoofs, L., Holman, G. M., Hayes, T. K., Nachman, R. J., & De Loof, A. (1990d). Locustatachykinin III and IV: two additional insect neuropeptides with homology to peptides of the vertebrate tachykinin family. *Regulatory Peptides,* 31, 199–212.

Schoofs, L., Holman, G. M., Hayes, T. K., Nachman, R. J., & De Loof, A. (1991a). Isolation and identification of locustapyrokinin, a myotropic peptide from *Locusta migratoria. General and Comparative Endocrinology,* 81, 97–104.

Schoofs, L., Holman, G. M., Hayes, T. K., Nachman, R. J., & De Loof, A. (1991b). Isolation, identification and synthesis of locustamyoinhibiting peptide (LOM-MIP), a novel biologically active neuropeptide from *Locusta migratoria. Regulatory Peptides,* 36, 111–119.

Schoofs, L., Holman, G. M., Hayes, T. K., Tips, A., Nachman, R. J., van de Sande, A., & De Loof, A. (1990e). Isolation, identification and synthesis of locustamyotropin (LOM-MT), a novel biologically active insect peptide. *Peptides,* 11, 427–433.

Schoofs, L., Holman, G. M., Proost, P., van Damme, J., Hayes, T. K., & De Loof, A. (1992). Locustakinin, a novel myotropic peptide from *Locusta migratoria:* isolation, primary structure and synthesis. *Regulatory Peptides,* 37, 49–57.

Schooley, D. A., Baker, F. C., Tsai, L. W., Miller, C. A., & Jamieson, G. C. (1984). Juvenile hormones 0, I, and II exist only in Lepidoptera. In *Biosynthesis, Metabolism and Mode of Action of Invertebrate Hormones,* ed. J. Hoffmann & M. Porchet, pp. 373–383. Heidelberg: Springer-Verlag.

Schüle, R., & Evans, R. M. (1991). Cross-coupling of signal transduction pathways: zinc finger meets leucine zipper. *Trends in Genetics,* 7, 377–381.

Schultes, N. P., & Szostak, J. W. (1990). Decreasing gradients of gene conversion on both sides of the initiation site for meiotic recombination at the *ARG4* locus of yeast. *Genetics,* 126, 813–822.

Schulz, H.-J., & Traut, W. (1979). The pachytene complement of the wildtype and a chromosome mutant strain of the flour moth, *Ephestia kuehniella* (Lepidoptera). *Genetica,* 50, 61–66.

Schulz-Aellen, M.-F., Roulet, E., Fischer-Lougheed, J., & O'Shea, M. (1989). Synthesis of a homodimer neurohormone precursor of locust adipokinetic hormone studied by in vitro translation and cDNA cloning. *Neuron,* 2, 1369–1373.

Schüpbach, T. (1987). Germ line and soma cooperate during oogenesis to establish the dorsoventral pattern of egg shell and embryo in *Drosophila melanogaster. Cell,* 49, 699–707.

Schüpbach, T., & Wieschaus, E. (1986). Maternal-effect mutations altering the anterior-posterior pattern of the *Drosophila* embryo. *Roux's Archives of Developmental Biology,* 195, 302–317.

Schwabe, J. W. R., & Rhodes, D. (1991). Beyond zinc fingers: steroid hormone receptors have a novel structural motif for DNA recognition. *Trends in Biochemical Sciences,* 16, 291–296.

Schwalm, F. E. (1988). *Insect Morphogenesis.* Basel: Karger.

Schwangart, F. (1905). Zur Entwicklungsgeschichte der Lepidopteren. *Biologisches Zentrallblatt,* 25, 677–690.

Schwartz, J.-L, Garneau, L., Masson, L., & Brousseau, R. (1991). Early response of cultured lepidopteran cells to k-endotoxin from *Bacillus thuringiensis:* involvement of calcium and anionic channels. *Biochimica et Biophysica Acta,* 1065, 250–260.

Schwartz, L. M., Kosz, L., & Kay, B. K. (1990). Gene activation is required for developmentally programmed cell death. *Proceedings of the National Academy of Sciences, U.S.A.,* 87, 6594–6598.

Schwartz, L. M., Myer, A., Kosz, L., Engelstein, M., & Maier, C. (1990). Activation of polyubiquitin gene expression during developmentally programmed cell death. *Neuron,* 5, 411–419.

Schwartz, L. M., & Truman, J. W. (1982). Peptide and steroid regulation of muscle degeneration in an insect. *Science,* 215, 1420–1421.

Schwartz, L. M., & Truman, J. W. (1983). Hormonal control of rates of metamorphic development in the tobacco hornworn *Manduca sexta. Developmental Biology,* 99, 103–114.

Schwartz, L. M., & Truman, J. W. (1984). Hormonal control of muscle atrophy and degeneration in the moth *Antheraea polyphemus. Journal of Experimental Biology,* 111, 13–30.

Schwarz-Sommer, Z., Leclercq, L., Gobel, E., & Saedler, H. (1987). *Cin4,* an insert altering the structure of the A1 gene in *Zea mays,* exhibits properties of non-viral retrotransposons. *The EMBO Journal,* 6, 3873–3880.

Schweitzer, E. S., Sanes, J. R., & Hildebrand, J. G. (1976). Ontogeny of electroantennogram responses in the moth, *Manduca sexta. Journal of Insect Physiology,* 22, 955–960.

Scott, M. P., Tamkun, J. W., & Hartzell, G. W. III. (1989). The structure and function of the homeodomain. *Biochimica et Biophysica Acta,* 989, 25–48.

Sedlak, B. J., & Gilbert, L. I. (1976). Epidermal cell development during the pupal-adult metamorphosis of *Hyalophora cecropia. Tissue and Cell,* 8, 637–648.

Sedlak, B. J., & Gilbert, L. I. (1979). Correlations between epidermal cell structure and endogenous hormone titers during the fifth larval instar of the tobacco hornworm, *Manduca sexta. Tissue and Cell,* 11, 643–653.

Segraves, W. A. (1991). Something old, some things new: the steroid receptor superfamily in *Drosophila. Cell,* 67, 225–228.

Segraves, W. A., & Hogness, D. S. (1990). The E75 ecdysone-inducible gene responsible for the 75B early puff in *Drosophila* encodes two new members of the steroid receptor superfamily. *Genes and Development,* 4, 204–219.

Segraves, W. A., & Woldin, C. (1993). The E75 gene of *Manduca sexta. Insect Biochemistry and Molecular Biology,* 23, 91–97.

Sehl, A. (1931). Fürchung und Bildung der Keimanlage bei der Mehlmotte *Ephestia kühniella* Zell. nebst einer algemeinen Ubersicht uber den Verlauf der Embryonalentwicklung. *Zeitschrift für Morphologie und Ökologie der Tiere,* 20, 533–598.

Sehnal, F. (1972). Action of ecdysone on ligated larvae of *Galleria mellonella* L. (Lepidoptera): induction of development. *Acta entomologica bohemoslovaca,* 69, 143–155.

Sehnal, F., Maroy, P., & Mala, J. (1981). Regulation and significance of ecdysteroid titre fluctuations in lepidopterous larvae and pupae. *Journal of Insect Physiology,* 8, 535–544.

Sen, R., & Baltimore, D. (1986). Inducibility of κ immunoglobulin enhancer-binding protein NF-κB by a posttranslational mechanism. *Cell,* 47, 921–928.

Seperack, P., Slatkin, M., & Arnheim, N. (1988). Linkage disequilibrium in human ribosomal genes: implications for multigene family evolution. *Genetics,* 119, 943–949.

Sevala, V. L., & Davey, K. G. (1989). Action of juvenile hormone on the follicle cells of *Rhodnius prolixus:* evidence for a novel regulatory mechanism involving protein kinase C. *Experientia,* 45, 355–356.

Shapiro, D. J., Blume, J. E., & Nielsen, D. A. (1987). Regulation of mRNA stability in eukaryotic cells. *BioEssays,* 6, 221–226.

Sharp, S. J., & Garcia, A. D. (1988). Transcription of the *Drosophila melanogaster* 5S RNA gene requires an upstream promoter and four intragenic sequence elements. *Molecular and Cellular Biology,* 8, 1266–1274.

Shaw, G., & Kamen, R. (1986). A conserved AU sequence from the 3′ untranslated region of GM-CSF mRNA mediates selective mRNA degradation. *Cell,* 46, 659–667.

Shea, M. J., King, D. L., Conboy, M. J., Mariani, B. D., & Kafatos, F. C. (1990). Proteins that bind to *Drosophila* chorion cis-regulatory elements: a new C_2H_2 zinc finger protein and a C_2C_2 steroid receptor-like component. *Genes and Development,* 4, 1128–1140.

Shemshedini, L., Lanoue, M., & Wilson, T. G. (1990). Evidence for a juvenile hormone receptor involved in protein synthesis in *Drosophila melanogaster. Journal of Biological Chemistry,* 265, 1913–1918.

Shemshedini, L., & Wilson, T. G. (1990). Resistance to juvenile hormone and an insect growth regulator in *Drosophila* is associated with an altered cytosolic juvenile hormone-binding protein. *Proceedings of the National Academy of Sciences, U.S.A.,* 87, 2072–2076.

Shepherd, G. M. (1992). Modules for molecules. *Nature,* 358, 457–458.

Sheppard, P. M. (1961). Some contributions to population genetics resulting from the study of the Lepidoptera. *Advances in Genetics,* 10, 165–216.

Shields, O. (1988). Mesozoic history and neontology of Lepidoptera in relation to Trichoptera, Mecoptera, and Angiosperms. *Journal of Paleontology,* 62, 251–258.

Shimada, T., Ebinuma, H., & Kobayashi, M. (1986). Expression of homeotic genes in *Bombyx mori* estimated from asymmetry of dorsal closure in mutant/normal mosaics. *Journal of Experimental Zoology,* 240, 335–342.

Shimada, T., Fujiwara, H., Hasegawa, T., Maekawa, H., Kobayashi, M., & Fujii, H. (1992). Detection of DNA polymorphism in *Bombyx mori* using polymerase chain reaction. *Proceedings of the XIXth International Congress of Entomology, Beijing,* 637.

Shimura, K. (1988). The structure, synthesis and secretion of fibroin in the silkworm, *Bombyx mori. Sericologia,* 28, 457–479.

Shirk, P. D., Bean, D. W., & Brookes, V. J. (1990). Ecdysteroids control vitellogenesis and egg maturation in pharate adult females of the Indian meal moth, *Plodia interpunctella. Archives of Insect Biochemistry and Physiology,* 15, 183–199.

Shirk, P. D., Dahm, K. H., & Röller, H. (1976). The accessory sex glands as the repository for juvenile hormone in male cecropia moths. *Zeitschrift für Naturforschung,* 31c, 199–200.
Silbergleid, R. E., Shepherd, J. G., & Dickinson, J. L. (1984). Eunuchs: the role of apyrene sperm in Lepidoptera? *American Naturalist,* 123, 255–265.
Sim, G.-K., Kafatos, F. C., Jones, C. W., Koehler, M. D., Efstratiadis, A., & Maniatis, T. (1979). Use of a cDNA library for studies on evolution and developmental expression of the chorion multigene families. *Cell,* 18, 1303–1316.
Simcox, A., & Sang, J. H. (1983). When does determination occur in *Drosophila* embryos? *Developmental Biology,* 97, 212–221.
Simpson, S. P. (1989). Detection of linkage between quantitative trait loci and restriction fragment length polymorphisms using inbred lines. *Theoretical and Applied Genetics,* 77, 815–819.
Sinclair, A. H., Berta, P., Palmer, M. S., Hawkins, J. R., Griffiths, B. L., Smith, M. J., Foster, J. W., Frischauf, A.-M., Lovell-Badge, R., & Goodfellow, P. N. (1990). A gene from the human sex-determining region encodes a protein with homology to a conserved DNA-binding motif. *Nature,* 346, 240–244.
Singer, M. F. (1982a). Highly repeated DNA sequences in mammalian genomes. *International Review of Cytology,* 76, 67–112.
Singer, M. F. (1982b). SINEs and LINEs: highly repeated short and long interspersed sequences in mammalian genomes. *Cell,* 28, 433–434.
Singer, M. F., & Skowronski, J. (1985). Making sense out of LINEs: long interspersed repeat sequences in mammalian genomes. *Trends in Biochemical Sciences,* 10, 119–122.
Singh, H., Clerc, R. G., & LeBowitz, J. H. (1989). Molecular cloning of sequence-specific DNA binding proteins using recognition site probes. *Bio/Techniques,* 7, 252–261.
Singh, H., LeBowitz, J. H., Baldwin, Jr., A. S., & Sharp, P. A. (1988). Molecular cloning of an enhancer binding protein: isolation by screening of an expression library with a recognition site DNA. *Cell,* 52, 415–423.
Skeiky, Y. A. W., & Iatrou, K. (1990). Silkmoth chorion antisense RNA. Structural characterization, developmental regulation and evolutionary conservation. *Journal of Molecular Biology,* 213, 53–66.
Skeiky, Y. A. W., & Iatrou, K. (1991). Synergistic interactions of silkmoth chorion promoter-binding factors. *Molecular and Cellular Biology,* 11, 1954–1964.
Sklar, P. B., Anholt, R. R. H., & Snyder, S. H. (1986). The odorant-sensitive adenylate cyclase of olfactory receptor cells. *Journal of Biological Chemistry,* 261, 15538–15543.
Sklar, V. E. F., Jaehning, J. A., Gage, L. P., & Roeder, R. G. (1976). Purification and subunit structure of deoxyribonucleic acid-dependent ribonucleic acid polymerase III from the posterior silk gland of *Bombyx mori. Journal of Biological Chemistry,* 251, 3794–3800.
Slightom, J. L., Bechl, A. E., & Smithies, O. (1980). Human fetal Gγ- and Aγ-globin genes: complete nucleotide sequences suggest that DNA can be exchanged between these duplicated genes. *Cell,* 21, 627–638.
Smith, G. E., Summers, M. D., & Fraser, M. J. (1983). Production of human beta interferon in insect cells infected with a baculovirus expression vector. *Molecular and Cellular Biology,* 3, 2156–2165.
Smith, G. P. (1976). Evolution of repeated DNA sequences by unequal crossover. *Science,* 191, 528–535.

Smith, G. P., Kunes, S. M., Schultz, D. W., Taylor, A., & Triman, K. L. (1981). Structure of chi hotspots of generalized recombination. *Cell,* 24, 429–436.

Smith, H. R., & Lautenschlager, R. A. (1978). *Gypsy Moth Handbook–Predators of the Gypsy Moth,* Agriculture Handbook No. 534. Washington, DC: U.S. Department of Agriculture.

Snodgrass, R. E. (1935). *Principles of Insect Morphology.* London and New York: McGraw-Hill.

Spies, G. A., Karlinsey, J. E., & Spence, K. D. (1986a). Antibacterial hemolymph proteins of *Manduca sexta. Comparative Biochemistry and Physiology,* B, 83, 125–133.

Spies, G. A., Karlinsey, J. E., & Spence, K. D. (1986b). The immune proteins of the darkling beetle, *Eleodes* (Coleoptera: Tenebrionidae). *Journal of Invertebrate Pathology,* 47, 234–235.

Spoerel, N., Nguyen, H. T., & Kafatos, F. C. (1986). Gene regulation and evolution in the chorion locus of *Bombyx mori.* Structural and developmental characterization of four eggshell genes and their flanking regions. *Journal of Molecular Biology,* 190, 23–35.

Spoerel, N. A., Nguyen, H. T., Eickbush, T. H., & Kafatos, F. C. (1989). Gene evolution and regulation in the chorion complex of *Bombyx mori:* Hybridization and sequence analysis of multiple developmentally middle A/B chorion gene pairs. *Journal of Molecular Biology,* 209, 1–19.

Spoerel, N. A., Nguyen, H. T., Towne, S., & Kafatos, F. C. (1993). Negative and positive regulators modulate the activity of a silkmoth chorion gene during choriogenesis. *Journal of Molecular Biology,* 230, 151–160.

Spradling, A. C., deCicco, D. V., Wakimoto, B. T., Levine, J. F., Kalfayan, L. J., & Cooley, L. (1987). Amplification of the X-linked *Drosophila* chorion gene cluster requires a region upstream from the *s38* chorion gene. *The EMBO Journal,* 6, 1045–1053.

Spradling, A. C., Waring, G. L., & Mahowald, A. P. (1979). *Drosophila* bearing the *ocelliless* mutation underproduce two major chorion proteins both of which map near this gene. *Cell,* 16, 609–616.

Sprague, K. U. (1975). The *Bombyx mori* silk proteins: characterization of large polypeptides. *Biochemistry,* 14, 925–931.

Sprague, K. U., Hagenbüchle, O., & Zuniga, M. C. (1977). The nucleotide sequence of two silk gland alanine tRNAs: implications for fibroin synthesis and for initiator tRNA structure. *Cell,* 11, 560–570.

Sprague, K. U., Larson, D., & Morton, D. (1980). 5′ Flanking sequence signals are required for activity of silkworm alanine tRNA genes in homologous in vitro transcription systems. *Cell,* 22, 171–178.

Sprague, K. U., Ottonello, S., Rivier, D. H., & Young, L. S. (1987). Control of tRNA gene transcription. In *RNA Polymerase and the Regulation of Transcription,* ed. W. S. Reznikoff, R. R. Burgess, J. E. Dahlberg, C. A. Gross, M. T. Record, Jr., & M. P. Wickens, pp. 195–207. New York: Elsevier.

Sridhara, S. (1985). Evidence that pupal and adult cuticular proteins are coded by different genes in the silkmoth *Antheraea polyphemus. Insect Biochemistry,* 15, 333–340.

Sroka, P., & Gilbert, L. E. (1971). Studies on the endocrine control of post-emergence ovarian maturation in *Manduca sexta. Journal of Insect Physiology,* 17, 2409–2419.

St. Johnston, D., & Nüsslein-Volhard, C. (1992). The origin of pattern and polarity in the *Drosophila* embryo. *Cell,* 68, 201–219.

Stairs, G. R. (1960). On the embryology of the spruce budworm, *Chorisoneura fumiferana* (Clem.) (Lepidoptera, Tortricidae). *Canadian Entomologist,* 62, 147–154.

Stairs, G. R. (1989). Effects of a nuclear polyhedrosis virus isolate from *Malocosoma disstria* on *Lymantria dispar* larval growth pattern. *Journal of Invertebrate Pathology,* 53, 247–250.

Starratt, A. N., & Brown, B. E. (1975). Structure of the pentapeptide proctolin, a proposed neurotransmitter in insects. *Life Science,* 17, 1253–1256.

Starratt, A. N., Dahm, K. H., Allen, N., Hildebrand, J. G., Payne, T. L., & Röller H. (1979). Bombykal, a sex pheromone of the sphinx moth *Manduca sexta. Zeitschrift für Naturforschung,* 34c, 9–12.

Steinbrecht, R. A. (1980). Cryofixation without cryoprotectants. Freeze substitution and freeze etching of an insect olfactory receptor. *Tissue and Cell,* 12, 73–100.

Steinbrecht, R. A. (1987). Functional morphology of pheromone-sensitive sensilla. In *Pheromone Biochemistry,* ed. G. D. Prestwich & G. J. Blomquist, pp. 353–384. Orlando, FL: Academic Press.

Steinbrecht, R. A. (1992). Experimental morphology of insect olfaction. Tracer studies, X-ray microanalysis, autoradiography and immunocytochemistry with silkmoth antennae. *Microscopy Research and Technique,* 22, 336–350.

Steinbrecht, R. A., & Gnatzy, W. (1984). Pheromone receptors in *Bombyx mori* and *Antheraea pernyi.* I. Reconstruction of the cellular organization of the sensilla trichodea. *Cell and Tissue Research,* 235, 25–34.

Steiner, H. (1982). Secondary structure of the cecropins: antibacterial peptides from the moth *Hyalophora cecropia. FEBS Letters,* 137, 283–287.

Steiner, H., Hultmark, D., Engström, Å., Bennich, H., & Boman, H. G. (1981). Sequence and specificity of two antibacterial proteins involved in insect immunity. *Nature,* 292, 246–248.

Steinmetz, M., Stephan, D., & Lindahl, K. F. (1986). Gene organization and recombination hotspots in the murine major histocompatibility complex. *Cell,* 44, 895–904.

Steitz, J. A., Black, D. L., Gerke, V., Parker, K., Kramer, A., Frendewey, D., & Keller, W. (1988). Functions of abundant U-snRNPs. In *Structure and Function of Major and Minor snRNPs,* ed. M. L. Birnstiel, pp. 115–154. Heidelberg: Springer.

Stephens, J. C., Cavanaugh, M. L., Gradie, M. I., Mador, M. L., & Kidd, K. K. (1990). Mapping the human genome: current status. *Science,* 250, 237–244.

Stephens, J. M. (1959). Immune response of some insects to some bacterial antigens. *Canadian Journal of Microbiology,* 5, 203–228.

Stephens, J. M. (1962). Bactericidal activity of the blood of actively immunized wax moth larvae. *Canadian Journal of Microbiology,* 8, 491–499.

Stern, H., & Hotta, Y. (1987). The biochemistry of meiosis. In *Meiosis,* ed. P. B. Moens, pp. 303–331. Orlando, FL: Academic Press.

Steward, R. (1989). Relocalization of the *dorsal* protein from the cytoplasm to the nucleus correlates with its function. *Cell,* 59, 1179–1188.

Stewart, L. M. D., Hirst, M., Ferber, M. L., Merryweather, A. T., Cayley, P. J., & Possee, R. D. (1991). Construction of an improved baculovirus insecticide containing an insect-specific toxin gene. *Nature,* 352, 85–88.

Stewart, W. B., Kauer, J. S., & Shepherd, G. M. (1979). Functional organization of rat olfactory bulb analysed by the 2-deoxyglucose method. *Journal of Comparative Neurobiology,* 185, 715–734.

Strong, R. G. (1984). The egg of *Hofmannophila pseudospretella* (Oecophoridae): fine structure of the chorion. *Journal of the Lepidoptera Society,* 38, 202–208.

Struhl, G. (1981). A homeotic mutation transforming leg to antenna in *Drosophila. Nature,* 292, 635–638.

Struhl, G. (1984). Splitting the bithorax complex of *Drosophila. Nature,* 308, 454–457.

Strunnikov, V. A. (1983). *Control of Silkworm Reproduction, Development and Sex,* transl. P. E. Chernilovskaya & V. V. Kuznetsov, ed. I. A. Zakharov. Moscow: MIR.

Stuart, J., Brown, S. J., Beeman, R. W., & Denell, R. E. (1991). A deficiency of the homeotic complex of the beetle *Tribolium. Nature,* 350, 72–74.

Sturtevant, A. H. (1915). No crossing over in the female of the silkworm moth. *American Naturalist,* 49, 42–44.

Stutz, F., Gouilloud, E., & Clarkson, S. G. (1989). Oocyte and somatic tyrosine tRNA genes in *Xenopus laevis. Genes and Development,* 3, 1190–1198.

Stys, P., & Bilinski, S. (1990). Ovariole types and the phylogeny of hexapods. *Biological Reviews,* 65, 401–429.

Sugawara, K. (1990). Comparison of chorion multigene families between *Bombyx mori* and *Bombyx mandarina.* Master's thesis, Kyushu University.

Sugumaran, M. (1988). Molecular mechanisms for cuticular sclerotization. *Advances in Insect Physiology,* 21, 179–231.

Sullivan, H. S., Young, L. S., White, C. N., & Sprague, K. U. (1994). Silkgland-specific $tRNA^{Ala}$ genes interact more weakly than constitutive $tRNA^{Ala}$ genes with silkworm TFIIIB and polymerase III fractions. *Molecular and Cellular Biology,* 14, 1806–1814.

Summers, M. D. (1989). Recombinant proteins expressed by baculovirus vectors. In *Concepts in Viral Pathogenesis,* vol. 3, ed. A. L. Notkins & M. B. A. Oldstone, pp. 77–86. New York: Springer-Verlag.

Summers, M. D. (1990). Baculovirus directed foreign gene expression. In *Insect Neuropeptides,* ed. J. Menn, T. Kelly & E. Mosler, *American Chemical Society Symposium Series,* 453, 237–251. Washington, DC: ACS.

Sun, S.-C., Åsling, B., & Faye, I. (1991). Organization and expression of the immunoresponsive lysozyme gene in the giant silk moth, *Hyalophora cecropia. Journal of Biological Chemistry,* 266, 6644–6649.

Sun, S.-C., & Faye, I. (1992a). *Cecropia* immunoresponsive factor, an insect immunoresponsive factor with DNA-binding properties similar to nuclear-factor κ-B. *European Journal of Biochemistry,* 204, 885–892.

Sun, S.-C., & Faye, I. (1992b). Affinity purification and characterization of CIF, an insect immunoresponsive factor with NF-kappa-B-like properties. *Comparative Biochemistry and Physiology,* B, 103, 225–233.

Sun, S.-C., Lindström, I., Boman, H. G., Faye, I., & Schmidt, O. (1990). Hemolin: an insect-immune protein belonging to the immunoglobulin superfamily. *Science,* 250, 1729–1732.

Sun, S.-C., Lindström, I., Lee, J.-Y., & Faye, I. (1991). Structure and expression of the attacin genes in *Hyalophora cecropia. European Journal of Biochemistry,* 196, 247–254.

Suomalainen, E. (1969). On the sex chromosome trivalent in some Lepidoptera females. *Chromosoma* (Berl.), 28, 298–308.
Suzuki, T., Matsuno, K., Takiya, S., Ohno, K., Ueno, K., & Suzuki, Y. (1991a). Purification and characterization of an enhancer binding protein of the fibroin gene. I. Complete purification of fibroin factor 1. *Journal of Biological Chemistry,* 266, 16935–16941.
Suzuki, T., & Suzuki, Y. (1988). Interaction of composite protein complex with the fibroin enhancer sequence. *Journal of Biological Chemistry,* 263, 5979–5986.
Suzuki, T., Takiya, S., Matsuno, K., Ohno, K., Ueno, K., & Suzuki, Y. (1991b). Purification and characterization of an enhancer binding protein of the fibroin gene. II. Functional analyses of fibroin factor 1. *Journal of Biological Chemistry,* 266, 16942–16947.
Suzuki, Y. (1977). Differentiation of the silk gland a model system for the study of differential gene action. In *Results and Problems in Cell Differentiation,* vol. 8, ed. W. Beermann, pp. 1–44. Berlin, Heidelberg, and New York: Springer-Verlag.
Suzuki, Y., & Adachi, S. (1984). Signal sequences associated with fibroin gene expression are identical in fibroin-producer and -nonproducer tissues. *Development, Growth and Differentiation,* 26, 139–147.
Suzuki, Y., & Brown, D. D. (1972). Isolation and identification of the messenger RNA for silk fibroin from *Bombyx mori. Journal of Molecular Biology,* 63, 409–429.
Suzuki, Y., Gage, L. P., & Brown, D. D. (1972). The gene for silk fibroin in *Bombyx mori. Journal of Molecular Biology,* 70, 637–649.
Suzuki, Y., & Giza, P. E. (1976). Accentuated expression of silk fibroin genes in vivo and in vitro. *Journal of Molecular Biology,* 107, 183–206.
Suzuki, Y., Obara, T., Takiya, S., Hui, C.-c., Matsuno, K., Suzuki, T., Suzuki, E., Ohkubo, M., & Tamura, T. (1990a). Differential transcription of the fibroin and sericin-1 genes in cell-free extracts. *Development, Growth and Differentiation,* 32, 179–187.
Suzuki, Y., & Suzuki, E. (1974). Quantitative measurements of fibroin messenger RNA synthesis in the posterior silk gland of normal and mutant *Bombyx mori. Journal of Molecular Biology,* 88, 393–407.
Suzuki, Y., Takiya, S., Hara, W., Obara, T., Suzuki, T., & Hui, C.-c. (1987). Developmental regulation of the tissue-specific genes and the homeotic genes in *Bombyx mori.* In *Gunma Symposia on Endocrinogy,* vol. 24, ed. K. Iwai, pp. 13–26. Tokyo and Utrecht: Center for Academic Publications and VNU Science Press.
Suzuki, Y., Takiya, S., Suzuki, T., Hui, C.-c., Matsuno, K., Fukuta, M., Nagata, T., & Ueno, K. (1990b). Developmental regulation of silk gene expression in *Bombyx mori.* In *Molecular Insect Science,* ed. H. H. Hagedorn, J. G. Hildebrand, M. G. Kidwell, & J. H. Law, pp. 83–89. New York: Plenum.
Suzuki, Y., Tsuda, M., Takiya, S., Hirose, S., Suzuki, E., Kameda, M., & Ninaki, O. (1986). Tissue-specific transcription enhancement of the fibroin gene characterized by cell-free systems. *Proceedings of the National Academy of Sciences, U.S.A.,* 83, 9522–9526.
Swimmer, C., Fenerjian, M. G., Martínez-Cruzado, J. C., & Kafatos, F. C. (1990). Evolution of the autosomal chorion cluster in *Drosophila.* III. Comparison

of the *s18* gene in evolutionarily distant species and heterospecific control of chorion gene amplification. *Journal of Molecular Biology,* 215, 225–235.
Szostak, J. W., Orr-Weaver, T. L., Rothstein, R. J., & Stahl, F. W. (1983). The double-strand break repair model for recombination. *Cell,* 33, 25–35.
Szostak, J. W., & Wu, R. (1980). Unequal crossing over in the ribosomal DNA of *Saccharomyces cerevisiae. Nature,* 284, 426–430.
Tabuchi, H., & Hirose, S. (1988). DNA supercoiling facilitates formation of the transcription initiation complex on the fibroin gene promoter. *Journal of Biological Chemistry,* 263, 15282–15287.
Taghert, P. H., & Truman, J. W. (1982). Identification of the bursicon-containing neurons in abdominal ganglia of the tobacco hornworm, *Manduca sexta. Journal of Experimental Biology,* 98, 385–401.
Takahashi, H., Komano, H., & Natori, S. (1986). Expression of the lectin gene in *Sarcophaga peregrina* during normal development and under conditions where the defence mechanism is activated. *Journal of Insect Physiology,* 32, 771–779.
Takami, T. (1942). Experimental studies on the embryo formation in *Bombyx mori. Zoological Magazine,* 54, 337–343. In Japanese with English summary.
Takami, T., & Kitazawa, T. (1960). Normal stages of the embryonic development in the silkworm, *Bombyx mori. Technical Bulletin of the Sericultural Experiment Station,* 75, 1–31. In Japanese.
Takasaki, T. (1947). On *Kp*-multiple allelic series, with special reference to a new mutant. *Journal of Sericultural Science of Japan,* 16, 42–43. In Japanese.
Takasaki, T. (1962). Studies on the second linkage group of the silkworm. Chapter II, Ph.D. thesis, Kyushu University. In Japanese.
Takei, F., Oyama, K., Kimura, K., Hyodo, S., Mizuno, S., & Shimura, K. (1984). Reduced level of secretion and absence of subunit combination in the fibroin synthesized by a mutant silkworm *Nd*(2). *Journal of Cell Biology,* 99, 2005–2010.
Takei, F., Yoshimi, K., Kikuchi, A., Mizuno, S., & Shimura, K. (1987). Further evidence for importance of the subunit combination of silk fibroin in its efficient secretion from the posterior silk gland cells. *Journal of Cell Biology,* 105, 175–180.
Takiya, S., Hui, C.-c., & Suzuki, Y. (1990). A contribution of the core-promoter and its surrounding regions to the preferential transcription of the fibroin gene in posterior silk gland extracts. *The EMBO Journal,* 9, 489–496.
Takiya, S., & Suzuki, Y. (1989). Factors involved in preferential transcription of the fibroin gene. *European Journal of Biochemistry,* 179, 1–9.
Talbot, W. S., Swyryd, E. A., & Hogness, D. S. (1993). Drosophila tissues with different metamorphic responses to ecdysone express different ecdysone receptor isoforms. *Cell,* 73, 1323–1337.
Tamaki, Y. (1985). Sex pheromones. In *Comprehensive Insect Physiology, Biochemistry and Pharmacology,* vol. 9, ed. G. A. Kerkut & L. I. Gilbert, pp. 145–191. Oxford: Pergamon.
Tamura, T. (1977). Deficiency in xanthine dehydrogenase activity in the *og* and *og*[t] mutants of the silkworm, *Bombyx mori. Journal of Sericultural Science of Japan,* 46, 113–119. In Japanese with English summary.
Tamura, T. (1983). Deficiency of xanthine dehydrogenase activity in the *oq* oily mutant of the silkworm, *Bombyx mori. Japanese Journal of Genetics,* 58, 165–168.

Tamura, T., & Akai, H. (1990). Comparative ultrastructure of larval hypodermal cell in normal and oily *Bombyx* mutants. *Cytologia,* 55, 519–530.

Tamura, T., Inoue, H., & Suzuki, Y. (1987). The fibroin genes of *Antheraea yamamai* and *Bombyx mori* are different in their core regions but reveal a striking sequence similarity in their 5′ ends and 5′ flanking regions. *Molecular and General Genetics,* 207, 189–195.

Tamura, T., Kanda, T., Takiya, S., Okano, K., & Maekawa, H. (1990). Transient expression of chimeric CAT genes injected into early embryos of the domesticated silkworm *Bombyx mori. Japanese Journal of Genetics,* 65, 401–410.

Tamura, T., & Sakate, S. (1975). Granules in the meconium of *og*-mutant of *Bombyx mori. Journal of Sericultural Science of Japan,* 44, 487–490. In Japanese with English summary.

Tamura, T., & Sakate, S. (1983). Relationship between the expression of oily character and uric acid incorporation in the larval integument of various oily mutants of the silkworm, *Bombyx mori. Bulletin of the Sericultural Experiment Station of Japan,* 28, 719–740. In Japanese with English summary.

Tanada, Y., & Hess, R. T. (1976). Development of a nuclear polyhedrosis virus in midgut cells and penetration of the virus into the hemocoel of the army worm. *Journal of Invertebrate Pathology,* 23, 325–332.

Tanaka, M. (1970). Embryonic development of the rice webworm, *Ancylolomia japonica* Zeller. II. From yolk segmentation to gnathal segment formation. *New Entomology,* 19, 65–71. In Japanese with English summary.

Tanaka, M. (1985). Early embryonic development of *Amata fortunei* (Lepidoptera, Amatidae). In *Recent Advances in Insect Embryology in Japan,* ed. H. Ando & K. Miya, pp. 139–156. Tsukuba, Japan: ISEBU.

Tanaka, M. (1987). Differentiation and behaviour of primordial germ cells during the early embryonic development of *Parnassius glacialis* Butler, *Luehdorifa japonica* Leech and *Byasa* (Atrophaneura) *alcinous alcinous* Klug (Lepidoptera: Papilionidae). In *Recent Advances in Insect Embryology in Japan and Poland,* ed. H. Ando & Cz. Jura, pp. 255–266. Tsukuba, Japan: ISEBU.

Taneja, R., Gopalkrishnan, R., & Gopinathan, K. P. (1992). Regulation of glycine tRNA gene expression in the posterior silk glands of the silkworm *Bombyx mori. Proceedings of the National Academy of Sciences, U.S.A.,* 89, 1070–1074.

Tanksley, S. D., & Hewitt, J. (1988). Use of molecular markers in breeding for soluble solids content in tomato–a re-examination. *Theoretical and Applied Genetics,* 75, 811–823.

Tata, J. R., Kawahara, A., & Baker, B. S. (1991). Prolactin inhibits both thyroid hormone-induced morphogenesis and cell death in cultured amphibian larval tissues. *Developmental Biology,* 146, 72–80.

Tayade, D. S. (1989). Genetic architecture of economic traits in some strains of mulberry silkworms *Bombyx mori* L. I. Combining ability analysis. *Sericologia,* 29, 43–60.

Taylor, B. A. (1976). Genetic analysis of susceptibility to isoniazid-induced seizures in mice. *Genetics,* 83, 373–377.

Taylor, B. A. (1978). Recombinant inbred strains: Use in gene mapping. In *Origins of Inbred Mice,* ed. H. C. Morse, pp. 423–438. New York: Academic Press.

Taylor, H. M., & Truman, J. W. (1974). Metamorphosis of the abdominal ganglia of the tobacco hornworm, *Manduca sexta*. *Journal of Comparative Physiology,* 90, 367–388.

Tazima, Y. (1964). *The Genetics of the Silkworm.* London and Englewood Cliffs, NJ: Logos Press and Prentice Hall.

Tazima, Y. (1989). Alteration of food habit of the domesticated silkworm, *Bombyx mori. Sericologia,* 29, 437–453.

Tazima, Y., Doira, H., & Akai, H. (1977). The domesticated silkmoth, *Bombyx mori.* In *Handbook of Genetics,* vol. 3, ed. R. C. King, pp. 63–124. New York: Plenum.

Tazima, Y., & Ohnuma, A. (1986). Further studies on non-preference mutations in the silkworm. 2. Genetic analyses of mutant strains and their interrelationship. *Reports of the Silk Science Research Institute,* 34, 1–16. In Japanese with English summary.

Tazima, Y., & Ohnuma, A. (1987). Further studies on non-preference mutations in the silkworm. (4) A non-preference mutant strain, D5, which is presumably controlled by cytoplasmic factor(s). *Reports of the Silk Science Research Institute,* 35, 1–6. In Japanese with English summary.

Tazima, Y., Ohnuma, A., & Tanaka, Y. (1987). Further studies on non-preference mutations in the silkworm. (5) Modifiers that affect the expression of non-preference character. *Reports of the Silk Science Research Institute,* 35, 7–16. In Japanese with English summary.

Tazima, Y., Ohnuma, A., & Tanaka, Y. (1988). Further studies on non-preference mutations in the silkworm. (7) Selective ingestion of artificial diet observed for strains having positive or null modifiers of non-preference mutation. *Reports of the Silk Science Research Institute,* 36, 11–16. In Japanese with English summary.

Tear, G., Akam, M., & Martinez-Arias, A. (1990). Isolation of an *abdominal-A* gene from the locust *Schistocerca gregaria* and its expression during early embryogenesis. *Development,* 110, 915–925.

Teicher, M. H., Stewart, W. B., Kauer, J. S., & Shepherd, G. M. (1980). Suckling pheromone stimulation of a modified glomerular region in the developing rat olfactory bulb revealed by the 2-deoxyglucose method. *Brain Research,* 194, 530–535.

Teitelbaum, Z., Lazarovici, P., & Zlotkin, E. (1979). Selective binding of the scorpion venom insect toxin to insect nervous system. *Insect Biochemistry,* 9, 343–346.

Telfer, W. H. (1954). Immunological studies of insect metamorphosis. II. The role of a sex-limited blood protein in egg formation by the *Cecropia* silkworm. *Journal of General Physiology,* 37, 539–558.

Telfer, W. H. (1960). The selective accumulation of blood proteins by the oocytes of saturniid moths. *Biological Bulletin,* 118, 338–351.

Telfer, W. H. (1961). The route of entry and localization of blood proteins in the oocytes of saturniid moths. *Journal of Biophysical and Biochemical Cytology,* 9, 747–759.

Telfer, W. H. (1975). Development and physiology of the oocyte nurse cell syncytium. *Advances in Insect Physiology,* 11, 223–319.

Telfer, W., Keim, P., & Law, J. (1983). Arylphorin, a new protein from *Hyalophora cecropia.* Comparisons with calliphorin and manducin. *Insect Biochemistry,* 13, 601–613.

Telfer, W. H., & Kunkel, J. G. (1991). The function and evolution of insect storage hexamers. *Annual Review of Entomology,* 36, 205–228.

Telfer, W. H., & Massey, H. C. (1987). A storage hexamer from *Hyalophora* that binds riboflavin and resembles the apoprotein of hemocyanin. In *Molecular Entomology,* ed. J. H. Law, pp. 305–314. New York: Alan R. Liss.

Temin, H. M. (1989). Retrons in bacteria. *Nature,* 339, 254–255.

Teshima, T., Ueki, Y., Nakai, T., Shiba, T., & Kikuchi, M. (1986). Structure determination of lepidopteran, self-defense substance produced by silkworm. *Tetrahedron,* 42, 829–834.

Thanos, D., & Maniatis, T. (1992). The high mobility group protein HMG I(Y) is required for NF-κB-dependent virus induction of the human IFN-β gene. *Cell,* 71, 777–789.

Thiem, S. M., & Miller, L. K. (1989). A baculovirus gene with a novel transcription pattern encodes a polypeptide with a zinc finger and a leucine zipper. *Journal of Virology,* 63, 4489–4497.

Thiem, S. M., & Miller, L. K. (1990). Differential gene expression mediated by late, very late and hybrid baculovirus promoters. *Gene,* 91, 87–94.

Thireos, G., & Kafatos, F. C. (1980). Cell-free translation of silkmoth chorion mRNAs: identification of protein precursors, and characterization of cloned DNAs by hybrid-selected translation. *Developmental Biology,* 78, 36–46.

Thummel, C. S. (1990). Puffs and gene regulation–molecular insights into the *Drosophila* ecdysone regulatory hierarchy. *BioEssays,* 12, 561–568.

Thummel, C. S., Burtis, K. C., & Hogness, D. S. (1990). Spatial and temporal patterns of E74 transcription during *Drosophila* development. *Cell,* 61, 101–111.

Tinsley, T. W., & Harrap, K. A. (1977). Viruses of invertebrates. *Comprehensive Virology,* 12, 1–101.

Tjia, S. T., zu Altenschildesche, G. M., & Doerfler, W. (1983). *Autographa californica* nuclear polyhedrosis virus (AcNPV) DNA does not persist in mass cultures of mammalian cells. *Virology,* 125, 107–117.

Tobe, S. S., & Stay, B. (1985). Structure and regulation of the corpus allatum. *Advances in Insect Physiology,* 18, 305–432.

Toh, H., Hayashida, H., & Miyata, T. (1983). Sequence homology between retroviral reverse transcriptase and putative polymerases of hepatitus B virus and cauliflower mosaic virus. *Nature,* 305, 827–829.

Tojo, S., Betchaku, T., Ziccardi, V. J., & Wyatt, G. R. (1978). Fat body protein granules and storage proteins in the silkmoth, *Hyalophora cecropia. Journal of Cell Biology,* 78, 823–838.

Tojo, S., Kiguchi, K., & Kimura, S. (1981). Hormonal control of storage protein synthesis and uptake by the fat body in the silkworm *Bombyx mori. Journal of Insect Physiology,* 27, 491–497.

Tojo, S., Morita, M., Agui, N., & Hiruma, K. (1985). Hormonal regulation of phase polymorphism and storage protein fluctuation in the common cutworm *Spodoptera litura. Journal of Insect Physiology,* 31, 283–292.

Tolbert, L. P., & Oland, L. A. (1989). A role for glia in the development of organized neuropilar structures. *Trends in Neurosciences,* 12,70–75.

Tolbert, L. P., & Sirianni, P. A. (1990). Requirement for olfactory axons in the induction and stabilization of olfactory glomeruli in an insect. *Journal of Comparative Neurobiology,* 298, 69–82.

Tolias, P. P., & Kafatos, F. C. (1990). Functional dissection of an early *Drosophila* chorion gene promoter: expression throughout the follicular epithelium is under spatially composite regulation. *The EMBO Journal,* 9, 1457–1464.

Tomalski, M. D., Kutney, R., Bruce, W. A., Brown, M. R., Blum, M. S., & Travis, J. (1989). Purification and characterization of insect toxins derived from the mite, *Pyemotes tritici. Toxicon,* 27, 1151–1167.

Tomalski, M. D., & Miller, L. K. (1991). Insect paralysis by baculovirus-mediated expression of a mite neurotoxin gene. *Nature,* 352, 82–85.

Tomalski, M. D., Wu, J., & Miller, L. K. (1988). The location, sequence, transcription and regulation of a baculovirus DNA polymerase gene. *Virology,* 167, 591–600.

Tomita, T., & Wada, Y. (1989). Multifactorial sex determination in natural populations of the housefly (*Musca domestica*) in Japan. *Japanese Journal of Genetics,* 64, 373–382.

Toyama, K. (1902). Contribution to the study of silkworm. I. On the embryology of the silkworm. *Bulletin of the College of Agriculture, Tokyo Imperial University,* 5, 73–118. In Japanese.

Toyama, K. (1909). *Embryology of the Silkworm, Bombyx mori.* Tokyo: Maruyamasha. In Japanese.

Toyama, K. (1912). Maternal inheritance and Mendelism. *Journal of Genetics,* 2, 352–405.

Trager, W. (1935). Cultivation of virus Grasserie in silkworm tissue culture. *Journal of Experimental Medicine,* 61, 501–514.

Trask, B. J. (1991). Fluorescence in situ hybridization: Applications in cytogenetics and gene mapping. *Trends in Genetics,* 7, 149–154.

Traut, W. (1976). Pachytene mapping in the female silkworm, *Bombyx mori* L. (Lepidoptera). *Chromosoma* (Berl.), 58, 275–284.

Traut, W. (1977). A study of recombination, formation of chiasmata and synaptonemal complexes in female and male meiosis of *Ephestia kuehniella* (Lepidoptera). *Genetica,* 47, 135–142.

Traut, W. (1987). Hypervariable Bkm DNA loci in a moth, *Ephestia kuehniella:* does transposition cause restriction fragment length polymorphism? *Genetics,* 115, 493–498.

Traut, W., Epplen, J. T., Weichenhan, D., & Rohwedel, J. (1992). Inheritance and mutation of hypervariable $(GATA)_n$ microsatellite loci in a moth, *Ephestia kuehniella. Genome,* 35, 659–666.

Traut, W., & Mosbacher, G. C. (1968). Geschlechtschromatin bei Lepidopteren. *Chromosoma* (Berl.), 25, 343–356.

Traut, W., & Rathjens, B. (1973). Das W-chromosom von *Ephestia kuehniella* (Lepidoptera) und die ableitung des geschlechtschromatins. *Chromosoma* (Berl.), 41, 437–446.

Traut, W., & Scholz, D. (1978). Structure, replication and transcriptional activity of the sex-specific heterochromatin in a moth. *Experimental Cell Research,* 113, 85–94.

Trenczek, T., & Bennich, H. (1987). Antibacterial proteins in insect hemocytes. *Scandinavian Journal of Immunology,* 26, 332.

Trenczek, T., & Faye, I. (1988). Synthesis of immune proteins in primary cultures of fat body from *Hyalophora cecropia. Insect Biochemistry,* 18, 299–312.

Tripoulas, N. A., & Samols, D. (1986). Developmental and hormonal regulation of sericin RNA in the silkworm, *Bombyx mori*. *Developmental Biology*, 116 328–336.

Trost, J. T., & Goodman, W. G. (1986). Hemolymph titers of the biliprotein, insecticyanin, during development of *Manduca sexta*. *Insect Biochemistry*, 16, 353–358.

Truman, J. W. (1983). Programmed cell death in the nervous system of an adult insect. *Journal of Comparative Neurology*, 216, 445–452.

Truman, J. W. (1985). Hormonal control of ecdysis. In *Comprehensive Insect Physiology, Biochemistry and Pharmacology*, vol. 8, ed. G. A. Kerkut & L. I. Gilbert, pp. 413–440. Oxford: Pergamon.

Truman, J. W. (1988). Hormonal approaches for studying nervous system development in insects. *Advances in Insect Physiology*, 21, 1–34.

Truman, J. W. (1989). Hormonal regulation of axon terminal reorganization during metamorphosis of the moth *Manduca sexta*. *Society for Neuroscience, Abstracts*, 15, 65.

Truman, J. W. (1990). Neuroendocrine control of ecdysis. In *Molting and Metamorphosis*, ed. E. Onishi & H. Ishizaki, pp. 67–82. Tokyo and Berlin: Japan Society Press and Springer-Verlag.

Truman, J. W., & Reiss, S. E. (1976). Dendritic reorganization of an identified motoneuron during metamorphosis of the tobacco hornworm moth. *Science*, 192, 477–479.

Truman, J. W., & Reiss, S. E. (1988). Hormonal regulation of the shape of identified motoneurons in the moth, *Manduca sexta*. *Journal of Neuroscience*, 8, 765–775.

Truman, J. W., & Riddiford, L. M. (1970). Neuroendocrine control of ecdysis in silkmoths. *Science*, 167, 1624–1626.

Truman, J. W., & Riddiford, L. M. (1971). The role of the corpora cardiaca in the behavior of saturniid moths. II. Oviposition. *Biological Bulletin*, 140, 8–14.

Truman, J. W., & Riddiford, L. M. (1974). Physiology of insect rhythms. III. The temporal organization of the endocrine events underlying pupation of the tobacco hornworm. *Journal of Experimental Biology*, 60, 371–382.

Truman, J. W., Riddiford, L. M., & Safranek, L. (1973). Hormonal control of cuticle coloration in the tobacco hornworm, *Manduca sexta*: basis of an ultrasensitive bioassay for juvenile hormone. *Journal of Insect Physiology*, 19, 195–203.

Truman, J. W., Riddiford, L. M., & Safranek, L. (1974). Temporal patterns of response to ecdysone and juvenile hormone in the epidermis of the tobacco hornworm, *Manduca sexta*. *Developmental Biology*, 39, 247–262.

Truman, J. W., Rountree, D. B., Reiss, S. E., & Schwartz, L. M. (1983). Ecdysteroids regulate the release and action of eclosion hormone in the tobacco hornworm, *Manduca sexta* (L). *Journal of Insect Physiology*, 29, 895–900.

Truman, J. W., & Schwartz, L. M. (1984). Steroid regulation of neuronal death in the moth nervous system. *Journal of Neuroscience*, 4, 274–280.

Truman, J. W., Talbot, W. S., Fahrbach, S. E., & Hogness, D. S. (1994). Ecdysone receptor expression in the CNS correlates with stage-specific responses to ecdysteroids during *Drosophila* and *Manduca* development. *Development*, 120, 219–234.

Tsitilou, S. G., & Kafatos, F. C. (1989). Nonuniform evolution of duplicated, developmentally controlled chorion genes in a silkmoth. *Journal of Molecular Evolution,* 29, 396–406.

Tsuchida, K., Kawooya, J. K., Law, J. H., & Wells, M. A. (1992). Isolation and characterization of two follicle-specific proteins from eggs of *Manduca sexta. Insect Biochemistry and Molecular Biology,* 22, 89–98.

Tsuchida, K., Nagata, M., & Suzuki, A. (1987). Hormonal control of ovarian development in the silkworm, *Bombyx mori. Archives of Insect Biochemistry and Physiology,* 5, 167–177.

Tsuda, M., Hirose, S., & Suzuki, Y. (1986). Participation of the upstream region of the fibroin gene in the formation of transcription complex in vitro. *Molecular and Cellular Biology,* 6, 3928–3933.

Tsuda, M., & Suzuki, Y. (1981). Faithful transcription initiation of fibroin gene in a homologous cell-free system reveals an enhancing effect of 5′ flanking sequence far upstream. *Cell,* 27, 175–182.

Tsuda, M., & Suzuki, Y. (1983). Transcription modulation in vitro of the fibroin gene exerted by a 200-base-pair region upstream from the "TATA" box. *Proceedings of the National Academy of Sciences, U.S.A.,* 80, 7442–7446.

Tsujimoto, Y., & Suzuki, Y. (1979a). Structural analysis of the fibroin gene at the 5′ end and its surrounding regions. *Cell,* 16, 425–436.

Tsujimoto, Y., & Suzuki, Y. (1979b). The DNA sequence of *Bombyx mori* fibroin gene including the 5′ flanking, mRNA coding, entire intervening and fibroin protein coding regions. *Cell,* 18, 591–600.

Tsujimoto, Y., & Suzuki, Y. (1984). Natural fibroin genes purified without using cloning procedures from fibroin-producing and -nonproducing tissues reveal indistinguishable structure and function. *Proceedings of the National Academy of Sciences, U.S.A.,* 81, 1644–1648.

Tsujita, M. (1955). Relation between lethality of *H-* and *Kp*-recombinants (E-pseudoallelic group) and environmental conditions in silkworm. *Japanese Journal of Genetics,* 30, 252–256.

Tsujita, M. (1961). Maternal effect of $+^{lem}$ gene on pterine reductase of *Bombyx mori. Japanese Journal of Genetics,* 36, 337–346.

Tsujita, M. (1963). Manifestation mechanism of yellow larval color in the silkworm, with special regard to *d-lem* gene. *Japanese Journal of Genetics,* 38, 48–60.

Tsujita, M., & Sakaguchi, B. (1955). Genetical and biochemical studies of yellow lethal larvae in the silkworm (1). On the nature of pterin obtained from the yellow lethal strain. *Japanese Journal of Genetics,* 30, 83–88. In Japanese with English summary.

Tsujita, M., & Sakaguchi, M. (1958). Studies on the maternal inheritance of the lethal yellow in the silkworm, *Bombyx mori.* II. Ovarian transplantations. *Japanese Journal of Genetics,* 33, 210–215.

Tsujita, M., & Sakurai, S. (1963). The association of a specific protein with yellow pigments (dihydropterin) in the silkworm, *Bombyx mori. Japanese Journal of Genetics,* 38, 97–105.

Tsujita, M., & Sakurai, S. (1964a). Genetical and biochemical studies of the chromogranules in the larval hypodermis of the silkworm, *Bombyx mori.* (I) Relation of uric acid to chromogranules. *Journal of Sericultural Science of Japan,* 33, 389–393. In Japanese with English summary.

Tsujita, M., & Sakurai, S. (1964b). Relationship between chromogranules and uric acid in hypodermal cells of silkworm larvae. *Proceedings of the Japan Academy,* 40, 561–565.

Tsujita, M., & Sakurai, S. (1965a). Purification of the three specific soluble chromoproteins from chromogranules in hypodermal cells of the silkworm larva. *Proceedings of the Japan Academy,* 41, 225–229.

Tsujita, M., & Sakurai, S. (1965b). Amino acid analysis of the three specific chromoproteins purified from chromogranules in hypodermal cells of the silkworm larva. *Proceedings of the Japan Academy,* 41, 230–235.

Tsujita, M., & Sakurai, S. (1966). Development of chromogranules in the larval skin of the silkworm. *Proceedings of the Japan Academy,* 42, 960–965.

Tsujita, M., & Sakurai, S. (1967a). Genetic variations in shape and size of pteridine granules of a normal silkworm strain and several other strains with transparent skin. *Proceedings of the Japan Academy,* 43, 997–1002.

Tsujita, M., & Sakurai, S. (1967b). Pteridine granules in hypodermal cells of the silkworm larva causing non-transparency of larval skin. *Proceedings of the Japan Academy,* 43, 991–996.

Tsujita, M., & Sakurai, S. (1969). Incorporation of ^{14}C-phenylalanine or ^{14}C-tyrosine into the hypodermal cuticle of the silkworm. I. Lethal lemon larvae. *Proceedings of the Japan Academy,* 45, 943–948.

Tsujita, M., & Sakurai, S. (1971). Genetic and biochemical studies of lethal albino larvae of the silkworm, *Bombyx mori. Japanese Journal of Genetics,* 46, 17–31.

Tublitz, N., Brink, D., Broadie, K. S., Loi, P. K., & Sylwester, A. W. (1991). From behavior to molecules: an integrated approach to the study of neuropeptides. *Trends in Neuroscience,* 14, 254–259.

Tublitz, N. J., & Sylwester, A. W. (1990). Postembryonic alteration of transmitter phenotype in individually identified peptidergic neurons. *Journal of Neuroscience,* 10, 161–168.

Tumlinson, J. H., Brennan, M. M., Doolittle, R. E., Mitchell, E. R., Brabman, A., Mazomenos, B. E., Baumhover, A. H., & Jackson, D. M. (1989). Identification of a pheromone blend attractive to *Manduca sexta* (L.). males in a wind tunnel. *Archives of Insect Biochemistry and Physiology,* 10, 255–271.

Turner, J. R. G. (1979). Genetic control of recombination in the silkworm. I. Multigenic control of chromosome 2. *Heredity,* 43, 273–293.

Tzertzinis, G., Malecki, A., & Kafatos, F. C. (in press). BmCFI a *Bombyx mori* RXR type receptor related to the *Drosophila* ultraspiracle. *Journal of Molecular Biology.*

Udupa, S. M., & Gowda, B. L. V. (1988). Heterotic expression in silk productivity of different crosses of silkworm, *Bombyx mori* L. *Sericologia,* 28, 395–399.

Ueda, H., & Hirose, S. (1990). Identification and purification of a *Bombyx mori* homologue of FTZ-F1. *Nucleic Acids Research,* 18, 7229–7234.

Ueda, H., Mizuno, S., & Shimura, K. (1985). Sequence polymorphisms around the 5′-ends of the silkworm fibroin H-chain gene suggesting the occurrence of crossing-over between heteromorphic alleles. *Gene,* 34, 351–355.

Ueda, H., Mizuno, S., & Shimura, K. (1986). Transposable genetic element found in the 5′-flanking region of the fibroin H-chain gene in a genomic clone from the silkworm *Bombyx mori. Journal of Molecular Biology,* 190, 319–327.

Ueno, K., Hui, C.-c., Fukuta, M., & Suzuki, Y. (1992). Molecular analysis of the deletion mutants in the E homeotic complex of the silkworm *Bombyx mori*. *Development,* 114, 555–563.

Ullu, E., & Tschudi, C. (1984). *Alu* sequences are processed 7SL RNA genes. *Nature,* 312, 171–172.

Underwood, D. C., Knickerbocker, H., Gardner, G., Condliffe, D. P., & Sprague, K. U. (1988). Silk gland-specific $tRNA^{Ala}$ genes are tightly clustered in the silkworm genome. *Molecular and Cellular Biology,* 8, 5504–5512.

Vaidya, V. G. (1968). Embryology of *Papilio polytes* L. (Lepidoptera, Papilionidae). I. Early embryology. *Journal of Animal Morphology and Physiology,* 13, 141–152.

Vail, P. V., & Jay, D. L. (1973). Pathology of a nuclear polyhedrosis virus of the alfalfa looper in alternate hosts. *Journal of Invertebrate Pathology,* 21, 198–204.

Valgeirsdottir, K., Traverse, K. L., & Pardue, M-L. (1990). HeT DNA: a family of mosaic repeated sequences specific for heterochromatin in *Drosophila melanogaster. Proceedings of the National Academy of Sciences, U.S.A.,* 87, 7998–8002.

Van Arsdell, S. W., Denison, R. A., Bernstein, A. M., Weiner, A. M., Manser, T., & Gesteland, R. F. (1981). Direct repeats flank three small nuclear RNA pseudogenes in the human genome. *Cell,* 26, 11–17.

van den Berg, M. J., & Ziegelberger, G. (1991). On the function of the pheromone binding protein in the olfactory hairs of *Antheraea polyphemus. Journal of Insect Physiology,* 37, 79–85.

Van Hofsten, P., Faye, I., Kockum, K., Lee, J.-Y., Xanthopoulos, K. G., Boman, I. A., Boman, H. G., Engström, Å., Andreu, D., & Merrifield, R. B. (1985). Molecular cloning, cDNA sequencing, and chemical synthesis of cecropin B from *Hyalophora cecropia. Proceedings of the National Academy of Sciences, U.S.A.,* 82, 2240–2243.

Varmus, H., & Brown, P. (1989). Retroviruses. In *Mobile DNA,* ed. D. E. Berg & M. M. Howe, pp. 53–108. Washington DC: American Society of Microbiology.

Veletza, V. (1988). An 18/401-like chorion gene pair in the silkmoth *Antheraea pernyi.* Ph.D. thesis, Harvard University.

Vinson, C. R., LaMarco, R. L., Johnson, P. F., Landschulz, W. H., & McKnight, S. L. (1988). In situ detection of sequence-specific DNA-binding activity specified by a recombinant bacteriophage. *Genes and Development,* 2, 801–806.

Viswanathan, S., & Dignam, J. D. (1988). Seryl-tRNA synthetase from *Bombyx mori:* purification and properties. *Journal of Biological Chemistry,* 263, 535–541.

Viswanathan, S., Dignam, S. S., & Dignam, J. D. (1988). Control of the levels of alanyl-, glycyl- and seryl-tRNA synthetases in the silkgland of *Bombyx mori. Developmental Biology,* 129, 350–357.

Vlak, J. M., Klinkenberg, F. A., Zaal, K. J. M., Usmany, M., Klinge-Roode, E. C., Geervliet, J. B. F., Roosien, J., & van Lent, J. W. M. (1988). Functional studies on the p10 gene of *Autographa californica* nuclear polyhedrosis virus using a recombinant expressing a p10-β-galactosidase fusion gene. *Journal of General Virology,* 69, 765–776.

Vogt, R. G. (1987). The molecular basis of pheromone reception: its influence on behavior. In *Pheromone Biochemistry,* ed. G. D. Prestwich & G. L. Blomquist, pp. 385–431. Orlando, FL: Academic Press.

Vogt, R. G., Köhne, A. C., Dubnau, J. T., & Prestwich, G. D. (1989). Expression of pheromone binding proteins during antennal development in the gypsy moth *Lymantria dispar. Journal of Neuroscience,* 9, 3332–3346.

Vogt, R. G., & Lerner, M. R. (1989). Two groups of odorant binding proteins in insects suggest specific and general olfactory pathways. *Society for Neuroscience, Abstracts,* 15, 1290.

Vogt, R. G., Prestwich, G. D., & Lerner, M. R. (1991). Odorant-binding-protein subfamilies associate with distinct classes of olfactory receptor neurons in insects. *Journal of Neurobiology,* 22, 74–84.

Vogt, R. G., Prestwich, G. D., & Riddiford, L. M. (1988). Sex pheromone receptor proteins: visualization using a radiolabeled photoaffinity analog. *Journal of Biological Chemistry,* 8, 3952–3959.

Vogt, R. G., & Riddiford, L. M. (1981a). Pheromone deactivation by antennal proteins of Lepidoptera. In *Insect Development and Behaviour,* ed. F. Sehnal, A. Zabza, J. J. Menn, & B. Cymborowski, pp. 955–967. Wroclaw: University of Wroclaw Press.

Vogt, R. G., & Riddiford, L. M. (1981b). Pheromone binding and inactivation by moth antennae. *Nature,* 293, 161–163.

Vogt, R. G., & Riddiford, L. M. (1986). Pheromone reception: a kinetic equilibrium. In *Mechanisms in Insect Olfaction,* ed. T. L. Payne, M. C. Birch, & C. E. J. Kennedy, pp. 201–208. London and New York: Oxford University Press and Clarendon Press.

Vogt, R. G., Riddiford, L. M., & Prestwich, G. D. (1985). Kinetic properties of a pheromone degrading enzyme: the sensillar esterase of *Antheraea polyphemus. Proceedings of the National Academy of Sciences, U.S.A.,* 82, 8827–8831.

Vogt, R. G., Rybczynski, R., Cruz, M., & Lerner, M. R. (1993). Ecdysteroid regulation of olfactory protein expression in the developing antenna of the tobacco hawk moth, *Manduca sexta. Journal of Neurobiology,* 24, 581–597.

Vogt, R. G., Rybczynski, R., & Lerner, M. R. (1990). The biochemistry of odorant reception and transduction. In *Chemosensory Information Processing,* ed. D. Schild, *NATO ASI Series,* vol. H39, pp. 33–76. Berlin and Heidelberg: Springer-Verlag.

Vogt, R. G., Rybczynski, R., & Lerner, M. R. (1991). Molecular cloning and sequencing of general odorant-binding proteins GOBP1 and GOBP2 from the tobacco hawk moth *Manduca sexta:* comparisons with other insect OBPs and their signal peptides. *Journal of Neuroscience,* 11, 2972–2984.

Volkman, L. E., & Goldsmith, P. A. (1983). In vitro survey of *Autographa californica* nuclear polyhedrosis virus interaction with nontarget vertebrate host cells. *Applied and Environmental Microbiology,* 45, 1085–1093.

von Wettstein, D., Rasmussen, S. W., & Holm, P. B. (1984). The synaptonemal complex in genetic segregation. *Annual Review of Genetics,* 18, 331–413.

Voytas, D. F., & Ausubel, F. M. (1988). A *copia*-like transposable element family in *Arabidopsis thaliana. Nature,* 336, 242–244.

Wade, D., Boman, A., Wahlin, B., Drain, C. M., Andreu, D., Boman, H. G., & Merrifield, R. B. (1990). All-D amino acid-containing channel-forming antibiotic peptides. *Proceedings of the National Academy of Sciences, U.S.A.*, 87, 4761–4765.

Wakimoto, B. T., & Kaufman, T. C. (1981). Analysis of larval segmentation in lethal genotypes associated with the Antennapedia gene complex in *Drosophila melanogaster. Developmental Biology,* 81, 51–64.

Waku, Y., Sumimoto, K.-I., & Eguchi, M. (1968). Cytoplasmic inclusion body in the epidermal cells of normal and transparent silkworm larvae. *Journal of Insect Physiology,* 14, 1319–1323.

Wall, C. (1973). Embryonic development of two species of *Chesias* (Lepidoptera, Geometridae). *Journal of Zoology,* 169, 65–84.

Walter, P., & Blobel, G. (1982). Signal recognition particle contains a 7S RNA essential for protein translocation across the endoplasmic reticulum. *Nature,* 299, 691–698.

Walther, C., Zlotkin, E., & Rathmayer, W. (1976). Action of different toxins from the scorpion *Androctonus australis* on a locust nerve-muscle preparation. *Journal of Insect Physiology,* 22, 1187–1194.

Wang, X., Ooi, B. G., & Miller, L. K. (1991). Baculovirus vectors for multiple gene expression and for occluded virus production. *Gene,* 100, 131–137.

Wang, X.-Y., Cole, K. D., & Law, J. H. (1989). The nucleotide sequence of a microvitellogenin gene from the tobacco hornworm, *Manduca sexta. Gene,* 80, 259–268.

Wang, Y. X., Marec, F., & Traut, W. (1993). The synaptonemal complex complement of the wax moth, *Galleria mellonella. Hereditas,* 118, 113–119.

Wang, Y., & Xu, J. (1990). Synaptonemal complex analysis of W·+P normal and W·Ze Zebra strains of silkworm (*Bombyx mori*). *Acta Genetica Sinica,* 17, 349–353.

Warren, J. T., & Gilbert, L. I. (1986). Ecdysone metabolism and distribution during the pupal-adult development of *Manduca sexta. Insect Biochemistry,* 16, 65–82.

Webb, B. A., & Riddiford, L. M. (1988a). Synthesis of two storage proteins during larval development of the tobacco hornworm, *Manduca sexta. Developmental Biology,* 130, 671–681.

Webb, B. A., & Riddiford, L. M. (1988b). Regulation of expression of arylphorin and female-specific protein mRNAs in the tobacco hornworm, *Manduca sexta. Developmental Biology,* 130, 682–692.

Webb, N. R., & Summers, M. D. (1991). Expression of proteins using recombinant baculoviruses. *Technique,* 2, 173–188.

Weeks, J. C. (1987). Time course of hormonal independence for developmental events in neurons and other cell types during insect metamorphosis. *Developmental Biology,* 124, 163–176.

Weeks, J. C., & Ernst-Utzschneider, K. (1989). Respecification of larval proleg motoneurons during metamorphosis of the tobacco hornworm, *Manduca sexta:* segmental dependence and hormonal regulation. *Journal of Neurobiology,* 20, 569–592.

Weeks, J. C., & Truman J. W. (1985). Independent steroid control of the fates of motoneurons and their muscles during insect metamorphosis. *Journal of Neuroscience,* 5, 2290–2300.

Weiner, A. J., Scott, M. P., & Kaufman, T. C. (1984). A molecular analysis of *fushi tarazu,* a gene in *Drosophila melanogaster* that encodes a product affecting embryonic segment number and cell fate. *Cell,* 37, 843–851.

Weiner, A. M., Deininger, P. L., & Efstratiadis, A. (1986). Nonviral retroposons: genes, pseudogenes, and transposable elements generated by the reverse flow of genetic information. *Annual Review of Biochemistry,* 55, 631–661.

Weirich, G. F., & Culver, M. G. (1979). S-adenosylmethionine: juvenile hormone acid methyltransferase in male accessory reproductive glands of *Hyalophora cecropia* (L). *Archives of Biochemistry and Biophysics,* 198, 175–181.

Weismann, I. (1935). Untersuchungen über den weiblichen genitalapparat, das Ei und die embryonal Entwicklung des Apfelwinckers *Carpocaspa (Cydia) pomonella* L. *Mitteilungen der Schwiezerischen Entomologischen Gesellschaft,* 16, 370–377.

Weiss, P., & Ferris, W. (1956). The basement lamella of amphibian skin: Its reconstruction after wounding. *Journal of Biophysical and Biochemical Cytology,* 2, 275–282.

Weith, A., & Traut, W. (1980). Synaptonemal complexes with associated chromatin in a moth, *Ephestia kuehniella Z. Chromosoma* (Berl.), 78, 275–291.

Weith, A., & Traut, W. (1986). Synaptic adjustment, non-homologous pairing, and non-pairing of homologous segments in sex chromosome mutants of *Ephestia kuehniella* (Insecta, Lepidoptera). *Chromosoma* (Berl.), 94, 125–131.

Weller, S. J., Friedlander, T. P., Martin, J. A., & Pashley, D. P. (1992). Phylogenetic studies of ribosomal RNA variation in higher moths and butterflies (Lepidoptera: Ditrysia). *Molecular Phylogenetics and Evolution,* 1, 312–337.

Welsh, J., & McClelland, M. (1990). Fingerprinting genomes using PCR with arbitrary primers. *Nucleic Acids Research,* 18, 7213–7218.

Wellauer, P. K., Dawid, I. B., & Tartof, K. D. (1978). X and Y chromosomal ribosomal DNA of *Drosophila:* comparison of spacers and insertions. *Cell,* 14, 269–278.

Weyer, U., Knight, S., & Possee, R. D. (1990). Analysis of very late gene expression by *Autographa californica* nuclear polyhedrosis virus and the further development of multiple expression vectors. *Journal of General Virology,* 71, 1525–1534.

Weyer, U., & Possee, R. D. (1988). Functional analysis of the p10 gene 5′ leader sequence of the *Autographa californica* nuclear polyhedrosis virus. *Nucleic Acids Research,* 16, 3635–3653.

Weyer, U., & Possee, R. D. (1989). Analysis of the promoter of the *Autographa californica* nuclear polyhedrosis virus p10 gene. *Journal of General Virology,* 70, 203–208.

Whalley, P. (1986). A review of the current fossil evidence of Lepidoptera in the Mesozoic. *Biological Journal of the Linnean Society,* 28, 253–271.

Wharton, D. A. (1978). The trichurid egg-shell: evidence in support of the Bouligand hypothesis of helicoidal architecture. *Tissue and Cell,* 10, 647–658.

White, J. H., Dimartino, J. F., Anderson, R. W., Lusnak, K., Hilbert, D., & Fogel, S. (1988). A DNA sequence conferring high postmeiotic segregation frequency to heterozygous deletions in *Saccharomyces cerevisiae* is related to sequences associated with eucaryotic recombination hotspots. *Molecular and Cellular Biology,* 8, 1253–1258.

Whitehouse, H. L. K. (1982). *Genetic Recombination, Understanding the Mechanisms.* New York: Wiley.

Wicker, C., Reichhart, J.-M., Hoffmann, D., Hultmark, D., Samakovlis, C., & Hoffmann, J. (1990). Characterization of a *Drosophila* cDNA encoding a

novel member of the diptericin family of immune peptides. *Journal of Biological Chemistry,* 265, 22493–22498.
Wickham, T. J., Davis, T., Granados, R. R., Hammer, D. A., Schuler, M. L., & Wood, H. A. (1991). Baculovirus defective interfering particles are responsible for variations in recombinant protein production as a function of multiplicity of infection. *Biotechnology Letters,* 13, 483–488.
Wigglesworth, V. B. (1934). The physiology of ecdysis in *Rhodnius prolixus* (Hemiptera). II. Factors controlling moulting and metamorphosis. *Quarterly Journal of Microscopical Science,* 77, 121–222.
Wigglesworth, V. B. (1936). The function of the corpus allatum in the growth and reproduction of *Rhodnius prolixus* (Hemiptera). *Quarterly Journal of Microscopical Science,* 79, 91–121.
Wigglesworth, V. B. (1940). The determination of characters at metamorphosis in *Rhodnius prolixus* (Hemiptera). *Journal of Experimental Biology,* 17, 201–222.
Williams, A. F., & Barclay, A. N. (1988). The immunoglobulin superfamily – domains for cell surface recognition. *Annual Review of Immunology,* 6, 381–405.
Williams, B. D., Schrank, B., Huynh, C., Shownkeen, R., & Waterston, R. H. (1992). A genetic mapping system in *Caenorhabditis elegans* based on polymorphic sequence-tagged sites. *Genetics,* 131, 609–624.
Williams, C. M. (1947). Physiology of insect diapause. II. Interaction between the pupal brain and prothoracic glands in the metamorphosis of the giant silkworm *Platysamia cecropia. Biological Bulletin,* 93, 89–98.
Williams, C. M. (1952). Physiology of insect diapause. IV. The brain and prothoracic glands as an endocrine system in the *Cecropia* silkworm. *Biological Bulletin,* 103, 120–138.
Williams, C. M. (1956a). The juvenile hormone of insects. *Nature,* 178, 212–213.
Williams, C. M. (1956b). Physiology of insect diapause. X. An endocrine mechanism for the influence of temperature on the diapausing pupa of the *Cecropia* silkworm. *Biological Bulletin,* 110, 201–218.
Williams, C. M. (1958). The juvenile hormone. *Scientific American,* 198, 67–75.
Williams, C. M. (1959). The juvenile hormone. I. Endocrine activity of the corpora allata of the adult *Cecropia* silkworm. *Biological Bulletin,* 116, 323–338.
Williams, C. M. (1961). The juvenile hormone. II. Its role in the endocrine control of molting, pupation, and adult development of the cecropia silkworm. *Biological Bulletin,* 121, 572–585.
Williams, C. M. (1967). Third-generation pesticides. *Scientific American,* 217, 13–17.
Williams, J. G. K., Kubelik, A. R., Livak, K. J., Rafalski, J. A., & Tingey, S. V. (1990). DNA polymorphisms amplified by arbitrary primers are useful as genetic markers. *Nucleic Acids Research,* 18, 6531–6535.
Willis, J. H. (1986). The paradigm of stage-specific gene sets in insect metamorphosis: time for revision! *Archives of Insect Biochemistry and Physiology,* Supplement 1, 47–57.
Willis, J. H. (1987). Cuticular proteins: the neglected component. *Archives of Insect Biochemistry and Physiology,* 6, 203–215.
Willis, J. H. (1989). Partial amino acid sequences of cuticular proteins from *Hyalophora cecropia. Insect Biochemistry,* 19, 41–46.

Willis, J. H., Binger, L. C., & Lampe, D. J. (1991). Isolation and characterization of the genes for two cuticular proteins that are regulated during metamorphosis. *Sericologia,* 31, Supplement, 13.

Willis, J. H., & Hollowell, M. P. (1976). The interaction of juvenile hormone and ecdysone: antagonistic, synergistic, or permissive? In *The Juvenile Hormones,* ed. L. I. Gilbert, pp. 270–287. New York and London: Plenum.

Willis, J. H., Regier, J. C., & Debrunner, B. A. (1981). The metamorphosis of arthropodin. In *Current Topics in Insect Endocrinology and Nutrition,* ed. G. Bhaskaran, S. Friedman, & J. G. Rodriguez, pp. 27–46. New York: Plenum.

Willott, E., Wang, X.-Y., & Wells, M. A. (1989). cDNA and gene sequence of *Manduca sexta* arylphorin, an aromatic amino acid-rich larval serum protein. *Journal of Biological Chemistry,* 264, 19052–19059.

Wilson, E. T., Condliffe, D. P., & Sprague, K. U. (1988). Transcriptional properties of *BmX,* a moderately repetitive silkworm gene that is an RNA polymerase III template. *Molecular and Cellular Biology,* 8, 624–631.

Wilson, E. T., Larson, D., Young, L. S., & Sprague, K. U. (1985). A large region controls tRNA gene transcription. *Journal of Molecular Biology,* 183, 153–163.

Wilson, M. E., Mainprize, T. H., Friesen, P. D., & Miller, L. K. (1987). Location, transcription and sequence of a baculovirus gene encoding a small arginine-rich polypeptide. *Journal of Virology,* 61, 661–666.

Wilson, P., & Keller, R. E. (1991). Cell rearrangement during gastrulation of *Xenopus:* direct observation of cultured explants. *Development,* 112, 289–300.

Witt, D. J., & Janus, C. A. (1976). Aspects of cross transmission to *Galleria mellonella* of a baculovirus from the alfalfa looper, *Autographa californica. Journal of Invertebrate Pathology,* 27, 65–69.

Witten, J. L., & Levine, R. B. (1991). Cellular specificity of steroid influences on process outgrowth of identified neurons in culture. *Society for Neuroscience, Abstracts,* 17, 1320.

Witten, J. L., & Truman, J. W. (1990). Stage specific expression of FMRFamide-like immunoreactivity in motoneurons of the tobacco hornworm, *Manduca sexta,* is mediated by steroid hormones. *Society for Neuroscience, Abstracts,* 16, 633.

Wolbert, P., & Schafer, F.-G. (1991). Macromolecular changes during metamorphosis of the integument. In *Physiology of the Insect Epidermis,* ed. K. Binnington & A. Retnakaran, pp. 169–184. East Melbourne: CSIRO Publications.

Wolfgang, W. J., & Riddiford, L. M. (1981). Cuticular morphogenesis during continuous growth of the final instar larva of a moth. *Tissue and Cell,* 13, 757–772.

Wolfgang, W. J., & Riddiford, L. M. (1986). Larval cuticular morphogenesis in the tobacco hornworm, *Manduca sexta,* and its hormonal regulation. *Developmental Biology,* 113, 305–316.

Wolfgang, W. J., & Riddiford, L. M. (1987). Cuticular mechanics during larval development of the tobacco hornworm. *Journal of Experimental Biology,* 128, 19–33.

Wong, Y.-C., Pustell, J., Spoerel, N., & Kafatos, F. C. (1985). Coding and potential regulatory sequences of a cluster of chorion genes in *Drosophila melanogaster. Chromosoma* (Berl.), 92, 124–135.
Woodhead, A. P., Stay, B., Seidel, S. L., Khan, M. A., & Tobe, S. S. (1989). Primary structure of four allatostatins: neuropeptide inhibitors of juvenile hormone synthesis. *Proceedings of the National Academy of Sciences, U.S.A.*, 86, 5997–6001.
Woodworth, C. W. (1889). Studies on the embryological development of *Euvanessa antiopa.* In *The Butterflies of Eastern United States and Canada*, ed. S. H. Scudder, pp. 95–104. Cambridge: S. H. Scudder.
Wu, C.-I., True, J. R., & Johnson, N. (1989). Fitness reduction associated with the deletion of a satellite DNA array. *Nature,* 341, 248–251.
Wu, J., & Miller, L. K. (1989). Sequence, transcription and translation of a late gene of the *Autographa californica* nuclear polyhedrosis virus encoding a 34.8K polypeptide. *Journal of General Virology,* 70, 2449–2459.
Wyatt, G. R. (1990). Developmental and juvenile hormone control of gene expression in locust fat body. In *Molecular Insect Science,* ed. H. H. Hagedorn, J. G. Hildebrand, M. G. Kidwell, & J. H. Law, pp. 163–172. New York: Plenum.
Wyatt, G. R., & Kalf, G. F. (1957). The chemistry of insect hemolymph. II. Trehalose and other carbohydrates. *Journal of General Physiology,* 40, 833–847.
Wyatt, S. S. (1956). Culture in vitro of tissue from the silkworm, *Bombyx mori* L. *Journal of General Physiology,* 39, 841–852.
Wyatt, S. S., & Wyatt, G. R. (1971). Stimulation of RNA and protein synthesis in silkmoth pupal wing tissue by ecdysone in vitro. *General and Comparative Endocrinology,* 16, 369–374.
Wysoki, C. J., & Meridith, M. (1987). The vomeronasal system. In *Neurobiology of Taste and Smell,* ed T. E. Finger & W. L. Silver, pp. 125–150. New York: Wiley.
Xanthopoulos, C. G., Lee, J.-Y., Gan, R., Kockum, K., Faye, I., & Boman, H. G. (1988). The structure of the gene for cecropin B, an antibacterial immune protein from *Hyalophora cecropia. European Journal of Biochemistry,* 172, 371–376.
Xiong, Y., Burke, W. D., & Eickbush, T. H. (1993). Pao, a highly divergent retrotransposable element from *Bombyx mori* containing long terminal repeats with tandem copies of the putative R region. *Nucleic Acids Research,* 21, 2117–2123.
Xiong, Y., Burke, W. D., Jakubczak, J. L., & Eickbush, T. H. (1988). Ribosomal DNA insertion elements *R1Bm* and *R2Bm* can transpose in a sequence specific manner to locations outside the 28S genes. *Nucleic Acids Research,* 16, 10561–10573.
Xiong, Y., & Eickbush, T. H. (1988a). The site-specific ribosomal DNA insertion element *R1Bm* belongs to a class of non-long-terminal-repeat retrotransposons. *Molecular and Cellular Biology,* 8, 114–123.
Xiong, Y., & Eickbush, T. H. (1988b). Functional expression of a sequence-specific endonuclease encoded by the retrotransposon *R2Bm. Cell,* 55, 235–246.
Xiong, Y., & Eickbush, T. H. (1988c). Similarity of reverse transcriptase-like sequences of viruses, transposable elements, and mitochondrial introns. *Molecular Biology and Evolution,* 5, 675–690.

Xiong, Y., & Eickbush, T. H. (1990). Origin and evolution of retroelements based upon their reverse transcriptase sequences. *The EMBO Journal,* 9, 3353–3362.

Xiong, Y., Sakaguchi, B., & Eickbush, T. H. (1988). Gene conversions can generate sequence variants in the late chorion multigene families of *Bombyx mori. Genetics,* 120, 221–231.

Yajima, H. (1960). Studies on embryonic determination of the harlequin-fly, *Chironomus dorsalis.* I. Effects of centrifugation and of its combination with constriction and puncturing. *Journal of Embryology and Experimental Morphology,* 8, 198–215.

Yamaguchi, K., Kikuchi, Y., Takagai, T., Kikuchi, A., Oyama, F., Shimura, K., & Mizuno, S. (1989). Primary structure of the silk fibroin light chain determined by cDNA sequencing and peptide analysis. *Journal of Molecular Biology,* 210, 127–139.

Yamamoto, K., Chadarevian, A., & Pelligrini, A. (1988). Juvenile hormone action mediated in male accessory glands of *Drosophila* by calcium and kinase C. *Science,* 239, 916–919.

Yamamoto, T. (1984). Studies on the mutant "black-striped pupal wing" sensitive to temperature in *Bombyx mori. Journal of Sericultural Science of Japan,* 53, 501–505. In Japanese with English summary.

Yanagawa, H., Watanabe, K., & Nakamura, M. (1988). Composition of artificial diets for the original strains of the silkworm, *Bombyx mori,* by applying a linear programming method. *Bulletin of the Sericultural Experiment Station of Japan,* 30, 569–588. In Japanese with English summary.

Yanofsky, M. F., Porter, S. G., Young, C., Albright, L. M., Gordon, M. P., & Nester, E. W. (1986). The virD operon of *Agrobacterium tumefaciens* encodes a site specific endonuclease. *Cell,* 47, 471–477.

Yao, T.-P., Forman, B. M., Jiang, Z., Cherbas, L., Chen, J. D., McKeown, M., Cherbas, P., & Evans, R. M. (1993). Functional ecdysone receptor is the product of *EcR* and *Ultraspiracle* genes. *Nature,* 366, 476–479.

Yao, T.-P., Segraves, W. A., Oro, A. E., McKeown, M., & Evans, R. M. (1992). *Drosophila* ultraspiracle modulates ecdysone receptor function via heterodimer formation. *Cell,* 71, 63–72.

Yaoita, V., & Brown, D. D. (1990). A correlation of thyroid hormone receptor gene expression with amphibian metamorphosis. *Genes and Development,* 4, 1917–1924.

Yazawa, M., Hirao, T., Arai, N., & Yagi, S. (1991). Feeding and gustatory responses in the "polyphagous strains" of the silkworm, *Bombyx mori* L. *Journal of Sericultural Science of Japan,* 60, 363–371. In Japanese with English summary.

Yokoyama, T. (1968). A scientist and a link of the chain uniting Thailand and Japan. *Japan Agricultural Research Quarterly,* 4, 30–33.

Yoshinaga, S. K., Boulanger, P. A., & Berk, A. J. (1987). Resolution of human transcription factor TFIIIC into two functional components. *Proceedings of the National Academy of Sciences, U.S.A.,* 84, 3585–3589.

Young, B. S., Pession, A., Traverse, K. L., French, C., & Pardue, M. L. (1983). Telomere regions in *Drosophila* share complex DNA sequences with pericentric heterochromatin. *Cell,* 34, 85–94.

Young, L. S., Dunstan, H. M., Witte, P. R., Smith, T. P., Ottonello, S., & Sprague, K. U. (1991). A class III transcription factor composed of RNA. *Science,* 252, 542–546.

Young, L. S., Rivier, D. H., & Sprague, K. U. (1991). Sequences far downstream from the classical tRNA promoter elements bind RNA polymerase III transcription factors. *Molecular and Cellular Biology,* 11, 1382–1392.

Young, L. S., Takahashi, N., & Sprague, K. U. (1986). Upstream sequences confer distinctive transcriptional properties on genes encoding silkgland-specific $tRNA^{Ala}$. *Proceedings of the National Academy of Sciences, U.S.A.,* 83, 374–378.

Yuki, S., Ishimaru, S., Inouye, S., & Saigo, K. (1986). Identification of genes for reverse transcriptase-like enzymes in two *Drosophila* retrotransposons, *412* and *gypsy:* a rapid detection method of reverse transcriptase genes using YXDD box probes. *Nucleic Acids Research,* 14, 3017–3030.

Zavortink, M., & Sakonju, S. (1989). The morphogenetic and regulator functions of the *Drosophila Abdominal-B* gene are encoded in overlapping RNAs transcribed from separate promoters. *Genes and Development,* 3, 1969–1981.

Zeh, D. W., Zeh, J. A., & Smith, R. L. (1989). Ovipositors, amnions and eggshell architecture in the diversification of terrestrial arthropods. *Quarterly Review of Biology,* 64, 147–168.

Zheng, L., Collins, F. H., Kumar, V., & Kafatos, F. C. (1993). A detailed map for the X chromosome of the malaria vector, *Anopheles gambiae. Science.* 261, 605–608.

Ziegler, I. (1961). Genetic aspects of ommochrome and pterin pigments. *Advances in Genetics,* 10, 349–403.

Zlotkin, E. (1983). Insect selective toxins derived from scorpion venoms: an approach to insect neuropharmacology. *Insect Biochemistry,* 13, 219–236.

Zlotkin, E., Rochat, H., Kopeyan, C., Miranda, F., & Lissitzky, S. (1971). Purification and properties of insect toxin from the venom of the scorpion *Androconus australis* Hector. *Biochimie,* 53, 1073–1078.

Zraket, C. A., Barth, J. L., Heckel, D. G., & Abbott, A. G. (1990). Genetic linkage mapping with restriction fragment length polymorphisms in the tobacco budworm, *Heliothis virescens.* In *Molecular Insect Science,* ed. H. H. Hagedorn, J. G. Hildebrand, M. G. Kidwell, & J. H. Law, pp. 13–20. New York: Plenum.

Index